Resonance Ionization Spectroscopy 1990

Sponsors

CNR, Comitato Nazionale Scienze Chimiche
ENEA
Instituto Nazionale Fisica Nucleare
Environment Institute, JRC Ispra
dB Electronics
Elicam
ISA Italia
Laser Optronic
Microcontrole
VG Instruments
Atom Sciences Inc
US Department of Energy
Chemical Sciences Department, Pacific Northwest Laboratory
EG&G Energy Measurements
Los Alamos National Laboratory
Martin Marietta Energy Systems
Perkin Elmer
National Institute of Standards & Technology
University of Tennessee

Advisory Group

*J E Parks, Co-Chairman (University of Tennessee, USA)
*N Omenetto, Co-Chairman (CEC, Joint Research Centre Ispra, Italy)
E Arimondo (Università di Pisa, Italy)
G Bekov (USSR Academy of Sciences, USSR)
*P Benetti (Università di Pavia, Italy)
*P Camus (CNRS Orsay, France)
*G Goldstein (US Department of Energy, USA)
*H-J Kluge (Universität Mainz, FRG)
P Knight (Imperial College of Science & Technology, UK)
Y Y Kuzyakov (Moscow State University, USSR)
B Lehmann (University of Bern, Switzerland)
A L'Huillier (CEN Saclay, France)
*T Lucatorto (National Inst. of Standards & Technology, USA)
*C M Miller (Los Alamos National Laboratory, USA)
*J C Miller (Oak Ridge National Laboratory, USA)
*K Niemax (ISAS Dortmund, FRG)
*G Rossi (CEC, Joint Research Centre Ispra, Italy)
H Rubinsztein-Dunlop (Chalmers University of Technology, Sweden)
*J C Travis (National Inst. of Standards & Technology, USA)
*T J Whitaker (Pacific Northwest Laboratory, USA)

*Member of Program Committee

Resonance Ionization Spectroscopy 1990

Proceedings of the Fifth International Symposium on Resonance Ionization Spectroscopy and its Applications held at the Congress Centre Villa Ponti, Varese, Italy, 16–21 September 1990

Edited by J E Parks and N Omenetto

Institute of Physics Conference Series Number 114
Institute of Physics, Bristol, Philadelphia and New York

CODEN IPHSAC 114 1–474 (1991)

British Library Cataloguing in Publication Data

International Symposium on Resonance Ionization
Spectroscopy and Its Applications
(5th 1990 Varese, Italy)
Resonance ionization spectroscopy: 1990.
— (Institute of Physics conference series)
I. Title II. Parks, J.E. III. Omenetto, N.
IV. Series
535.8

ISBN 0-85498-046-6

Library of Congress Cataloging-in-Publication Data are available

Published under The Institute of Physics imprint by IOP Publishing Ltd
Techno House, Redcliffe Way, Bristol BS1 6NX, England
335 East 45th Street, New York, NY 10017-3483, USA
US Editorial Office: 1411 Walnut Street, Philadelphia, PA 19102, USA

Printed in Great Britain by Galliard (Printers) Ltd, Great Yarmouth, Norfolk

Preface

The Fifth International Symposium on Resonance Ionization Spectroscopy and its Applications was held in Varese, Italy, 16–21 September 1990. The Symposium was organized by an International Advisory Group through a Program Committee and was co-ordinated jointly by the Institute of Resonance Ionization Spectroscopy of the University of Tennessee (USA) and the Commission of the European Communities, Joint Research Centre, Ispra Site, Italy.

Interest in RIS and its applications continues to grow, and RIS is expanding into a more diverse and mature field of study. This maturity was evident in this meeting both in the basic science and understanding of RIS processes and in the number of new and improved applications and techniques. The application of RIS techniques to molecular detection problems made remarkable progress since the last meeting two years ago. Subtle effects pertaining to isotopic discrimination received more theoretical attention, and there now seems to be good understanding of these effects, which can lead to correction procedures and/or methods to avoid isotopic effects. RIS applications were presented in which significant, real world problems were addressed, demonstrating its capability to solve problems that previously could not be accurately solved by other more traditional techniques. The interest in Resonance Ionization Spectroscopy and its Applications is made even more apparent by the growth in the number of sponsors who provided partial financial support.

This symposium had three notable additions which highlighted the success of the program. First, a conscientious effort to involve graduate students in the meeting succeeded in attracting over 30 graduate students from the US, Europe, and Asia. Second, a short course on Resonance Ionization Spectroscopy was offered and over 60 participants attended on each night of the two-night course. Third, excellence awards were presented for outstanding work in RIS performed by a graduate student and for an outstanding poster presentation. Roland Hergenröeder was presented the Excellence Award for RIS Research and Development Performed by a Graduate Student, and Mohamed El-Maazawi, also a graduate student, won the Excellence Award for RIS Research and Development Presented in a Poster Session. While also recognizing the many excellent examples of work in RIS presented at the meeting, we congratulate Mr Hergenröeder and Mr El-Maazawi for the work they presented.

In an attempt to promote efficiency and cost-effectiveness, members of the Program Committee were organized into two subcommittees, one US based and one European based. These committees functioned well, each co-ordinating its activities with frequent communication between their co-chairmen. The conscientious work of our colleagues on these committees, P Benetti, P Camus, G Goldstein, H-J Kluge, T Lucatorto, C M Miller, J C Miller, K Niemax, G Rossi, J C Travis, and T J Whitaker, is greatly appreciated.

The Local Organizing Committee in Varese and the staffs of the Joint Research Centre and the Congress Centre Villa Ponti in Varese are to be commended for the excellent planning and arrangements they made in hosting this very successful symposium. The work of the session chairmen of the symposium in helping with the review of the papers in this proceedings is gratefully acknowledged.

The support provided by the following sponsors is also gratefully acknowledged: CNR, Comitato Nazionale Scienze Chimiche; ENEA; Istituto Nazionale Fisica Nucleare; Environment Institute, JRC Ispra; dB Electronics; Elicam; ISA Italia; Laser Optronic; Microcontrole; VG Instruments; Atom Sciences Inc; US Department of Energy, Office of Health and Environmental Research; Battelle Pacific Northwest Laboratories, Molecular Sciences Research Center; EG&G Energy Measurements; Los Alamos National Laboratory; Martin Marietta Energy Systems; Perkin Elmer, Physical Electronics Division; National Institute of Standards and Technology; and the Institute of Resonance Ionization Spectroscopy and the Science Alliance of the University of Tennessee.

Finally, we wish to express our appreciation to all the authors for having submitted their contributions on time and to the publishers for their collaborative effort and efficiency. We are confident that basic studies, as well as more applied research, in the field of Resonance Ionization Spectroscopy will result in a constant increase in the number of analytical chemists and spectroscopists involved in the development and application of this technique.

N Omenetto
J E Parks

Contents

Section 6: Environmental Trace Analysis

Section 8: State Selected Chemistry

Section 9: New Laser Sources and Techniques

Section 10: Ultra-high Resolution and Isotopic Selectivity

Section 11: Surface and Bulk Analysis

Inst. Phys. Conf. Ser. No 114: Section 1
Paper presented at RIS 90, Varese, Italy, 16–21 September 1990

Resonance ionization mass spectroscopy of radioactive isotopes

H.-Jürgen Kluge

Institut für Physik, Universität Mainz D-65 Mainz, Germany

A survey is given on spectroscopic techniques applied to the determination of nuclear ground state properties of short-lived isotopes as produced at on-line isotope separators.

1. INTRODUCTION

The majority of our information on nuclear ground state properties comes from atomic spectroscopy: The hyperfine structure (HFS) of atomic levels gives access to a determination of the nuclear spin I, the nuclear magnetic dipole moment μ_I, and the spectroscopic quadrupole moment Q_s. The isotope shift (IS) in the optical transitions of different isotopes enable to calculate the change in the mean square charge radius $\delta\langle r^2\rangle^{A,A'}$ between isotopes with massnumber A and A'. In case of radioactive isotopes, only optical spectroscopy provides the latter information due to its by far higher sensitivity as compared to other techniques.

It should be pointed out that these nuclear properties are model-independent. Furthermore, the data on short-lived isotopes far from stability, with an unusual ratio of proton to neutron number, enable stringent tests of nuclear models and concepts developed from the knowledge of the properties of stable or long-lived nuclei. Finally, there is the chance to discover new phenomena by the study of nuclei far away from stability which are not present or not detectable in the region of known and well studied nuclei.

2. PRODUCTION OF ISOTOPES

Proton-induced spallation and fission are the most powerful nuclear reactions to produce nuclei far away from stability. Since these reactions are not selective for the production of one isotope only, mass separation of the reaction products is mandatory (Ravn and Allardyce 1989). This provides isotopic selectivity. Elemental selectivity is aimed for by especially designed target/ion source systems. Only in case of heavy-ion induced reactions the selectivity of these processes might make a mass separator unnecessary.

The on-line mass separator ISOLDE at Geneva/CERN is the facility with the widest spectrum of produced isotopes and – in most cases – highest yields. The isotopes (up to 39) of about 70 elements are available with intensities up to 10^{11} mass separated ions per second (Kluge 1986).

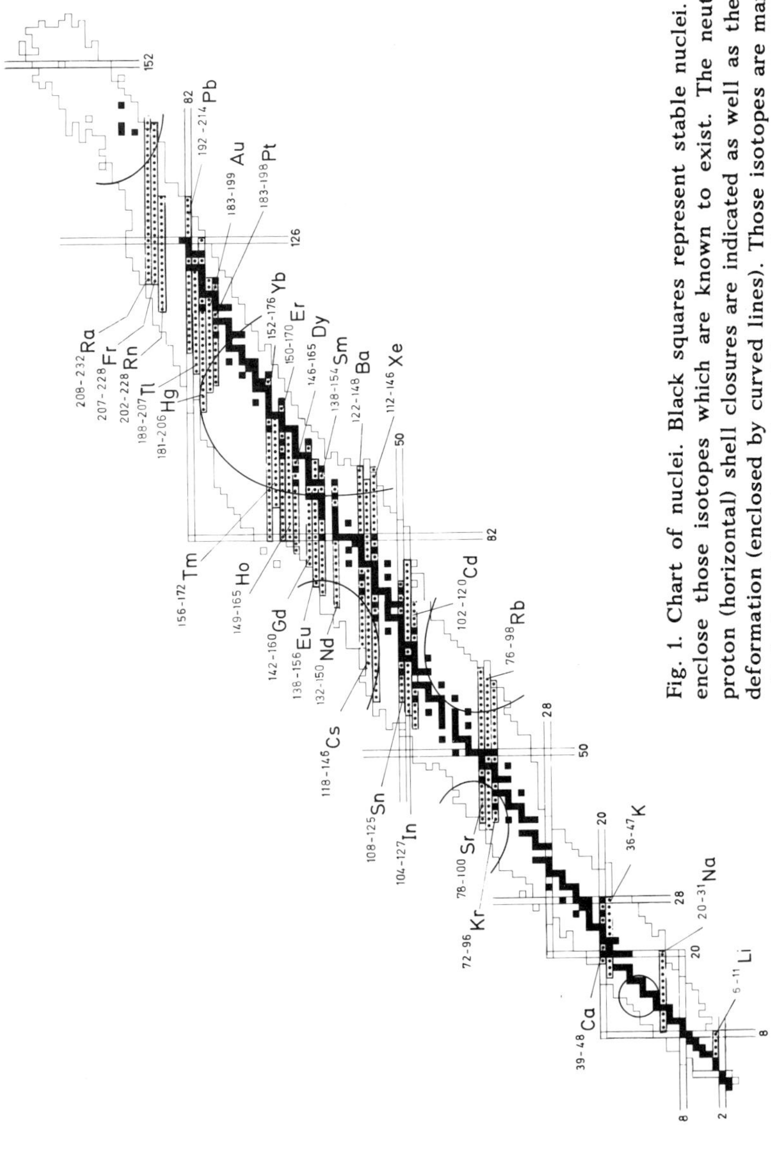

Fig. 1. Chart of nuclei. Black squares represent stable nuclei. The boundary line enclose those isotopes which are known to exist. The neutron (vertical) and proton (horizontal) shell closures are indicated as well as the regions of strong deformation (enclosed by curved lines). Those isotopes are marked (dotted areas) which have been investigated by optical spectroscopy in long isotopic chains leading far off stability.

3. OPTICAL SPECTROSCOPY OF SHORT-LIVED ISOTOPES

Optical spectroscopy of short-lived isotopes started some 20 years ago and provided an impressive set of data (Otten 1989). Today, charge radii (Aufmuth *et al* 1987), spins, and moments (Raghavan 1989) have been measured for about 30 elements in long isotopic chains for about 500 nuclear ground or isomeric states (Fig.1). Several techniques have been applied:

* Radioactive detection of optical pumping (RADOP) (Bonn *et al* 1976)
* Fluorescence spectroscopy in cells (Kluge 1985, Sprouse *et al* 1990)
* Fluorescence spectroscopy on collimated atomic beams (Jacquinot *et al* 1979, Meisel *et al* 1987)
* Laser-induced optical pumping (LIOP) of collimated atomic beams (Jaquinot *et al* 1979, Thibault 1985)
* Collinear spectroscopy (CS) with fluorescence detection (Otten 1981, Neugart 1985, 1987)
* Collinear spectroscopy with fluorescence/atom coincidence (Eastham *et al* 1986)
* Collinear spectroscopy with detection by collisional ionization (Neugart *et al* 1986) or charge exchange (Silverans *et al* 1988)
* Collinear spectroscopy with detection by RADOP (Arnold *et al* 1987)
* Collinear spectroscopy with detection by resonance ionization spectroscopy (Aseev *et al* 1990, Schulz *et al* 1990)
* Resonance ionization (mass) spectroscopy (RIS, RIMS) (Barzakh 1990, Passler *et al* 1990)
* Resonance ionization mass spectroscopy with pulsed laser induced desorption (RIMS/PLID) (Passler *et al* 1990, Le Blanc *et al* 1990)

The references given above are by no means complete. For further references and details the reader is refered to some recent reviews (Kluge 1987, Neugart 1987, Otten 1989) and the proceedings of the conference 'Lasers in Nuclear Physics' (Bemis and Carter 1982).

3. ON–LINE RESONANCE IONIZATION SPECTROSCOPY

Resonance ionization spectroscopy (RIS) was first applied to the study of short–lived isotopes at Leningrad. These experiments are described in the contribution to this conference by Barzakh *et al* (1990). The combination of RIS with collinear spectroscopy is subject of the contributions by Aseev *et al* (1990) and Schulz *et al* (1990). Gold and platinum isotopes were studied at ISOCELE/Orsay (Le Blanc *et al* 1990) and at ISOLDE/CERN (Passler *et al* 1990).

Figure 2 shows those elements (marked with an open circle) which have been investigated at on-line isotope separators. In addition, those elementes are marked by black triangls, where RIS studies of stable or long-lived isotopes have been reported.

4. COMPARISON OF TECHNIQUES

There is no technique which is superior to all other in respect to the investigation of short-lived isotopes. Table 1 gives some guide lines and states some criteria. However, the choice of an appropriate technique depends on the element under investigation (chemical reactivity, excitation scheme and so on), the nuclear decay properties (half life, asymmetry of beta decay etc.), the availability as on-line mass separated ion beam and, finally, on the yield of the isotope under investigation and its isobaric contamination.

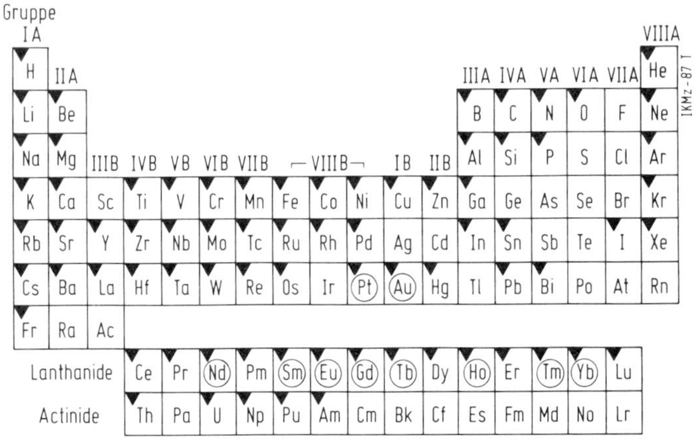

Fig.2 Periodic table of elements. The black triangles indicate those elements where resonance ionization spectroscopy experiments have already been performed. Those elements are marked by open circles where nuclear ground state properties have been studied in long isotopic chains at accelerators.

Conceptionally, fluorescence spectroscopy in cells or in collimated thermal atomic are the most simple techniques. Collinear spectroscopy with fluorescence detection is the most general technique since this scheme is tailored for the investigation of ions delivered by an on-line isotope separator. Ions as well as atoms (obtained by charge exchange in the beam) can be investigated. If there is no resonance transition in the ionic level system suitable for excitation by a cw-laser, a ground or metastable state might be populated in the atomic system by charge exchange. In this way the isotope of allmost any element produced in the ion source of an on-line isotope separator is accessible for investigation. This technique has also the highest resolving power due to the velocity bunching effect. This reduction in Doppler width near to the natural linewidth leads also to high efficiency: Since the total area under the resonance curve remains unchanged, the peak intensity of the resonance rises by the same factor by which the line width is reduced.

Resonance ionization (mass) spectroscopy is well suited when the isotope under investigation is only available as daughter in the nuclear decay of the mass-separated ion beam. For example, refractory elements will not leave the target matrix of an ISOLDE-type separator although they are produced in the nuclear reaction. Quite generally, those techniques are more sensitive where the optical resonance is detected via a detection of charged particles, γ-rays, or fast atoms instead of the photons of fluorescecnce. This is due to the small solid angle of photon collection, the limited quantum efficiency of photomultipliers and their background caused by radioactivity and dark current. As shown in table 1, several techniques avoid photon detection among which RIMS is the one with lowest background. For example, less than 1 background event per 1000 laser shots where observed in the RIMS and RIMS/PLID experiment on Au and Pt (Wallmeroth *et al* 1987, 1989, Krönert *et al* 1988). Still, the overall efficiency of 10^{-4} (defined als the ratio of ions detected in resonance to the number of atoms collected from the the implanted ions) limits the minimum yield to be of the order of 10^{4} atoms per second.

Table 1. Comparison of different on-line techniques for study of short-lived isotopes. Outstanding performance is marked by **bolt**, good one by standard, and less good performance by *tilted* letters. The numbers given as minimum yield are those cited in the literature or are calculated for a nuclear half life of some seconds. They are not absolute but strongly depend on the isotope under investigation.

Technique	Applicability	Example	Resolution [MHz]	Minimum yield [s^{-1}]	Background/ detected part.
RADOP in cells	volatile elements	alkalis Hg	*10^3*	10^3	asymmetry effect/ β, γ
Fluorescence spectroscopy in cells	inert elements	Hg, Au, Yb	*10^3*	10^3	*stray light*/ photons
Fluorescence spectroscopy in collimated beams	**general**	alkaline earths	10^2	*10^8*	**dark current**/ photons
LIOP in collimated atomic beams	surface ionizable elements	alkalies	10^2	10^6	opt. pumping effect/atoms **or ions**
CS with detection of fluorescence	**general, direct use of ion beam**	many elements	**10**	10^4	radioactivity/ photons
CS with detection by collisional ionization	high-lying metastable state	noble gasses	**10**	**10^2**	isobars/ **fast ions**
CS with detection by charge exchange	metastable state of ions	Sr	**10**	**10^2**	isobars/ **fast atoms**
CS with coincidence of fluorescent photon and atom	*weak optical pumping required*	Sr, Ba	**10**	**10^2**	isobars/ **fast atoms or ions**
CS with detection by RADOP	*dedicated problems*	^{11}Li	**10**	**10^2**	asymmetry effect/ β, γ
CS with RIS detection	general, direct use of ion beam	Yb	10^2	10^6	isobars/ **fast ions**
RIS or RIMS	**general, daughters, accumulation**	rare earths	10^2	10^4	**residual-gas ions/ ions**
RIS/PLID	**as above, refractory elements**	Au, Pt	*10^3*	10^4	**residual-gas and desorption ions/ ions**

5. CONCLUSION AND OUTLOOK

Resonance ionization (mass) spectroscopy is a rather general, very sensitive technique for the investigation of short-lived isotopes which is unique in case the isotope under investigation can only be obtained as daughter isotope at an on-line isotope separator. Still important improvements are possible like the combination which detection of nuclear decay after resonant ionization (Backe *et al* 1990). Higher efficiencies of the order of some 10 percent are feasible in a laser ion source (Mishin 1990, F. Ames *et al* 1990, Fedoseev *et al* 1990). Still this techniques is not yet exploited for the study of nuclear ground state properties at on-line isotope separators but only tested until now in on-line experiments as efficient, selective ion source (Fedoseev *et al* 1990).

It can be concluded that resonance ionization (mass) spectroscopy will continue to play an important role at the existing and the new isotope production facilities like the PS-Booster-ISOLDE at CERN and the IGISOL-systems in Japan.

This work was supported by the Bundesminister für Forschung und Technologie under grant number 06HZ188-I.

REFERENCES

Ames F *et al* 1990 contribution to this conference
Arnold E *et al* 1987 *Phys. Lett.* **B197** 311
Aseev S A *et al* 1990 contribution to this conference
Aufmuth P *et al* 1987 *At. Data Nucl. Data Tables* **37** 455
Backe H *et al* 1990 contribution to this conference
Barzakh A E 1990 contribution to this conference
Bemis C E and Carter H-K 1982 (editors) Proceedings: *Lasers in Nuclear Physics*, Harwood Academic Publishers, New York
Bonn J *et al* 1976 *Z. Phys.* **A276** 187
Eastham D A *et al* 1986 *J. Phys. G: Nucl. Phys.* **12** L205
Fedoseev V N *et al* 1990 contribution to this conference
Jaquinot P. *et al* 1979 *Rep. Progr. Phys.* **42** 773
Kluge H-J 1985 *Hyperfine Interactions* **24** 69
Kluge H J 1986 *ISOLDE User's Guide.* CERN, Geneva, Yellow Report 86--05
Kluge H J 1987 *Hyperfine Interactions* **37** 347
Krönert U *et al* 1988 *Z. Phys.* **A331** 521
Le Blanc F *et al* 1990 contribution to this conference
Meisel G *et al* 1987 *Hyperfine Interactions* **38** 723
Mishin V I 1990 contribution to this conference
Neugart R 1985 *Hyperfine Interactions* **24** 159
Neugart R *et al* 1986 *Nucl. Instr. Meth.* **B17** 354
Neugart R 1987 in: *Progress in Atomic Spectroscopy D* (Ed.: H.-J. Beyer and H. Kleinpoppen). Plenum Press, New York p. 75
Otten E W 1981 *Nucl. Phys.* **A354** 471c
Otten E W 1989 *Treatise on Heavy Ion Science 8* ed D A Bromley (New York: Plenum Press) p 515
Passler G *et al* 1990 contribution to this conference
Raghavan P 1989 *At. Data Nucl. Data Tables* **42** 89
Ravn H L and Allardyce B W 1989 *Treatise on Heavy Ion Science 8* ed D A Bromley (New York: Plenum Press) p. 363
Thibault C 1985 *Hyperfine Interactions* **24** 95
Schulz *et al* 1990 contribution to this confererence
Silverans R E *et al* 1988 *Phys. Rev. Lett.* **60** 2607
Sprouse G D *et al* 1989 *Phys. Rev. Lett.* **63** 1463
Wallmeroth K *et al* 1987 *Phys. Rev. Lett.* **58** 516
Wallmeroth K *et al* 1989 *Nucl. Phys.* **A493** 224

Inst. Phys. Conf. Ser. No 114: Section 1
Paper presented at RIS 90, Varese, Italy, 16–21 September 1990

Detection of very rare isotopes by collinear resonance ionization of accelerated atoms

S.A. Aseyev, Yu.A. Kudryavtsev, V.S. Letokhov, V.V. Petrunin

Institute of Spectroscopy, USSR Academy of Sciences, 142092 Troitsk, Moscow Region, USSR

ABSTRACT: Results are presented of the work on the development of a method for detecting ultrarare radioactive isotopes, based on the collinear isotope-selective laser ionization of atoms in a fast beam. The rare isotope ^3He was detected at a selectivity of 10^9, and the isotope-selective ionization of krypton atoms was realized. A scheme is suggested for detecting the rare radioactive isotopes ^{85}Kr and ^{81}Kr.

1. INTRODUCTION

One of the problems in laser spectroscopy is to develop a detector for very rare radioactive isotopes contained in samples together with the abundant stable isotopes of the same elements (Letokhov 1981, 1987; Hurst 1988). The solution of this problem will help to solve many a problem in nuclear physics, geochemistry, environment protection, detecting past events, and so on.

The paper describes a fairly universal method of detecting rare isotopes, based on the collinear isotope-selective laser ionization of fast isotopic atoms in a beam. The idea of the method (Kudryavtsev, Letokhov 1982) is as follows. The atoms of all isotopes in the substance of interest are ionized and accelerated to an energy of eU_A in an ion source and then turned to neutral in a charge-exchange cell. The ions that failed to turn to neutral atoms are extracted from the beam by means of a filtering capacitor. The atoms of the isotope to be detected are then raised to a Rydberg state in an isotope-selective fashion in a field-free region. The excited atoms are ionized in the electric field of an ionizer, and the ions thus produced are deflected onto a detector. Since in the field-free region all the isotopic atoms of the same energy move with different velocities, there occurs an additional isotope shift in collinear excitation. The magnitude of this shift is given by

$$\Delta\nu^{is} = \pm(\nu_0/c)(2eU_A)^{1/2}[1/(M_1)^{1/2} - 1/(M_2)^{1/2}],$$

where M_1 and M_2 are the masses of two isotopes, ν_0 is the atomic transition frequency, and c is the velocity of light. The plus and minus signs correspond to the laser light directions with and against the atomic beam, respectively. The solid lines in Fig. 1 show the additional isotope shift (for $\Delta M = 1$) as a function of the isotope mass for the energies eU_A equal to 10 and 100 keV. The circles indicate the natural isotope shifts for resonance transitions of various elements. It can be seen that at an energy of a mere 10 keV the additional isotope shift for atoms in the middle of the periodic table is 100 times their natural shift. It is also essential that an artificial isotope shift exists in any transition of any element. Another important feature of the method is that the Doppler absorption linewidth in the collinear excitation of a fast atomic beam is reduced by a factor of $2(eU_A/kT)^{1/2}$ and can be commensurable with the radiative linewidth (Kaufman, 1976). This allows one to make a narrow-band laser radiation interact with all the atoms of the isotope being detected in the fast beam and achieve their highly selective excitation. The method makes it possible to effect a stepwise excitation to Rydberg states of high-ionization-potential atoms by means of the existing dye lasers, provided the atoms possess high-lying metastable states to be populated in the course of charge exchange. It is also necessary that the ionization potential of the atoms contained in the charge-exchange-cell be close to the electron binding energy in the metastable states.

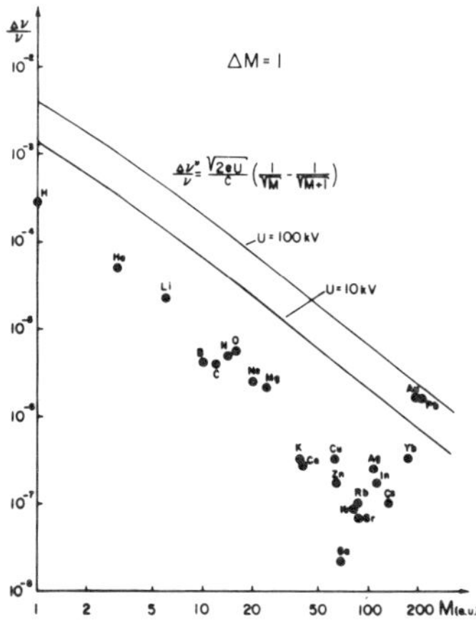

Fig. 1. Relative isotope shift of resonance lines of various elements (circles) and kinematic isotope shift of accelerated atoms 10 and 100 keV in energy as a function of atomic mass (solid curves) for $\Delta M = 1$

2. DETECTION OF THE ISOTOPE ^{3}He

The concentration of the rare isotope ^{3}He relative to that of the abundant isotope ^{4}He in various objects varies over a wide range. To illustrate, the relative concentration of ^{3}He in the atmosphere is equal to 1.4×10^{-6} and remains the same throughout the globe; it is 10^{-7}-10^{-8} in the commercial-grade helium gas obtained from natural gas deposits and comes to 10^{-9}-10^{-10} in some minerals.

The experiments on detecting the rare isotope ^{3}He (Kudryavtsev, Petrunin 1988) have shown that the capabilities of the technique are limited by the background noise signal due to collisions between the accelerated atoms of the abundant isotope ^{4}He and residual gas molecules. At a residual gas pressure in the vacuum system of 10^{-7} mm Hg, the level of the background noise signal corresponded to a ^{3}He concentration of 10^{-6}.

In the present work, to combat the background noise, use was made of an additional time-of-flight separation of the accelerated isotopic atoms in a pulsed beam. This made it possible to suppress greatly the noise signal and detect reliably the rare isotope ^{3}He in samples with its relative concentration as low as 10^{-9}.

The experimental setup is shown schematically in Fig. 2*a*. A beam of helium ions accelerated to an energy of eU_A = 3.9 keV is produced by a hot-cathode ion source and then focused with a single electrostatic lens and deflected by means of a deflector d_1 to free it from the neutral component. The deflected beam is then made to pass through a plane

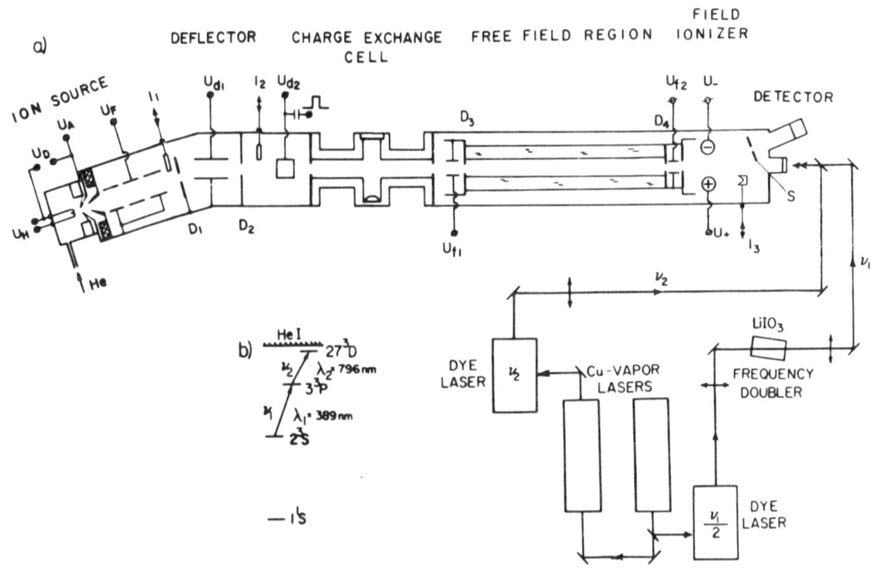

Fig. 2. (*a*) Schematic diagram of the experimental setup and (*b*) energy level diagram of the helium atom

capacitor d_2 used as an ion beam modulator. By applying a voltage of $U_{d2} = 50$ V across the capacitor d_2, the ions can be completely deflected from the axis of the setup. Modulation is effected by means of short square pulses returning the ion beam to the axis. The time of flight from the modulator to the detector is 4.5 μs for ^{3}He and 5.2 μs for ^{4}He. With the modulating pulse duration τ_p being as short as 0.5 μs, the isotopes are separated completely.

Downstream of the modulator, the ion beam enters a charge exchange cell containing potassium vapor at a temperature of 160°C, in which around 40% of the ions exchange their charge to become neutral atoms, some 3/4 of the atoms being produced in the 2^3S triplet state. The ions that failed to exchange their charge and the atoms produced in highly excited states are extracted from the beam by the electric field (9 kV/cm in strength) of a filtering capacitor f_1. Neutral atoms enter a 1 m long field-free region where they undergo two-step laser excitation from the 2^3S metastable state state to the 27^3D Rydberg state via the 3^3P intermediate level. Excitation is effected when a group of ^{3}He atoms (at $\tau_p = 0.5$. μs, the group is 25 cm long) is at the exit from the field- free region (diaphragm D_4 2 mm in diameter). The diagram of the energy levels of He used in excitation is presented in Fig. 2*b*.

The Rydberg atoms entering the electric field of the two cylinders ⊕ and ⊖ (usually $U_{f2} = 0$) are ionized and deflected by the same field onto an adjustable slit S with a secondary electron multiplier behind. Such ionization of the Rydberg atoms by the transverse electric field features dispersion with respect to n, i.e., the signals from different n^3D states are observed at different potentials U_+ and U_- of the cylinders.

The $^3He^+$ ions are detected during a time interval of τ_{det} = 0.6 μs by means of a gated counting system a delay time of τ_d = 4.5 μs after the modulating pulse is applied to the capacitor d_2. The background ions resulting from collisions involving ^{4}He atoms arrive at the detector a delay time of t_d = 5.2 μs after the modulating pulse.
The Rydberg states are excited with two pulsed dye lasers generating at frequencies ν_2 and $\nu_1/2$. Radiation at ν_1 is then obtained by frequency doubling in a $LiIO_3$ crystal. Radiations at ν_1 and ν_2 (λ_1 = 389 nm, λ_2 = 796 nm) are then brought together by means of a dielectric mirror and directed strictly counter to the accelerated atomic beam.

The dye lasers are pumped by two copper-vapor laser tubes, one being used as an oscillator and the other, as an amplifier. This ensures reliable locking of the dye lasers in step with each other. The repetition frequency of the laser pulses is 8.6 kHz, and their duration is some 18 ns.

Figure 3 presents the counting rate of ^{3}He ions as a function of the isotope concentration in the gas. As can be seen, the function is a straight line, and the results of our measuremenets of ^{3}He isotopic concentrations in air and commercial-grade helium gas are in full agreement with the literature data. The counting rate of photoions of a rare isotope at a concentration of C is $N = I\tau_p f \eta C$, where I is the continuous atomic beam current, τ_p is the beam current pulse duration, f is the laser pulse repetition frequency, and η is the efficiency of exciting the atoms and detecting the photoions produced. For f = 8.6 kHz, $I = 2.3\times10^{11}$ s^{-1}, τ = 0.5 μs, $C = 10^{-6}$, and η = 0.01, the ^{3}He photoion counting rate is N = 10 counts per second. The background noise signal corresponds to a ^{3}He concentration of 10^{-9} at a vacuum of 10^{-7} Torr in the system. The efficiency of detecting the rare isotope atoms, determined as the ratio between the counted number N of atoms to the total number of atoms in the beam, equal to $C\times I$, amounted to 4.3×10^{-5}. The detection efficiency can be improved tenfold, to amount to 4×10^{-4}, by enhancing the laser pulse energy. The background noise signal in that case will correspond to a ^{3}He concentration of 10^{-10}.

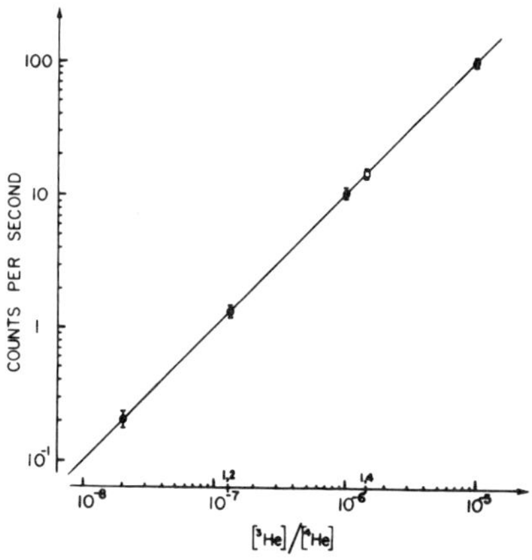

Fig. 3. Photoion signal from ^{3}He as a function of ^{3}He concentration. The square indicates helium gas extracted from the atmosphere

3. ISOTOPE-SELECTIVE IONIZATION OF KRYPTON ATOMS

The detection of the rare radioactive krypton isotopes ^{81}Kr ($T_{1/2}$ = 2.1× 10^5 years, $C = 5\times10^{-13}$) and ^{85}Kr ($T_{1/2}$ = 10.7 years, $C = 5\times10^{-11}$) is of interest in environment protection, as well as in the study of hydrological processes occurring on a time scale of hundreds of thousands of years.

At present, there exists a method for detecting rare krypton isotopes, based on a combination of a repeated filtration of the rare isotope of interest by means of a mass filter with subsequent laser resonance photoionization detection of Kr atoms (Thonnard *et al.* 1987).

Our experiment on the isotope-selective ionization of krypton atoms was conducted on the setup described above. Metastable krypton atoms in the $1S_5$ state were prepared by quasiresonant charge exchange between krypton ions and potassium atoms (Fig. 4). The first-step wide-band dye laser radiation of $\Delta\nu_1 = 1\ cm^{-1}$ propagating counter to the atomic beam raised all the krypton isotopes from the $1S_5$ to the $2P_6$ level, $\lambda_1 = 7602$ Å. The laser was pumped with the yellow line of a copper vapor laser. The second-step radiation ν_2 was the green line of the copper laser, $\lambda_2 = 5106$ Å. This line resonated with the krypton atoms on the $2P_6$-$24d[3/2]_2$ transition, the atoms being 2.7 keV in energy and the light propagating in the same direction as the atomic beam. To reduce the bandwidth $\Delta\nu_2$ to some 0.03 cm^{-1}, use was made of two Fabry-Perot etalons 10 and 30 mm in free spectral range. To tune to resonance with different isotopes in the second-step transition, the atomic beam was retarded by applying a decelerating voltage across the charge-exchange cell, the ionizer voltage being corrected in step. The experimental ionization spectrum of krypton isotopes is shown in Fig. 5*a*. A 1-GHz change in frequency corresponds to a 31-eV change in the atomic energy. Figure 5*b* shows the theoretical spectrum of all even Kr isotopes and the odd isotope ^{85}Kr, along with their relative concentrations. The width of the spectrum obtained is determined by the laser linewidth $\Delta\nu_2$ and amounts to 850 MHz.

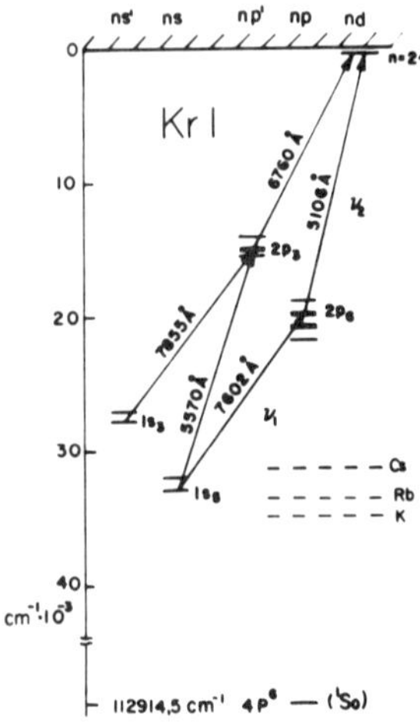

Fig. 4. Energy level diagram of the Kr atom

As in the case of the isotope He, rare krypton isotopes can be detected in conjunction with a preliminary mass separation, so that the collisional background noise can be reduced. To implement the method, use should be made of a pulsed narrow-band radiation in both excitation steps, which can be obtained by amplifying the radiation of a single-mode CW dye laser in a pulsed amplifier pumped by a copper vapor laser. Let us estimate the selectivity and sensitivity of the method using the $1S_5 \dashrightarrow 2P_3 \dashrightarrow 24d$ transitions whose hyperfine splitting constant for the ^{85}Kr isotope are known. The theoretical absorption spectrum for the first-step transition $1S_5 \dashrightarrow 2P_3$, $\lambda_1 = 5570$ Å is shown in Fig. 6. The natural isotope shift in this transition is 50 MHz. When the atoms are accelerated to 10 keV, the isotope shift increases to 1.7 GHz. Because of the hyperfine level splitting of ^{85}Kr ($I = 9/2$), its strongest line $F = 13/2 \dashrightarrow F = 11/2$ (extreme right) is 1.9 GHz distant from the nearest ^{86}Kr line. The laser excitation selectivity in this step is $S_1 = 2.4 \times 10^5$ (at Γ = 3 MHz). The selectivity S_2 in the second step will be of the same order of magnitude as S_1, and the selectivity of the two-step excitation process, $S = S_1 \times S_2$, will be no less than 10^{10}. Should the preliminary mass separation ensure an enrichment of $S_M = 10^5$, the total selectivity will be high

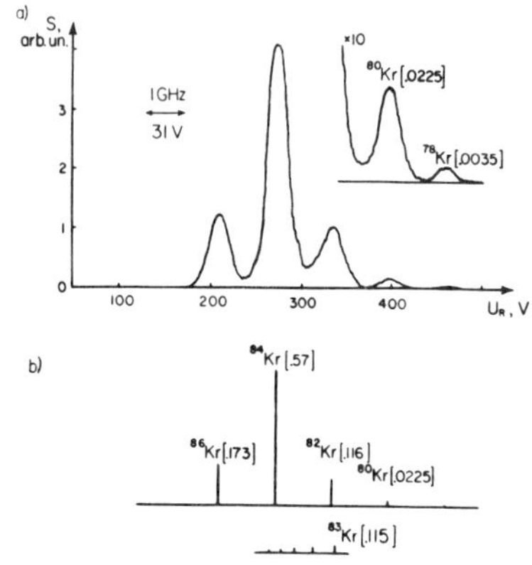

Fig. 5. (*a*) Photoion signal as a function of the decelerating voltage; (*b*) theoretical absorption spectrum of Kr isotopes for the $2P_6 \rightarrow 24d[3/2]$ transition

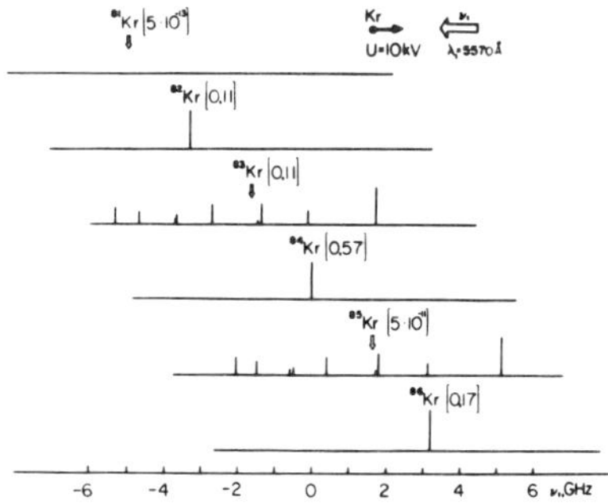

Fig. 6. Theoretical absorption spectrum of 10-keV isotopic Kr atoms for the $1S_5 \rightarrow 2P_3$ transition

enough to detect both ^{85}Kr and ^{81}Kr.

Let us now consider the sensitivity of the method. The counting rate for the ^{85}Kr atoms is $N = \eta_{c-e} \times \eta_{ex} \times \eta_{det} \times I_{Kr} \times \eta_{1S_5} \times C \times \tau_{det} \times f$, where η_{c-e} = 0.5 is the charge-exchange efficiency, η_{1S_5} = 5/12 the proportion of atoms in the $1S_5$ state, η_{ex} = 0.16 the excitation efficiency, η_{det} = 0.8 the detection efficiency, I_{Kr} the total krypton ion current, C the concentration of the isotope being detected, $\tau_{det} = L/v = 7$ μs is the detection time equal to the time it takes for the atoms to traverse the field-free region, and $f = 10^4$ Hz is the laser pulse repetition frequency. The counting rate for ^{85}Kr at a concentration of 5×10^{-11} is 0.1 counts/s at an ion current of $I_{Kr} = 1.1\times10^{12}$ ions/s (180 nA). The counting rate for the background ions produced by collision is N_{bg} = 10^{-3} counts/s at a vacuum of 10^{-9} Torr. Thus, the signal/noise ratio will be equal to 10^2.

The experimental results on the detection of ^{3}He and the estimates presented point to the possibility of using the collinear laser photoionoization method in conjunction with preliminary mass selection for detecting other isotopes (^{90}Sr, ^{3}H) of interest to environmental sciences.

REFERENCES

Hurst G S and Payne M G 1988 *Principles and Applications of Resonance Ionization Spectroscopy* (Bristol: Adam HGilger Ltd.)
Kaufman S L 1976 *Opt. Comm.* **17** 309
Kudryavtsev Yu A and Letokhov V S 1982 *Appl. Phys.* **B29** 219
Kudryavtsev Yu A and Petrunin V V 1988 *Sov. Phys. JETP* **67** 691
Letokhov V S 1981 *Comm. At. Mol. Phys.* **10** 257
Letokhov V S 1987 *Laser Photoionization Spectroscopy* (Orlando: Academic Press)
Thonnard N, Willis R D, Wright M C and Davis W A 1987 *Nucl. Instr. Meth.* **B29** 398

Paper presented at RIS 90, Varese, Italy, 16–21 September 1990

Laser resonance ionization spectroscopy of radioactive isotopes of rare-earth elements

G D Alkhazov[1], A E Barzakh[1], V P Denisov[1], V N Fedoseyev[2], V S Letokhov[2], V I Mishin[2], S K Sekatsky[2]

[1]Leningrad Nuclear Physics Institute, USSR Academy of Sciences, SU-188350, Gatchina, Leningrad district, USSR
[2]Institute of Spectroscopy, USSR Academy of Sciences SU-142092, Troitsk, Moscow district, USSR

ABSTRACT: Application of the resonance ionization spectroscopy for measurements of the electromagnetic moments and isotopic changes of nuclear charge radii of radioactive nuclei at the mass-separator IRIS and the directions of further development of our facility are described. A short rewiew of nuclear physics results obtained at this installation is presented.

1. INTRODUCTION

Investigation of nuclear ground states has a fundamental significance for nuclear physics . Of particular value is experimental information on long isotopic chains which enables to study the isotopic changes of nuclear shapes and sizes. Atomic spectroscopy is an unequalled means to obtain such information (magnetic dipole and electric quadrupole moments, isotopic changes in mean square charge radii $\Delta\langle r^2\rangle$). These values are derived from isotope shifts (IS) and hyperfine structure (hfs) of spectral lines. Since the late 70s when the first encouraging results have been received, a good many different techniques have been proposed and put into operation. The principal goal of this methodological development is to increase the sensitivity for studying the nuclei as far from stability as possible . Among the very successful methods we could mention laser-excited fluorescence in resonance cells, collinear laser spectroscopy and some others (Otten 1989). Our method - resonance ionization spectroscopy (RIS)- proves to be one of the most efficient and promising technique for the IS and hfs investigations of radioactive atoms. The RIS method was put into effect on-line with the mass-separator IRIS and synchrocyclotron at LNPI in collaboration with Institute of spectroscopy (Troitsk) in 1984 (Fedoseyev et al 1984). By now approximately 100 ground and long lived isomeric states of the rare-earth nuclei have been studied: 132-144 Nd, 138-147 Sm, 138-159 Eu, 146-160 Gd, 147-159 Tb, 152-165 Ho, 156-172 Tm (Alkhazov et al 1987, 1988, 1989a).
In this paper we shall describe briefly our experimental set-

up (Sec.2), outline the directions of further development and refinement of our RIS facility (Sec.3) and present a short review of the most interesting nuclear physics results derived from the data obtained at LNPI (Sec.4).

2. EXPERIMENTAL SET-UP

2.1 RIS for nuclear physics measurements

The principle of the experimental method is the following. The atoms under study are excited sequentially from level to level into autoionization or Rydberg states by laser radiation of several frequencies. The ions arising from the decay of these states are detected by channel electron multiplier. In the case of Rydberg states the ionization is produced by pulsed electric field (10 kV/cm). For the rare-earth elements three-step excitation is utilized. Some examples of corresponding photoionization schemes are presented in fig.1. The laser radiation frequencies for the second and the third excitation steps are kept in resonance with the atomic transitions chosen and the linewidths of these radiations are sufficiently broad to ensure overlapping both the IS's and hfs of the corresponding atomic lines. For the first excitation step the narrow-band laser is used, its frequency being scanned over a suitable range. Photoions appear when the scanned frequency coincides with the first-step atomic transition frequency. The optical spectra represent the numbers of detected photoions versus the scanned frequency (Fig.2).

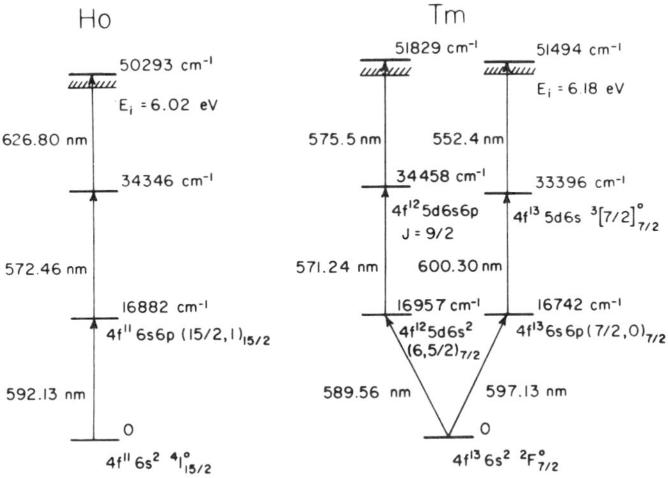

Fig.1 Three-step photoionization schemes for Ho and Tm atoms.

2.2 Experiment

The isotopes under investigation were produced in uranium and tantalum targets by 1 GeV protons from LNPI synchrocyclotron. The mass of the tantalum target was near 100g, uranium target contained up to 40g of uranium. After mass-separation, 30 keV ion beam of radioactive isotopes was directed into collector - a tantalum tube with a plugged end (50 mm long and 5 mm in diameter). When this tube was heated, the implanted ions were evaporated and emerged in the opposite direction as a collimated atomic beam (Fig.3). This beam was intersected at right angles by 3 dye laser beams tuned to the frequencies of 3 consecutive atomic transitions. To increase the efficiency of photoionization the laser beam crossed the atomic beam several times using successive reflection from two plane mirrors. Some of the laser radiation was split off and directed into a separate vacuum chamber to obtain the hfs spectra of stable reference isotopes.

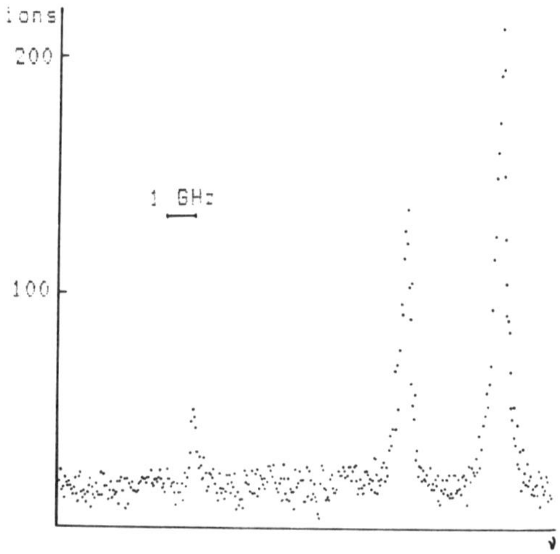

Fig.2. Hfs spectrum of ^{151}Tb (I=1/2)

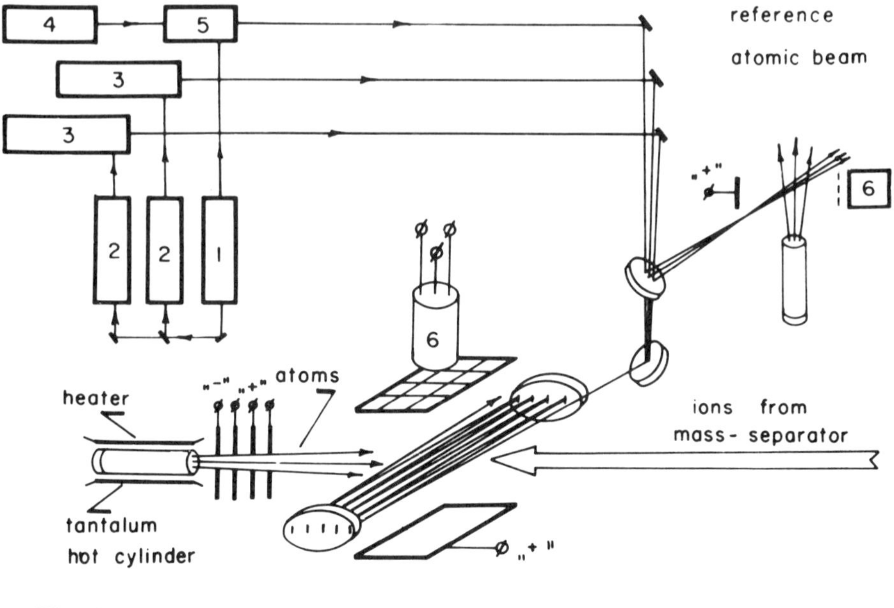

Fig.3. Experimental set-up.
1 - Cu-vapor laser, 2 - Cu-vapor amplifiers, 3 - dye lasers, 4 -Ar^{+} and CW dye lasers, 5 - dye amplifier, 6 - channel electron multiplier.

The dye lasers were pumped by three pulsed Cu-vapor lasers,

one of them being operated as an oscillator and the other two as amplifiers. The laser pulse repetition rate was 10 kHz. For the first excitation step a CW dye laser pumped by an Ar laser was used. The output from this CW dye laser was amplified in a pulsed laser amplifier pumped by one of the Cu-vapor lasers. The linewidth of the amplified radiation was 30-50 MHz. The observed width of the optical lines in our experiments was determined primarily by the residual Doppler broadening due to the angular divergence of the atomic beam. The achieved resolution of 100-200 MHz was usually sufficient to analyse successfully the hfs of the isotopes in question. The overall efficiency, i.e. the ratio between the number of photoions recorded at the resonance frequencies of the excitation lasers and the number of the mass-separated ions entering the collector per second, was as a rule 10^{-4}-10^{-5}. This efficiency permitted to carry out experiments with the mass-separated ion fluxes down to 10^{4}-10^{5} pps.

3. GENERAL OUTLOOK FOR THE FURTHER DEVELOPMENT OF THE EXPERIMENTAL INSTALLATION.

It is obvious that the sensitivity of the experimental installation may be increased by lowering the background. Time-of-flight mass-spectrometer used in the RIMS experiments (Wallmeroth et al 1987) considerably reduces the ion background. Similar mass-spectrometer for our experiment will be put into operation in the near future.

The second direction of the development of our facility is the application of RIS at the entrance of the mass-separator. To provide a sufficiently high resolution (similar to that at the exit of the mass-separator) it is necessary to collimate atomic beam from the target. This beam is intersected by 3 dye laser beams at right angles (Fig.4). The ions obtained after photoionization can be transported through the mass-separator to the detector. This eliminates the losses caused by low efficiencies of the surface or discharge ionization in the usual ion sources (10^{-3}-10^{-1}) . Besides, the background conditions are essentially improved in this case due to the reduction of the overall ion flux in the mass-separator. These advantages

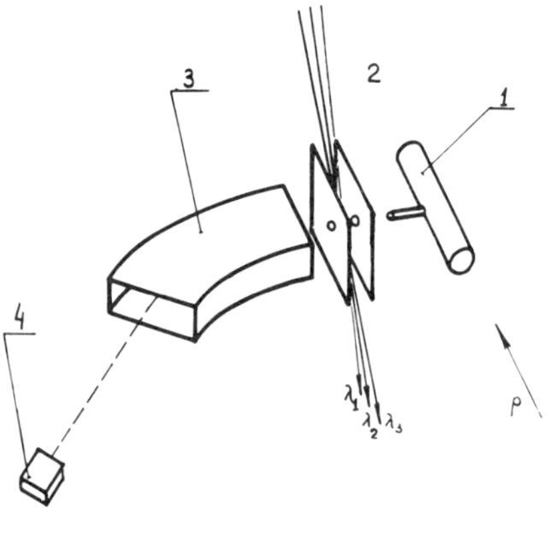

Fig.4. Schematic diagram of RIS experiment at the entrance of the mass-separator: 1-target, 2-laser radiations, 3 - mass-separator, 4-detector

have been confirmed in the course of the preliminary experiments with Yb isotopes.

Recently a higly efficient laser ion source has been proposed and put into operation at LNPI (Alkhazov et al 1989b). In this ion source the laser beam is directed immediatly in the output opening of the target instead of the perpendicular geometry of the beams crossing as it has been just described. It results in a very high ionization efficiency (10^{-1}-10^{-2}) but suffers from a rather poor resolution. Nevertheless a number of acute nuclear physics problems can be proposed to be solved with the aid of this method bearing in mind its high sensitivity. Such experiments are now in progress.

4. SOME NUCLEAR PHYSICS RESULTS FROM RIS INVESTIGATIONS AT THE LNPI.

"Critical" points on the nuclide chart, i.e. the regions where drastic alteration of some nuclear characteristics occurs, are of particular interest for nuclear physics. Our experimental efforts were concentrated in the vicinity of two of them: near the magic neutron number N=82 and near "critical" neutron number N=90. The problem of importance was the investigation of the influence of the third particular point - the proton subshell closure at Z=64 - on the changes of nuclear properties near the above mentioned critical points. At the same time the even-Z nuclei of the shape transition region near N=90 have been studied at CERN by collinear laser spectroscopy (Otten 1989). At the LNPI odd-Z nuclei and some even-Z Gd isotopes of this region have been investigated. Eu isotopes have been studied by both groups. As a result it was ascertained that the abrupt change of deformation at N=88-90 (which displays itself as a jump of the $\Delta\langle r^2\rangle$ values) occurs whenever the proton number Z of the corresponding nuclei are not far from semi-magic proton number Z=64 ($|Z-64|<5$) (fig.5). It is worth mentioning that laser spectroscopy data are also sensitive to the degrees of freedom other than pure quadrupole deformation. After the thorough analisys of these data some conclusions on the octupole deformation in neutron-rich Eu isotopes and nonaxiality in neutron-deficient Tm isotopes can be drawn.

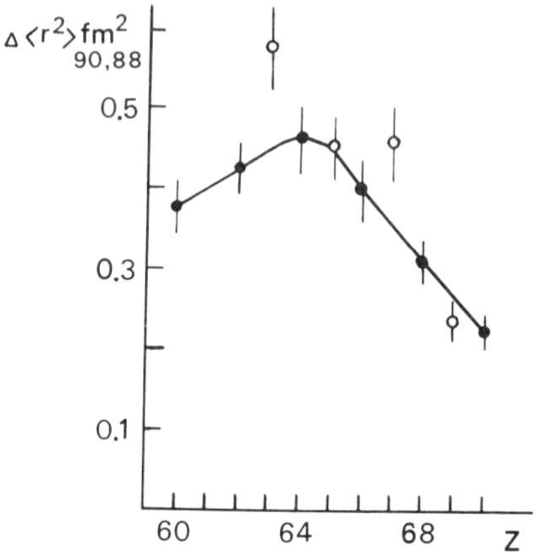

Fig.5. Value of $\Delta\langle r^2\rangle$ for nuclei with N=88 and N=90 as a as a function of Z for even-Z (full circles) (Otten 1989) and odd-Z (open circles) nuclei.

Isotopic dependencies of $\Delta\langle r^2\rangle$ for nuclei with different Z display a striking similarity at 82<N<88, while at N<82 and at N>88 one can observe a strong Z-dependence of the isotopic changes of nuclear charge radii (Fig.6) . This difference seems to be connected with the different influence of the subshell Z=64 on the growth of the deformation for the nuclei on both sides of the magic neutron number N=82 through the strong n-p interaction.

A significant role of the proton subshell Z=64 attracts a marked attention to the properties of the "magic" nucleus ^{146}Gd. With the aid of the measurement of $\Delta \langle r^2 \rangle$ for the Gd isotopic chain from stable ^{160}Gd to radioactive ^{146}Gd we have succeded to find the mean square charge radius of the "double-magic" ^{146}Gd, the values of $\langle r^2 \rangle$ for the stable Gd isotopes from mesic experiments being used.

To summarise, RIS proves to be one of the most efficient and promising technique for the IS's and hfs investigations of the radioactive atoms. The results obtained by this method promote our understanding of nuclear structure and allow new interesting and unexpected effects to be revealed.

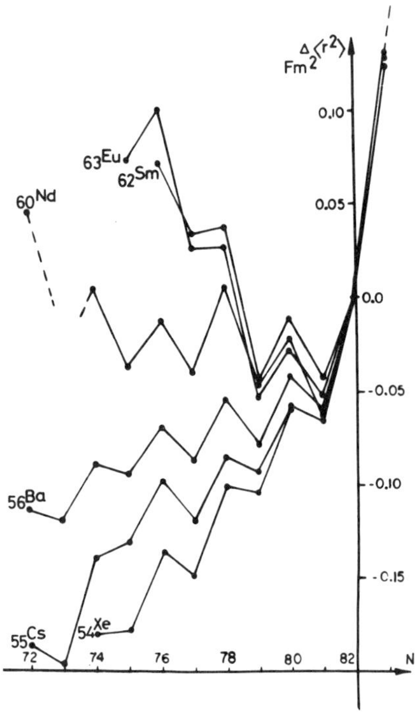

Fig.6. Isotopic dependencies of $\Delta \langle r^2 \rangle$ for Ba, Cs, Xe (Otten 1989) and for Nd,Sm,Eu nuclei with N<82.

References.

Alkhazov G D et al 1987 Preprint LNPI N 1309
Alkhazov G D et al 1988 Nucl.Phys. A477 37
Alkhazov G D et al 1989a Nucl.Phys A504 549
Alkhazov G D et al 1989b Preprint LNPI N 1492
Fedoseyev V N et al 1984 Opt.Comm. 52 24
Otten E W 1989 Tretise on heavy-ion science v 8, ed D A Bromley (New-York: Plenum) pp 517-638
Wallmeroth K et al 1987 Phys.Rev.Lett. 58 1516

Inst. Phys. Conf. Ser. No 114: Section 1
Paper presented at RIS 90, Varese, Italy, 16–21 September 1990

Radioactive decay detected resonance ionization spectroscopy in a buffer gas cell (RADRIS) at ^{208}Tl

W.Lauth, H.Backe, M.Dahlinger, I.Klaft, A.Scherrer, P.Schwamb, G.Schwickert, N.Trautmann*, U.Othmer*

Institut für Physik and (*) Institut für Kernchemie der Gutenberg Universität, Postfach 3980, D-65 Mainz, Federal Republic of Germany

ABSTRACT: An ultrasensitive laser spectroscopic method is being developed to perform hyperfine spectroscopy of heavy-ion induced reaction products far off stability. It is based on resonance ionization in a buffer gas cell combined with radioactive decay detection of the ionization process. The feasibility of such a scheme has been demonstrated in a laboratory experiment using the β-active isotope ^{208}Tl ($T_{1/2}$ = 183 s). A sensitivity $S = 1.3 \cdot 10^{-3}$ of the two step resonance ionization process via the 7s $^2S_{1/2}$- and $n = 17$ Rydberg-levels has been determined, utilizing a pulsed excimer-dye-laser combination of high repetition rate. The previously unknown nuclear magnetic moment $\mu(^{208}\text{Tl}) = 0.29(2)\mu_n$ of the ($\pi 3s^{-1}_{1/2} \otimes \nu 2g_{9/2}$)$_{I=5}$ ground state configuration of ^{208}Tl was deduced from the hyperfine splitting of the 6p $^2P_{1/2}$ – 7s $^2S_{1/2}$ transition.

Optical hyperfine spectroscopy with modern tunable narrow band lasers has been proven to be a very effective tool for the investigation of ground state properties of unstable nuclides far off stability. Nuclear spins, electromagnetic moments and the change of the rms radius have been measured for long chains of isotopes, see e.g. the recent review article by E.W. Otten [1]. Most of the results were obtained Doppler-free in collinear geometry of the laser and a fast atomic beam. A flux of about $10^4 - 10^6$ atoms per second is required for such a fluorescence light spectroscopy experiment. The nuclides are usually produced by spallation reactions induced e.g. by a high energy proton beam. However, of potential interest are also short-lived isotopes far off stability produced by heavy ion induced nuclear reactions. We particularly mention in this respect transuranium nuclides which cannot be produced by spallation reactions, and refractory elements for which atomic beams cannot be prepared. An ultrasensitive method is required for the hyperfine spectroscopy of such isotopes since the production rate in heavy ion reactions of interest lies often well below 10^4/s. Following this scheme, progress has been made recently by G.Sprouse et al. [2] who studied isotope shifts of ytterbium isotopes via fluorescence light spectroscopy in an optical cell on-line at the Stony Brook tandem-linac heavy-ion-accelerator facility with a production rate as low as 900/s.

In this contribution we report on our approach to tackle the above outlined problem. We will show that our RAdioactive Decay detected Resonance Ionization Spectroscopy (RADRIS) in an optical buffer gas cell also may open new prospects for on line laser spectroscopy at low energy heavy ion accelerators. This method combines both, the very sensitive resonance ionization with pulsed lasers and the low background detection of the resonantly ionized species by means of their nuclear radiation. In addition, the utilization of a high resolution α- or γ-spectrometer permits even the identification of a specific isotope or of an isomeric state. Advantage is taken of the facts that (i) about a fraction of 15 % of the ions recoiling after a nuclear reaction from the target into the buffer gas neutralizes while the slowing down process and (ii) that the resulting atoms remain stored for about 50 ms in the laser illuminated domain before diffusing out. This offers the possibility of several attempts to resonantly ionize the atoms if a modern pulse laser system with a repetiton rate greater than 100/s is employed. Consequently, a very high resonance ionization efficiency of the order of unity may be achieved with laser beams of mJ/cm^2 pulse energy flux.

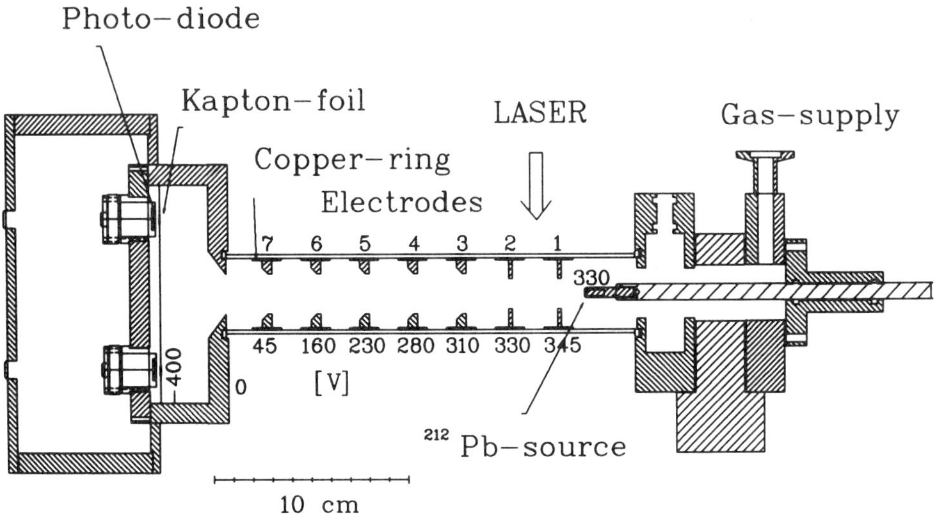

Figure 1: Experimental set-up

In order to prove the feasibility of such a scheme we carried out a laboratory RADRIS-experiment using the β-active isotope ^{208}Tl ($T_{1/2}$ = 183s). The experimental set-up is shown in fig.1. The ^{212}Pb-source was prepared by chemical precipitation of the lead daughter from the equilibrium activity of a 10 μCi ^{228}Th mother substance in aqueous solution. The ^{212}Pb source with an activity of typically 10^5 Bq is β-active and decays with a half life time of 10.6 h to ^{212}Bi. The ^{208}Tl recoil ions which leave the source after α-decay of ^{212}Bi with an efficiency of up to 100 % are slowed down in the argon buffer gas at a pressure of 28 mbar and come to rest in the domain just between the first electrode and the source. All ^{208}Tl recoils which remained positively charged after the slowing down process (approximately 85 %) will be pushed back to the source electrode by applying electrical potentials on the electrodes as indicated in fig.1. Part of the remainder of (neutral) ^{208}Tl atoms diffuses through the aperture of the first electrode the fraction of which was determined by solving numerically the diffusion equation. These atoms can be resonantly ionized by the laser beams which illuminate a part of the region between the first and the second electrode. The created ^{208}Tl ions are transported in the electric field generated by the electrode system just in front of that of the eight β-counters which was put at high voltage. As β-counters Hamamatsu photodiodes of the typ S-3590-04 have been used with an area of 1×1 cm^2 and a thickness of 300 μm.

The resonance ionization scheme is shown in the insert of fig.2. Two photons with a wave-length of 377.57 nm and 452.1 nm have to be absorbed to excite the ^{208}Tl atoms from the 6p $^2P_{1/2}$ ground state via the 7s $^2S_{1/2}$ intermediate state into the n = 17 Rydberg state from which the actual ionization process occurs radiationless by collisions with the buffer gas atoms. The laser beams were prepared with two Lambda Physik FL 2001 lasers by using the dyes BBD (λ = 377.57 nm) and Coumarin 47 (λ = 452.1 nm). Both lasers were equipped with intracavity etalons (finesse F^* = 25) to reduce the band-width to about $\Delta\nu$ = 1.2 GHz. The dye lasers were pumped with a Lambda Physics EMG 104 MSC excimer laser at a wave-length of 308 nm. The experiments were performed at a repetition rate of 200 Hz.

During the course of the experiment the frequency and band width stability of the laser was checked as a matter of routine by scanning a reference cell signal. For that purpose a buffer gas quartz cell was built which was loaded with a small sample of natural thallium

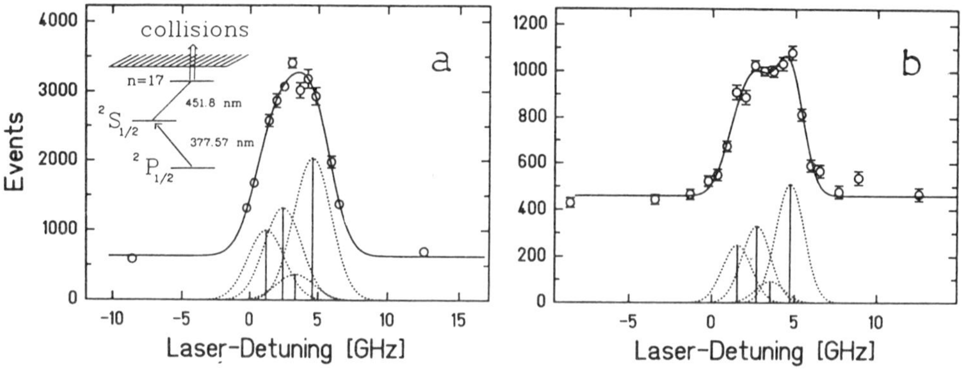

Figure 2: Resonanzionization signal of ^{208}Tl, (a) high intensity scan with 0.9 μJ/cm^2 (b) low intensity scan with 0.08 μJ/cm^2 of the 6p $^2P_{1/2} \rightarrow$ 7s $^2S_{1/2}$ transition.

and kept at a temperature of about 600 K. The charge which originated from resonance ionization of the thallium vapour atoms in an individual laser pulse was collected on two planeparallel stainless steel electrodes (area 5 cm×1 cm, distance 1 cm) and integrated by means of a charge sensitive preamplifier. The electrical field strength was choosen at the low value of 50 V/cm in order to avoid Stark shifts and -splittings of the Rydberg levels.

The choice of the $n = 17$ Rydberg state was the result of careful investigations of the resonance ionization signal with this buffer gas reference cell. During the course of these experiments the laser was operated without intracavity etalons at a band width of about 4 – 6 GHz. The pulse energy flux of the $\lambda = 452.1$ nm beam amounted to about 700 μJ/cm^2. We observed a strong increase of the resonance ionization signal with decreasing main quantum number n of the Rydberg level. At $n = 17$ the signal turned out to be a factor of about 40 larger as the observed signal when the laser frequency was tuned close to the ionization limit. At still smaller n-values the signal is expected to decrease since the ionization propability owing to buffer gas collisions drops off steeply [3]. However, for $n = 17$, Niemax et al. reported an ionization probability of still 50 % for rubidium in 16 mbar neon.

In fig.2 a high and a low intensity scan of the $^2P_{1/2} - {}^2S_{1/2}$ transition is presented keeping the $\lambda = 452.1$ nm transition tuned on resonance with the highest possible energy flux of 1 mJ/cm^2. The high intensity scan with a pulse energy flux of 0.9 μJ/cm^2 exhibits a large sensitivity, the low intensity scan with 0.08 μJ/cm^2 exhibits good resolution.

The sensitivity S of RADRIS, defined as the ratio of the numbers of detected β-particles and ^{208}Tl recoils implanted into the buffer gas, turned out to be $S = 1.3 \cdot 10^{-3}$. This experimental result is in accord with the product of (i) the neutralisation efficiency assumed to be the same as for ^{242}Am in argon [4] $\epsilon_{neut} = 0.13$, (ii) the calculated diffusion efficiency $\epsilon_{diff} = 0.12$, (iii) the measured transport efficiency of the resonantly ionized ^{208}Tl atoms $\epsilon_{trans} = 0.7$, (iv) the measured detection efficiency of the β-detectors $\epsilon_{det} = 0.2$ and finally (v) the resonance ionization efficiency for which, to be consistent with the experimental result, the large value $\epsilon_{RIS} = 0.64$ must be assumed. The biggest losses in the sensitivity result from the neutralisation and the diffusion efficiencies which, probably, can not be avoided in a simple manner.

The previously unknown nuclear magnetic moment of ^{208}Tl was extracted from the broadened structure of the low intensity scan, see fig.2b. Four hyperfine components contribute to the observed line profile which originate from transition between the F = 9/2, 11/2 hyperfine levels of the 6p $^2P_{1/2}$ and the 7s $^2S_{1/2}$ terms. A nuclear spin I = 5 was assumed for the ($\pi 3s^{-1}_{1/2} \otimes \nu 2g_{9/2}$) ground state configuration of ^{208}Tl in accord with Nordheim's rule [5]. The assumption that I = 4 gives a magnetic moment totally at variance with the extreme single particle model. In the analysis procedure the ratio of the hyperfine constants $A_{^2S_{1/2}} / A_{^2P_{1/2}} = 0.576$ was kept constant at the experimental value of ^{205}Tl, i.e. a possible hyperfine anomaly was neglected. In principle, the intensity ratios of the four hyperfine components are given by the 6j coefficients of the transition matrix elements. However, the ionization probabilities of the F = 9/2 and the F = 11/2 hyperfine levels of the $^2S_{1/2}$ term depend also on the frequency tuning of the second laser beam. This fact was taken also into consideration in the analysis procedure. The hyperfine components were approximated by Gauss-functions as suggested by scans in the reference cell under comparable experimental conditions. The five parameter fit yielded a hyperfine constant $A_{^2P_{1/2}}(^{208}\text{Tl}) = 0.377(26)$ GHz from which the nuclear moment $\mu(^{208}\text{Tl}) = 0.29(2)\ \mu_n$ was deduced assuming $A_{^2P_{1/2}}(^{205}\text{Tl}) = 21.311$ GHz [6] and $\mu(^{205}\text{Tl}) = 1.638\ \mu_n$ [7].

Finally, we compare our experimental value of the magnetic moment with the outcome of the extreme single particle model. In the stretched configuration resulting in I = 5 the magnetic moment turns out to be the sum of the magnetic moments $\mu(^{207}\text{Tl}) = 1.876(5)\ \mu_n$ [8] and $\mu(^{209}\text{Pb}) = -1.4735(16)\ \mu_n$ [9] of ^{207}Tl and ^{209}Pb, respectively, which yields $\mu_{s.p.}(^{208}\text{Tl}) = 0.403\ \mu_n$. The experimental value is somewhat smaller which most probably is caused by configuration mixings. The hyperfine anomaly was estimated to be in the order of at most 0.02 μ_n.

Prospects are quite good to study with this RADRIS method nuclear moments and isotope shifts of short lived nuclides on-line at heavy-ion-accelerator facilities. In principle, only the radioactive ^{212}Pb source has to be replaced by the heavy-ion target. We do not expect major complications associated with the passage of the heavy ion beam through the buffer gas. However, it could be advantageous to seperate the reaction products from the primary beam prior to the implantation into the buffer gas cell. The feasibility of such a possibility has been demonstrated again in an laboratory experiment with ^{208}Tl ions from a ^{212}Pb source. After post acceleration under high vacuum conditions with the aid of a suitable electrostatic field ($\Delta U = 90$KV) the ^{208}Tl recoil ions were implanted into the buffer gas cell. A clear resonance ionization signal has been observed. The sensitivity amounted to $S = 5 \cdot 10^{-4}$.

This work has been supported by the Bundesministerium für Forschung und Technologie under contract 06 MZ 188 I.

[1] E. W. Otten : *Treatise on Heavy-Ion Science,* Vol.8, p.517, ed. D. A. Bromley, Plenum Publishing Corporation, 1989

[2] G. D. Sprouse et. al. : *Phys. Rev. Lett.* 63 (1989)1463

[3] K. Niemax : *Appl. Phys. B* 32 (1983)59

[4] S. Ziegler : Diploma thesis, Mainz (1987)

[5] L. W. Nordheim : *Phys. Rev.* 78 (1950)294

[6] A.A. Radzig and B.M. Smirnov : in *Reference Data on Atoms, Molecules and Ions,* Vol. 31, Springer Series in Chemical Physics, Springer-Verlag, Berlin, Heidelberg, and New York, 1985

[7] C. Ekström, G. Wannberg, and Y.S. Shishodia : *Hyp. Int.* 1 (1976)437

[8] R. Neugart et. al. : *Phys. Rev. Lett.* 55 (1985)1559

[9] M. Anselment et. al. : *Nucl. Phys. A* 451 (1986)471

Charge radii and shape transitions in short-lived Hg, Au and Pt isotopes

G. Passler, St. Becker, G. Bollen, M. Gerber, Th. Hilberath, H.-J. Kluge, U. Krönert, A. Venugopalan+ and the ISOLDE Collaboration++
Inst. f. Physik, Univ. Mainz, Postfach 3980, D-6500 Mainz, F. R. Germany
+on leave from Bhabha Atomic Research Center, Bombay, India
++CERN, CH-1211 Geneva, Switzerland

ABSTRACT: The hyperfine structure and the isotope shift of very neutron deficient Au and Pt isotopes has been determined at the on-line isotope separator ISOLDE at CERN by resonance ionization mass spectroscopy combined with pulsed laser induced desorption of the implanted radioactive sample. The changes of the mean square charge radii were determined for ^{184}Au ($T_{1/2}$ = 53 sec) and ^{183}Au ($T_{1/2}$ = 42 sec) as well as for 15 isotopes of Pt in the range between ^{198}Pt (stable) and ^{183}Pt ($T_{1/2}$ = 6.5 min). The magnetic moment of ^{183}Pt was found to be μ_I = +0.51(3) μ_N

1. INTRODUCTION

Nuclear spins and moments and the changes of the nuclear charge radii are key input parameters for nuclear model calculations. These quantities can be determined by measuring the hyperfine structure (HFS) and the isotope shift (IS) in optical transitions with the help of tunable lasers [1]. At the isotope separator ISOLDE at CERN/Geneva these investigations can be performed on-line with the production of the isotopes, which are available in a broad isotopic range with high purity and intensity [2]. Since 1977 investigations of more than 300 short-lived isotopes have been made resulting in an enormous increase of our knowledge about nuclear structure [1]. Especially the region near Z = 80 and N = 104 has attracted a lot of attention: In 1972 the phenomena of nuclear shape transition, and later on shape coexistence and shape staggering were discovered in the neutron-deficient Hg isotopes [3]. Because these effects critically depend on shell and pairing energies, we have performed similar systematic studies in the neighbouring elements Au and Pt.

2. EXPERIMENTAL SET-UP AND PERFORMANCE

Fig. 1 shows the experimental set-up which had been installed at the new on-line isotope separator ISOLDE-3 at CERN/Geneva. Since at that time no efficient target-ion source system was available for the direct production of gold or platinum isotopes, the nuclei were obtained as daughters in the decay of isobaric Hg isotopes. Resonance ionization mass spectroscopy (RIMS) in combination with pulsed-laser induced desorption (PLID) [4, 5] was applied in order to investigate these isotopes. The mass-separated Hg^+-beam from the isotope separator was focused onto a target wheel made of graphite. After the selected isotope had decayed into Au or Pt this wheel was turned by 180° and the implanted atoms were evaporated by means of PLID using 10 ns light pulses (λ = 532 nm) of a Nd:YAG laser. The atoms desorbed were photo-ionized using tunable dye lasers pumped by a second Nd:YAG laser. In the case of gold a three-colour, three-step resonant excitation to an autoionizing state (λ_1 = 267 nm, λ_2 = 407 nm, λ_3 = 592 nm) was used whereas in the case of platinum the last ionizing step was accomplished by the frequency-doubled light of the Nd:YAG pump laser (λ_1 = 266 nm, λ_2 = 451 nm, λ_3 = 532 nm). A time-of flight (TOF) spectrometer with a two-stage acceleration region, a field-free drift tube, a 40° ion reflector and a microchannel plate detector served for mass-selective detection of the photo ions and hence reduction of background.

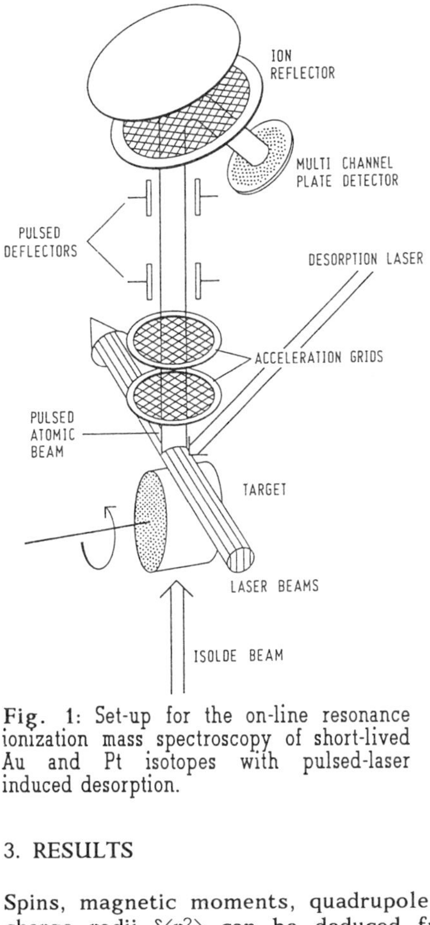

Fig. 1: Set-up for the on-line resonance ionization mass spectroscopy of short-lived Au and Pt isotopes with pulsed-laser induced desorption.

The overall detection efficiency ε_{exp} of the RIMS apparatus, which is defined as the ratio of the number of ions detected in resonance to the number of atoms implanted into the target, was determined to be $\varepsilon_{exp} = 10^{-5}$ for the case of Au in satisfactory agreement with the calculated value [5]. This is an increase of the efficiency by three orders of magnitude in comparison to a continous evaporation of the Au atoms as used in an earlier experiment on heavier Au isotopes [6]. During the run at ISOLDE an efficiency was observed which was two orders of magnitude lower than stated above. For the case of Au this reduction is due to the limited spatial overlap of the beamspot of the desorption laser with the area of the implanted ion beam (20%), the losses due to radioactive decay in the case of short-lived isotopes (50%), and the bad performance of the dye laser used for the third excitation step during the beam time. For platinum, a similar efficiency was found. Here, the desorption efficiency might have been lower and/or the third ionization step into the continuum far from saturation. A background counting rate of typically one event per 1000 laser pulses was observed in test as well as during the on-line experiments.

3. RESULTS

Spins, magnetic moments, quadrupole moments and changes of the mean square charge radii $\delta\langle r^2\rangle$ can be deduced from the HFS splitting and the IS in the first excitation step, which is in both cases a 6s → 6p transition (Au: 6s $^2S_{1/2}$ → 6p $^2P_{1/2}$, Pt: 5d^9 6s $^3D_3$ → 5d^{9}6p 7_4). Fig. 2 shows the resonances in the ion signal of ^{183}Pt, the most neutron-deficient Pt isotope investigated by us. Since only two resonances are observed, the nuclear spin is determined to be I = 1/2. The data points of Fig. 2 are corrected for the time constant of the desorption, which is of the order of 30 s (300 laser shots). The measured intensity ratio of the two HFS components of R_{exp} = 1.4(5) favours the choice of a positive sign of the magnetic moment (R_{theo} = 1.3 for $\mu_I > 0$, R_{theo} = 0.77 for $\mu_I < 0$). Hence the splitting of the two HFS components yields a magnetic moment of μ_I = + 0.51(3) μ_N, which is typical for a ν 1/2$^-$[521] orbital at large prolate deformation.
The most striking data obtained are the changes of the mean square charge radii of the Hg, Au and Pt nuclei in the mid-shell region around N = 104 compiled in Fig. 3. In the case of Hg the well-known nuclear shape transition from slightly oblate (A > 185) to strongly prolate deformation (A = 185, 183, 181) is shown as

well as the shape coexistence in ^{185}Hg and the huge odd-even staggering in the region A = 181 to 186 [3]. A similar shape transition from a slightly oblate to a strong prolate shape was also observed in gold. The deformation changes from $|\beta_2|$= 0.16 in ^{187}Au to $|\beta_2|$= 0.25 in ^{186}Au and ^{185}Au [6]. The RIMS/PLID method enabled us to investigate two even more neutrondeficient isotopes ^{184,183}Au. From an analysis of the IS a similar strong deformation is obtained for these isotopes [7]. This behaviour is due to the $\pi h_{9/2}$ intruder orbital which stabilizes the strong deformation of the isotopes below A = 186.

In the case of platinum the influence of the proton shell closure at Z = 82 is diminished drastically. As can be seen from Fig. 3 by comparison with the slope of the spherical droplet prediction [8] the nuclear deformation starts to build up already in the heavy isotopes.

A smooth change in the ground state deformation from predominantly oblate shape for A > 188 to predominantly prolate shape for A < 188 is expected from nuclear decay spectroscopy [9].

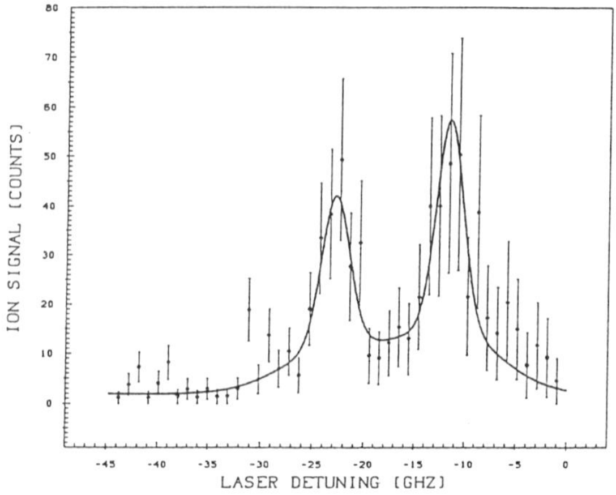

Fig. 2: Resonance signal of laser desorbed ^{183}Pt atoms (I = 1/2, $T_{1/2}$ = 6.5 min) versus the laser frequency in the first excitation step as obtained by resonance ionization mass spectroscopy. The solid line represents a fit to the data points. For this signal $3{\cdot}10^9$ ions of ^{183}Hg were implanted into the target.

No evidence for such a shape transition in Pt can be found from Fig. 3. The explanation is that IS measurements are only sensitive to a change in the size of β_2, since $\delta\langle r^2\rangle$ is depending on $\langle\beta_2^2\rangle$. The spectroscopic quadrupole moments are much more decisive for a shape transition as long as the coupling of the valence neutron to the nuclear core is known [10]. Such measurements have recently been performed by nuclear orientation [11] and RIMS/PLID experiments [12] very similar to ours. These experiments yield a change in the ground state deformation between ^{187}Pt and ^{185}Pt from oblate to prolate shape. Since the IS values show no evidence for any discontinuity in this region, it can be concluded that the Pt nuclei near A = 186 have minima in the energy-deformation curve on the prolate and oblate side at nearly identical absolute deformation, which might be connected by the γ-degree of free-

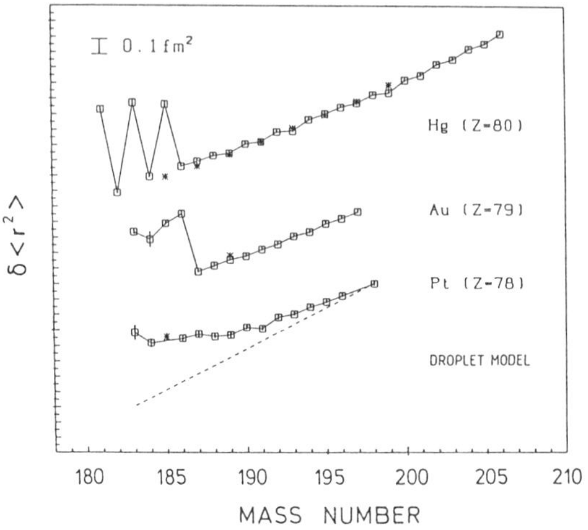

Fig. 3: Changes of the mean-square charge radii of Hg, Au, and Pt isotopes as a function of the mass number. The ground state data points are connected by eye-guiding lines. The isomeric states are indicated by asterisks. The values for Hg and Au are taken from Refs. (3) and (7, 12). The dashed line indicates the change of the radius of a spherical nucleus according to the droplet model (8).

dom. An analysis of all the spectroscopic information on the Pt nuclei by the Lund particle-plus-triaxial rotor model yields in fact a γ-soft transition from oblate to prolate shape around A = 186.

Supported by the Bundesministerium für Forschung und Technologie under grant number 06 HZ-188-I

REFERENCES

[1] E. W. Otten, *Treatise on Heavy-Ion Physics,* Vol. 8: *Nuclei far from Stability,* D. A. Bromley, ed.; Plenum Press, New York (1989), page 515

[2] ISOLDE Users Guide, H.-J. Kluge, ed.; *CERN Yellow Report 86-05,* CERN, Geneva (1986)

[3] G. Ulm, S. K. Bhattacherjee, P. Dabkiewicz, G. Huber, H.-J. Kluge, T. Kühl, H. Lochmann, E. W. Otten, K. Wendt, S. A. Ahmad, W. Klempt and R. Neugart, *Z. Phys.* A325, 247 (1986) and references therein

[4] H.-J. Kluge, *Hyperfine Interact.* 24, 69 (1985) and *Hyperfine Interact.* 37, 347 (1987)

[5] U. Krönert, St. Becker, Th. Hilberath, H.-J. Kluge, C. Schulz and the ISOLDE Collaboration, *Appl. Phys.* A44, 339 (1987)

[6] K. Wallmeroth, G. Bollen, A. Dohn, P. Egelhof, J. Grüner, F. Lindenlauf, U. Krönert, J. Campos, A. Rodriguez Yunta, M. J. G. Borge, A. Venugopalan, J. L. Wood, R. B. Moore, H.-J. Kluge and the ISOLDE Collaboration, *Phys. Rev. Lett.* 58, 1516 (1987) and
K. Wallmeroth, G. Bollen, A. Dohn, P. Egelhof, U. Krönert, M. J. G. Borge, J. Campos, A. Rodriguez Yunta, K. Heyde, C. de Coster, J. L. Wood, H.-J. Kluge and the ISOLDE Collaboration, *Nucl. Phys.* A493, 224 (1989)

[7] U. Krönert, St. Becker, G. Bollen, M. Gerber, Th. Hilberath, H.-J. Kluge, G. Passler and the ISOLDE Collaboration, *Z. Phys.* A331, 521 (1988)

[8] W. D. Myers, K.-H. Schmidt, *Nucl. Phys.* A410, 61 (1983)

[9] M. Finger, R. Foucher, J. P. Husson, J. Jastrzebski, A. Johnson, G. Astner, B. R. Erdal, A. Kjelberg, P. Patzelt, H. Hoglund, S. G. Malmskog and R. Henck, *Nucl. Phys.* A188, 369 (1972)

[10] I. Ragnarsson, *Proc. of the Int. Symp. on Future directions in studies of nuclei far from stability (Nashville 1979),* J. H. Hamilton, E. H. Spejewski, C. R. Bingham, E. F. Zganjar, eds.; North-Holland Publ. Comp., Amsterdam (1980), page 367

[11] R. Eder, C. J. Ashworth, I. Berkes, D. E. Brown, U. Dämmrich, I. S. Grant, A. G. Griffiths, E. Hagn, R. Hassani, P. Herzog, S. Ohya, M. Massaq, C. Richard-Serre, J. Rikovska, K. Schlösser, N. Severijns, N. J. Stone, W. Vanderpoortren, J. Vanhaverbeke, L. Vanneste, T. Woelfle, E. Zech and the ISOLDE Collaboration, *Hyp. Int.* 43, 469 (1988)

[12] J. K. P. Lee, G. Savard, J. E. Crawford, G. Thekkadath, H. T. Duong, J. Pinard, S. Liberman, F. LeBlanc, P. Kilcher, J. Obert, J. Oms, J. C. Putaux, B. Roussiere and J. Sauvage, *Phys. Rev.* C38, 2985 (1988)

Resonance ionization spectroscopy on a fast atomic beam

Ch. Schulz, E. Arnold, W. Borchers, W. Neu, R. Neugart, M. Neuroth,
E. W. Otten, M. Scherf, K. Wendt,
Institut für Physik, Universität Mainz, Mainz, W. Germany
P. Lievens, Leuven University, Leuven, Belgium
Yu.A. Kudryavtsev, V.S. Letokhov, V.I. Mishin, V.V. Petrunin
Institute of Spectroscopy, USSR Academy of Sciences, Troitzk, USSR
and the ISOLDE Collaboration, CERN, Geneva, Switzerland

ABSTRACT: Resonance Ionization Spectroscopy in collinear geometry has been successfully applied to a fast beam of *Yb* atoms. The atoms were excited stepwise into a Rydberg state by pulsed laser light, ionized in an electrical field and deflected onto a secondary electron detector. The efficiency was $1 \cdot 10^{-5}$ detected ions per incoming atom on a background from collisional ionization of $1 \cdot 10^{-8}$. The technique has been exploited for the measurement of hyperfine structures and isotope shifts of unstable *Yb* isotopes, in particular ^{157}Yb, ^{159}Yb and ^{175}Yb.

1. Introduction

Owing to their high resolution and sensitivity, laser-spectroscopic techniques are well established tools for the study of nuclear moments and radii of unstable isotopes. In the rare earth region two methods have provided most of the present results: collinear laser spectroscopy on fast beams (Neugart 1987) and resonance ionization spectroscopy (RIS) on thermal atomic beams (Balykin et al 1980, Alkhazov et al 1988). Both methods have their inherent advantages. While collinear spectroscopy combines high resolution with a large excitation efficiency, the RIS scheme mainly offers the detection of optical resonances with an extremly low background. Therefore the combination of both techniques (Kudryavtsev et al 1988) promises a considerable improvement in sensitivity. The present contribution reports on an application of this techique to the neutron deficient *Yb* isotopes.

2. Experimental Setup

The experiment was carried out at the ISOLDE isotope separator facility at CERN. The unstable Yb isotopes are produced among other rare earth elements in the spallation reaction $^{181}Ta(p,4p+xn)^{178-x}Yb$. After ionization in a hot surface ionization source, acceleration to $60\,keV$ and mass separation, the isobaric contamination in the beam exceeds the amount of Yb for most isotopes by several orders of magnitude. This beam is superimposed on the laser beams by means of an electrostatic deflector and then neutralized by charge exchange collisions with alkali atoms in a vapour cell. The choice of caesium as appropriate alkali metal gives a population of approximately 5% for the metastable $6s6p\,^3P_0$ which is used as the initial state for the two step excitation ladder. Downstream, in the interaction region (l = *100cm*), the atoms are excited via the $6s7s\,^3S_1$ ($\lambda = 648.9nm$) into one of the $6s21p\,^3P_{0,1,2}, {}^1P_0$ ($\lambda = 575.6nm$) Rydberg states. From here the atoms are ionized in an electrical field of about $4\,kV/cm$. Pulsed amplification of cw laser radiation is employed for inducing the transition into the

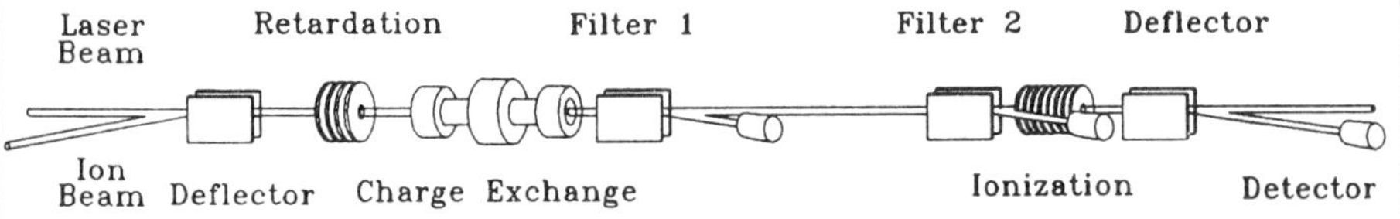

Figure 1: Experimental Setup

Rydberg state. This maintains the resolution of collinear spectroscopy and maximizes the spectral overlap between the laser radiation ($\delta\nu \approx 70\,MHz$) and the small natural linewidth ($\delta\nu < 50\,KHz$).

Optical pumping losses in this excitation scheme are mainly due to the decay of the intermediate $6s7s\,^3S_1$ state into the states $6s6p\,^3P_2,^3P_1$ ($\tau_2 \approx 15nsec$). The Rydberg states have a lifetime of $\tau_3 \approx 5\mu sec$, which is somewhat larger than the time of flight through the interaction region.

With this experimental setup (Fig.1) one has two possible sources of background: (*i*) Rydberg atoms are produced in the charge exchange process and ionized together with the Rydberg atoms produced by laser excitation. (*ii*) Ions are directly produced in collisions with rest gas atoms. The former ones are removed by ionization and deflection in the electrical field ($5\,kV/cm$) of a capacitor (Filter 1) situated in front of the excitation region. Ions produced in the excitation region are deflected to the wall by a second capacitor (Filter 2) at $500\,V$. Thus only the ions produced in the ionization region can reach the detector. The ionizer itself is constructed as an acceleration lens with eleven diaphragms at a distance of $4\,mm$. The potential on the diaphragms increases slowly along the beam direction to a peak value of $7\,kV$ and then decreases sharply on the remaining diaphragms. The Rydberg atoms are ionized mainly between the diaphragms 8 and 9, where the field strength on the axis is sufficiently high ($4\,kV/cm$). This arrangement has two important advantages: *(i)* the ionization of laser excited Rydberg atoms is restricted to a volume of $1.5\,cm^3$, which reduces the background from collisionally produced ions, and *(ii)* the ionized Rydberg atoms are accelerated after the ionization. The resulting energy difference of $3\,keV$ between the "Rydberg ions" and the background can be exploited to deflect only the Rydberg ions onto the secondary electron detector. The background count rate is further minimized by keeping the ionization region at ultra high vacuum of $2 \cdot 10^{-9}\,mBar$.

For the remaining background count rate, one can give the following estimate,

$$N_{back} = \sigma_{coll} \cdot N_T \cdot I_P \cdot L_T \cdot \eta_L \approx 40/sec,$$

with a cross section for collisional ionization at $60\,keV$ beam energy of $\sigma_{coll} \approx 10^{-15}\,cm^2$, a number of target particles of $N_T = 6 \cdot 10^7/cm^3$, a current of projectile particles of $I_P = 3 \cdot 10^{10}/s$ (corresponding to a beam of $5\,nA$), a length of the ionization region of $L_t = 1cm$. The duty cycle η_L of the copper vapour laser is 3×10^{-2}, given by the time of flight through the interaction region ($3 \times 10^{-6}s$) and the repetition rate of the copper vapour laser ($10^4\,s^{-1}$).

3. Results and evaluation of data

The apparatus was tested with measurements on the stable isotopes $^{168,170-174,176}Yb$. The ion count rate was recorded versus the acceleration voltage (Doppler tuning) at fixed frequencies of the dye lasers. The best results were obtained with the transtion into the fine structure component $6s21p\,^3P_2$. On the weakest of the stable isotopes, ^{168}Yb with the spin $I = 0$, an overall detection efficiency of $\epsilon_{exp} = 1.5 \cdot 10^{-5}$ (22 ions/s from a beam of $1.5 \cdot 10^6\,Yb$ atoms/s at a signal-to-noise ratio of $S/N = 20$ and an integration time of $0.8\,s$ per channel) was measured. The efficiency is given by

$$\epsilon_{theo} = \eta_L \cdot \epsilon_{pop} \cdot \epsilon_{exc} \cdot \epsilon_{ion} \cdot \epsilon_{trans},$$

where η_L is the duty cycle of the laser, ϵ_{pop} the population of the $6s6p\,^3P_0$ state, ϵ_{exc} the overall excitation efficiency, ϵ_{ion} the ionization efficiency of the Rydberg atoms and ϵ_{trans} the transmission of the ion optics. With reasonable values ($\eta_L = 3\%$, $\epsilon_{pop} = 5\%$, $\epsilon_{exc} = 2.5\%$, $\epsilon_{ion} = 80\%$, $\epsilon_{trans} = 70\%$) one obtains $\epsilon_{theo} = 2 \cdot 10^{-5}$, which agrees with the experimental value. For isotopes with $I \neq 0$, the efficiency is reduced by about one order of magnitude, due to the splitting into HFS sublevels.

Measurements on radioactive *Yb* isotopes were performed on the even isotopes in the sequence $^{156-166}Yb$ and on the odd isotopes $^{157,159,175}Yb$. The sensitivity is demonstrated by the first measurement of a hyperfine structure in ^{157}Yb. Fig. 2 gives a spectrum of this isotope (production rate $\approx 5 * 10^7\,atoms/s$, isobaric contamination $\approx 10^{10}\,atoms/s$), which was recorded with an integration time of $0.8\,s$ per channel.

The analysis of the data has to take into acount that the fine structure splitting between the $6s21p\,^3P_2$ and the $6s21p\,^1P_1$ Rydberg state is only $4.5\,GHz$, so that the hyperfine structures of both states overlap. This leads to a strong perturbation of the HFS sublevels ($F = I+1, I, I-1$), which has to be analyzed by a full diagonalization of the fine and hyperfine structure interaction matrix (Rinneberg 1989). The analysis of the hyperfine structure of ^{157}Yb yields the spin $I = 7/2$ and the magnetic moment $\mu = -0.639(8)\,\mu_N$. This fits well into the trend of $f_{7/2}$ neutron states above the shell closure at $N = 82$. For ^{159}Yb, the analysis confirms the spin assignment $I = 5/2$ and

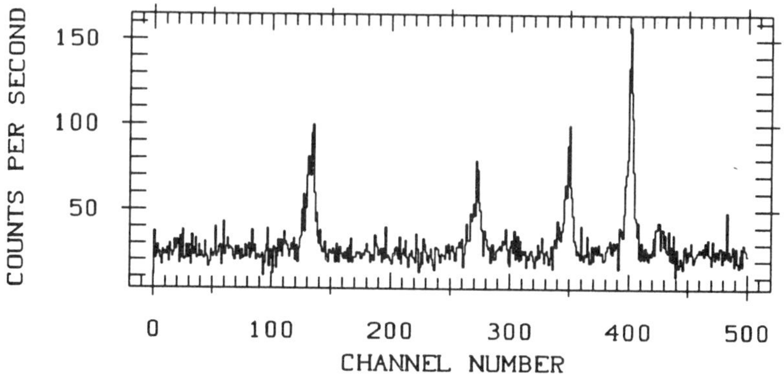

Figure 2: Spectrum of ^{157}Yb

the magnetic moment obtained from earlier measurements in the intercombination line $6s^2\,{}^1S_0 \rightarrow 6s6p\,{}^3P_1$ (Neugart et al 1983). The data taken on ^{175}Yb yield the first precise value for the magnetic moment of this isotope with spin $I = 7/2$, $\mu = 0.766(8)\mu_N$. The measurement of the isotope shift yield values for the changes in mean square charge radii $\delta <r^2>$, which, combined with earlier results (Clark et al 1979, Buchinger et al 1982, Sprouse et al 1989), now cover the even isotopes $^{152-176}Yb$ and the odd isotopes $^{157-175}Yb$.

4. Conclusion

The combination of collinear laser spectroscopy with resonance ionization detection has been shown to be a competitive technique for the investigation of unstable isotopes. It offers major advantages for beams with high isobaric contamination which require efficient background suppression.

The ionization efficiency is reduced mainly by the duty cycle of the copper vapour laser and the population of the metastable state. It could be increased by the use of a pulsed ion source (conventional or laser ion source) and with an excitation scheme that starts from a state which is more efficiently populated in the charge exchange process. Altogether a gain in sensitivity of 10 - 100 seems feasible, so that spectroscopy of 10^4 *atoms/sec* with a background contamination of 10^9 *atoms/sec* for an isotope with $I=0$ is in reach.

This work has been funded by the German Federal Minister for Research and Technology (BMFT) under the contract number 06 Mz 188I.

Alkhazov G D, Barzakh A E, Denisov V P, Ivanov V S, Chubukov I Ya, Buyanov N B, Letokhov V S, Mishin V I, Sekatsky S K, Fedoseev V N *Prc. 5th.Int. Conf. on Nuclei far from Stability, ed Towner I S, AIP Conf. Proc., Vol 164 (New York 1984) 115*

Balykin V I, Bekov G I, Letokhov V S, Mishin V I 1980 *Usp. Fiz. Nauk* **132** *293*

Buchinger F, Mueller A C, Schinzler B, Wendt K, Ekström C, Klempt W, Neugart R 1982 *Nucl. Instr. and Meth.* **202** *(1982) 159*

Clark D L, Cage M E, Lewis D A, Greenlees G W 1979 *Phys. Rev.* **A20** *(1979) 239*

Kudryavtsev Yu A, Lethokov V S, Petrunin V V 1988 *Optics Comm.* **68** *25*

Neugart R 1987 *Progress in Atomic Spectroscopy ed Beyer H J, Kleinpoppen H (New York: Plenum Press) 75*

Neugart R, Wendt K, Ahmad S A, Klempt W, Ekström C 1983 *Hyp. Int.* **15/16** *(1983) 181*

Rinneberg H 1987 *Progress in atomic spectroscopy ed Beyer H J, Kleinpoppen H (New York: PLenum Press) 157*

Sprouse G D, Das J, Lauritsen T, Schecker J, Berger A, Billowes J, Holbrow C H, Mahnke H E, Rolston S L 1989 *Phys. Rev. Lett.* **63** *(1989) 1463*

Inst. Phys. Conf. Ser. No 114: Section 2
Paper presented at RIS 90, Varese, Italy, 16–21 September 1990

Anomalous lineshapes and signals in laser-enhanced ionization spectrometry in flames—the role of 'scattered' light, dynamical Stark effects and two-step versus two-photon excitations

Ove Axner and Sten Sjöström

Department of Physics, Chalmers University of Technology, S-412 96 Göteborg, Sweden

ABSTRACT: Anomalous contributions to the signal in Laser-Enhanced Ionization spectrometry in flames are investigated. It is found that a significant part of the one-step LEI signal (> 90 %) can be attributed to ionization of atoms positioned outside the interaction region out-traced by the laser beam due to excitation of atoms by "scattered" light. This is especially significant when small beam diameters and high light intensities are being used. Furthermore, dynamical Stark effects and two-photon excitations can significantly affect the signals. Several examples are given.

1. INTRODUCTION

Laser-Enhanced Ionization Spectrometry (LEI) in flames is a laser-based spectroscopic technique which previously has been shown to be well suited for ultra-sensitive trace element analysis (Travis 1982, Travis 1984, Axner 1989a, Axner 1989b). Detection limits in the ppt-ppb region for traces in aqueous solutions have been demonstrated for a number of elements. In short, the LEI technique relies upon the basic principle that the ionization rate of atoms (through collisions, associative ionization, or photoionization) can be increased by exciting atoms by resonant laser light to high lying states. This enhanced ionization is detected by applying an electric field across the interaction region.

Although several attempts to model the LEI signal strengths have been presented in the literature [Axner 1984, Omenetto 1986] there is, however, still no good theory which accurately or even qualitatively (better than within one order of magnitude) can predict LEI signal strengths. In this work we give some explanations to why several of the previous models fail by presenting several "anomalous" contributions to the LEI signals. These contributions consist of: (i) the influence of "scattered" light; (ii) the influence of dynamical Stark effects (primarily splitting or broadening of levels); and (iii) the existence of two-photon excitations.

One of the reasons for the high sensitivity of LEI in flame is the important property that virtually all of the produced charges can be detected, i.e. there is a high charge collection efficiency in a large part of the flame. This unique property of the LEI technique has the potential of introducing significant and unexpected contributions to the measured signal. As is shown below, a significant part of the one-step LEI signal (>> 90 % in certain cases) can be attributed to ionization of atoms positioned outside the "true" interaction region, i.e. the "geometrically confined laser-atom interaction volume", due to their excitation by "scattered" light (the word "scattered" is here used in its widest meaning, i.e. it refers to scattered, diffracted as well as fluorescence light). This phenomenon is most significant when small beam diameters and high light intensities are being used.

Furthermore, virtually all of the previous models for LEI have described the excitation processes of atoms in flames by the rate-equation formalism. With such a description phenomena such as dynamical Stark effects and two-photon excitations can not be accounted for. By using a density matrix formulation of the light atom interactions all these phenomena can at least qualitatively be described.

2. EXPERIMENTAL

The experimental set-up has been described previously and will only shortly be summarized below [Axner 1990a, Axner 1990b]. The light was produced by an excimer pumped dye laser system consisting of a pump laser operating on XeCl at 308 nm and one or two dye-lasers. All experiments were performed on Sr using the 460.733 nm and 554.333 nm transitions (from the $5s^2\ ^1S_0$ ground state to the $5s5p\ ^1P_1$ state at E = 21698.48 cm^{-1} and, in the two-step experiments, further to the $5s6d\ ^1D_2$ state at 39733.11 cm^{-1}). The laser beams were directed through the flame at a height of 20 mm unless otherwise is stated. Pulse energies between a few µJ and some mJ were used together with beam diameters ranging from 2 to 8 mm. The pulse duration was around 25 ns. A commercially available 5 cm long single-slot burner head was used. The flame was fed with acetylene and air in approximate stoichiometric proportions. The signal-collecting electrical field was established by using a split-cathode arrangement and a voltage of 2500 V was applied between the pair of electrodes and the burner head. The electrodes were positioned at a height of 30 mm above the burner head with an approximate separation of 24 mm. The signal produced in the flame was amplified by a current-to-voltage amplifier placed in close proximity to the flame.

3. RESULTS AND DISCUSSION

3.1. The Influence of "Scattered" Light upon the LEI-signal.

First, a number of one-step LEI scans were measured. Figure 1 displays signals from a set of scans between 460.4 and 461.2 nm in 3 ppm Sr for a beam diameter of 2 mm and for various laser light intensities. As can be seen from the figure a distinct peak is clearly visible in all of the scans. Furthermore, even though the intensity of the light is increased by a factor of 40 between the lowest and highest intensities (Fig. 1a and Fig. 1d, respectively) the peak signal is only increased by a factor of 2. This is normally assumed to be due to optical saturation. However, during optical saturation conditions the measured signal is expected to become broader the more the transition is saturated. A simple theory for the behaviour of saturated transitions predicts that the width of the measured peaks should increase as the square-root of the light intensity, i.e. the widths of the curves should differ by roughly a factor of 6. This is indeed not the case in Fig. 1. Consequently, the assumption that the behaviour of the signal is due to optical saturation must be revised.

Another experiment performed is displayed in Fig. 2. In this case, all the three scans were measured using the same beam intensity but various diameters of the beam. The most striking observation about this set of scans is that the signal obtained for a beam diameter of 2 mm is also found on the other two scans with the only difference that for larger beam diameters there are also some "broad wings" appearing.

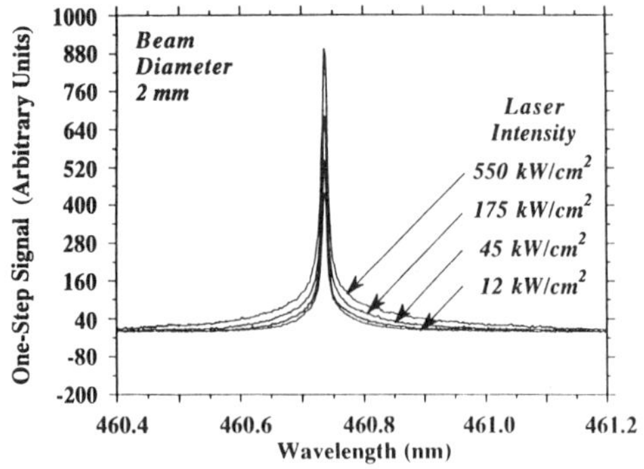

Fig. 1. *A set of one-step LEI scans in Sr for various beam intensities.*

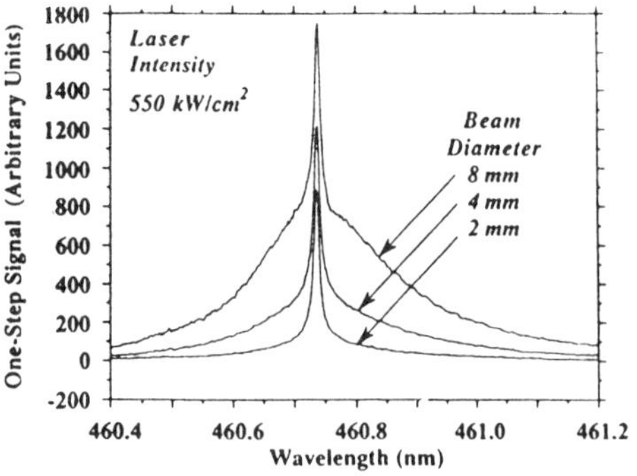

Fig. 2. *A set of one-step LEI scans in Sr for for various beam diameters.*

From these figures it is possible to gain some insight into the true physical origin of the peaks and hence also offer a possible explanation of their somewhat strange and unexpected behaviour. The "broad wings" of the uppermost curve in Fig. 2 can be explained as originating from atoms in the "true" interaction region, i.e. the region out-traced by the laser beam (in this case a volume with a circular area with a diameter of 8 mm and a length of 5 cm). Since the intensity in this part of the flame is significantly higher than what is needed for reaching saturation, the signal becomes wide, in agreement with the above-mentioned predictions. The narrow superimposed peak, on the other hand, which is not broadened by the intensity increase in Fig. 1, can be explained as originating from excitations of atoms which are positioned outside the "true" interaction region by weak "scattered" laser light. Due to the high charge collection efficiency in a large part of the flame this narrow signal from atoms outside the "true" interaction region might fully dominate the "true" signal. This is in particular the case for smaller beam areas, since then the number of atoms outside the "true" interaction region might be significantly larger than those inside, which more than compensates for the fact that they are exposed to much weaker light intensities. A consequence of this is thus that, when high laser light intensities are used in conjunction with small apertures, *the signal from the "true" interaction region might be **insignificant** as compared to that from scattered laser light.*

Additional experiments were performed to support the above mentioned assumption of signals originating from other parts of the flame than the "true" interaction region. In one investigation, a two-step experiment was performed with spatially non-overlapping beams (the second step was positioned at a height of 24 mm above the burner head while the first beam still was positioned at 20 mm - both beams had a diameter of 2 mm). From such an experiment it was found that the LEI signal indeed was enhanced when the second step was applied. In addition, upon closer examination of the signal shapes, it was found that it was only the "narrow" peak of the signal which was enhanced (approximately 10 times). This supports the assumption that the "narrow" peak originates from excitation of atoms outside the "true" interaction region.

When investigating possible mechanisms responsible for the weak exciting light outside the "true" interaction region, one finds (at least) three possibilities:

i) The light might be diffracted from an aperture positioned in front of the flame;

ii) The light might originate from laser light scattering from particles or flame molecules in the heavily illuminated interaction region in the flame; or

iii) The light might originate from fluorescence light emitted from the excited analyte atoms in the interaction region.

Already from the previously described experiments (Fig. 1) it is possible to draw the conclusion that the "scattered" laser light can not originate mainly from diffraction of the aperture [point i)]. The reason for this is that the signal from the "scattered" light does not increase significantly when the laser light intensity is increased.

In order to determine whether the laser light outside the "true" interaction region originates from emitted fluorescence light from Sr atoms in the interaction region [point iii)], a one-step experiment was performed in which the number density of strontium atoms in the flame was increased by aspirating 1, 3 and 10 ppm solutions. Such a study gave as a result that the "scattered" signal shapes were independent of concentration and that the signal strengths increased virtually linearly with concentration. Although the shape of a curves of growth obtained in this way can vary significantly due to several reasons, it is believed that in the range of concentrations studied above a quadratic dependence should have been obtained if the scattered laser light was mainly due to emitted fluorescence from Sr atoms. This is because both the fluorescence and the absorption process leading to ionization increase linearly with the number density of strontium, resulting in an overall quadratic dependence upon concentration. Hence, our experiments give no evidence for such a process as described in point iii) above.

In order to study the influence of scattered laser light from particles and molecules in the flame an experiment was performed in which the scattered laser light was investigated as a function of different flame compositions. Since the flame temperature and hence the atomization of the Sr atoms is significantly affected by the flame composition, the two-step signal (non-overlapping beams), which thus depends completely on the scattered light process, was compared to the one-step signal in order to correct for the differences in atomization throughout the experiment. It was found that the ratio of the two-step signal to the one-step signal increased significantly with increasing richness of the flame (approx. by a factor of 3). This indicates that the light scattering process originates from molecules or particles in the flame which increase in number density when the flame is run fuel-rich. Consequently, the scattering process is found to originate from some sort of saturable process from so far unknown molecules or particles in the flame.

The true process behind this finding has not yet been found. However, the effect of scattered light on one-step LEI signals has been demonstrated and shown to drastically change the signal behaviour in certain cases.

3.2. The Influence of Dynamical Stark Effects and Two-Photon versus Two-Step Transitions in LEI Spectrometry

It is a common opinion among flame spectroscopists that the rate-equation formalism is adequate for a satisfactory description of most of the phenomena associated with interactions between atoms and light in flames. The reason for this is the fact that the high collision rates in flames destroy much of the coherence of the system. It has previously been shown that in media with high collision rates the more general density-matrix formalism approaches the rate-equation formalism for one-step excitations (Daily 1977). However, as is further demonstrated in this paper, the above-mentioned argument is valid only for one-step excitations and in some limited cases for multi-step excitation.

A few examples of situations when the rate-equation approach is clearly insufficient for even qualitatively describing the experimentally measured lineshapes obtainable at the intensities modern pulsed dye-laser systems easily can produce are presented below. An adoption of the density-matrix formalism for the situations encountered in LEI in flames has recently been presented by the authors (Axner 1990b). The theory is able to predict virtually all main features of the "anomalous" lineshapes which have been found experimentally: two-photon peaks, power- and saturation-broadening including Stark splitting of peaks, etc.

In general, it is possible to obtain two kinds of excitations of atoms in two-color LEI, namely: i) those when the atoms first are excited to the intermediate level by absorption of one photon and then, a short time later, are excited to the uppermost level by absorption of a second photon, generally termed ***two-step excitations;*** and ii) those when the atoms simultaneously absorb two photons and consequently directly are being excited from the ground state to the uppermost state, termed ***two-photon excitations.*** This is displayed in Fig. 3 in which the second step laser was tuned over the upper resonance for various wavelength settings of the first step laser. In Fig. 3a, the first step laser was held at its resonance wavelength, while it was detuned 50, 100 and 200 pm in the other three curves, respectively. In this figure, it is clear that there are both two-step and two-photon excitations - the two-step signals appear at the resonance wavelength for the scanning laser (even though the fixed laser might be slightly detuned from the resonance wavelength) while the two-photon signals, on the other hand, appear at a wavelength for which the sum of the energy of the two photons corresponds to the energy difference between the uppermost and the lowermost levels. Hence, the position of this two-photon peak, when one laser is scanned, is dependent on the detuning of the other lasers. When both lasers are at resonance, both types of excitations occur simultaneously.

Figure 4 displays a similar situation although with the difference that the first step laser is scanned and that the second step laser is held fixed. In this case, the uppermost curve (curve a), which was obtained when the second laser was held at resonance, is found to consist of two peaks, symmetrically positioned around a minimum approximately at the resonance

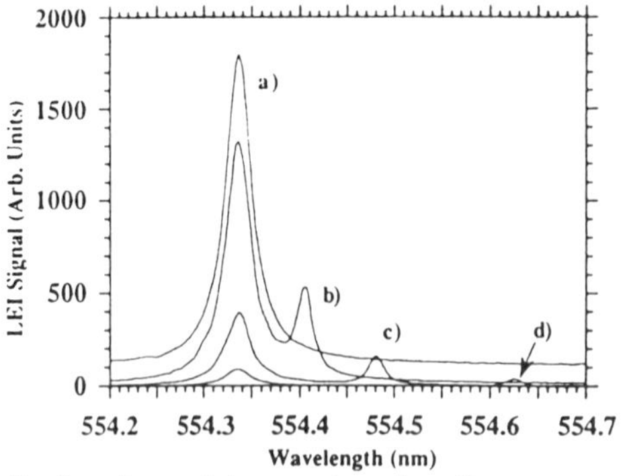

Fig. 3. Scans of the second step laser for various detunings of the first step. The laser intensities were 20 kW/cm².

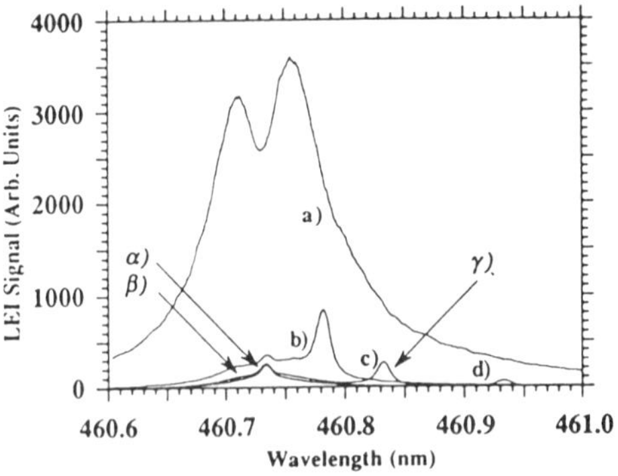

Fig. 4. Scans of the first step laser for various detunings of the second step. The laser intensities were 20 kW/cm².

wavelength. In addition, the signal in Fig. 4a is much broader (> 100 pm) than all of the previous curves (10 - 30 pm). The three other curves (b, c, and d), when the second laser was detuned, show both a two-step and a two-photon peak as in the scan above - a two-photon signal at various wavelengths depending on the detuning of the second step laser (labeled γ in curve c) and a two-step signal at the resonance wavelength (labeled α and β in curve c). The labelling of one of the curves is done for clarifying reasons since it is found that this two-step signal in turn consists of two different contributions - one from the atoms outside the "true" interaction region due to "scattered" light, i.e. the "narrow" banded peak, denoted (α), and one from the atoms which are present in the geometrically confined laser interaction volume in the flame, the ordinary two-step signal, denoted (β). Hence, it is here clear that the LEI signal at resonance has contributions from ***three*** different origins. Consequently, it is easy to understand that a proper modelling of the LEI signal has been difficult to obtain.

The reason for the split of the signal in Fig. 4a is that dynamical Stark effects and power broadening appear as a consequence of the interaction between the atoms and strong laser light fields. These kinds of effects can show up as broadenings, splittings or shiftings of the transitions. Generally, the Rabi flop frequency of a two-level system, which determines the dynamical Stark splitting of a level, is proportional to the square-root of the intensity of the light and can for the case of Sr be calculated to be approximately 7 and 8 GHz for a laser intensity of 100 kW/cm^2, for the two transitions, respectively. Such splittings can be probed, for example, by tuning a weak "probe" laser (the second step) across the transition. In general, this Stark split is too small to be seen directly in flames where total broadening mechanisms may account for some tens of GHz (pm).

However, when the first step laser is tuned and the second step laser is positioned at resonance, as in Fig. 4a, another type of situation occurs. As the strong first step laser is tuned towards the resonance wavelength, the population of the intermediate state increase (and hence the two-step LEI signal) until a situation occurs when either power-broadening starts to broaden the transition so that the transition probability per unit frequency interval is diminished or, when the Rabi-flop frequency becomes large enough, a significant dynamical Stark splitting of the levels appears. Since the second step laser is positioned at its resonance wavelength, it merely acts as a probe of the population at the centre of the intermediate level. A broadening or a splitting of the intermediate level diminishes the population of that level in the energy interval where it is probed by the second laser (at the unperturbed energy of the state). Thus, since fewer excited atoms are accessible for further excitation by the second step laser when both lasers are on "resonance" than when the strong first step laser is slightly detuned a dip in the LEI signal results when the first step laser is scanned across the resonance. This behaviour corresponds well to what is found in the experiments (Fig. 4a). The width of this dip is roughly given by the detuning needed in order to diminish the excitation and deexcitation rates so that also the broadening/split of the intermediate level decreases. This width can therefore be considerably larger than the actual induced Stark split.

In order to investigate the behaviour of the broad double peak at resonance as a function of laser light intensity four first step scans were measured for various first step pulse energies (see Fig. 5). The scans show that the width of the double peak increases significantly as a function of first step laser intensity. (The four curves correspond to intensities of 20, 80, 300 and 1100 kW/cm^2, respectively) In fact, for the highest intensity, the width of the dip in the centre of the curve is as large as approximately 0.2 nm. It is also worth noticing that the signal at resonance in fact *decreases* as a function of laser light intensity of the first step laser due to this effect. Hence, when a saturation curve is constructed, these kind of effects will cause *the saturation curve to bend over,* i.e. have a negative derivative.

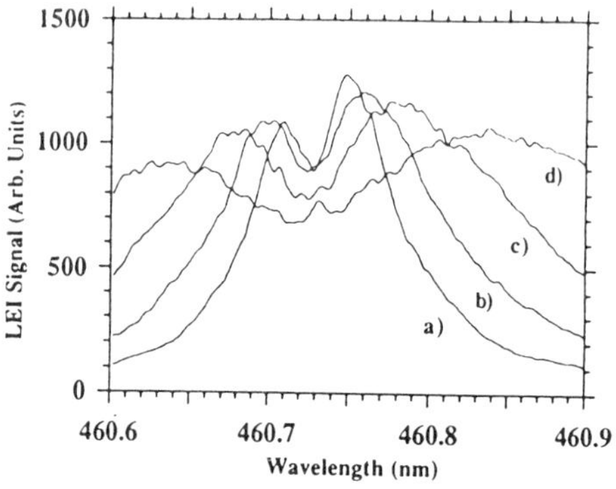

Fig. 5. Scans of the first step laser for various first step intensities.

4. CONCLUSIONS

In conclusion we have showed that when high intensity light (compared to that needed to saturate the transition) is used in one-step LEI spectroscopy *a significant part of the signal (>> 90 % in certain cases) can be attributed to ionization of atoms outside the "true" interaction region (out-traced by the laser beam) due to scattered laser light.* This behaviour is especially significant when small beam diameters are being used since then the number of atomized atoms outside the "true" interaction region is considerably larger than the number of atoms in the interaction region. This implies, among other things, that saturation curves (signal versus intensity) might be significantly distorted and irrelevant enhancement factors measured. Secondly, we found that power-broadening, dynamical Stark splitting, two-photon peaks etc might influence the lineshapes and signal strengths in two-colour LEI in flames considerably already at the moderate laser intensities which are easily obtained with a normal pulsed dye-laser system today without any focussing of the laser beams. Step-wise excitations of atoms in flames by pulsed laser light can give rise to two different types of peaks; i) two-step signals - when the atoms first absorb one photon from the first laser and then, a short time later, absorb a second photon from the second laser; and ii) two-photon signals - when the atoms simultaneously absorb two photons, one from each laser. Dynamical Stark effects and power broadening might, furthermore, distort the LEI signals considerably by producing double peaks (with dips several Å wide) and saturation curves which might "bend over".

REFERENCES

Axner O, Berglind T, Heully J - L, Lindgren I and Rubinsztein-Dunlop H 1984 *Appl. Phys.* **55** 3215
Axner O and Rubinsztein-Dunlop H 1989a *Spectrochim. Acta* **44B** 835
Axner O, Rubinsztein-Dunlop H and Sjöström S 1989b *Microchimica Acta* **III** 197
Axner O and Sjöström S 1990a *Applied Spectrosc.* **44** 864
Axner O and Sjöström S 1990b *Submitted to Spectrochim. Acta B*
Daily J W 1977 *Appl. Opt.* **16** 2322
Omenetto N, Smith B W, Hart L P 1986 *Fresenius Z. Anal. Chem.* **324** 683
Travis J C, Turk G C and Green R B 1982 *Anal. Chem.* **54** 1006A
Travis J C, Turk G C, DeVoe J R, Schenck P K and van Dijk C A 1984 *Prog. Analyt. Atom. Spectrosc.* **7** 199.

Paper presented at RIS 90, Varese, Italy, 16–21 September 1990

Analytical description of oscillatory structure in photodetachment cross section for negative ion placed into uniform electric field

V D Kondratovich[+] and V N Ostrovsky[++]

[+]Vinnitsa Polytechnical Institute, Vinnitsa 286021, USSR
[++]Institute of Physics, Leningrad State University, Leningrad 198904,USSR

ABSTRACT: The structure in the photodetachment cross section for a negative ion placed into a uniform electric field is calculated analytically in the framework of semiclassical approximation.

1. INTRODUCTION

The nature of oscillatory structure in the photodetachment cross section for a negative ion placed into a uniform electric field can be understood in the terms of classical electron trajectories. Outside the atomic core in the uniform electric field the motion possesses important property: in the classically axcessible region an arbitrary pair of points in space can be connected by two different classical trajectories with the same energy. In application to the photodetachment process one of this points is the negative ion which is a source of photoelectrons.Electrons ejected in two different directions pass the same final point in space. In quantum mechanics this results in interference effects. The role of such effects is well known in semiconductors theory (see e.g. Anselm 1978). For the negative atomic ions in the uniform electric field the total photodetachment cross section was calculated originally by Fabrikant (1980). An oscillatory structure in the cross section was obtained. The principal contribution to it is given by the interference of the photoelectrons ejected along the uniform field axis but in the opposite directions. A similar conclusion was drawn later by Reinhardt (1983) but no quantitative results were obtained by this author.

2. ANALYTICAL EXPRESSION FOR CROSS SECTION STRUCTURE

It was noted by Kondratovich and Ostrovsky (1984) that due to the steep rise with the energy of the cross section background the observed structure is not oscillatory but has a form of rounded steps. Recently this behaviour was observed in experiments by Bryant et al (1987). The theoretical interpretation of the cross section structure by Bryant et al (1987) lies in the framework of the papers cited above but no analytic expressions for the cross section were obtained. Quite recently Du and Delos (1988), Rau and Wong (1988) and Fabrikant (1990) proposed analytical description of the effect.

In fact the analytic expressions are contained in the earlier paper by the authors (Kondratovich and Ostrovsky 1984) where the photoionisation of neutral atom in the uniform electric field was analysed within general framework of semiclassical approximation. The latter problem is much more complicated due to the Coulomb field of atomic core (for instance generally more than two electron trajectories interfer). However when the charge of Coulomb centre Z tends to zero the photodetachment limit is achieved. This somewhat non-straightforward approach allows to obtain analytical expressions for the cross section in various approximations. The simplest of them corresponds to the case $\nu \gg 1$ ($\nu = k^3/F$, F is the electric field strength, $k=(2E)^{1/2}$, E is the photoelectron energy, atomic units are used):

$$\sigma_m(E) = \frac{MF\nu}{12\pi}\left[1 + \frac{12(m+1)^{m+4}\ln\nu}{(8e^2\nu)^{m+1}}\cos\left(\frac{2}{3}\nu + m\pi\right)\right], \tag{1}$$

$$M = [2^{21/2}\pi E_0^{5/2}]/[c(E_0+E)^3]$$

Here E_0 is the electron affinity energy of the atom, $E+E_0$ is the photon energy, c is the light velocity, m=0,1 is the azimuthal quantum number in the electron final state. The cross section structure depends strongly on the light polarisation. Namely, the oscillations are significantly dumped (by the factor $2/e^2\nu<<1$) in the case of σ polarisation (i.e. when m=1) as compared with π polarisation of light (m=o). Experimentally oscillations were observed only in the latter case. In the narrow region around field-free threshold (E=o) the more general analytical expression should be applied (Kondratovich and Ostrovsky 1984,1990).

The expression (1) somewhat differs from the analytical results mentioned in the Introduction. Most apparently the distinction can be seen in the amplitude of the oscillating term namely as a presence of logarithmical term. Next distinction is a phase shift

$$\gamma = - \tan^{-1} \left(\frac{\pi(m+1)/2}{m+(m+1)\ln[8\nu/(m+1)]} \right), \quad m=0,1 \qquad (2)$$

omitted in (1) because $\gamma \to 0$ when $\nu \to \infty$. These distinctions are in fact the small corrections and can be less than the accuracy of the approximation. However some ideas stated in the paper of Bryant (1988) could be helpful to define the physical meaning of these corrections. In the paper of Bryant (1988) the recently reported ripples in the H^- photodetachment cross section near threshold in an external electric field are interpreted as interference fringes in a dual beam interferometer like a Michelson one. It is stated that the phase correction to the argument of cos in (1) is related with an additional retard of the photoelectron caused by tunneling. I.e. from classical point of view, the photoelectron turns out in the classically forbidden region of motion (not at the border of classically accessible region).

The amplitude of oscillations can be interpreted as a function of the photoelectron's coherence length and of non-complet overlapping of the wave functions corresponding to the electrons escaping the atom opposit to the electric field direction and to the electrons leaved the atom along the electric field and next rejected by the potential wall. The length of coherence of rejected electrons differs from the initial length because the time of "quantum delay", related with the phase shift (2), depends on the energy of electron. The logarithmical dependance transfers in the amplitude of oscillations.

3. DISCUSSION

The agreement between the theory and experiment is good for all photon energies where the cross section structure was observed sufficiently well. Only relative experimental cross sections are available now. In contradiction with the other authors (see e.g. Bryant et al 1987) we normalise the experimental data to the theory in the high energy region since near the threshold the electron-core interaction in the final state is expected to be more important. The latter interaction was not taken into account in all versions of theoretical calculations (see references above) except by Fabrikant (1989).

REFERENCES

Anselm A I 1978 *Introduction to the Theory of Semiconductors* (in Russian, Moscow: Nauka)

Bryant H C 1988 *Atomic Spectra and Collisions in External Fields* ed C W Clark, M H Nayfeh and K T Taylor (New York: Plenum) pp 177-85

Du M L and Delos J B 1988 Phys.Rev.A **38** 5609

Fabrikant I I 1980 JETP **52** 1045

-- 1989 Phys.Rev.A **40** 2373

-- 1990 J.Phys.B: At.Mol.Opt.Phys. **23** 1139

Kondratovich V D and Ostrovsky V N 1984 J.Phys.B: At.Mol.Phys. **I7** 2011

-- 1990 J.Phys.B: At.Mol.Opt.Phys. **23** 21

Rau A R P and Wong H Y 1988 Phys.Rev.A **37** 632

Reinhardt W P 1983 J.Phys.B: At.Mol.Phys. **I6** L635

Reduction of chemical interferences in a graphite furnace by laser-assisted atomization

Madelen Norberg and Sten Sjöström

Department of Physics, Chalmers University of Technology, S-412 96 Göteborg, Sweden

ABSTRACT: Chemical vapor phase interferences in a graphite furnace due to matrices can be reduced by laser-induced photo-dissociation of analyte-containing molecules. The laser-induced fluorescence signal from indium was reduced by a factor of 40 when 3% NaCl was added to the sample. This signal suppression was due to creation of InCl and by photo-dissociation of InCl 50 % of this reduction could be retained. Similar results has been obtained for Sn (SnO). Diagnostic studies can also be performed by using this method.

1. INTRODUCTION

Laser-induced fluorescence in a graphite furnace is an ultra sensitive method for trace element analysis (Butcher 1988, Sjöström 1990). The detection limits of several elements are in the order of fg or, by means of concentrations, in the ppt-region. However, chemical interferences due to matrices in the sample and in particular those which results in non-complete atomization may cause severe losses of analyte signal (Frech 1990). In atomic absorption spectrometry in a graphite furnace several techniques have been used in order to reduce this problem, for example matrix modifiers, platforms, rapid heating and chemical extraction techniques.

We present here a technique for reduction of chemical interferences in the graphite furnace originating from molecular formation by performing laser-assisted atomization (LAA). The laser-assisted atomization is done by using a second laser prior to the detection laser in order to photodissociate molecules containing analyte atoms. The laser-induced photo-dissocation is followed by detection of the analyte atoms by laser-induced fluorescence spectrometry. This method has been used in order to increase atomization of 4 ppb indium in a matrix of 3% NaCl and to tin which forms oxides in the vapor phase.

This method can also be used in order to investigate diagnostic properties in the graphite furnace such as molecular recombination and atomization efficiencies. Photo-dissociation has previously been used as a method for determination of cross-sections (Silver 1986) and for reaction kinetics studies (Fontijn 1988).

2. EXPERIMENTAL

The experimental set-up for photo-dissociation in conjunction with laser-induced fluorescence is schematically shown in figure 1. The laser system is composed of two excimer lasers pumping two dye-lasers which are producing the detection laser light and the photo-dissociation laser light separately. The detection laser can be delayed a certain time (up to milliseconds in steps of ten nanoseconds) with respect to the dissociation laser. The length of the laser pulses are 20 ns. The area of the laser beams is about 3 mm^2 and they overlap perfectly when entering the graphite furnace where the sample is atomized. The fluorescence light from the analyte atoms is collected by a photo-multiplier through a monochromator.

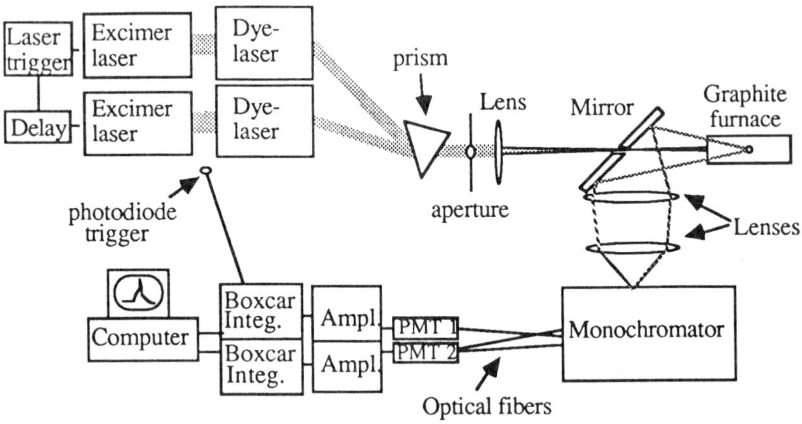

Figure 1. Experimental set-up for laser-induced fluorescence in conjunction with laser-assisted atomization.

3. RESULTS

The first sample investigated was 4 ppb of indium in a matrix of 3% NaCl at a temperature of 1700 K. When detecting the indium atoms by laser-induced fluorescence we found that the signal suppression due to the NaCl-matrix was 40 times as compared to indium in pure water. This signal suppression was due to formation of InCl and probably also indium-oxides. By photo-dissociation of InCl using a repulsive state at 38000 cm^{-1}, which corresponds to a laser dissociating wavelength of 265 nm, the signal was enhanced by a factor of 20. The signal recovery compared to indium in pure water was therefore 50 %. The temperature in this experiment was deliberately held low in order to encourage molecular formation. Measurements at a temperature of 2500 K, at which the sample is supposed to be almost completely atomized, were also performed. At that temperature there was a signal suppression of 15 when 3% NaCl was added to the In concentration. An enhancement of 5 due to photo-dissociation of InCl could be achieved meaning that the signal recovery was here about 30 %.

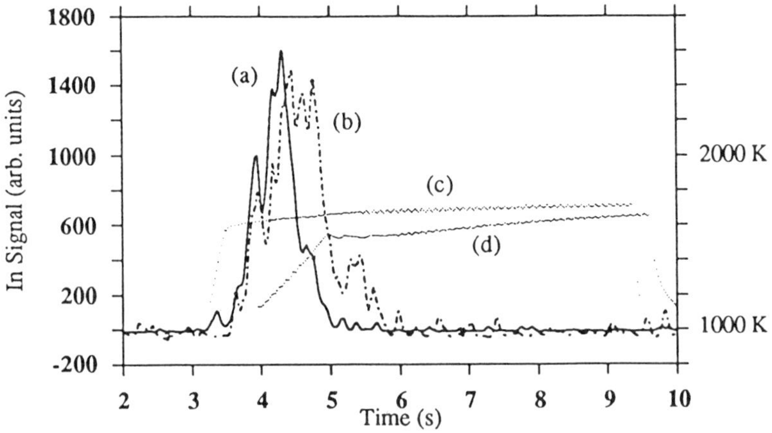

Figure 2. Indium signal shape with (a) and without (b) photo-dissociation of InCl. Curves (c) and (d) show the tube and the cup temperatures, respectively. Curve (b) is amplified 20 times.

The LIF-signal shape with and without photo-dissociation is shown in figure 2. As mentioned there are also indium-oxides, mainly In_2O, that exist in vapor-phase in the furnace. In order to investigate if dissociation of these indium-oxides contribute to the enhanced In-signal the same experiment as for InCl was performed for indium in pure water, that is, the indium signals with and without photo-dissociation were compared. With the furnace temperature kept low (1700 K) the signal increased a factor of two and could therefore have contributed slightly to the signal-enhancement in the NaCl-matrix case. With the higher temperature, 2500 K, no enhancement due to photo-dissociation was noticed and the atomization for indium in pure water was therefore complete.

This technique has also proved to be successful when detecting tin. Tin requires high temperatures in the furnace for complete atomization (3000 K). When the temperature is lower (1700 K), tin exists mainly in the form of SnO and the analytical signal from Sn atoms is almost completely suppressed. By using laser-assisted atomization this suppressed Sn-signal could be enhanced by a factor of 70 at 1700 K.

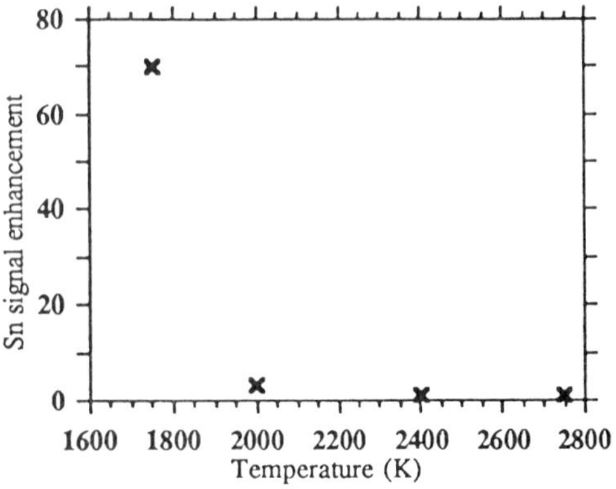

Figure 3. Enhanced Sn signal due to laser-assisted atomization vs temperature.

We believe that this signal enhancement is due to photodissocation of SnO. The wavelength of the dissociation laser was 230 nm. The energy carried by these photons are, however, not enough fordirect photo-dissociation of groundstate molecules. The remaining energy is supplied from thermally populated vibrational and rotational energy levels. At high temperatures, 2800 K, the thermal dissociation dominates and there is only a slight contribution from the photo-dissociation process. The enhanced Sn-signal due to photo-dissociation of SnO as a function of temperature is shown in figure 3.

Although the above mentioned experiments were performed at reduced temperatures in the furnace, it is shown that the laser-assisted technique is especially important when elements which are not readily atomized even at the highest temperature in the furnace are detected. It is also suitable for such samples that need addition of matrix modifiers or the use of chemical extraction techniques in order to reduce interferences. With the technique of laser-assisted atomization the risk of contaminating the sample by adding these chemicals is avoided and atomization could be just as efficient.

In general the technique of using one laser in order to prepare a sample in a desired manner and another laser for detection of the products with laser-induced fluorescence can also be used for diagnostic investigations in the graphite furnace. Atomization efficiencies in the graphite furnace can be determined by comparing the analyte signal with and without the photo-dissociation process. The problem is to know whether all molecules are dissociated or not. This requires a high laser intensity and also knowledge about what molecules are present in the vapor phase. In the experiments described above the laser pulse energy was between 200 and 400 μJ. No saturation of the analyte fluorescence signal due to depletion of molecules was noticed. This means that a much higher laser intensity is needed in order to dissociate all molecules. Even if the fluorescence signal is saturated meaning that there are no molecules left to dissociate it might be a false message since we still do not know whether there might be other molecules that are not effected by the dissociation laser.

It is also possible to study molecular recombination processes during the atomization. This can be done by varying the time delay between the dissociation laser and the probe laser.

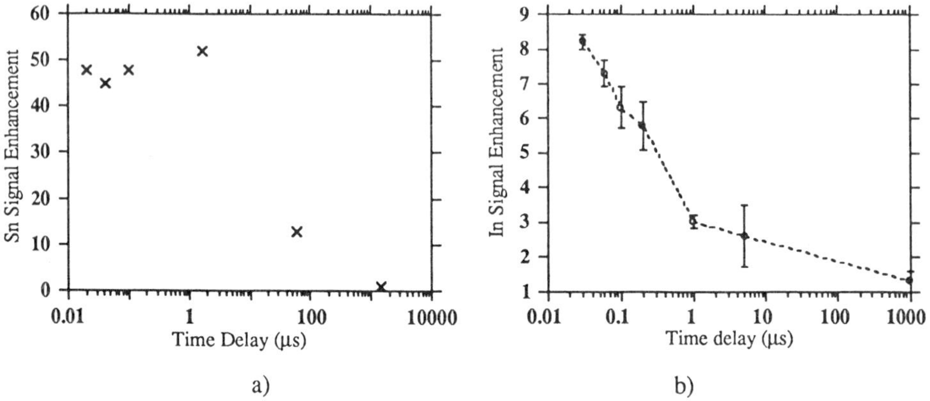

Figure 4. Signal enhancement due to laser-assisted atomization as a function of time delay between the dissociation laser and the detection laser. The signal decrease indicates molecular recombination of a) SnO and b) InCl.

Figure 4 shows the enhanced signals due to photo-dissociation of InCl and SnO, respectively as a function of time delay between the two lasers. The signal enhancement is largest just after the dissociation laser pulse. The delay between the lasers is therefore necessary in order to avoid background scattering. These fast recombination rates also mean that if the detection laser pulse was longer than the 20 ns we have in this experiment, recombination of analyte atoms to molecules within the laser pulse would affect the signal. When the analyte signal decreases the analyte atoms recombine into molecules. We see that after one μs the In-signal enhancement due to photo-dissociation is only one third of the original signal enhancement. For Sn-atoms the total recombination of SnO occurs within approximately one ms.

4. CONCLUSIONS

Laser-assisted atomization in a graphite furnace by laser-induced photo-dissociation of molecules has shown to be a powerful technique in trace-element analysis in order to increase sensitivity for elements in samples containing interfering matrices.

Laser-induced fluorescence spectrometry in a graphite furnace together with laser-induced photo-dissociation also shows a great potential as a method for diagnostic studies. Since laser-induced fluorescence is such a sensitive method it gives a possibility to study systems containing low concentrations of analyte atoms which is the normal situation in trace element analysis.

REFERENCES

Butcher D J , Dougerthy J P , Preli F R , Walton A P , Wei G-T , Irwin R L and Michel R G 1988 *J. Anal. Atom. Spectrom.* **3** 1059

Fontijn A 1988 *Spectrochim. Acta* **43B** 1075

Frech W and Baxter D C 1990 *Spectrochim. Acta* **45B** 867

Silver J A , Worsnop D R , Freedman A and Kolb C E 1986 *J. Chem. Phys.* **84** 4378

Sjöström S 1990 *Laser Spectroscopic Techniques in Flame and Graphite Furnace for Trace Element Analysis* (Dissertation thesis, University of Göteborg)

Inst. Phys. Conf. Ser. No 114: Section 2
Paper presented at RIS 90, Varese, Italy, 16–21 September 1990

The National Institute of Standards and Technology resonance ionization spectroscopy/resonance ionization mass spectrometry data service

E. B. Saloman

Electron and Optical Physics Division, National Institute of Standards and Technology, Gaithersburg, Maryland 20899, USA

ABSTRACT: A resonance ionization spectroscopy data service has been established at the National Institute of Standards and Technology. It publishes formatted data sheets which provide the atomic and laser physics information to apply RIS/RIMS to a given element. A first group of ten data sheets has been published. It covers As, B, Cd, C, Ge, Au, Fe, Pb, Si, and Zn. A second group has been submitted for publication. It covers Al, Ca, Cs, Cr, Co, Cu, Kr, Mg, Hg, and Ni. More will be produced in the future.

1. INTRODUCTION

The techniques of Resonance Ionization Spectroscopy (RIS) and Resonance Ionization Spectroscopy followed by Mass Spectrometry (RIMS) have been demonstrated in several state-of-the-art laser laboratories as analytical techniques of high elemental sensitivity. For example, Bekov (1989) and co-workers have demonstrated detection limits as low as .001 ppt (one part in 10^{15}) for Pt and Rh in sea water. Hurst and co-workers (1977) have demonstrated single atom detection and also the counting of 1000 atoms of one Kr isotope which was mixed with 2×10^5 other Kr isotopes and 1×10^{10} atoms of He (Hurst *et al* 1985).

These demonstrations present the RIS/RIMS technique as a potentially very powerful tool for monitoring environmental pollutants, performing trace impurity analysis in semiconductors, alloys, and other materials, measuring radioactive contamination, studying biological processes, as well as many other areas. For these techniques to meet their potential, they must be made available to analytical chemists and other scientists who do not have the most comprehensive knowledge of atomic structure and laser physics. The task is made more difficult for these scientists because much of the information needed to apply RIS and RIMS exists in scattered data bases. Much of the relevant data has been collected at NIST for seven decades. NIST, in consultation with the RIS/RIMS user community, has begun a RIS data service to provide RIS schemes and atomic data for application to RIS/RIMS elemental analysis. It is the object of this data service to organize the available information and supplement it with calculations (where the many gaps exist) to provide the needed information. The output of this data service is a set of formatted data sheets which concisely provides the needed information on appropriate RIS schemes for each element.

2. DATA SHEETS

The format of the data sheets has been designed in an iterative fashion through consultation with the RIS community. A preliminary version of the data sheets was presented at the Fourth International Symposium on Resonance Ionization Spectroscopy and its Applications in 1988 (Saloman 1989). Comments were solicited from the participants at that symposium and from other members of the RIS community. These comments showed considerable diversity in the kinds of RIS schemes believed to be the most important to include. As a result an expanded data sheet format was designed which usually includes several RIS schemes for each element. Wherever possible, experimentally proven schemes are used.

Calculations and data collection have been carried out and the first group of ten data sheets (for the elements As, B, Cd, C, Ge, Au, Fe, Pb, Si, and Zn) has been published (Saloman 1990) in Spectrochimica Acta B. A second group has been submitted for publication (Saloman 1991). It covers Al, Ca, Cs, Cr, Co, Cu, Kr, Mg, Hg, and Ni. Data sheets for additional elements will be produced in the future.

The sheets list the element, its ionization potential, an introduction to the RIS/RIMS studies carried out on it, its stable isotopes, isotope shifts and hyperfine structure, RIS schemes, atomic energy levels, lifetimes, oscillator strengths, laser schemes, atom sources, and both data and RIS/RIMS references. Also included are the results of calculations of excited state photoionization cross sections by Hartree-Fock techniques. Suggestions as well as reprints, preprints, and reports of RIS/RIMS work are solicited to allow us to include the information in future data sheets.

The table summarizes the schemes included in the data sheets completed so far. For each element completed, the table lists the reference containing that data sheet. Reference #1 is (Saloman 1990) and reference #2 is (Saloman 1991). Then the first ionization potential of that element is given. Next the table specifies the types of RIS schemes for which data is provided for that element. The notation of the RIS schemes will be explained by a series of examples which may be generalized to obtain cases not specifically stated: $\omega_1+\omega_1$ is a one-color two-photon RIS process in which the first photon excites the atom to a resonance level which is then photoionized by a second photon of the same color; $\omega_1+\omega_2$ is a two-color two-photon process in which the photoionizing photon has a different color than the excitation photon; $\omega_1+\omega_2+\omega_3$ is a three-color process involving two resonance steps and concluding with photoionization; $\omega_1+\omega_2^{AI}$ is a two-color process in which the first photon excites the atom to a resonance level which is then excited to an autoionizing state by the second photon; $\omega_1+\omega_2^{R}$ is a two-color process in which the first photon excites the atom to a resonance level and the second photon excites the atom from this level to a high-lying Rydberg state which is subsequently ionized by collisions or by an electric-field pulse; and $2\omega_1+\omega_1$ is a one-color process in which the first step is a two-photon transition to a resonance level followed by photoionization by a third photon of the same color.

Summary of Completed Data Sheets for the RIS/RIMS Data Service

Symbol of Element	Ref. #	Ionization Potential (cm^{-1})	Types of RIS Schemes covered in the data sheet for this element		
Al	2	48278	$\omega_1+\omega_1$ $\omega_1+\omega_2+\omega_3$	$\omega_1+\omega_2$ $\omega_1+\omega_2+\omega_3^{AI}$	$\omega_1+\omega_2^{R}$
As	1	78950	$2\omega_1+\omega_1$	$2\omega_1+\omega_2^{AI}$	
Au	1	74409	$\omega_1+\omega_1$ $\omega_1+\omega_2+\omega_3$	$\omega_1+\omega_2$ $\omega_1+\omega_2+\omega_3^{AI}$	$2\omega_1+\omega_1$
B	1	66928	$\omega_1+\omega_2+\omega_3$		
C	1	90820	$2\omega_1+\omega_1$		
Ca	2	49306	$\omega_1+\omega_2+\omega_1$ $\omega_1+\omega_2+\omega_3^{AI}$	$\omega_1+\omega_2+\omega_2$ $2\omega_1+\omega_1$	$\omega_1+\omega_2+\omega_3$ $2\omega_1+\omega_2^{AI}$
Cd	1	72540	$\omega_1+\omega_1$ $\omega_1+\omega_2+\omega_1$ $2\omega_1+\omega_2$	$\omega_1+\omega_2$ $\omega_1+\omega_2+\omega_2$	$\omega_1+\omega_2+\omega_3$ $2\omega_1+\omega_1$
Co	2	63564	$\omega_1+\omega_1$	$\omega_1+\omega_2+\omega_3$	
Cr	2	54576	$\omega_1+\omega_1$	$\omega_1+\omega_2$	$\omega_1+\omega_2^{AI}$
Cs	2	31406	$\omega_1+\omega_1$	$\omega_1+\omega_2+\omega_3$	$2\omega_1+\omega_2$
Cu	2	62317	$\omega_1+\omega_1$ $2\omega_1+\omega_1$	$\omega_1+\omega_2$	$\omega_1+\omega_2^{AI}$
Fe	1	63737	$\omega_1+\omega_1$	$\omega_1+\omega_2$	$2\omega_1+\omega_1$
Ge	1	63713	$\omega_1+\omega_1$ $\omega_1+\omega_2+\omega_3$	$\omega_1+\omega_2+\omega_1$	$\omega_1+\omega_2+\omega_2$
Hg	2	84184	$2\omega_1+\omega_1$ $\omega_1+\omega_2+\omega_3$	$\omega_1+\omega_2^{AI}$ $\omega_1+\omega_2+\omega_3^{AI}$	$\omega_1+\omega_2+\omega_2$
Kr	2	112914	$\omega_1+\omega_2$	$\omega_1+\omega_2+\omega_3$	$2\omega_1+\omega_1$
Mg	2	61671	$\omega_1+\omega_1$ $2\omega_1+\omega_1$	$\omega_1+\omega_2^{AI}$	$\omega_1+\omega_2+\omega_3$
Ni	2	61619	$\omega_1+\omega_1$	$\omega_1+\omega_2+\omega_3$	
Pb	1	59820	$\omega_1+\omega_1$	$2\omega_1+\omega_1$	$\omega_1+\omega_2+\omega_3$
Si	1	65748	$\omega_1+\omega_1$	$\omega_1+\omega_2+\omega_3$	
Zn	1	75769	$\omega_1+\omega_1$ $\omega_1+\omega_2+\omega_3$	$\omega_1+\omega_2+\omega_1$ $2\omega_1+\omega_1$	$\omega_1+\omega_2+\omega_2$ $2\omega_1+\omega_2$

3. RIS SCHEMES

A number of different types of RIS schemes are included in these data sheets since the choice of the best one to use depends on the application. Preference is given to experimentally proven schemes. One-color schemes are usually the simplest to set up since only a single laser is required. Schemes involving two-photon transitions to resonance levels require higher power levels which may produce interferences but can be used in atoms with higher ionization potential and are amenable to Doppler-free techniques when isotope selectivity is required (Clark *et al* 1984). Three-color schemes involving two resonance steps allow for increased elemental sensitivity by permitting the optimization of the power levels of each step in the RIS process. Excitation to an autoionizing state is often a means of increasing the efficiency of the ionization process over that of photoionization as is excitation of a high-lying Rydberg state followed by ionization by an electric field pulse (Bekov 1989).

4. ACKNOWLEDGEMENTS

I wish to thank G. S. Hurst, J. E. Parks, J. C. Travis, and T. B. Lucatorto for their encouragement of this work. This project is supported in part by the U. S. Department of Energy Office of Health and Environmental Research under contract DE-AI05-86ER-60447.

5. REFERENCES

Bekov G I 1989 *Lasers in Atomic, Molecular, and Optical Physics* ed V S Letokhov (Singapore, World Scientific) pp 339-48

Clark C W, Fassett J D, Lucatorto T B and Moore L J *Resonance Ionization Spectroscopy 1984* eds G S Hurst and M G Payne (Bristol, Institute of Physics) pp 107-117

Hurst G S, Nayfeh M H and Young J P 1977 *Appl. Phys. Lett.* **30** 229

Hurst G S *et al.* 1985 *Rep. Prog. Phys.* **48** 1333

Saloman E B 1989 *Resonance Ionization Spectroscopy 1988* eds T B Lucatorto and J E Parks (Bristol, Institute of Physics) pp 255-8

Saloman E B 1990 *Spectrochim. Acta* **45B** 37

Saloman E B 1991 submitted for publication

Stark-induced resonances and giant shifts in multiphoton ionization of atoms

P. Agostini, P. Breger, E. Mevel and G. Petite
Service de Physique des Atomes et des Surfaces
CEN Saclay 91191 Gif sur Yvette Cedex
A. Antonetti, J. P. Chambaret
LOA, ENSTA Ecole Polytechnique 91120 Palaiseau

I. INTRODUCTION

When an atom is submitted to an electromagnetic field in the optical or near infrared frequency domain, only the electrons on the outermost shell are responding to the perturbation. In low intensity fields, the electrons will, in general, oscillate at the field frequency and re-radiate the energy absorbed from one mode into another mode (Raleigh scattering). As the field strength is increased, the electron motion becomes more complicated and in particular, harmonics of the incident frequency may be generated and multiphoton transitions may be observed. At intensities of the order of the Terawatt per cm^2 transitions involving more than 10 photons are easily observed and this number increases to the hundreds when intensity reaches 10^{14}-10^{16} $W.cm^{-2}$. Under such irradiation most atoms ionize by Multiphoton Ionization (MPI). The electrons which are released in an MPI process have a kinetic energy E which is given by the energy conservation :

$$E = Nh\nu - E_I \qquad (1)$$

where N is the smallest integer which makes E positive, hν is the photon energy and E_I is the ionization potential.

As it is well known by now, such an MPI process usually (as soon as the number of photons is larger than five or so) goes along with the so-called Above-Threshold Ionization (ATI) process in which K extra-photons are absorbed, thus changing eq.(1) into:

$$E = (N + K)h\nu - E_I \qquad (2)$$

The energy spectrum of the photoelectrons then displays a series of peaks starting at the energy given by eq. (1) and spaced by the photon energy.
Both MPI and ATI are quantum processes which may become resonant if the energy of an integer number of photons (say m) matches the energy of an atomic transitions allowed by electric dipole selection rules:

$$mh\nu = E_R - E_g \qquad (3)$$

where E_g and E_R are respectively the ground state and the resonant state energies. The increase of the ionization rate when such a condition is met are well known by every one attending this meeting.

Another unvoidable consequense of the interaction of a strong electromagnetic field and atoms is the distortion of the electron trajectories and the resulting shifts and broadenings of the atomic energy levels. At $10^{14} W/cm^2$ such shifts may theoretically reach several tens of eV, depending on the state and the frequency. Such intensities are usually produced by focusing intense pulsed lasers and, during the rise of the pulse one or several atomic levels may be scanned through resonance as soon as the intensity becomes enough for the shift to compensate for the static detuning, ie without field. Such "Stark-induced" resonances had been predicted long ago[1] but were thought to be unobservable

II.2 Freeman resonances.

The mechanism that produces the substructures in the photoelectron energy spectra (Fig. 2) has been proposed by Freeman et al.[2]. In essence, it is the following: MPI and ATI take place at intensities at which most atomic levels are strongly shifted. For instance, a Rydberg state, close to the ionization threshold, has a second order Stark shift which may be calculated in the high-frequency approximation (HFA) since the photon energy is much larger than the separation between neighbouring states. In this case, it is well known[8] that the shift is just given by E_Q. Such shifts are sufficient to induce resonances, ie to make the level energy equal to the energy of an integer number of photons, and consequently to boost the ionization rate at this particular intensity. Since, in short pulses, the corresponding photoelectron energy is shifted by $-E_Q$, a peak appears in the energy spectrum at $E - E_Q$ where E is given by eq.(2). The intensity axis is thus mapped onto the energy axis. In the simplest case where the Stark shift of the ground state can be neglected and the shift of the m-photon resonant state is given by E_Q, the energy of the resonant state is related to the energy of the outgoing electron by:

$$E_R = E + E_I - (n\text{-}m)h\nu \qquad (5)$$

where n is the total number of photons absorbed. Naturally, the resonances must obey the selection rules and their position is independent of the pulse peak intensity, in first approximation, as clear from eq.(5). Only the amplitudes of the substructures depend on the peak intensity: as it is increased, more resonances are posible and appear as substructures at lower energies. In the special case of xenon at 620 nm the observed resonances are identified to 6-photons resonances with f and p states and are consequently suppressed with circular polarization[10]. However, for these states in xenon, it was very likely that the high-frequency approximation (HFA) would breakdown as discussed in the next section.

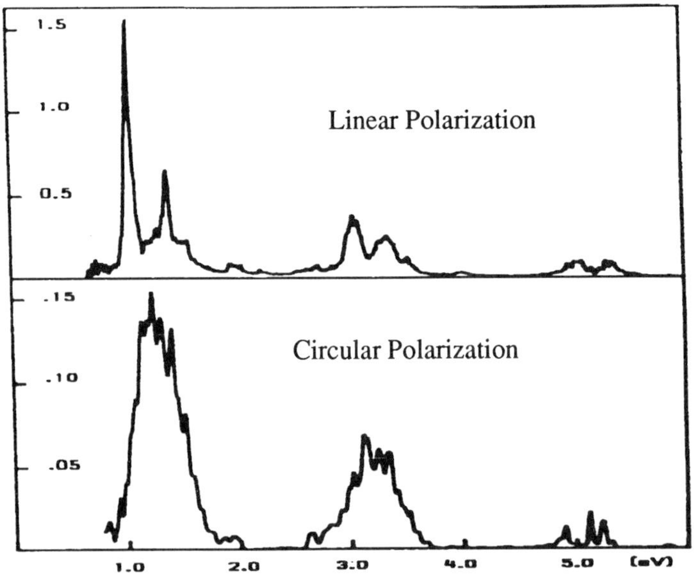

Figure 2. ATI spectra of xenon at 620 nm 100fs pulses in linear and circular polarization[7].

due to the averaging over the intensity distribution in the interaction region. It is recently that it was discovered[2] that they were responsible for new structures appearing in the electron energy spectra when using ultrashort pulses. These are commonly known as Freeman resonances by now. They have been subsequently investigated by a number of groups[3,4,5,6] for various atoms and proved to be valuable for studying the distortion of atomic structure under strong irradiation.

II. CHARACTERISTICS OF PHOTOELECTRON SPECTRA IN SHORT PULSES.

II.1 Ponderomotive shift.

Experimental investigation of the dependence of ATI photoelectron spectra on the pulse duration[7] have shown that the kinetic energies of the outgoing electrons could be affected or not during the escape of the particles from the laser beam. If the pulse duration is long compared to the electron exit time ($\Phi/(\sqrt{(2E/m)}$ where Φ is the beam diameter and E the electron initial kinetic energy as given by eq. (2)) then all the quiver energy gained by a free electron in the oscillatory motion forced by the electromagnetic field is converted into drift energy. This happens because of the work of the gradient force which derives from the ponderomotive potential which is nothing else than the quiver kinetic energy. Since the same quiver energy must be added to the ionization potential to obtain the effective ionization potential in the field, the electron is produced with less drift energy but regains whatever is missing in its way out of the field. Calling E_0 one energy given by eq.(2) and E_Q the quiver energy, the measured kinetic energy is, in this case:

$$E = E_0 - E_Q + E_Q = E_0 \tag{4}$$

The situation is completely different in the opposite case of pulses short compared to the exit time of the electron and turns off before the electron has travelled enough for the gradient force to produce some work. Then the quiver energy is lost either partially (Fig.1) or totally and the measured kinetic energy is, in the limit of ultrashort pulses, just $E_0 - E_Q$. As a consequence the spectra taken in short pulses appear to be shifted towards low energies by the amount E_Q. At short enough pulse durations the spectra are dominated by this shift and the Freeman resonances.

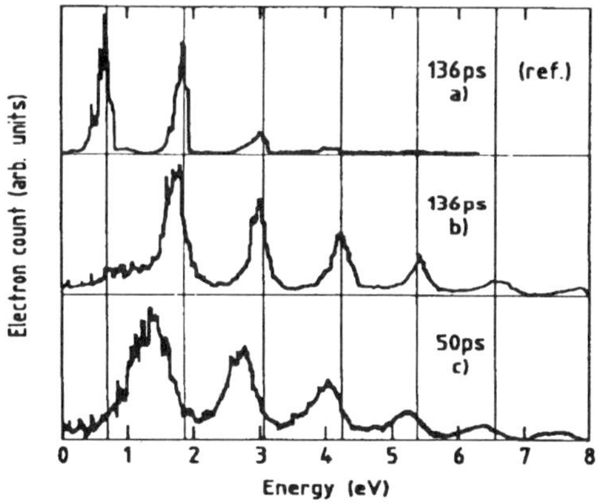

Figure 1. Photoelectron energy spectra from MPI of xenon at 1064 nm displaying the ponderomotive shift for short pulses[7].

III. AC-STARK SHIFTS IN Xe AND HIGH-FREQUENCY APPROXIMATION

There are several reasons for such a breakdown. Even if the second order calculation is acceptable, the HFA is valid only to the extent that the ***Ap*** part of the interaction Hamiltonian is neglected. Taking the corresponding term into account somehow changes the assignements of the resonances[10]. Another strong reason for the breakdown of HFA is the possible coupling of the Rydberg state with states about one photon above (or below) the state under consideration. In xenon, around 620nm, there are such states between the 3/2 and the 1/2 ionization thresholds: 9s', 8d', 10s', 9d', 11s' etc. In order to calculate exactly the shifts (to second order) we have used wavefunctions derived from a quantum-defect analysis of Rydberg series of xenon[10,11]. Figure 3 shows, as an example, the calculated shift of the state 7p[1/2] state as a function of the photon energy (dashed line). Indeed the shift takes values significantly different from E_Q (shown as a dotted line) around the frequencies corresponding to the autoionizing resonances. It was therefore interesting to check if this could be observed in the experiment. Changing the wavelength also changes the static detuning of the various states and, therefore, the intensity needed to shift them into resonance. The energy of the substructure in the spectrum is changed accordingly. This method provides the only way to investigate ac-Stark shifts at intensities in the range of 10^{13}W.cm^{-2}. The main conclusion of the experiment[3] is that all identified states shift linearly with intensity with a slope very close to the ponderomotive value, which seems to imply that the HFA is valid throughout the investigated wavelength range. Most probably the reason for this is the averaging over the laser bandwidth due to the short pulse duration necessary to observe the substructures (solid line on fig.3) wich smoothes out the variations due to the autoionizing states.

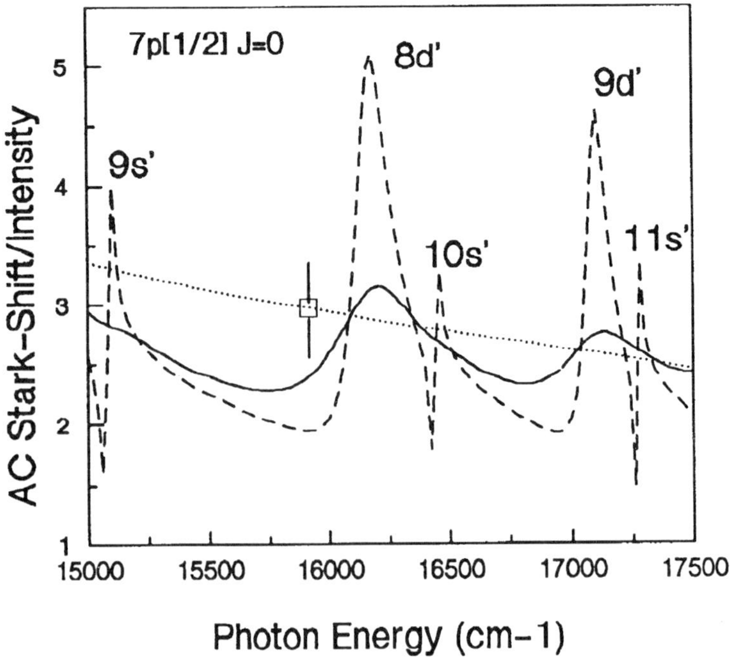

Figure 3. Second order Stark-shift of the 7p[1/2] J=0 state.

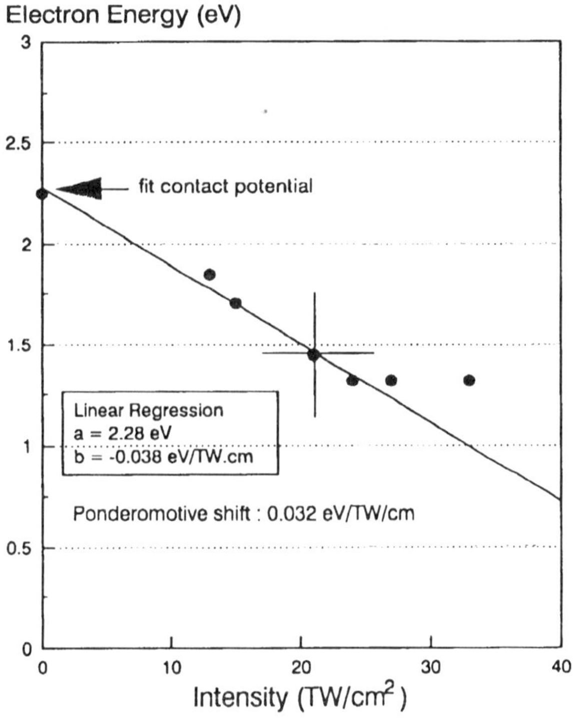

Figure 4. Ponderomotive shift as a function of intensity in MPI of argon with 620nm 100fs pulses.

IV. DIRECT MEASUREMENT OF THE THRESHOLD SHIFT IN ARGON

As discussed above, in the limit of the HFA, the shift of Rydberg states is just E_Q $(0.9410^{-13} I\lambda^2 \; eV/Wcm^{-2}/\mu m^2)$ where I is the laser intensity and λ the wavelength. This value was experimentally measured at low intensity[12]. However, recent attempts to extend this mesurement to higher intensities[9] have yielded a value about 50% smaller. This experiment and other in progress[13,14] have in common to be two-color experiments: one beam is a high intensity one and induces the shift while the other one is a weak one and probes the level energy. As a consequense they pause the delicate problem of the superposition of the two beams in space and time in spite of careful consideration of this question in these works. Electron spectrometry in short pulses with circularly polarized light provides a direct method to measure the shift in one-color experiments. In circular polarization all important resonances are forbidden and the electron energies are determined by eq.(2) as discussed in section II.1. A careful measurement of the electron energies and the absolute laser intensity in principle yields the value of E_Q. Such a measurement is currently in progress in 8-photon MPI of argon around $10^{13} W.cm^{-2}$. Preliminary results are shown in fig.4. In spite of still important error bars, the experimental slope (least-square fit) is not significantly different from the theoretical value. This measurement must be refined, mostly by improving the background pressure in the spectrometer in order to get better spectra and signal to noise ratio. If confirmed, it will bring some relief in the multiphoton community since there is, for the moment no explanation for the current discrepancy.

V. Summary

Under strong field conditions, all MPI transitions become multi-resonant whatever the light frequency. The resonances are induced by strong Stark shifts on the atomic energy levels. This fact had been known for long time and was observed as new structures in the photoelectron spectra from MPI, ATI transitions. The experimental requirement is to use pulses short enough to prevent changes of the electron kinetic energies by the work of the gradient force. These so-called Freeman resonances have been observed in xenon, argon, hydrogen and helium.Most observed shifts have been found so far to be linear functions of intensity and very close to the ponderomotive shift although very important shifts have been observed (>1 eV) This simple behaviour is not expected at such intensities and in some cases might be due to the lack of resolution of the method. However, electron spectrometry with strong short pulses is the only experimental approach to study ac-Stark shifts at high intensity. Finally, a direct measurement of the ponderomotive shift using circular polarization to suppress the resonances yields a value close to the theoretical prediction while previous two-color investigations had reported much smaller values.

This work was supported in part by the EEC grant No. SCI-1103.

REFERENCES.

[1] Y. Gontier, M. Trahin and N. K. Rahman Phys. Rev. Lett **34**, 779 (1975).
[2] R. R. Freeman, P. H. Bucksbaum, H. Milchberg, S. Darack, D. Schumacher and G. M. Geusic, Phys. Rev. Lett. **59**, 1092 (1987).
[3] P. Agostini, P. Breger, A. L'Huillier, G. Petite, A. Migus and A. Antonetti Phys. Rev. Lett. **63**, 2208 (1989).
[4] H. Rottke, B. Wolff, M. Brickwedde, D. Feldmann and K. H. Welge Phys. Rev. Lett. **64**, 404 (1990).
[5] M. Perry, K. Kulander and A. Szoke Phys. Rev. Lett. (1989).
[6] H. P. Helm private commmunication.
[7] P.Agostini, J. Kupersztych,, L. A. Lompré, G. Petite and F. Yergeau Phys. Rev. **A36,** 4111 (1987).
[8] P. Avan, C. Cohen-Tannoudji, J. Dupont-Roc, C. Fabre J. Phys. (Paris)**37**, 993 (1976).
[9] D. Normand, L. A. Lompré, A. L'Huillier, J. Morellec, M. Ferray, J. Lavancier, G. Mainfray and C. Manus J. Opt. Soc. Am B 6, 1513 (1989).
[10] P. Agostini, A. Antonetti, P. Breger, M. Crance, A. Migus, H. G. Muller and G. Petite J. Phys. B **22**, 1971 (1989).
[11] A. L'Huillier, X. Tang and P. Lambropoulos Phys. Rev. A **39**, 1112 (1989).
[12] S. Liberman, J. Pinard and A. Taleb Phys. Rev. Lett. **50**, 888 (1983).
[13] T. McIllrath et al. private comm.
[14] M. White and L. DiMauro private comm.

Paper presented at RIS 90, Varese, Italy, 16–21 September 1990

RIS studies of autoionization in calcium

Q. Li†, T.J. McIlrath†, E.B. Saloman and T.B. Lucatorto

National Institute of Standards and Technology, Gaithersburg, MD 20899, USA
†Also with IPST, University of Maryland, College Park, MD 20742, USA

ABSTRACT: Autoionizing levels belonging to configurations of $3p^6\,3dnl$ are being studied as part of an effort to find efficient ionization pathways applicable to ultrasensitive isotopic analysis and laser isotope separation. The studies are performed in a RIMS apparatus with a time-of-flight mass spectrometer and simple thermal filament atomization source. Estimates of two-photon transition rates and excited state photoionization cross sections show that the $4s^2\,{}^1S_0 \rightarrow\rightarrow 3d5s\,{}^1D_2 \rightarrow 3d7p\,{}^1P_1$ (autoionizing) ionization pathway might be efficient enough to allow effective isotopically selective ionization with cw Ti:sapphire lasers.

1. INTRODUCTION

Biomedical research on bone demineralization caused by certain diseases and by weightlessness requires relatively large amounts of the enriched rare isotopes of calcium and the capability for the sensitive detection of these isotopes. The naturally occuring isotopes (all nonradioactive) and their abundances are shown below.

^{40}Ca	^{42}Ca	^{43}Ca	^{44}Ca	^{46}Ca	^{48}Ca
96.941%	0.647%	0.135%	2.086%	0.004%	0.187%

Presently, ^{48}Ca is the most suitable for biomedical use because its eight-amu difference from ^{40}Ca, the major isotope, makes separation and detection less difficult. Enriched ^{48}Ca costs about $250/mg, a factor that limits the research efforts on calcium metabolism.

Ti:sapphire, a new solid state laser material, is capable of high power tunable cw generation in a broad wavelength region centered at around 8250Å. (Powers in excess of 15W have been achieved (Boss *et al* 1989).) High efficiency doubling of the Ti:sapphire laser has also been achieved, with powers in excess of 150mW reported (Pease *et al* 1990). These developments are particularly relevant to the application of resonance ionization in Ca since the energy level structure provides suitable ionization pathways using radiation around 4200Å and 8400Å.

2. EXPERIMENTAL APPARATUS

Fig. 1 shows a layout of the experimental apparatus. A time-of-flight mass spectrometer was constructed with a simple three-zone ion flight path: a weak field extraction region, an acceleration region, and a 50cm drift region. With proper voltages a mass resolution better than 1 part in 100 was attained. Ion detection with single ion sensitivity was provided by a channel plate multiplier. Signal acquisition was either by a digital oscilloscope, which gave simultaneous information on ion yields for all mass channels, or by a boxcar integrator which provided the ion yield for a single mass channel. Atomization of Ca was achieved by heating metallic Ca on a thermal filament.

For most of the work reported here we used a dye laser pumped by an injection seeded Nd:YAG laser doubled to 5320Å. Dye laser outputs were about 20mJ near 6000Å (doubled output about 5mJ near 3000Å) and 10mJ in the region near 8400Å (doubled output about 1mJ near 4200Å). The unfocussed beam had a diameter of about 1.4mm, and the pulse duration was about 5ns

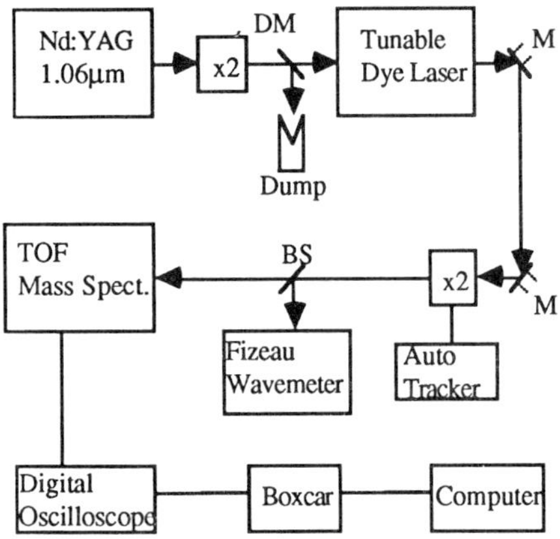

Fig. 1

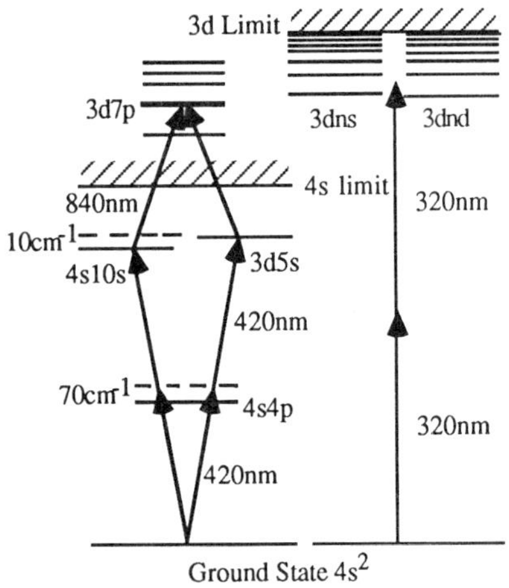

Fig. 2a Fig. 2b

3. MEASUREMENTS

Two types of resonance ionization pathways were studied, both terminating on a 3dnl-type autoionizing level: the two-photon resonant, three-photon ionization processes with either the $3d5s^1D_2$ or $4s10s^1S_0$ intermediate levels (Fig. 2a) and the direct two-photon excitation of an autoionizing level (Fig. 2b). The latter type is not suitable for cw ionization and will be reported on elsewhere (Qi Li *et al* 1991).

Both $3d5s^1D_2$ and $4s10s^1S_0$ pathways are well suited for cw sub-Doppler ionization because the location of the $4s4p^1P_1$ level provides strong enhancement to the two-photon transition. (We note that our calculations show the oscillator strength of the $4s4p^1P_1 \rightarrow 3d5s^1D_2$ transition to be relatively large because of the significant configuration interaction for the $3d5s^1D_2$ level.) In addition, further enhancement is obtained by tuning close to the broad autoionizing $3d7p^1P_1$ level, first observed by photoabsorption from the ground state (Garton and Codling 1965). The wavelengths for the two steps, ~ 4220Å and ~ 8430Å, were the second harmonic and fundamental from our dye laser.

Figure 3 shows the ionization yield as a function of blue and infrared laser pulse energies for the $3d5s^1D_2$ pathway. We note that there is clear evidence of saturation of both the two-photon rate and photoionization rate. We have attempted to extract from the data of Figure 3 the parameters that characterize the dynamics of the resonant ionization process so that estimates of cw laser ionization efficiencies can be made. An exact solution of the equations of motion for a

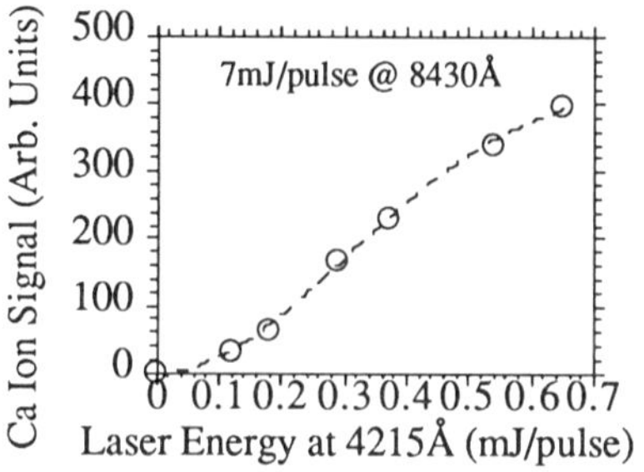

Fig. 3a

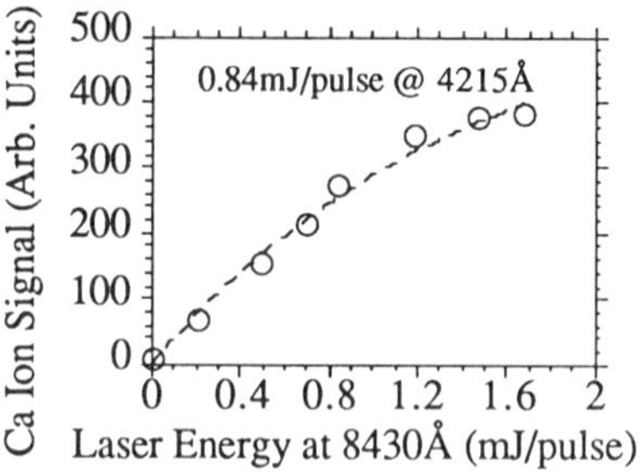

Fig. 3b

density matrix for our pulsed laser system would be a complicated task. Instead we used a formula (Eberly 1979) derived by J. Eberly from an extended two-level rate model that is appropriate for high-intensity, near resonant processes. The formula when tuned to exact two-photon resonance (neglecting the small AC Stark shifts) is:

$$R = \frac{\Omega^2}{4}\frac{\sigma_c I_{IR}}{\hbar\omega}\frac{\Gamma_{od}/\Gamma_d}{(\Gamma_{od}/2)^2+\Omega^2\,\Gamma_{od}/2\Gamma_d} \qquad (1)$$

where R is the total ionization rate, Ω is the two-photon Rabi rate, σ_c is the photoionization cross section out of the excited state, and Γ_d and Γ_{od} are the diagonal and off-diagonal relaxation rates respectively. Specifically for our collisionless environment:

$$\Gamma_d = A + \sigma_c I_{IR}/\hbar\omega \quad \text{and}$$
$$\Gamma_{od} = A + \sigma_c I_{IR}/\hbar\omega + \Gamma_{laser}$$

where A is the spontaneous radiative decay rate for the resonantly excited state, Γ_{laser} is the laser bandwidth of the resonant laser, and I_{IR} is the infrared laser intensity. To further simplify the analysis, a rectangular 5ns laser pulse was assumed. We used formula (1) to fit the data in Figs. 3a and 3b with σ_c and K, the constant of proportionality defined by $\Omega = KI_{uv}$, as fitting parameters. The best fit, shown as the dashed curves in Figs. 3a and 3b, occurs with

$$\sigma_c = 85 \times 10^{-18}\ \text{cm}^2 \quad \text{and}$$
$$K = 1.04 \times 10^3\ \text{cm}^2/\text{J}$$

4. DISCUSSION

We may express K in terms of atomic parameters (for a linearly polarized laser)

$$K = \frac{4\pi c e^2}{10^7 \hbar^2}\left|\sum_i \frac{\langle g|z|i\rangle\langle i|z|e\rangle}{\omega_{gi} - \omega_{laser}}\right|$$

where $\hbar(\omega_{gi} - \omega_{laser})$ is the energy difference between the laser photon energy and the energy of the intermediate state i. Given that the $4s4p\,^1P_1$ intermediate level dominates, we may replace the sum by a single term and evaluate the matrix elements by using the f-value for the $4s^2\,^1S_0 \rightarrow 4s4p\,^1P_1$ as tabulated (Wiese *et al* 1972) and for the $4s4p\,^1P_1 \rightarrow 3d5s\,^1D_2$ as calculated by us using the code developed by R. Cowan (Cowan 1981). The value obtained is

$$K_{theory} = 4.3 \times 10^3\ \text{cm}^2/\text{J}$$

which is four times larger than the experimental value.

Similarly, we have calculated the cross section for the photoionization process

$$3d5s^1D_2 \rightarrow 3d7p^1P_1 \rightarrow Ca^+ + e$$

and obtain the result

$$\sigma_{c(theory)} = 11 \times 10^{-18}\ cm^2$$

The fact that Formula 1) with two adjustable parameters was able to fit both Figs. 3a and 3b indicates the functional dependences of R on Ω and σI are realistic. The discrepancies between the experimental and theoretical values are probably due to inaccuracies in our simplified characterization of the laser parameters on the experimental side and the difficulties in calculating accurate oscillator strengths and photoionization cross sections on the theoretical side.

Efficient cw resonant ionization requires that the two-photon transition be saturated or $\Omega \sim 1/\tau$, (where τ is the lifetime of the $3d5s^1D_2$ level 81ns (Hansen 1983), and that the photoionization rate be comparable to the inverse of the atom's resident time in the laser beams, i.e. $\sigma I/\hbar\omega \sim 1/T_r$. In the cw case we chose to increase the resonant enhancement by the $4s4p^1P_1$ intermediate level through a two-color (4228Å and 4202Å), two-photon scheme which is detuned 500 linewidths from the $^1P_1$ level. As a conservative estimate of Ω as a function of cw intensity, we have chosen parameters connected with the smaller (i.e. experimental) value for K. To estimate the photoionization rate, we have chosen the experimental value for σ_c on the basis that smaller values had marked effects on the fits, especially when one increased Ω towards the theoretical value. With these two assumptions and with $T_r = 10^{-5}$ sec, we get that efficient ionization requires

$$I_{420} = 5\ W/cm^2$$
$$I_{840} = 300W/cm^2$$

We expect that such intensities should be achievable with Ti:sapphire lasers in the near future.

Similar measurements were made for the $4s10s^1S_0$ pathway. Here there was almost no saturation in the yield as a function of 4216Å energy (up to 1mJ) and only a slight sign of saturation in yield vs IR energy (up to 2mJ). This evidence of reduced overall ionization efficiency was also reflected in the theoretical calculation of K and σ_c which were over an order of magnitude smaller than the 3d5s pathway.

Work partially supported by the DoE office of Health and Environmental Research, Contract No. DE-AI05-86ER60447 and the National Science Foundation.

REFERENCES:

Boss I L, Brown P G, Bonanno R E, Delos-Sartos K, Enbert G V, Hockel R P, Hayman C A, and Paisner J A 1989, Postdeadline paper 20, QELS '89, Baltimore

Pease B W, Johnson T J, and Bischel W K 1990, Paper CTUN 2, CLEO '90, Anaheim

Garton W R S and Codling K 1965, Proc. Phys. Soc. 86, 1067; see also: Brown C M, Tilford S G, and Ginter M L 1973, J. Opt. Soc. Amer. **63**, 1454

Eberly J H 1979, Phys. Rev. Lett. **42**, 1049

Wiese W L, Smith M W, and Miles B M 1972, *Atomic Transition Probabilities Vol II*, NSRPS publication NBS 22

Cowan R D 1981,*The Theory of Atomic Structure and Spectra* (University of California Press)

Li Q, Lucatorto T B, McIlrath T J 1991 to be published

Hansen W 1983, J. Phys. B **16**, 2309

Inst. Phys. Conf. Ser. No 114: Section 2
Paper presented at RIS 90, Varese, Italy, 16–21 September 1990

Theory of two-photon-resonant three-photon ionization by broad-band radiation applied to the determination of isotopic abundances

Takashi Nakajima and P. Lambropoulos

The University of Southern California, Los Angeles, CA 90089-0484

We present a theoretical model for two-photon-resonant three-photon ionization by a broad-band laser. We carry out a specific calculation for ^{9}Be and ^{10}Be under the condition that the laser bandwidth is sufficiently broad to overlap all hyperfine levels of the state (2s4d) $^1D_2$. Our calculation shows that the ionization rates for ^{9}Be and ^{10}Be could depend strongly on the laser intensity.

1. INTRODUCTION

One of the potential applications of resonance ionization spectroscopy (RIS) is to use this technique to determine isotopic abundances (Clark et al. 1985, Fairbank et al. 1989). To obtain reliable experimental isotopic abundances, however, it is necessary to know the behavior of the ionization rates under various conditions. In recent papers (Lambropoulos and Lyras (1989), Lyras et al. (1990)), we have discussed the theory of one-photon resonant two-photon ionization and doubly resonant three-photon ionization, respectively. This paper is the extention of our previous work, with emphasis on the application to Be, where two-photon resonant three-photon ionization has been employed.

Most of the elements have isotopes, with different nuclear spins(I), which cause the hyperfine splittings (hfs) of atomic energy levels. For example, I=0 for ^{10}Be and $I = \frac{3}{2}$ for ^{9}Be. One might expect that the hyperfine splittings will not affect the ionization rates. -i.e., the different isotopes will be equally ionized, - as long as the laser bandwidth is large enough to overlap all the hfs levels, and the laser pulse duration is not long compared with the hfs coupling time. But this expectation is not always correct.

Let Δ_{hfs} be the hyperfine splittings. Then Δ_{hfs}^{-1} gives the hyperfine coupling time. Because of this hyperfine coupling, a redistribution of the populations among the hyperfine multiplets occurs, if the laser pulse duration T_L is longer than Δ_{hfs}^{-1}, which is often the case in experiments. Other factors that destroy the coherence by broad-band laser excitation are the Rabi frequency Ω and the ionization width Γ. One might think that if the Rabi oscillation is faster than Δ_{hfs}^{-1}, the effect of the hyperfine coupling can be reduced. As we see later, however, this is not the case.

Fast Rabi oscillation means that the electron comes back to the excited state fast. Thus the total time spent in the upper level does not change. As Γ becomes larger, on the other hand, the ionization occurs faster, and the hyperfine coupling is less likely to occur. To take all of these factors into account at the same time, we need to solve the density matrix equations.

2. Theoretical formulation

We consider the scheme $Be(2s^2)^1S_0 \to\to (2s4d)^1D_2 \to$continuum. For the even isotope, we denote the initial state $(2s^2)^1S_0$ as $|0\rangle$, and the resonant state $(2s4d)^1D_2$ as $|1\rangle$. The possible magnetic sublevel is $M_J=0$ only, since the initial state $|0\rangle$ has $M_J=0$, and for linearly polaraized light the selection rule demands $\Delta M_J=0$.

As for the odd isotope, the situation is more complicated. We introduce the $|(J\ I)F\ M_F\rangle$ representation. Since we assume light linearly polarized, the magnetic quantum number M_F will be conserved for the whole process.

The possible initial states are $|(0\ \frac{3}{2})\ \frac{3}{2}\ \frac{1}{2}\rangle$ and $|(0\ \frac{3}{2})\ \frac{3}{2}\ \frac{3}{2}\rangle$. We do not need to consider $M_F = \frac{-1}{2}$ and $\frac{-3}{2}$ due to the symmetry.) So, we need to set-up 2 independent sets of density matrix equations for $M_F = \frac{1}{2}$ and $M_F = \frac{3}{2}$ in the case of the odd isotope. Also the resonant state $2s4d^1D_2$ splits into 4 hfs levels $(2s4d)^1D_2(F = \frac{1}{2}, \frac{3}{2}, \frac{5}{2}$ and $\frac{7}{2})$. We denote the initial state $(2s^2)^1S_0|(0\ \frac{3}{2})\ \frac{3}{2}\ \frac{1}{2}\rangle$, and the resonant states $(2s4d)^1D_2\ |(2\ \frac{3}{2})\frac{1}{2}\ \frac{1}{2}\rangle$, $|(2\ \frac{3}{2})\ \frac{3}{2}\ \frac{1}{2}\rangle$, $|(2\ \frac{3}{2})\ \frac{5}{2}\ \frac{1}{2}\rangle$ and $|(2\ \frac{3}{2})\ \frac{7}{2}\ \frac{1}{2}\rangle$ as $|0\rangle, |1\rangle, |2\rangle, |3\rangle$, respectively, for $M_F = \frac{1}{2}$. Similarly, we denote $2s^2\ ^1S_0\ |(0\ \frac{3}{2})\ \frac{3}{2}\ \frac{3}{2}\rangle$, $2s4d\ ^1D_2\ |(2\ \frac{3}{2})\ \frac{3}{2}\ \frac{3}{2}\rangle$, $|(2\ \frac{3}{2})\ \frac{5}{2}\ \frac{3}{2}\rangle$, and $|(2\ \frac{3}{2})\ \frac{7}{2}\ \frac{3}{2}\rangle$ as $|0\rangle$, $|2\rangle$, $|3\rangle$ and $|4\rangle$, respectively, for $M_F = \frac{3}{2}$.

(Note that no state $|1\rangle$ exists for $M_F = \frac{3}{2}$).

Using these notations, the density matrix equations for the odd isotope may be written as follows:

$$\frac{d}{dt}\sigma_{00} = \sum_{k=1}^{4} I_m(\Omega_{ok}\ \sigma_{ko}) \tag{1}$$

$$\frac{d}{dt}\sigma_{\ell\ell} = -\Gamma_\ell\sigma_{\ell\ell} - I_m(\Omega_{o\ell}\ \sigma_{\ell o}) \qquad (\ell = 1,2,3,4) \tag{2}$$

$$\left[\frac{d}{dt} - i\Delta_\ell + \frac{1}{2}\Gamma_\ell + 2\gamma_L \frac{\beta^2}{\Delta_\ell^{\,2} + \beta^2}\right]\sigma_{\ell o}$$

$$= \frac{1}{2} i\, \Omega_{\ell o}(\sigma_{\ell\ell} - \sigma_{oo}) + \frac{1}{2} i \sum_{\substack{k=1\\ k\neq\ell}}^{4} \Omega_{ko}\, \sigma_{\ell k} \qquad (\ell = 1,2,3,4) \qquad (3)$$

$$\left[\frac{d}{dt} + i\omega_{k\ell} + \frac{1}{2}(\Gamma_k + \Gamma_\ell)\right] \sigma_{k\ell}$$

$$= \frac{-1}{2} i\, \Omega_{ko}\sigma_{o\ell} + \frac{1}{2} i\, \Omega_{o\ell}\sigma_{ko} \qquad (k=1,2,3;\ \ell=2,3,4;\ k\neq\ell) \qquad (4)$$

where $\Omega_{o\ell}$ are the Rabi frequencies coupling $|0\rangle$ to $|\ell\rangle$, Γ_ℓ the ionization width of state $|\ell\rangle$, $\Delta_\ell = \omega - \omega_\ell$, and $\omega_{12} = \omega_1 - \omega_2$. ($\omega$ is the laser frequency). The details of this formulation have been given elsewhere (Dixit et al. 1981, Dixit and Lambropoulos, 1983).

We now need to calculate the atomic parameters Ω's and Γ's to use in the density matrix equations. We can not show here the calculation because of space limitations. We simply mention that single-channel quantum defect theory has been used to calculate the radial parts of the matrix elements, after the separation from the angular parts.

Having determined the atomic parameters, we proceed to solve the density matrix equations with laser pulse duration (FWHM, Gaussian) T_L = 5 nsec, laser bandwidth γ_L=0.8 cm^{-1}, and cutoff parameter β=1.6. We employ the hyperfine splittings for the $|2s3d\ ^1D_2\rangle$ state, since the data for the $|2s4d\ ^1D_2\rangle$ state were not available to us. Then, the maximum hyperfine splittings $\Delta_{hfs} = \omega_4 - \omega_1$=37.5 MHZ , must be somewhat of an overestimate.

We list here the values of the relevant atomic parameters (all in units of rad/sec) as obtained in our calculations: (1) (Odd isotope, $M_F = \frac{1}{2}$. Ω_{10} = 5.04I, Ω_{20} = -5.04I, Ω_{30} = -3.30I, Ω_{40} = 8.08I, Γ_1 = 2.41I, Γ_2 = 2.41I, Γ_3 = 2.65I, Γ_4 = 3.02I); (2) (Odd isotope, $M_F = \frac{3}{2}$. Ω_{20} = Ω_{30} = - 8.08J, Ω_{40} = 6.02I, Γ_2 = 2.41I, Γ_3 = 2.47I, Γ_4 = 2.78I); (even isotope Ω_{10} = 11.27I, Γ_1 = 3.27I); where the laser intensity I must be inserted in Watts/cm².

Using these parameters, and setting the initial conditions as $\sigma_{oo}(t=0)=1$ for each isotope, the total ionization is calculated through $P(t) = 1 - \sum_i \sigma_{ii}(t)$

3. Results and discussion

We calculate the ionization P for the even isotopes, and $P'_{m=\frac{1}{2}}$ and $P'_{m=\frac{3}{2}}$ for the odd isotopes $m=\frac{1}{2}$ and $m=\frac{3}{2}$, respectively. We assume that all of the magnetic sublevels in the ground state are thermally populated. We plot the quantity $R=2(P'_{m=\frac{1}{2}}+P'_{m=\frac{3}{2}}-P)/(P'_{m=\frac{1}{2}}+P'_{m=\frac{3}{2}}+P)$ (%) as a function of laser intensity. (Fig. 1)

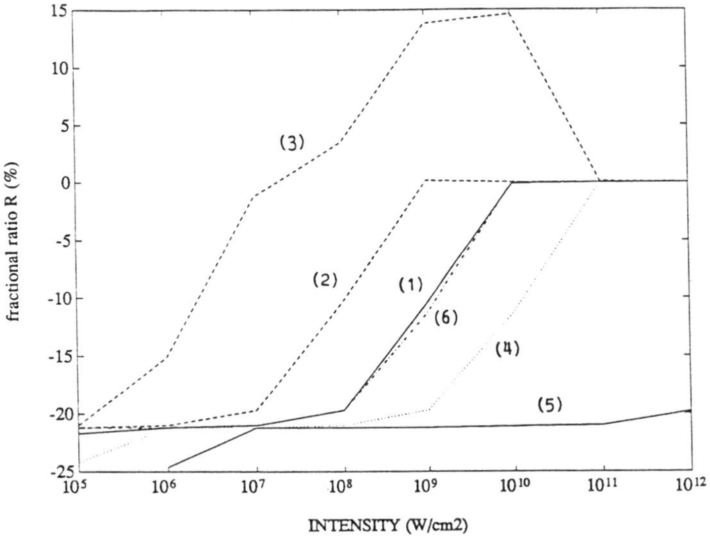

Fig.1
Line (1);result for Be with the original Ω's and Γ's
(2);result for Be with the original Ω's and 10x Γ's
(3);result for Be with the original Ω's and 500x Γ's
(4);result for Be with the original Ω's and 0.1x Γ's
(5);result for Be with the original Ω's and 0.0001x Γ's
(6);result for Be with the 10xΩ's and original Γ's
All other parameters used in the calculations are given in the text.

At the low intensity range, the curve is flat, and R has negative values. It means that the even isotope is easier to be ionized than the odd one. As the intensity is increasing, R begins to be larger, and has a value -1.7% at 10^{10} W/cm^2, where the ionization rate P is about 0.96. At the higher intensity range, R is 0. We can explain this behavior qualitatively as follows: At the low intensity range, the rate of ionization is much smaller than the hyperfine coupling time. So, the coherence is destroyed by the hyperfine coupling and R takes a nonzero value. Around the intensity 10^8 W/cm^2, the rate of ionization and the rate of hyperfine coupling begin to compete, and at the high intensity, R reaches zero since all of the atoms ionize regardless of their nuclear spin state. More qualitatively, the behavior of R is decided by the combination of many factors, such as T_L, γ_L, Γ, Ω and Δ_{hfs}.

To illustrate the role of the ionization width in relation to Δ_{hfs}, we also plot in Figure 1 the results of some calculations obtained by changing the magnitudes of Γ's (but the ratios among the Γ's are kept constant). The larger the magnitudes of the Γ's, at the lower intensity, the larger R becomes. In the calculations described above, the spontaneous decay has been neglected by the assumption that the spontaneous decay rate τ^{-1} is much smaller than the ionization width (i.e. $\tau^{-1} >> \Gamma$). But this assumption is not valid when the intensity is low and the ionization width Γ is small. To demonstrate the role of the spontaneous decay terms, which are not shown in the density matrix equations (1)~(4), we carried out calculations with these terms included. Since the decay from the excited states $|1\rangle$~$|4\rangle$ to $|0\rangle$ is not simple because of the cascade, etc., we performed the calculation by simiply assuming that all of the atoms in the states $|1\rangle$~$|4\rangle$ decay to the ground state with the rate τ^{-1}, where τ is the life-time of $|2a4d\ ^1D_2\rangle$ state in the fine structure scheme. The result is shown in Figure 2.

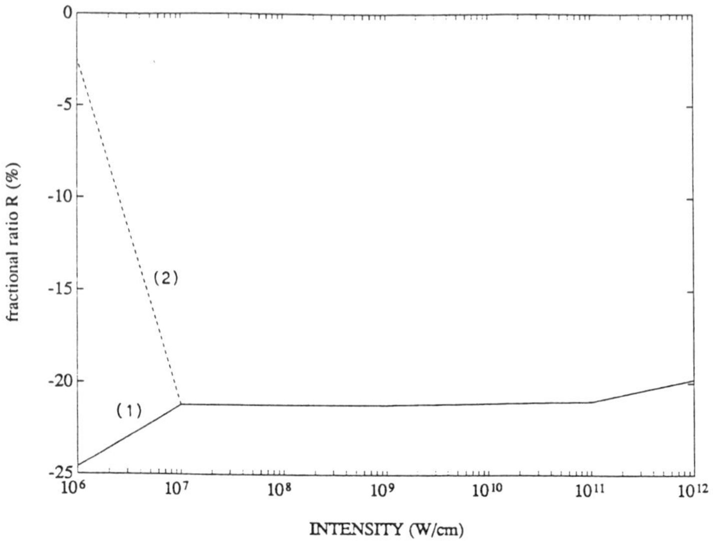

Fig.2
Line (1);result without the spontaneous decay terms. Same parameters with the line(5) of Fig.1 are used.
Line (2);result corresponding to the line(1) with the spontaneous decay terms.

The effect of the decay terms is large at the low intensity range when the Γ's are small, as we expected. It should be noted that the decay terms are negligible for the result with the Γ's as obtained by our atomic parameter calculations.

One may ask if the increase or the decrease of the Rabi frequency changes the fractional ratio R. A calculation performed with Ω's 10 times larger than those in Fig. 1 shows that this effect seems to be very small as was also the case for Sn (Lambropoulos 1989).

In summary, we have shown that the two-photon resonant 3-photon ionization of Be with broadband radiation will in general exhibit strong discrimination between the isotopes ^{9}Be and ^{10}Be. Our results at this time should be viewed as semiquantitative because the calculation of the atomic parameters through single-channel quantum defect theory is very approximate. It is the ionization widths that are most sensitive to this approximation since the energy range within the continuum reached by the third photon is not too far from the vicinity of autoionizing states (Clark et al. 1985). That is why we have explored the effect of a variety of values for the ionization widths. The conclusion at this point is that the dynamics of the process, imposed by the angular momentum of the states involved, is such that will lead to selective ionization under most circumstances. For the prediction of an accurate value of R and comparison with experiment, we need a more sophisticated calculation of the matrix elements which we expect to complete in the near future.

References

Clark C W, Fassett J D, Lucatorto T B, and Moore L J. 1985 J. Opt. Soc. B 2 891

Fairbank Jr W M, Spaar M T, Parks J E, and Hutchinson J M R 1989 Phys. Rev. A40 2195

Dixit S N, Lambropoulos P, and Zoller P 1981 Phys. Rev. A24 318

Dixit S N and Lambropoulos P 1983 Phys. Rev. A27 861

Lambropoulos P and Lyras A 1989 Phys. Rev. A40 2199

Lyras A, Zorman B, and Lambropoulos P 1990 Phys. Rev. A42 543

Inst. Phys. Conf. Ser. No 114: Section 2
Paper presented at RIS 90, Varese, Italy, 16–21 September 1990

Non-linearities in laser enhanced ionization spectroscopy by energy pooling

R. Hergenröder, A. Obrebski* and K. Niemax
Institut für Spektrochemie und angewandte Spektroskopie(ISAS)
Dortmund, Federal Republic of Germany
*Univesität Bielefeld, Fakultät für Physik, Bielefeld, FRG

ABSTRACT: Energy pooling collisions are effecting the direct proportionality between excited state number density and signal in LEI spectroscopy. Narrowing of spectral lines and unexpected intensity ratios can be observed. The effect was studied in the spectra of Ca and Ba intercombination lines.

1. INTRODUCTION

It is a basic assumption in quantitative laser enhanced ionisation (LEI) spectroscopy that the signal is linear proportional to the excited state number density. This is justified as long as the energy of the laser excited level is near (about kT) to the ionization limit. If the excited level is far from the ionization limit the thermal energy transfered through collisions cannot induce direct ionization, and other mechanism have to be taken into account. These processes do not necessary produce the linear correlation between the number of excited states and the ion signal which is required for quantitative measurements. Two processes - Penning ionization (Allegrini et al. 1976) and collisional energy pooling transfer (e.g. Huennekens and Gallagher 1983) - which can cause these effects were studied in our laboratory. In the present work we report on experiments dealing with the latter process.

2. EXPERIMENTS AND RESULTS

We have studied the Doppler limited LEI spectra of the barium and calcium intercombination lines (Ba :$6s^2\ ^1S_o$ -> $6s6p\ ^3P_1$, Ca :$4s^2\ ^1S_o$ -> $4s4p\ ^3P_1$) at 791.4 nm and at 657.3 nm, respectively. The spectra were measured in a thermionic diode filled with the relevant atomic vapour and neon as a buffer gas.

In the barium experiment the buffer gas pressure was 500 mTorr, the number density of barium atoms about 2×10^{12} cm^{-3} and the laser intensity of about 100 Wcm^{-1}, sufficiently to saturate the transition. Fig.1 shows the LEI spectrum of the barium intercombination line. The intensity ratio of the 'even' isotope components I_E to the 'odd' components I_O is 27 ± 3 (F=5/2). On the other hand, taking into account the saturation and the isotopic composition a ratio of 6.9 is expected. That means the intensity ratio was closer to $(I_E/I_O)^2$ than to (I_E/I_O). It can be attributed to the energy pooling process: 2 Ba($^3P_1$) -> Ba^{**}+Ba, (Ba^{**}:Rydberg atom). In this process the population density of the higher quantum state depends on the square of the number density of the laser excited state. The difference between the expected ratio (6.9^2) and the measured value can be attributed to the optical thickness of the strong component (65% absorption in the present case).

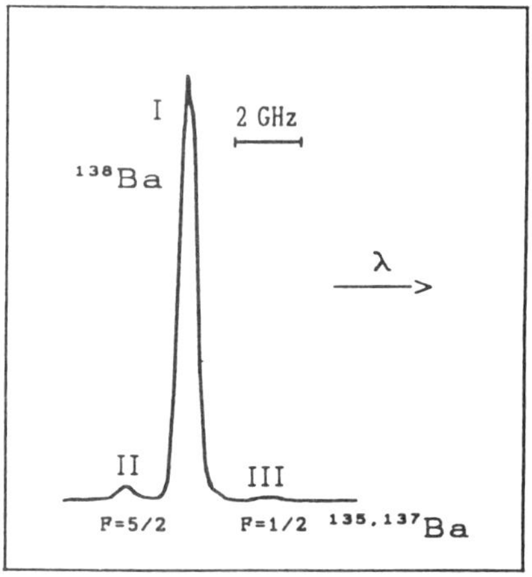

Fig. 1. LEI spectrum of the barium intercombination line.

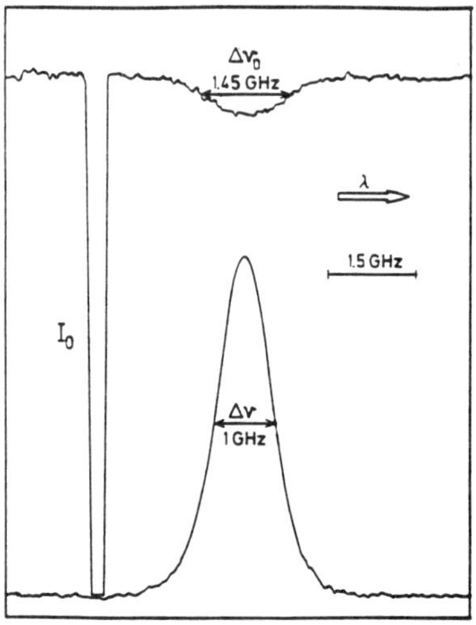

Fig. 2. LEI signal (lower trace) and absorption spectrum of the calcium line.

The calcium intercombination line is more suitable for investigations of energy pooling for different reasons. First the oscillator strength of the transition (Corliss and Bozman 1962) is very small (about $f=5x10^{-5}$). Therefore, even for relatively high calcium number densities the line is still optical thin. The lifetime is long, there is no ^3D_J state below the laser excited level as in the case of barium and optical pumping can be neglected. Natural abunded calcium mainly consists of the ^{40}Ca isotope (97%). For this reason the Doppler profil of the ^{40}Ca isotope is only slightly disturbed by the other isotopes. It should be mentioned that the energy of the laser excited level is three times less than the ionization energy of calcium (Sugar and Corliss 1985).Therefore direct ionization of excited atoms by thermal collisions is unlikely.

In the Ca experiment the buffer gas pressure was 400 mTorr. The temperature of the heat pipe was varied between 770 and 900 K which resulted in a calcium number density range of $5x10^{12}$ - $1.2x10^{14}$ cm^{-3} (Nesmeyanov 1963). The gas temperature was determined by the Doppler widths of the calcium absorption line and by measuring the integrated absorption (Unsöld 1968) taking into account the oscillator strength of the transition and the length of the absorption layer (L=10cm).

Fig.2 , which shows the absorption and the LEI spectrum of the investigated calcium intercombination line, demonstrates the influence of energy pooling effects on the line shape. Because the peak absorption of the line is only about 8%, the Doppler line widths of the absorption can be directly taken from the upper trace. Comparing the linewidth of the Doppler profil (1.45 GHz) and of the LEI signal (1GHz) the narrowing of the LEI line due to energy pooling collisions can be seen immediately. For further investigations of the line profil we plot the logarithm of the absorption coefficient k, converted to intensity and normalized to the peak value of the LEI signal, against the square of the detuning from the line centre (Fig.3). In the case of the absorption profil this is an straight line. From the Doppler formula $k=k_o\exp[-\Delta\nu^2/(0.36\Delta\nu_D^2)]$ we can determine the Doppler width. The LEI profiles were analysed applying the modified formula $I=I_o\exp[-m\Delta\nu^2/(0.36\Delta\nu_D^2)]$, where m gives the number of excited atoms which are involved in the energy pooling process. The LEI profil had to be corrected in the line centre, because the absorption could not be neglected. From the line centre to about 1.2 GHz detuning the slope of the LEI profile was about 2.7. This means that energy pooling collisions of two and three atoms are responsible for the observed narrowing of the line. The deviation from symmetry in the far wings is due to the less abundant isotopes which are shifted to the blue (Lorenzen et al. 1983).

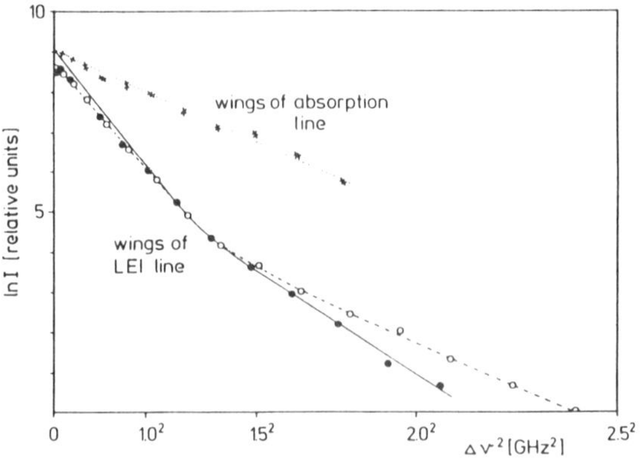

Fig. 3. (●) and (○) are the LEI red and blue wing data. (x) are red and blue wing data of the absorption line which is normalized to the LEI line. Full curve: LEI line wings corrected for optical thickness; dashed-dotted curve: deviation from line symmetry

There is another possibility to find out how many low-excited atoms are involved in the collisional energy pooling process. One had to vary the excited state number density. We did this by variation of the laser power at constant calcium number density ($N_{Ca}=1.2\text{x}10^{14}\text{cm}^{-3}$) and constant beam diameter. Fig.4 shows the LEI signals from the line centre of the ^{40}Ca line in a double logarithmic scale plotted against the laser power. The filled circles were taken from a measurement with an unfocused laser beam, the data given by open circles were measured with an expanded laser beam. As long as there is no optical saturation the number of laser excited atoms is proportional to the laser power. However, the number of collisional excited atoms scales with the square of the beam diameter. Fig.4 shows that in the range of high laser power the slope of the data obtained with the unfocused beam is approximately 2.6. With decreasing laser power the slope is smaller. With the expanded beam we measured a slope of about 2.4 in the region of high laser power. Together, this means that with increasing intensity energy pooling of two and three atoms is dominating. At lower densities the contribution of energy pooling with three atoms decreases. Within the limits of uncertainty the number of atoms

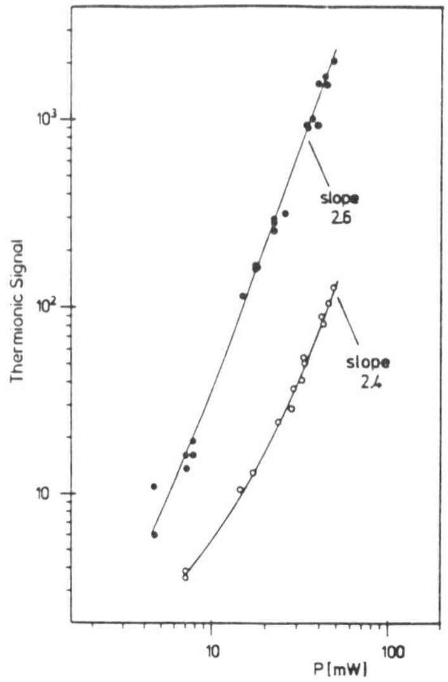

Fig. 4. LEI signal of the peak intensity of the Ca line versus the laser power. (●): unfocused laser beam; (○) expanded laser beam.

involved in the energy pooling process obtained from dependance of the laser power were in agreement with the values found in the analysis of the line profiles.

3. CONCLUSION

Our investigations show that energy pooling can strongly influence quantitative LEI spectroscopy of low lying states. Energy pooling processes give rise to non-linearities between laser excited number densities and ion signal, resulting in, for example, wrong line intensities and narrowed spectral lines.

Allegrini,M., Alzetta,G., Kopystynska,K., Moi,L., Orriols,G., Opt.Commun. 19, 96(1976)

Corliss,C., Bozman,W.R.: Experimental transition probabilities for spectral lines of seventy elements. Washington, D.C.: U.S. National Bureau of Standards (1962).

Huennekens,J., Gallagher,A., Phys.Rev. A27, 771(1983)

Lorenzen,C.-J., Niemax,K., Pendrill,L.R.: Phys.Rev. A28, 2051(1983)

Nesmeyanov,A.N.: Vapour pressure of the elements, Amsterdam: Elsevier(1963)

Sugar,J., Corliss,C.: J. Phys. Chem. Ref. Data 14 [Suppl.2], 51(1985)

Unsöld,A.,: Physik der Sternatmosphären. New York, Berlin, Heidelberg:Springer (1968)

Inst. Phys. Conf. Ser. No 114: Section 2
Paper presented at RIS 90, Varese, Italy, 16–21 September 1990

Theoretical study of the photoionization of a hydrogen atom in an external electric field

V D Kondratovich† and V N Ostrovsky‡

†Vinnitsa Polytechnical Institute, Vinnitsa 286021, USSR
‡Institute of Physics, Leningrad State University, Leningrad 198904, USSR

ABSTRACT: The present contribution is a synopsis of a review report summarising about ten years of theoretical investigation of the photoionisation of the hydrogen and hydrogenlike atoms in an external electric field.

1. INTRODUCTION

The Stark (or more justfully Lo Surdo - Stark) effect in hydrogen is one of fundamental reference problems in atomic physics. It is a typical example of the shape resonance resulting from the existence of barrier in the potential seen by the atomic electron with the tunnelling which leads to ionisation. The theory is simplified for the hydrogen atom as compared with the other species due to the separability of the dynamical variables in the parabolic coordinates. Enormous number of publications is devoted to the calculation of resonances positions and widths. However originally the object of study primarily was the resonances lying well below the top of the potential barrier what is related with the description of the optical spectral lines shift and broadening. Little was questioned about the manifestation of the resonances positioned above the barrier top.

The situation has changed with the advent of the high resolution experimental observations of the photoionisation cross section of the atom placed into uniform stationary electric field. The structure in the cross section energy dependence was revealed not only for the negative energies lying both below and above the potential barrier top but also in the positive energy region (relative to the zero-energy threshold). This development stimulated the resonance parameters calculation in the new energy regions as well as direct evaluation of the total Stark photoionisation cross section (Kondratovich and Ostrovsky 1980, 1982).

2. SEMICLASSICAL APPROXIMATION

The essential point of the theory is a construction of the final state electron wave function which includes the effect of Coulomb and uniform electric field since both of them are strong and none can be treated as a small perturbation. The semiclassical approximation used by the authors (Kondratovich and Ostrovsky 1980, 1982, 1984a,b, 1990a) is a natural device here since it has advantage of the analytical form of the description suitable for successive analysis (in contradistinction to the numerical studies (Luc Koenig and Bachelier 1980a,b). The semiclassical approximation was formulated in the uniform way what means that it is well applicable without essential loss of accuracy in the extremely vast range of the photoelectron energy and the electric field strength. Although the accuracy of the semiclassical approximation hardly can be improved by tractable methods, it is usually sufficient for the description of the experimental data even for high resolution experiments.

The set of quite simple analytical expressions was proposed which is suitable for the prompt evaluation of the resonance position and width in both below- and above-threshold regimes (for the general algorithm of calculations see Kondratovich and Ostrovsky 1990a). It should be stressed that the notion of the effective potential barrier top itself is not obvious in the multidimensional problems. Due to the peculiarities of the variables separation in the parabolic coordinates the effective barrier top for the electron escape to infinity depends on the motion partition between the coordinates. In the other words this implies that the barrier top position is a function of the resonance quantum numbers. The simple universal scaled diagrams were constructed which allow straightforward classification of every resonance in terms of its position relative to the effective barrier top (Kondratovich and Ostrovsky 1984).

3. RESONANCE PROFILES

The theory reproduces well also the Stark photoionisation cross section variation with the energy giving complete analytical description of the cross section modulation. This allows to argue, for example, that if the cross section oscillation mode discussed somewhat obscurely by Damburg (1987) exists then it should coincide with that obtained in the framework of the semiclassical approach, namely, the interference modulation (for more details see Kondratovich and Ostrovsky 1990a). One of the important universal conclusions is that the resonance lineshape is distinct from the symmetrical Lorentsian profile and can be described as a non-symmetrical Beutler-Fano contour which is characterised not only by the resonance position and width but also by the profile index (Kondratovich and Ostrovsky 1982, 1984a,b). Originally Fano parametrisation was constructed for the multichannel (or Feshbach) resonances but as it was demonstrated by the authors it is applicable also for the shape resonances which form alternative major type of resonances. In the Feshbach case the resonance asymmetry stems from the interference of direct and resonance transitions to the final state. Tentatively it can be stated that the same is the physical nature of the asymmetry in the shape resonance case although the partition of the resonance and nonresonance (background) contributions to the process is not so straightforward here. This formulation was applied recently by Alverz and Silverstone (1990a,b).

Mathematically the resonance position and width define the real and imaginary part of the point in the complex energy plane where the scattering matrix pole is situated. The photoionisation cross section is naturally represented as a sum of poles contributions and the resonance parameters (including the profile index) possess appealing meaning. Much less natural is the concept of Harmin (1985) who introduced an artificial energy dependence of width in order to describe the resonance profile.

Thus the cross section background can be described as a bulk contribution of very broad resonances. In order to make it more tractable it is converted to the compact analytical form using Watson transformation (the latter is employed in the scattering theory when the Regge representation is used) (Kondratovich and Ostrovsky 1984b). The background energy dependence is essential in many cases (see for example Kondratovich and Ostrovsky 1990a). For example, it is important in the description of the asymmetry of the resonance profiles observed by Bergeman *et al* (1984) in the superstrong electric field. The profile shape here differs significantly from the conventional Fano contour. Namely, the minimum on the low-energy side of the resonance is absent whereas on the high-energy the cross section rises and forms a plateau. The reason for this behaviour lies in the rapid growth of the background term within the resonance width. Generally, due to the finite resolution the experimental study of the resonance form is feasible only for sufficiently broad resonances, i.e. for those lying above the classical ionisation barrier. In this conditions the background is inevitably present.

4. RESONANCE AND INTERFERENCE EFFECTS

The analytical approach allows qualitative insight into the physical aspects of the problem. First of all it should be stressed that the cross section modulation is produced not only by the resonances but also by the interference effects. This is particularly clear if one turns to the process of the electron photodetachment from the negative ions where the modulation arises even when no resonance effects are included into the description. More definitely, this modulations appear even in the calculations when only the uniform electric field is taken into account in the photoelectron final state and all core effects are neglected (for the more detailed discussion see Kondratovich and Ostrovsky, this volume). Thus the physical separation of resonance and interference effects becomes an interesting problem. It is treated by the present authors (Kondratovich and Ostrovsky 1984a,b) in the following manner. For some parameter domains the cross section poles can be unambigousely classified as governing resonance or interference effects. Then looking for the smooth evolution of the poles position with variation of the parameters (energy and electric field) strength one can genetically classify poles in the other conditions. The energy scale can be subdivided into the regions where the resonance or interference effects prevail and where both of them are important. The general trend is that the significance of interference increases with the energy.

5. CLASSICAL ELECTRON TRAJECTORIES AND QUANTUM CROSS SECTIONS

The second important physical aspect of the problem is the interpretation of the cross section features in the terms of the ejected photoelectron classical trajectories. Consider the trajectories for the fixed energy which differ by the direction of electron ejection from the atomic nucleus. The prominent peculiarity of this family of trajectories is the existence of the limiting cycle (Kondratovich and Ostrovsky 1984b). The notion of limiting cycle is well known in the theory of differential equations. In our case the limiting cycle corresponds to the trajectory of electron ejected against direction of the force exerted on it by the uniform field. When the ejection direction tends to this limit the electron spends more time in the vicinity of the cycle. Thus the electron motion exhibits the resonance-type behaviour in the classical description. In the other terms the limiting cycle is the periodic unstable trajectory. Furthermore for the positive energy an infinite number of the trajectories exist which pass via the same point in space. The implications of this fact are especially appealing in the description of the differential Stark photoionisation cross sections (Demkov *et al* 1981, 1990b and this volume). The latter seems to be promising tool both for the fundamental study of the process peculiarities and for some practical applications.

REFERENCES

Alvarez G and Silverstone H J 1989a *Phys.Rev.A* **40** 90
-- 1990b *Phys.Rev.Lett.* **63** 1364
Bergeman T *et al* 1984 *Phys.Rev.Lett.* **53** 775
Damburg R J 1987 Preprint LAFI-108 Salaspils
Demkov Yu N, Kondratovich V D and Ostrovsky V N 1981 *JETP Lett.* **34** 403
Harmin D A 1985 *Phys.Rev.A* **31** 2984
Kondratovich V D and Ostrovsky V N 1980 *JETP* **52** 198
--1982 *JETP* **56** 719
--1984a *J.Phys.B: At.Mol.Phys.* **17** 1981
--1984b *J.Phys.B: At.Mol.Phys.* **17** 2011
--1990a *J.Phys.B: At.Mol.Opt.Phys.* **23** 21
--1990b *J.Phys.B: At.Mol.Opt.Phys.* **24** (in print)
Luc Koenig E and Bachelier 1980a *J.Phys.B: At.Mol.Phys.* **13** 1743
--1980b *J.Phys.B: At.Mol.Phys.* **13** 1769

Inst. Phys. Conf. Ser. No 114: Section 2
Paper presented at RIS 90, Varese, Italy, 16–21 September 1990

Differential cross section for photoionization of a hydrogen atom in a uniform electric field

V D Kondratovich† and V N Ostrovsky‡

†Vinnitsa Polytechnical Institute, Vinnitsa 286021, USSR
‡Institute of Physics, Leningrad State University, Leningrad 198904, USSR

ABSTRACT: The differential cross section for the photoionisation of an atom in a uniform electric field can be observed if the photocurrent is registered on the plane screen perpendicular to the uniform field axis. It manifests complicated structure due to both resonance and interference effects which are interpreted in terms of the classical electron trajectories. In certain experimental conditions the pattern can achieve a macroscopic size. The tentative applications of the differential cross section observation are discussed.

1. INTRODUCTION

For the atomic photoprocesses in the stationary uniform electric field in most cases the total cross sections are investigated. The study of the photocurrent distribution in space suggests some new possibilities. In order to visualize such a distribution one can detect the photocurrent on the plane screen perpendicular to the uniform field axis. The pattern observed can be named the differential cross section of the process. The important point stressed by Demkov *et al* (1981) is that the experimental conditions can be chosen so that the pattern on the screen would be of macroscopic size. The preceeding studies (Fabrikant 1980, 1990, Demkov *et al* 1981, Du 1989) were devoted to the electron photodetachment from the negative ion. In this case the differential cross section structure is interpreted exclusively in terms of interference. In the classically accessible region the electrons ejected from the point source in two different directions can reach the same final point on the screen (Fabrikant 1980). The interference of the related contributions gives the oscillatory (Airy-type) structure in the differential cross section which can be visualized as the rings on the screen.

2. RESONANCE AND INTERFERENCE EFFECTS IN PHOTOIONISATION

In the case of photodetachment the electron motion in the final state is infinite and qualitatively similar in all energy range. Unlike it, in the case of photoionisation the principal features of the photoelectron motion vary drastically with the energy region. The parameters of such a dependence are the classical ionisation threshold in the presence of an external electric field $E_c=-2(ZF)^{1/2}$ and the field-free ionisation potential of the atom I. Here Z is the effective charge of atomic core, F is the uniform electric field strength, the atomic units are used. The zero point of the photoelectron energy E=0 corresponds to the field-free ionisation limit. The calculations of present paper are performed for the case of hydrogen atom photoionised from a spherically symmetric initial state, so Z=1. Nevertheless we retain this parameter in the general formulae.

For the photoelectron energies $E<E_c$ only a finite motion is possible. The pattern on the screen can appear only as a result of tunnelling in the case of resonance on the quasidiscrete Stark level (i.e. non-classically). Then the differential cross section represents essentially the squared resonance wave function in the parabolic coordinate ξ. The motion along this coordinate is finite for all energies.
When $E>E_c$ the "bottleneck" is opened in the potential barrier around the atomic nucleus, so a part of photoelectrons can escape to infinity. In the differential and total cross sections this results in the appearence of a background. The background in the total cross section of photoionisation is described in detail in our previous works (Kondratovich and

Ostrovsky 1984, 1990). The differential cross section background is also formed by superimposed broad above-barrier resonances. They form a concentric pattern on the screen which radius ρ is a smooth function of energy. The pattern radius depends on the radius of the "bottleneck". It is less than the maximal possible radius of interference picture in the energy region $E_c<E<0$. The maximal radius appears for the narrow below-barrier resonances, in which the photoelectron escapes the core along the direction of the external electric field. Thus the effect of tunneling is appreciable well above the classical ionisation limit E_c.

The above mentioned most stable (narrow) resonances (with the parabolic quantum number $n_2=0$, as it was proved earlier, see Kondratovich and Ostrovsky 1984) emerge above barrier in the narrow energy region: $-F^{2/3}<E<F^{2/3}$. So for $E>F^{2/3}$ the radius of the background pattern is equal to the maximal possible one:

$$\rho^c_{max}=[p+(p^2+4Z/F)^{1/2}]^{1/2}(2z_o)^{1/2} \tag{1}$$

z_o being the distance between the photoionising atom and the screen, $p=E/F$ is the scale parameter for the classical motion in a uniform electric field. The stable above-barrier resonances with $n_2=0$ manifest themselves as an enhancement of the pattern intensity at the resonance energy.

In all region $E>0$ the pattern radius in the case of photoionisation is larger than in the photodetachment case. The pattern structure in the photoionisation case is more complicated because of interference of an infinite number of photoelectron classical trajectories leading to the same point in space. Two of these trajectories are of the same nature as the interfering trajectories in the case of photodetachment. They mainly contribute to the rainbow-type part of interference pattern which is due to the interference near the caustical surface. The electrons moving on the mentioned trajectories do not return to the vicinity of nucleus after ejection. Such a return is characteristic property of the other type of trajectories. These trajectories form two bundles embracing the Coulomb centre and generate the glory-type part of interference pattern. The serpent-like form of the trajectories indicates their connection with the resonances in the superposition of Coulomb and uniform fields. Only the glory-type interference exists in the region $E<0$. In the case $E>0$ two parts of pattern are superimposed. The radius of the glory-type pattern is the largest one since the multiple trajectory passages around the nucleus enhance the component of momentum perpendicular to the external electric field. The difference between the radia decreases when the energy increases.

3. DISCUSSION

The experimental observation of the differential cross section can be useful in some applications. The high sensitivity of the interference pattern to the photoelectron energy can be used tentatively in the measurements of the ionisation potentials or electron affinities. Due to Aharonov- Bohm effect (see e.g. Feynman *et al* 1965) the interference phase is influenced by the magnetic field applied transversally to the electric field, the phase increment being proportional to the magnetic flux through the loop bounded by the interfering trajectories. This property can be used e.g. for measurement of weak magnetic fields.

REFERENCES

Demkov Yu N, Kondratovich V D and Ostrovsky V N 1981 *JETP Lett.* **34** 403

Du M L 1989 *Phys.Rev.A* **40** 4983

Fabrikant I I 1980 *JETP* **52** 1045

--1990 *J.Phys.B: At.Mol.Opt.Phys.* **23** 1139

Feynman R P, Leighton R B and Sands M 1965 *The Feynman Lectures on Physics* (Reading: Addison-Wesley) vol.3 *(Quantum Mechanics)*

Kondratovich V D and Ostrovsky V N 1984 *J.Phys.B: At.Mol.Phys.* **17** 2011

--1990 *J.Phys.B: At.Mol.Opt.Phys.* **23** 21

Giant electric-field-induced linear dichroism

V D Kondratovich

Vinnitsa Polytechnical Institute, Vinnitsa 286021, USSR

ABSTRACT: The electric-field-induced anisotropic absorption in photoionising atomic vapour placed in an external electric field is theoretically investigated. The effect is characterized by giant values of dichroism and by specific dependance on the external electric field.

1. INTRODUCTION.

A linear dichroism can be related with different absorption in photoionising atomic vapour of two ortogonal components E_{ix} and E_{iz} of plane-polarized incoming light wave with vector of electric field strength E_i. It is supposed that the light wave propagate in the y axis direction and the external electric field is directed along z axis. For anisotropic absorption the ratio E_{ox}/E_{ix} differs from E_{oz}/E_{iz} (subscript o is used for the outcoming wave). Hense the vector E_o is rotated relative the vector of polarization of incoming wave E_i. So, anisotropic absorption can be characterized by dichroism constant D and the angle of rotation φ. For D we use the standard definition

$$D=(æ_z-æ_x)/(æ_z+æ_x) \qquad (1)$$

where $æ_z$ and $æ_x$ are absorption factors for two components of plane-polarized light.

The aim of this work is the study of anisotropic photoabsorption in the region of continuum above the classical ionisation limit in an external electric field $E_c=-2F^{1/2}$. Atomic units are used with few exceptions indicated below. F is the external electric field strength. The zero point of energy corresponds to the field-free ionisation threshold, so the frequency of the ionising light is equal $\omega=I+E$ where I is the ionisation potential of free atom. No significant ellipticity appears in the continuum region because of relatively short lifetime of an excited and next ionised atom.

2. DICHROISM AND ROTATION IN A VAPOUR OF ATOMIC HYDROGEN PLACED IN AN EXTERNAL ELECTRIC FIELD.

In the work of Kondratovich and Ostrovsky (1984b) the cross section of the one-photon ionisation in an electric field was represented in the region $E>E_c$ as a sum of two terms. The first one (denoted further as σ_s) of smooth dependance on energy, describes the contribution of the continuum part of Stark spectrum to the photoionisation cross section. For photoionisation by plane-polarized light from 1s initial state of a hydrogen atom the approximate analytical formulae are stated (Kondratovich and Ostrovsky 1984b and 1990).For the sake of brevity I do not reproduse here these formulae They show a strong influence of polarization of the ionising light on the energy dependance of continuum part σ_s.In the region E>0 this influence is absent except the case of interference oscillations.

The second term in the above mentioned expansion of the photoionisation cross section describes the contribution of sharp resonances. Here we fix attention on the case of non-resonant ionisation in the region of energy $E_c<E<0$. The non-resonant case is easy to be obtained by small variation of the light frequency or, what is more convenient, of the electric field strength. In the non-resonant case the cross section of photoionisation can be described by the above mentioned expressions which show a strong anisotropy of absorption.

The energy dependance of the continuum σ_s in the case $E_i \parallel F$ (denoted further as σ_{sz}) and in the case $E_i \perp F$ (denoted further as σ_{sx}) is characterized by equality of σ_{sz} and σ_{sx} at the energies $E=E_c$, $E=E_o=-(2F)^{1/2}=2^{-1/2}E_c$ and $E=0$. In the intervals $E_c<E<E_o$ and $E_o<E<0$ the difference $\Delta\sigma_s=\sigma_{sz}-\sigma_{sx}$ has different signs. The maximum value $|\Delta\sigma|$ has at energes $E_\pm = (1 \pm 3^{-1/2})^{1/2}E_o$. This value do not depends on the external electric field strength. Using the relation between the absorption factor æ and the cross section of photoionisation σ, $\text{æ} = n\sigma$ (n being the density of vapour), we obtain from formula (1) the approximate expression for the dichroism constant in the interval of energy $E_c<E<0$:

$$D=6Z_c(Z_c-1/2)/(2Z_c^2-Z_c+2) \qquad (2)$$

where $Z_c=E^2/4F$. The dichroism constant varies from the maximal value equal to unity at the energy $E=E_c$ to the zero value at $E=2^{-1/2}E_c$. The minimal value -0.2 is obtained at $E=E_c/2$. As the energy increases further, the dichroism constant also increases to the zero value at $E=0$. It is to be stressed that in the first approach the maximal and minimal values of the dichroism constant *do not depends* on the external electric field strength F. The latter only gives the scale in the dependance of D on the energy E throughthe relation $E_c =-2F^{1/2}$.

As it was mentioned in Introduction, the effect is accompanied by giant ellipticity-free rotation. The constant of rotation $\gamma=d\varphi/dl$ (l being the distance along the ray) has maximal value at $E=E_\pm$:

$$\gamma = \frac{d\varphi(E_\pm)}{dl} \approx \frac{1}{4}\Delta\sigma n = \pm \frac{2^{10}\pi^2(\alpha-1/2)^2 n}{3^{3/2}\alpha^7 c \exp(4/\alpha)} . \qquad (3)$$

It should be also stressed that the extremal values of the constant of rotation *are independent* on the external electric field strength. The extremal values of γ depend only on the density of vapour and on the quantum numbers of the initial state. For molar density $(2.6\times10^{19}cm^{-3})$ γ is about 100 radians/cm.

In the region E>0 the total cross section σ_z ($E_i \parallel F$) exhibits pronounced oscillations with the amplitude proportional to $F^{1/4}$ (Kondratovich and Ostrovsky 1984a) . These oscillations are due to the resonances with parabolic quantum numbers $n_2=0$ and m=0 shifted in this region by the external electric field. In the case $E_i \perp F$ no significant oscillations appear. So noticeable dichroism of resonance origin exists even in the energy region above field-free ionisation threshold E=0. For the energies E>I (I being the field-free ionisation potential) the oscillations of the interference origin in the cross section σ_z only remain. These oscillations of the constant depth of modulation of the cross section also causes an ellipticity-free dichroism.

The regions of spectrum unaccessible in the one-photon experiments with atomic hydrogen can be easily reached for hydrogenlike systems. A set of these systems includes hydrogenlike atoms and atoms with well separated in energy regions of one- and two-electron excitation and also the excitons in semiconductors. Dichroism of above mentioned interference type appears also in the case of photodetachment in an external electric field.

A strong interaction of light with photoionising systems might be used in an elaboration of logical elements for optical computers. For example, one of two or three crossed beams of light with different frequencies can govern the passage of other beams through a volume of vapour or semiconductor. An external electric field also could be used for the same goal.

REFERENCES

Kondratovich V D and Ostrovsky V N 1984a J.Phys.B: At.Mol.Phys. I7 1981-2010

--1984b J.Phys.B: At.Mol.Phys. I7 2011-38

--1990 J.Phys.B: At.Mol.Opt.Phys. **23** 21-43

Paper presented at RIS 90, Varese, Italy, 16–21 September 1990

Electron-nucleus resonance in a laser radiation field

F F Karpeshin and B A Zon

Leningrad State University, Institute of Physics, Petrodvorets, Leningrad, 198904, USSR (F F K) and

Voronezh State University, Voronezh, 394693, USSR (B A Z)

ABSTRACT: A way of significant acceleration (by several orders of magnitude) of a nuclear transition rate is offered through resonance interaction with the electronic shell. External radiation field serves for the compensation of the energy difference. Formation of vacancies during this process can in principle be used for generation of a coherent radiation of a much higher frequency in comparison with that of the externally applied field.

1. INTRODUCTION

Many authors endeavoured looking for ways of application of strong electromagnetic fields for the purpose of affecting particle and nuclear decay rate (see, e.g., Nikishov and Ritus 1964, Goldansky and Letokhov 1974, Zaretsky and Lomonosov 1985). Direct effect turned out to be quite negligible, but in some cases it can attain a feasible value through interaction with the electronic shell. We first offer a method of drastic acceleration of nuclear decay rate.

2. Electronic shell as a resonator for nuclear radiation.

The electronic shell is proposed to play a role of a resonator for nuclear radiation. Energy exchange between nuclei and their electronic shells occures by virtue of a new type of internal conversion. The electron having absorbed the nuclear transition energy is lifted up to a higher discrete level. - This is the discrete internal conversion (see Figures 1, 2). Note that during conventional internal conversion the electron is lifted up into the continuum.

It is convenient to introduce the <u>conversion factor</u>:

$$R = \frac{\text{Conversion probability}}{\text{Radiative transition probability}} \qquad (1)$$

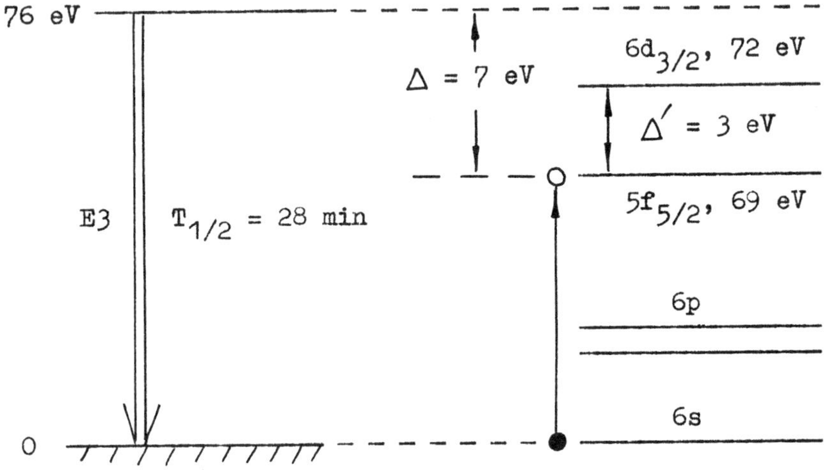

Figure 1. Nuclear (left) and electronic (right) transition schemes for ^{235}U.

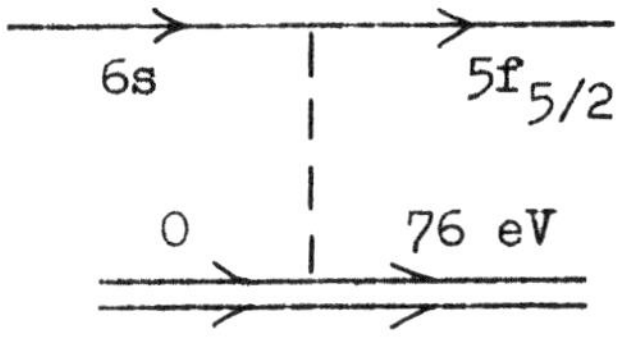

Figure 2. Feynman graph corresponding to the scheme of Figure 1.

The expression for R can be deduced in the Breit-Wigner form (Zon and Karpeshin 1990)

$$R(\tau L) = \alpha(\tau L) \frac{\Gamma / 2\pi}{\Delta^2 + (\Gamma/2)^2} \qquad (2)$$

Here τL stand for the multipole order. The total width $\Gamma \cong \Gamma_a$, Γ_a being the electronic hole width. α is the energy-dimensional discrete analog of the conventional internal conversion coefficient (ICC) (Zon and Karpeshin, 1990). It is calculated by means of the standard formulae.

3. Tuning of the electronic "resonator" by applying a laser radiation field.

As is known, in a laser radiation field energy of electronic transitions can be shifted by $\hbar\omega$, ω being the laser frequency. As a result, exact coincidence of the nuclear and electronic energies can be achieved.

Final-state electronic wave function in a laser radiation field can be presented in the form

$$|6d_{3/2}\rangle_1 = \left\{|6d_{3/2}\rangle + \exp(\pm i\hbar\omega t)\beta\,|5f_{5/2}\rangle\right\}\exp(-i\frac{E}{\hbar}t) \quad (3)$$

Here E is the energy of the atomic state $6d_{3/2}$, β is the admixture amplitude of atomic states $5f_{5/2}$ to $6d_{3/2}$ arising in a laser radiation field. Corresponding Feynman graph reads as follows:

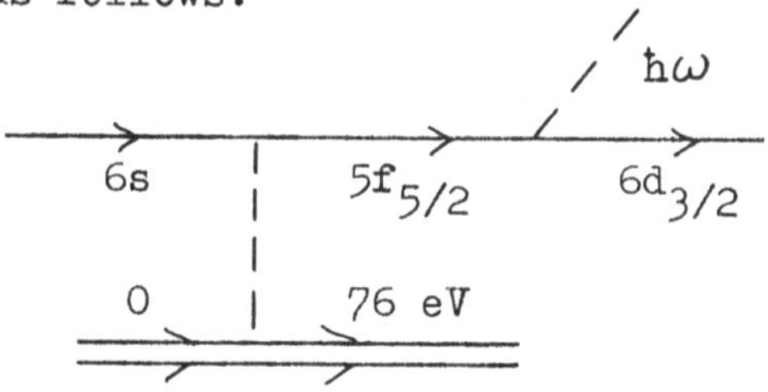

Figure 3. Feynman graph of the discrete conversion (Figures 1, 2) in a laser radiation field.

Using wave function (3) in a conventional expression for α in (2) and retaining only the nearest $5f_{5/2}$ component in the intermediate state, one readily obtains for the conversion factor the expression

$$R_1(\tau L) = |\beta|^2\alpha(\tau L)\,\frac{\Gamma/2\pi}{(\Delta - \Delta' - \hbar\omega)^2 + (\Gamma/2)^2} \,. \quad (4)$$

4. Numerical results for ^{235}U and further examples.

We use the next values given by Grechukhin and Soldatov (1983): $\Delta = 7$ eV, $\Delta' = 3$ eV and $\Gamma \cong 4 \cdot 10^{10}\ s^{-1}$. Therefore, $\hbar\omega = 4$ eV. The α value can be calculated by using the matrix element

$$\langle 6d_{3/2}|r^{-4}|5f_{5/2}\rangle = -60\ a^{-4},$$

a being the Bohr radius. The calculation gives $\alpha = 6 \cdot 10^{20}$ eV. Correspondingly, we obtain by means of eq. (2) the conversion factor

$$R = 2 \cdot 10^{12}.$$

This is nevertheless by 7 orders of magnitude smaller than the conventional ICC in the upper electronic shells 5f, 6p, 7s (Grechukhin and Soldatov 1976). But, after compensation in an externally applied field, one obtains by means of eq. (4) that in the case of exact resonance $|\Delta - \Delta' - \hbar\omega| \lesssim \Gamma$ the R value increases by $|\beta|^2 (\Delta/\Gamma)^2$ times. Accurate calculation of β allowing for broadening of Γ due to multiple ionization (Zon and Karpeshin 1990) shows that the optimum radiation intensity is about 10^{14} W/cm^2. At this intensity $\beta = 0.03$, and $R = 7 \cdot 10^{20}$. So, the effect in the case of ^{235}U can be observed at the level of at least two orders of magnitude.

Note, that formation of vacancies in the 6s shell could be used for generation of coherent radiation with a frequency of transition 6p$\longrightarrow$6s (about 30 eV).

There exists a number of other nuclides which can be considered as candidates for observing the effect, e.g., ^{99}Tc (2.12 keV E3 transition) and others. All the statements hold for these elements, besides that the Δ and Δ' values are of hundreds eV and larger. Hence, intense X-ray beams are to be used to search for the effect in these atoms.

References

Goldansky V I and Letokhov V S 1974 Zh. Eksp. & Teor. Fiz. 67 513 (Engl. transl. JETP)
Grechukhin D P and Soldatov A A 1976 IAE report No. 2706 (Moscow)
Grechukhin D P and Soldatov A A 1983 IAE report No. 3824/2 (Moscow)
Nikishov A I and Ritus B I 1964 Zh. Eksp. & Teor. Fiz. 46 1768 (Engl. transl. JETP)
Zaretsky D F and Lomonosov V V 1985 Yad. Fiz. 42 1383 (Engl. transl. Sov.J. Nucl. Phys.)
Zon B A and Karpeshin F F 1990 Zh. Eksp. & Teor. Fiz. 97 401 (Engl. transl. JETP)

Inst. Phys. Conf. Ser. No 114: Section 2
Paper presented at RIS 90, Varese, Italy, 16–21 September 1990

Doubly excited autoionization states of Ca and Sr 0-20000 cm^{-2} above the ionization limit

J O Gaardsted, T Andersen, H K Haugen and H Stapelfeldt

Institute of Physics, University of Aarhus, DK-8000 Aarhus C, Denmark

ABSTRACT: Doubly excited autoionization states of Ca and Sr are studied by multi-photon absorption through the subsequent decay to excited 4p and 5p ionic states, respectively. The observed resonances in the fluorescense signal due to doubly excited states are characterized. Sufficient atom density is obtained utilizing a vapor in a moderate-pressure regime (0.1-5 mTorr) such that collisional effects are negligible. The pulsed (12 ns duration) dye laser was scanned in the 490-540 nm and 560-575 nm ranges. Both linearly and circularly polarized light have been used.

1. INTRODUCTION

Atomic structure studied by absorption of a single photon is a well known technique which has also been applied to the $Ca(4s^2)$ and $Sr(5s^2)$ ground states. Since the ground state term is $^1S_0$, one is however restricted to odd J=1 states. The technique where more than one photon is absorbed in the same process (multi-photon absorption) allows studies of both odd and even states with more than one J value. For example, absorbing from a J=0 level the possible final states are:

2 phot. abs.	even	J=0,2
3 - " -	odd	J-1,3
4 - " -	even	J=0,2,4
etc.		

Since the appearence of pulsed lasers producing strong fields the multi-photon absorption process has become possible and thereby studies of many new states.

2. EXPERIMENTAL TECHNIQUE

We have studied doubly excited autoionization states of Ca and Sr via the fluorescense from excited ions which are formed after multi-photon absorption from the neutral atom ground state. The experimental technique is analogous to the that described earlier /1/, although some improvements have been made. In order to obtain sufficient atom density to facilitate the fluorescense experiments, a vapor - in contrast to an atomic beam - is utilized. This allows the search for weak channels, that is higher lying autoionization states. In the present work a crossed heat-pipe oven is operated in a non-heat-pipe regime with vapor pressures of ~0.1-5 mTorr. Since lifetimes of the excited atomic states are < 10 ns, collision processes can be neglected and we have demonstrated that the fluorescense signal is proportional to the density of atoms. Wavelength-resolved fluorescense from $Ca^+(4p)$ or $Sr^+(5p)$ J=1/2,3/2 is collected at 90° to the laser-beam direction.

The study has concentrated on the characterization of the observed resonances due to doubly excited states of the elements rather than the ionization process itself. The pulsed (12 ns duration) dye laser was scanned in the 490-540 and 560-575 nm ranges with intensities in the interaction region typically around 1-4 GW/cm^2 (4-photon) and 10-40 GW/cm^2 (5-photon), respectively. The influence of both linearly and circularly polarized light has been studied. A minimum of 3 photons is required to ionize the $Ca(4s^2)$ and $Sr(5s^2)$ ground states. Most of the observed resonances are reached after 3 photons, although 1 or 2 additional photons must be absorbed in order to populate $Ca^+(4p)$ or $Sr^+(5p)$.

3. RESULTS AND DISCUSSION

Up to the present point only preliminary results are obtained.

The Ca resonances observed in the 490-540 nm region (1700-7400 cm^{-1} below the $Ca^+(3d\ ^2D_{3/2})$ state) are dominated by the 3dnf series while the 3dnp series are weaker. The J=1 states in these series are well known from earlier single VUV-photon absorption studies.

The presence of the 3dnf series is allowed because the ground state is not pure $4s^2$ but contains $4p^2$. In this series the 3d4f-resonances are the strongest, probably due to almost resonant absorption of the 4th photon. The outgoing electron has an energy of only 5-18 meV when the residual ion is left in the $4p_{1/2}$ state.

In the 560-575 nm region 5 photons are required to populate the $Ca^+(4p)$ state. Only one resonance, at 565.5 nm, is observed in this region. This wavelength range has been studied previously utilizing an atomic beam, 50 ps laser pulses and Ca^+, Ca^{++} detection /2/. They observed a resonance (only in the Ca^{++} signal) at 564.6 nm which moved towards lower wavelengths with increasing laser intensity. The two resonances differ in wavelength, in width, in response to circ.pol. light, and in shift as fucntion of laser intensity. Clearly, these two resonances are not identical. None of the 3d5p J=1 states (all well known) appearing in this wavelength range fits with our wavelength, thus indicating a 4 or maybe 5 photon resonance.

Previously, Sr has been studied in the 560-575 nm region by multi-photon spectroscopy. This region contains doubly excited J=1 and J=3 states belonging to the 5p5d and 4d4f configurations. Also the 530-540 nm region is well studied and understood. Resonances here belong to the 4d5f configuration. The odd J=3 states in the 490-530 nm region have been studied recently by multi-photon spectroscopy /3/. The 4dnf, n≥6 and 4dnp, n≥8 are well represented in this region but also the 4dnh, n≥6 series has been observed. These series converge to 4d J=3/2,5/2 around 494 and 496 nm. Our spectra exhibit a dense series of resonances of which several coincide with those reported in ref. 3.

1. H K Haugen and A Othonos, Phys.Rev.A 38, 6448 (1988)
   and Phys.Rev.A 39, 3392 (1989)
2. P Agostini and C Petite, J.Phys.B 17, L811 (1984)
3. M Kompitsas, S Cohen, C A Nicolaïdes, O Robaux, M Aymar and P Camus, J.Phys.B 23, 2247 (1990)

Population trapping for above-threshold ionization induced by a train of ultra-short laser pulses—analytical solution

J Swiatlowski and W Leonski

Institute of Physics, A Mickiewicz University, Grunwaldzka 6, 60-780 Poznan, Poland.

ABSTRACT: We deal with an atom containing one discrete ground level and two continua driven by a train of ultra-short laser pulses. We find recurrential relations permitting the numerical calculation of the time-evolution of the ground state population for Lorentzian and Gaussian continua. The analytical solution for Lorentzian continua is also obtained.

1. THE MODEL

The atomic system under consideration involves one discrete ground level $|0\rangle$ and two orthogonal continua $|c_1\rangle$ and $|c_2\rangle$. The continua are of the same shape and have identical widths. This model resembles that of Deng and Eberly (1984) and Rzazewski et al (1989). However, we assume that the atom is driven by a train of ultra-short laser pulses instead of a field with well defined frequency. The pulses couple the ground level $|0\rangle$ to the lower continuum $|c_1\rangle$ and the two continua $|c_1\rangle$ and $|c_2\rangle$ mutually. The time interval between two successive pulses is T. A similar model involving only one continuum has been discussed by Rzazewki and Mostowski (1987). The case of a three level atom driven by a train of ultra-short laser pulses has been discussed by Greenland (1983). The Hamiltonian of our system can be written in the form

$$H = H_0 + \sum_{n=0}^{\infty} H_1 \delta(t-nT) , \tag{1}$$

where H_0 is the free atom Hamiltonian and H_1 describes the interaction between the atom and the field. The states $|0\rangle$, $|c_1\rangle$ and $|c_2\rangle$ are eigenstates of the free atom Hamiltonian H_0:

$$H_0|0\rangle = 0|0\rangle , \qquad H_0|c_j\rangle = E_j|c_j\rangle \quad (j = 1,2) , \tag{2}$$

where we assume the energy of the ground level equal to 0. We write the time evolution operator U transforming the wave function from its value Φ_n after the n-th pulse to its value $\Phi_{n+1}=U\Phi_n$ after the (n+1)-st pulse in the form

$$U = \exp(-iH_1) \exp(-iH_0T) \tag{3}$$

(in units of $h/2\pi=1$). The operator $U_0=\exp(-iH_0T)$ describes the free evolution of the system between two successive pulses whereas the operator $U_1=\exp(-iH_1)$ transforms the wave function describing the system just before the pulse to that just after it. We choose the operator U_1 in the form:

$$U_1 = \exp\,[iF(x_1+qx_2)]\ , \tag{4}$$

where F is a measure of the field strength. The operators x_1 and x_2 describe the transitions between the ground state $|0\rangle$ and the lower continuum $|c_1\rangle$, and between the two continua, respectively. The factor q permits a comparison of the strengths of these two couplings. The nonzero matrix elements for the Hermitian operators x_1 and x_2 are

$$\langle 0|x_1|c_1\rangle = (\langle c_1|x_1|0\rangle)^* = \varepsilon_1(E_1)\ , \tag{5a}$$
$$\langle c_1|x_2|c_2\rangle = (\langle c_2|x_2|c_1\rangle)^* = \varepsilon_1(E_1)^*\varepsilon_2(E_2)\ , \tag{5b}$$

where $\varepsilon_j(E_j)$ are functions describing the shape of the continua $|c_j\rangle$. In this paper we discuss two cases, the one involving Lorentzian-shape continua and the other involving continua of Gaussian shape.

2. NUMERICAL CALCULATIONS

To obtain the relations for the probability amplitudes we apply the procedure proposed by Rzazewski and Mostowski (1987). We decompose the wave function of the system into the eigenstates of H_0

$$\Phi_n = a_n|0\rangle + \int b_{1,n}(E_1)|c_1\rangle dE_1 + \int b_{2,n}(E_2)|c_2\rangle dE_2\ . \tag{6}$$

Acting with the time-evolution operator U on the wave function Φ_n and introducing new quantities $v_{j,k}$ (j =1,2) as follows

$$b_{j,n}(E_j) = i\varepsilon_j(E_j)^* \sum_{k=0}^{n} \exp[-i(n-k)E_jT]v_{j,k}\ , \tag{7}$$

we obtain (assuming that the atom is initially in its ground state) the following set of recurrential relations

$$\begin{aligned} a_{n+1} &= a_n[q^2+\cos(QF)]/Q^2 - S_1\sin(QF)/Q + iqS_2[\cos(QF)-1]/Q^2\ ,\\ v_{1,n+1} &= a_n\sin(QF)/Q + S_1[\cos(QF)-1] + iS_2\sin(QF)/Q\ ,\\ v_{2,n+1} &= -ia_nq[\cos(QF)-1]/Q^2 + iqS_1\sin(QF)/Q + q^2S_2[\cos(QF)-1]/Q^2\ , \end{aligned} \tag{8}$$

where

$$Q = (1+q^2)^{1/2}\ ; \qquad S_k = \sum_{j=0}^{n} B_k(n-j+1)v_{k,j} \quad (k=1,2)\ . \tag{9}$$

For the Lorentzian case the functions $B_k(n)$ are given by

$$B_k(n) = \exp[-n(iE_{Lk} + \Gamma_L)T]\ , \tag{10a}$$

whereas for the Gaussian case they take the form

$$B_k(n) = \exp(-inE_{Gk}T)\,\exp(-n^2\Gamma_G^2T^2/4)\ , \tag{10b}$$

where Γ_i and E_{ik} (i=L,G) determine the widths and centers of the continua.

In this work we are interested in the population of the ground level - this quantity determines the total ionization rate. The probability $P=|a_n|^2$ of finding electrons in the ground state can be calculated straightforwardly from the recurrential formulas (8). They are accessible to study even using standard microcomputers. Our considerations are restricted to simulation of the system for up to about 100 laser pulses. For longer times the numerical errors grow enormously.

3. ANALYTICAL SOLUTION

The functions $\varepsilon_k(E_k)$ are chosen in such a way that it becomes possible to find an analytical solution for the probability amplitudes. We introduce generating functions for a_n and $v_{k,n}$ (k=1,2)

$$A(z) = \sum_{n=0}^{\infty} z^n a_n \ ; \quad C(z) = \sum_{n=0}^{\infty} z^n v_{1,n} \ ; \quad D(z) = \sum_{n=0}^{\infty} z^n v_{2,n} \ . \tag{11}$$

Using the formulas (8) with the assumption that the atom is initially in its ground state we obtain the following system of equations for the generating functions

$$\begin{bmatrix} 1-z[k^2+\cos(QF)]/Q^2 & z\sin(QF)/Q & izq[1-\cos(QF)]/Q^2 \\ -z\sin(QF)/Q & 1+z[1-\cos(QF)] & -izq\sin(QF)/Q \\ -izq[1-\cos(QF)]/Q^2 & -izq\sin(QF)/Q & 1+zq^2[1-\cos(QF)]/Q^2 \end{bmatrix} \begin{bmatrix} A(z) \\ C(z)G_1(z) \\ D(z)G_2(z) \end{bmatrix} = \begin{bmatrix} 1 \\ 0 \\ 0 \end{bmatrix}$$

where (12)

$$G_j(z) = \sum_{n=0}^{\infty} z^n B_j(n) \quad (j=1,2) \ . \tag{12a}$$

The probability amplitude a_n of interest to us can now be obtained from the formula

$$n!a_n = \left. \frac{d^n W_A(z)}{dz^n W(z)} \right|_{z=0} , \tag{13}$$

where

$$W_A(z) = 1 + [1-\cos(QF)]\{zq^2G_2(z)/Q^2 + zG_1(z) + 2z^2k^2G_1(z)G_2(z)\} , \tag{13a}$$

$$W(z) = 1-z+[1-\cos(QF)]\{z[(G_1(z)+k^2G_2(z)/Q^2+1/Q^2)] + z^2[2k^2G_1(z)G_2(z)-k^2G_2(z)+(1-k^2)G_1(z)]/Q^2 - 2z^3k^2G_1(z)G_2(z)/Q^2\} . \tag{13b}$$

Although the series (12a) is rapidly convergent, no analytical formulas for the functions $G_k(z)$ can be derived in the Gaussian case. One can, however, calculate the probability amplitude an using the formula (13) and replacing the functions (12a) by a cut-off (finite) series. ⁿevertheless, it seems that for Gaussian continua it is preferable to use the recurrential relations (8).

For Lorentzian-type continua we obtain the following formulas for the functions $G_k(z)$ (k=1,2):

$$G_k(z) = g_k \ / \ (1-zg_k) \ , \tag{14}$$

where

$$g_k = \exp[-(iE_{Lk} + \Gamma_L)T] \ . \tag{14a}$$

The probability amplitude now becomes

$$a_n = \frac{V(z_1)(z_2-z_3)(z_2z_3)^{n+1} + V(z_2)(z_3-z_1)(z_1z_3)^{n+1} + V(z_3)(z_1-z_2)(z_1z_2)^{n+1}}{[z_1z_2(z_1-z_2)+z_2z_3(z_2-z_3)+z_1z_3(z_3-z_1)](z_1z_2z_3)^n} , \tag{15}$$

where

$$V(z) = W_A(z)(1-g_1z)(1-g_2z) \tag{15a}$$

and z_i (i=1,2,3) are the roots of the equation $W(z)(1-g_1z)(1-g_2z)=0$. The formula (15) allows to calculate the probability P of finding electrons in the ground level and the total ionization probability after any number of pulses.

4. RESULTS

Figure 1 shows the time-evolution of the probability P for finding electrons in the ground state for the above discussed kinds of continua: Lorentzian (i) and Gaussian (ii). We chose the laser field strength F equal to 0.3 and the factor q equal to 0.25. In both cases we assume that the two continua have identical width $G_k=0.1/2\pi$ (in units of T; k=L,G). Because the laser field is in the form of a train of ultra-short pulses, we are unable to define precisely the Rabi frequency or the detunings of the laser field. We introduce the "characteristic" frequency $\Omega = 2\pi/T$ playing the role of the Rabi frequency for systems driven by a harmonic field. We assume that the centers of the two continua are located in a way allowing the transitions to be resonant with the characteristic frequency Ω, i.e. we assume $\Omega = E_{k1} = 0.5E_{k2}$. We see that for both cases the field is strong enough to induce oscillations of P. For Lorentzian continua the oscillations are damped and the steady state value of the probability P is 0 - the atom becomes permanently ionized. This phenomenon resembles that observed for an atom with one continuum driven by a harmonic field: as it was shown by Agarwal et al (1984) for simple autoionization, damped Rabi oscillations occur.

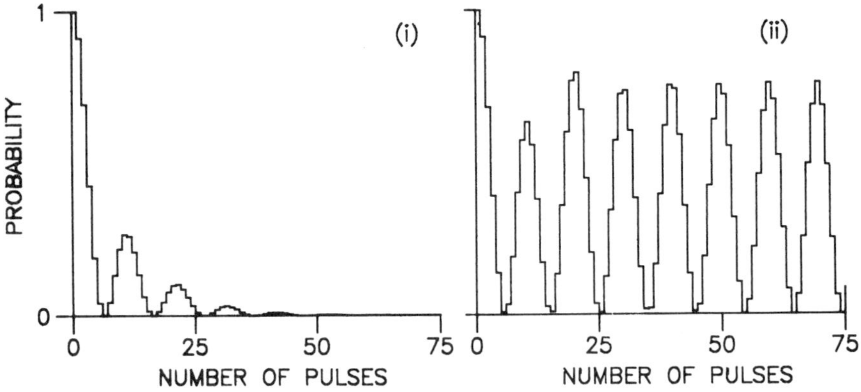

Fig 1. The time evolution of the ground state population for Lorentzian (i) and Gaussian (ii) continua.

For Gaussian continua we observe quite different behavior of the groundstate population. The probability P also oscillates; however, after several oscillations, the amplitude apparently becomes constant - we observe a kind of "population trapping" (Radmore and Knight (1984)). This is an effect similar to that discussed by Rzazewski and Mostowski (1987) for autoionization induced by a train of ultra-short pulses. The damping rate is very small and hardly detectable. This means that the system contains a long-living discrete level and, as a consequence, a sharp peak should occur in the photoelectron spectrum.

This work was supported by Research Project CPBP 01.07.

Agarwal G S Haan S L and Cooper J 1984 Phys.Rev.A 29 2552
Deng Z and Eberly J H 1985 J.Phys.B. 18 L287
Greenland P T 1983 J.Phys.B 16 2515
Radmore P M and Knight P L 1984 Phys.Lett. 102A 180
Rzazewski K and Eberly J H 1981 Phys.Rev.Lett. 43 408
Rzazewski K Li Wang and Haus J W 1989 Phys.Rev.A 40 3453
Rzazewski K and Mostowski J 1987 Phys.Rev.A 35 4414

Paper presented at RIS 90, Varese, Italy, 16–21 September 1990

Influence of quantum nature of electromagnetic field on the autoionizing photoelectron spectrum—SU (1,1) state vs. coherent state

W.Leonski

Nonlinear Optics Division, Institute of Physics, A.Mickiewicz University, Grunwaldzka 6, 60-780 Poznan, Poland

ABSTRACT: The atomic system under consideration comprises a ground level and a structurized continuum. The atom is irradiated by an external electromagnetic field of quantum nature. The analytical formula for the long-time photoelectron spectrum has been found and is valid for any continuum structure and for any initial quantum field state. We discuss two cases: (i) that for the field in SU(1,1) Holstein-Primakoff state, and (ii) that for a coherent state, and compared with that for initial Fock state.

1. INTRODUCTION

The atomic model under consideration is one of the simplest for ionization processes. It contains a single discrete ground level and one continuum. These atomic states are labeled by $|0\rangle$ and $|c\rangle$, respectively. They are mutually coupled by an external electromagnetic field. We shall define the nature of this field later. Moreover, we assume that the continuum $|c\rangle$ has its own structure, determined by a shape-function; for an atomic system with one single autoionizing level it is a so-called "Fano profile" (Fano 1961, Rzazewski and Eberly 1981). Obviously, for systems containing more autoionizing levels the function is of a more complicated form (for example, see Leonski et al (1987). Due to the assumption concerning the structure of the continuum, we shall apply the formalism referred to as "Fano diagonalisation" (Fano 1961, Rzazewski and Eberly 1981); thus, our result will be applicable for any kind of subsystem of atomic levels containing an arbitrary number of autoionizing states.

2. THE ATOMIC MODEL AND THE FIELD

Our aim in this paper is to show how the quantum nature and type of the quantum state of the electromagnetic field can act on and influence the ionization (for the model discussed here - autoionization) processes. Therefore, we define the following wave-function:

$$|\Phi(t)\rangle = \exp(-iE_L(n-1)t) \sum_{n=0}^{\infty} \left(g_n(t) \; |0,n\rangle + \int dE \; c_n(E,t)\,|c,n-1\rangle \right) , \qquad (1)$$

where $g_n(t)$ and $c_n(E,t)$ are time-dependent probability amplitudes, whereas E and E_L label the structurized continuum and photon energies in units of $h/2\pi=1$. The summation in the definition (1) extends over all n-photon Fock states from the vacuum state for n=0 to infinity. Although the wavefunction (1) resembles that first proposed by Rzazewski and Eberly (1981) the states introduced here are photon-electron states, whereas theirs are only electron states. Moreover, we define the Hamiltonian for our system in the following form:

$$H = E_0 |0,n\rangle\langle 0,n| + \int dE\, E|c,n-1\rangle\langle c,n-1| + a^+a\, E_L + \\ + (a \int dE\, F(E)\, |c,n-1\rangle\langle 0,n| + H.c.), \tag{2}$$

where E_0, E and E_L are the energies of the ground state, the continuum and the photon, respectively. The function F(E) occurring in (2) is the shape-function mentioned above, whereas a and a^+ are the usual boson annihilation and creation operators, respectively. Applying the Schrodinger equation and standard projection procedure we get the set of equations for the probability amplitudes in the form

$$i \frac{dg_n(t)}{dt} = (E_0+ E_L)\, g_n(t) + \int dE\, c_n(E,t)\, F(E)^*\, (n)^{1/2} + i\delta(t)Q_n \quad , \tag{3a}$$

$$i \frac{dc_n(E,t)}{dt} = F(E)\, (n)^{1/2} g_n(t) + E\, c_n(E,t) \tag{3b}$$

We have assumed here that for the initial moment of time t=0 the atom was in its ground state and the field was defined in such a way that the wave-function (1) can be written as

$$|\Phi(t=0)\rangle = \sum_{n=0}^{\infty} Q_n |0,n\rangle \tag{4}$$

where Q_n is determined by the expansion of the field states in usual n-photon Fock states. We refrain at this point from determining the exact form of Q_n, and shall do so later.

3. THE PHOTOELECTRON SPECTRUM

It would obviously be possible to find the analytical solution for the equations (3). However, since we are interested in the long-time photo-electron spectrum, we now proceed to define it. Namely, the spectrum can be written as:

$$W(E) = \lim_{t\to\infty} \sum_{n=0}^{\infty} |\, c_n(E,t)\, |^2 \quad . \tag{5}$$

From the form of the equation (3a) one sees that it is possible to write the photoelectron spectrum W(E) in the long-time limit as:

$$W(E) = \sum_{n=0}^{\infty} n \left| F(E)G(z) \right|^2_{z=-iE} \quad , \tag{6}$$

where G(z) is the Laplace transform of the time-dependent probability amplitude $g_n(t)$, and F(E) is the shape-function. For its simplicity, we have chosen F(E) as the "Fano profile" (Fano 1961) corresponding to atomic systems containing one single autoionizing level.

Here, we shall discuss two cases: (i) that corresponding to a coherent initial field state (eigen-state of the boson annihilation operator), and (ii) that for Holstein-Primakoff SU(1,1) coherent state. For the case (i) the quantity Q_n is determined by a Poisson distribution. Thus, the formula for the photoelectron spectrum is of the following form:

$$\frac{\pi W(E)}{\Gamma_0} = \sum_{n=0}^{\infty} \left| \frac{(n)^{1/2}(E - E_1)\,\langle n\rangle^n \exp(-|\langle n\rangle/2|^2)}{n!(\,(E - E_1 + i\Gamma)(E - E_0 - E_L + in\Gamma_0) - n\Gamma_0\Gamma\,(q - i)^2)} \right|^2 . \qquad (7)$$

The widths Γ_0 and Γ are radiative and autoionizing widths, respectively, whereas q is the asymmetry parameter first introduced by Fano (1961) and E_1 is the energy of the autoionizing level. The width Γ_0 is defined by a matrix element determining the transitions between the ground level and the continuum: $\Gamma 0 = \pi m_{0c}$. Moreover, the quantity $\langle n\rangle$ denotes the average numbers of photons. Applying the formula (7) we have found the photo-electron spectrum for strong field coupling. In addition, we assume that the asymmetry parameter is small (q=2), and the energy of the autoionizing level $E_1=1$. Figure 1 shows the spectrum for this case (solid line). We compare it with that induced by the field in Fock state (dashed line) for $\Omega=3$ (the Rabi frequency Ω has been defined in the same way as by Rzazewski and Eberly (1981) in their discussion of classical field spectra). The Fock state spectrum resembles that for classical field coupling for $\Omega=3$. The spectrum consists of two Autler-Townes peaks. One of them is very broad and practically disappears, whereas the other falls on the zero and is very sharp exhibiting "confluence of coherences" character. For the spectrum induced by the quantum coherent state ($\Omega=1$, but $\langle n\rangle=10$), the

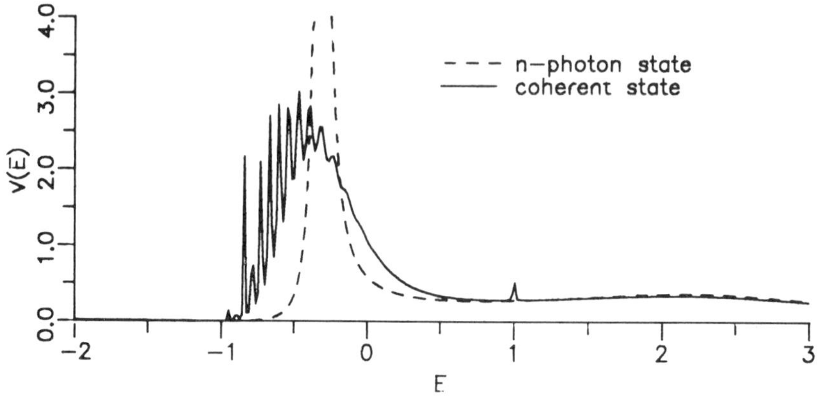

Fig.1. The photoelectron spectra $V(E)=\pi W(E)$ for Fock and coherent states.

sharp peak is replaced by a series of sharp peaks. This phenomenon has been referred to as "comb confluence of coherences" (Leonski and Buzek 1990). Moreover, these peaks are spread on the slope of some maximum. In addition the spectrum contains a single small, sharp peak elastic in nature. It results from the coupling by vacuum Fock state.

As the field is initially in Holstein-Primakoff SU(1,1) coherent state, the photoelectron spectrum is expressable in the form:

$$\frac{\pi W(E)}{\Gamma_0} = \sum_{n=0}^{\infty} \left| \frac{(n)^{1/2}(E - E_1)\ \mu n\ (1-|\mu|^2)^{-1/2}}{(E - E_1 + i\Gamma)(E - E_0 - E_L + in\Gamma_0) - n\,\Gamma_0\Gamma\,(q - i)^2} \right|^2 , \qquad (8)$$

where μ is a complex parameter referred to as "squeezing parameter" (Wodkiewicz and Eberly 1988, Buzek 1989). When $|\mu|$ tends to unity, the Holstein-Primakoff state tends to become a 100% squeezed state. From (9) it is apparent that the photoelectron spectrum depends on $|\mu|$. Therefore,

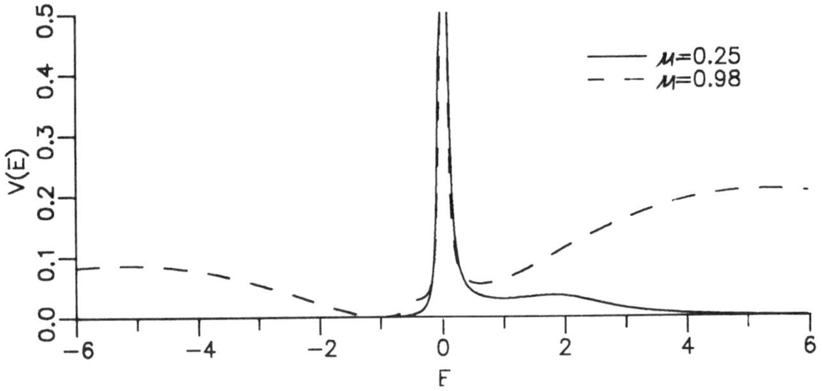

Fig.2 The spectra for various values of the squeezing parameter μ.

with the value of μ available one can measure the degree of squeezing of the field state from this spectrum. Figure 2 depicts the spectra for various values of the squeezing parameter. The remaining parameters are identical to those for the situation shown in Figure 1). When the squeezing parameter is far from unity (μ=0.25) the spectrum is of the same form as that for Fock states of the field. As μ increases, (μ=0.98), the spectrum undergoes radical transformation: instead of two, three peaks are now dominant. Two of them are very broad, whereas the third central peak remains high, although it loses its "confluence of coherences" character and is located on the slope of one of the broad peaks.

4. CONCLUSIONS

We have discussed the photoelectron spectra for two kinds of quantum field states. One of them is a coherent state (an eigen-state of boson annihilation operator), whereas the other is a SU(1,1) Holstein-Primakoff coherent state. We have shown that for the case of a coherent state the spectrum significantly differs from that for a Fock n-photon state, or equivalently, from that induced by a classical field, and produces a novel kind of phenomenon referred to as "comb confluence of coherences" (Leonski and Buzek 1990). The spectra induced by the SU(1,1) state are strongly dependent on the value of the squeezing parameter μ and differ in character from those produced by a coherent field. For μ sufficiently close to unity the spectrum can lose its confluence of coherences (Rzazewski and Eberly 1981) character and can produce an additional broad peak.

This research was supported by Research Project CPBP 01.07.

Buzek V 1989 Phys.Rev.**A39** 3196
Fano U 1961 Phys.Rev.**124** 1866
Leonski W and Buzek V 1990 J.Mod.Opt. (in press)
Leonski W Tanas R and Kielich S 1987 J.Opt.Soc.Am.**B4** 72
Rzazewski K and Eberly J H 1981 Phys.Rev.Lett.**48** 408
Wodkiewicz K and Eberly J H 1988 J.Opt.Soc.Am.B2 2683 and references quoted therein

Inst. Phys. Conf. Ser. No 114: Section 2
Paper presented at RIS 90, Varese, Italy, 16–21 September 1990

Intensity- and time-dependent spectra of doubly resonant ionization of barium

A.Wójcik and R.Parzyński

Institute of Physics, A.Mickiewicz University, Grunwaldzka 6, 60-780 Poznań, Poland

ABSTRACT: A model of ladder-type double optical resonance in natural barium including isotope structure of atomic beam and hyperfine structure of the middle level is presented. Ionisation spectra are evaluated for a wide range of laser pulse intensities, durations and detunings.

1. INTRODUCTION

Three monochromatic, coincident in time, square laser pulses are focused on an atomic beam of natural multi-isotope barium (^{138}Ba - 71.9%; ^{137}Ba - 11.2%; ^{136}Ba - 7.8%; ^{135}Ba - 6.5%; ^{134}Ba - 2.4%).First laser (probe) is tuned to resonance with $6s^2\,{}^1S_0 \rightarrow 6s6p\,{}^1P_1$ transition (535.5 nm), second laser (strong) drives the $6s6p\,{}^1P_1 \rightarrow 6p^2\,{}^3P_0$ transition (608.3 nm) and third laser (ionising) couples the upper level $6p^2\,{}^3P_0$ to continuum. Ionisation spectrum, i.e. total ionisation probability as a function of probe laser detuning δ is calculated, using Laplace - transformed Schrödinger equations for population probability amplitudes under the rotating wave and pole approximation. Because of the hyperfine structure of $6s6p\,{}^1P_1$ in odd mass number isotopes, it is necessary to consider an extended model of double optical resonance. This model allows the middle level to have some internal structure. Realistic spectra so obtained are multi-peaked due to isotope, hyperfine and Autler-Townes effects. Dependence of the ionisation spectrum on lasers intensities, detunings and pulse duration is investigated.

2. RESULTS

The figures attached show the evolution of the spectra with such parameters as:

- upper-resonance Rabi frequency Ω,
- upper-level ionisation damping γ,
- atom-photon interaction time t.

All parameters refer to ^{138}Ba.

Figure 1 presents the effect of upper-resonance Rabi frequency Ω. For small Ω ($\Omega<\Gamma$; $\Gamma=19$ MHz is the middle-level natural width) only elastic peaks are built up. However, at strong upper resonance ($\Omega>\Gamma$) the Ba elastic peak splits and Autler-Townes doublet is observed. In figure 2 the effect of upper-level ionisation damping is presented. If ionisation damping is small ($\gamma<\Omega$) well pronounced Autler-Townes doublet is created. If ionisation damping is increasing Autler-Townes components broaden, then overlap and finally become one peak at $\gamma > \Omega$. However, we observe narrowing of this peak on increasing ionisation damping more and more. In this condition ($\gamma \gg \Omega$) population is trapped

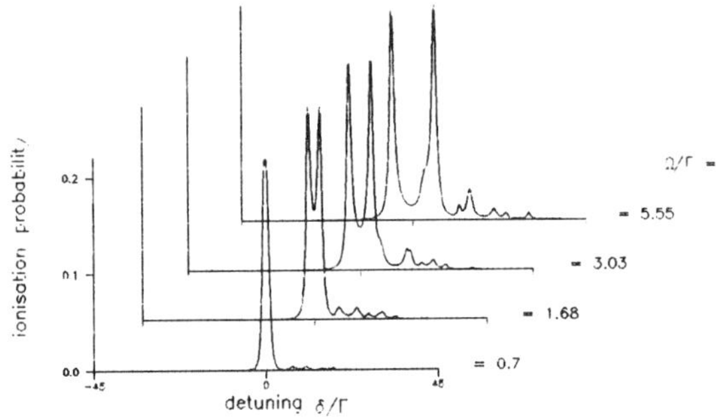

Figure 1. Autler-Townes effect.

by ground and middle level because extremely strong ionisation damping

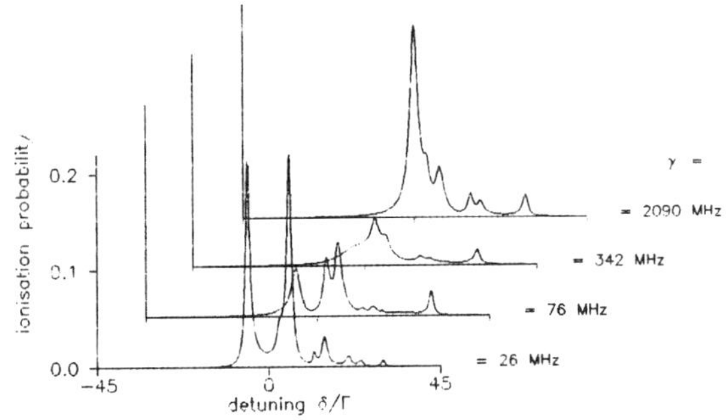

Figure 2. Peak narrowing.

decouples the upper level from the other ones. Figure 3. shows the number-of-peak reduction and flattening the spectrum by shortening the atom-photon interaction time t. If the interaction time is shorter

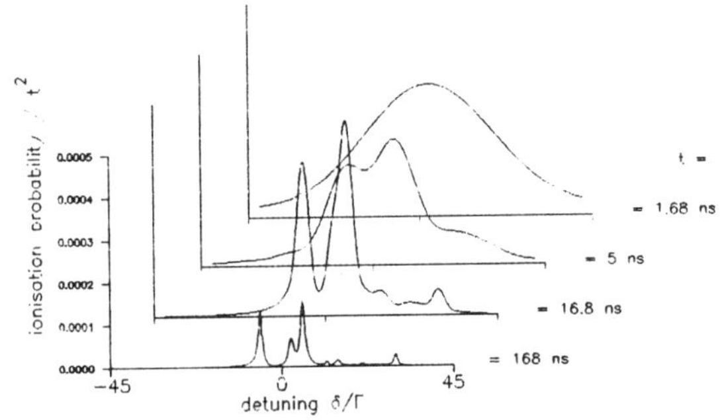

Figure 3. Number-of-peak reduction.

then the reciprocal of Rabi frequency and reciprocal of spacing between the lowest and highest sublevels of the middle level, neither the Autler-Townes doublet nor additional peaks arising from the middle-level structure have a chance to reveal themselves. In extreme case only a single broad peak is present in the spectrum. All spectra are obtained under the assumption of lower resonance Rabi frequency M = 1.5 MHz.

Inst. Phys. Conf. Ser. No 114: Section 2
Paper presented at RIS 90, Varese, Italy, 16–21 September 1990

Positive-energy resonances in Stark effect: Calculations via classical electron trajectories

A K Kazansky, V N Ostrovsky and D A Telnov

Institute of Physics, Leningrad State University,
198904 Leningrad, USSR

ABSTRACT: For the hydrogen atom placed into uniform electric field the positions and widths of resonances in the positive-energy region are calculated approximately from the analysis of unstable periodic classical electron trajectory. The results are in excellent agreement with the accurate quantum calculations.

1. INTRODUCTION

The photoionisation cross section for atoms placed into uniform electric field exhibits structure in the positive-energy region which was observed experimentally (see e.g. Freeman and Economou 1979). The appearance of the structure is due to the resonances with the width comparable to the energy. The positive-energy resonances cannot be interpreted as a shape resonances. In the early paper by Freeman and Economou (1979) the nature of resonances was tentatively related with the unstable classical trajectories of the atomic electron. In the present communication we show how this relation can be established on quantitative level giving quite precise results.

2. GENERAL THEORY

The recent consideration (Kazansky and Ostrovsky 1989) shows that individual unstable periodic classical trajectory generates a whole series of resonances in quantum mechanics. The width of resonance Γ is proportional to the well known Lyapunov index α which characterises the trajectory instability:

$$\Gamma = 2\alpha(2n_2+|m|+1) \qquad (1)$$

where m is azimuthal quantum number for rotations in the plane locally perpendicular to the cycle, n_2 is the integer quantum number related with the radial motion in the mentioned plane (the atomic units are used throughout the report). The meaning of Eq.(1) is clear: since α characterises time of life of the classical system in the vicinity of the cycle then due to Uncertainty Principle it is related with the quantum width Γ.

The resonance position E is given by the quantisation

condition applied to the periodic orbit

$$\oint pdq = 2\pi(n_1 + \gamma) \qquad (2)$$

where integration is performed over the cycle, n_1 is integer and the constant γ depends on the Maslov index for the trajectory. The quantum number n_1 (which fixes E) enumerates the resonance series. Within each of them the resonances are indexed by the quantum number n_2 and differ in the width (1). We assume here mathematically strict definition of the resonance as a position of S-matrix (or Green function) pole on the complex energy plane: $E_r = E - i\Gamma/2$ which is applicable in all cases (e.g. for overlapping or superbroad resonances). The series of resonances in question lie in the complex E-plane on the line perpendicular to the energy real axis.

3. APPLICATION TO ATOM IN ELECTRIC FIELD

Here the electron with the energy E moves in the Coulomb field of atomic nucleus with the charge Z and the uniform external field with the strength F. The unstable cycle corresponds to the motion of the electron along the axis of uniform field. The electron starts from the nucleus moving against the direction of the force exerted on it by the uniform field, reaches the turning point and then returns back to the nucleus. This oscillating motion is unstable since small deviation from it grows exponentially in time (for illustration see fig.1 in the paper by Kondratovich and Ostrovsky 1984). The factor before time in this exponent is essentially the Lyapunov index which can be calculated explicitly:

$$\alpha = \frac{2F(2E)^{-1/2} K(p)}{2R\mathbf{E}(p) + (1-R)\mathbf{K}(p)},$$

$$p = \sqrt{(1+R)/(2R)} \quad , \; R = \sqrt{1+4E^{-2}ZF} \qquad (3)$$

where $\mathbf{K}(p)$ and $\mathbf{E}(p)$ are the complete elliptic integrals. This expression seems to differ from that by Alber (1989). The quantisation condition (2) takes the form ($\gamma = 1/2$):

$$\left[\frac{64\,Z^3}{F\,[p^2(1-p^2)]^3}\right]^{1/4}\left[(2p^2-1)\mathbf{E}(p) + (1-p^2)\mathbf{K}(p)\right] =$$

$$= 3\pi(n_1 + (|m| + 1)/2) \qquad (4)$$

The quantum numbers n_1, n_2, m in this case coincide with the well known parabolic quantum numbers. Note that the resonance position in this approximation depends only on the quantum number n_1 whereas n_2-dependence of width is factorised in the universal form.

The approximation described above gives unexpectedly good agreement with the exact results in wide range of energy E and field strength F. As a typical example we give here the complex energies describing the resonances with the parabolic quantum numbers $(n_1,n_2,m) = (15,n_2,0)$, $n_2 = 0,1,2$.

Table 1. Complex energies for the resonance (15,0,0). M denotes the calculation method: a, the accurate quantum calculation according to Telnov (1989); b, the present classical trajectory method. The number in parentheses indicates the power of 10.

F	M	Re E_r	$-$Im E_r
0.7(-5)	a	0.122518331(-3)	0.630193353(-4)
	b	0.100760 (-3)	0.404264 (-4)
0.1(-4)	a	0.925585958(-3)	0.133464866(-3)
	b	0.918409 (-3)	0.131037 (-3)
0.2(-4)	a	0.338910451(-2)	0.295509405(-3)
	b	0.338891 (-2)	0.295274 (-3)
0.5(-4)	a	0.961097809(-2)	0.636492362(-3)
	b	0.961490 (-2)	0.636462 (-3)
0.1(-3)	a	0.182218932(-1)	0.107760366(-2)
	b	0.182303 (-1)	0.107761 (-2)
0.2(-3)	a	0.325943528(-1)	0.179464957(-2)
	b	0.326109 (-1)	0.179469 (-2)
0.5(-3)	a	0.668307254(-1)	0.348279272(-2)
	b	0.668712 (-1)	0.348291 (-2)
0.1(-2)	a	0.112325808	0.572698962(-2)
	b	0.112406	0.572725 (-2)
0.2(-2)	a	0.186225265	0.939871279(-2)
	b	0.186386	0.939929 (-2)
0.5(-2)	a	0.357985310	0.180570669(-1)
	b	0.358394	0.180588 (-1)

Table 2. Complex energies for the resonance (15,1,0) obtained from the numerical procedure by Telnov (1989). The ratios of the energy real (RR) and imaginary (RI) parts for the resonance (15,0,0) to those of the resonance (15,1,0) are also presented. The classical trajectory theory gives their values equal to 1 and 1/3, respectively

F	Re E_r	$-$Im E_r	RR	RI
0.7(-5)	0.172102487(-3)	0.233315464(-3)	0.711892	0.270103
0.1(-4)	0.940760595(-3)	0.410987638(-3)	0.983870	0.324742
0.2(-4)	0.336225718(-2)	0.883532897(-3)	1.00798	0.334463
0.5(-4)	0.953919941(-2)	0.190264301(-2)	1.00752	0.334531
0.1(-3)	0.180930490(-1)	0.321910753(-2)	1.00712	0.334752
0.2(-3)	0.323609014(-1)	0.535281470(-2)	1.00721	0.335272
0.5(-3)	0.663052549(-1)	0.103502262(-1)	1.00793	0.336494
0.1(-2)	0.111345051	0.169464744(-1)	1.00881	0.337946
0.2(-2)	0.184395997	0.276443920(-1)	1.00992	0.339986
0.5(-2)	0.353872502	0.525251811(-1)	1.01162	0.343779

Table 3. As in Table 2 for the resonance (15,2,0) compared with (15,0,0). The classical trajectory approach gives the values RR and RI equal to 1 and 1/5, respectively

F	Re E_r	$-$Im E_r	RR	RI
0.7(-5)	0.215320920(-3)	0.436388161(-3)	0.569003	0.144411
0.1(-4)	0.954926196(-3)	0.704303909(-3)	0.969275	0.189499
0.2(-4)	0.331076207(-2)	0.146309675(-2)	1.02366	0.201975
0.5(-4)	0.939945086(-2)	0.314914842(-2)	1.02250	0.202116
0.1(-3)	0.178455573(-1)	0.532312628(-2)	1.02109	0.202438
0.2(-3)	0.319239212(-1)	0.883123832(-2)	1.02100	0.203216
0.5(-3)	0.653712069(-1)	0.169938141(-1)	1.02233	0.204945
0.1(-2)	0.109690419	0.276900004(-1)	1.02403	0.206825
0.2(-2)	0.181494360	0.449195725(-1)	1.02607	0.209234
0.5(-2)	0.347905379	0.846800902(-1)	1.02897	0.213239

4. DISCUSSION

Note that the present approach is applicable only for the positive energy resonances since the cycle is unstable only for $E > 0$. Naturally, the resonance width is non-zero also for $E < 0$ but the decay cannot be interpreted in the classical terms here. Therefore it is understandable that the theory looses its accuracy when E approaches zero. The same occurs also for higher electric field strength.

The expressions (3), (4) can be obtained also via some specific analysis of the conventional semiclassical approximation (Kazansky et al 1990). The latter allows natural interpretation in terms of tunnelling in the case of shape resonances but becomes much less appealing in the positive-energy region.

The unstable trajectory remains the same in the case of potential with the non-Coulomb core. Thus the analysis in terms of classical trajectories is possible in this case although the specific analytic expressions (3), (4) do not hold. The general approach was applied recently also to the two-centre Coulomb problem (Du et al 1990).

REFERENCES

Alber G 1989 *Phys.Rev.A* **40** 1321
Du M L , Kazansky A K and Ostrovsky V N 1990 *Phys.Rev.A* **42** (in print)
Freeman R R and Economou N P 1979 *Phys.Rev.A* **20** 2356
Kazansky A K and Ostrovsky V N 1989 *Sov.Phys.-JETP* **68** 670
Kazansky A K , Ostrovsky V N and Telnov D A 1990 *J.Phys.B: At.Mol.Opt.Phys.* **23** (in print)
Kondratovich V D and Ostrovsky V N 1984 *J.Phys.B: At.Mol.Phys.* **17** 2011
Telnov D A 1989 *J.Phys.B: At.Mol.Opt.Phys.* **22** L399

Inst. Phys. Conf. Ser. No 114: Section 2
Paper presented at RIS 90, Varese, Italy, 16–21 September 1990

Method for determining the cross section for two-photon excitation of discrete levels

A.Yu. Elizarov and N.A. Cherepkov

A.F. Ioffe Physicotechnical Institute, 194021 Leningrad, USSR

ABSTRACT: Measurements of the absolute cross sections for the nonresonant two-photon excitation of discrete states of barium atoms are presented. They are based on the detection of an ion currents due to absorption of a third photon in two processes of both resonant and nonresonant two-photon excitation of the same discrete level.

The experimental apparatus [1] includes two dye lasers, pumped by the same Nd-YAG laser. The width of the output line is 3 cm^{-1}. The energy in the pulse is 0.1 mJ and the pulse length (τ) is 18 ns. The density of atoms in the beam can be varied from 10^8 to 10^{12} cm^{-3}. The region in which the laser light interacts with the atomic beam lies between the grids of the accelerating gap of a time-of-flight mass spectrometer.

The laser power was insufficient for nonresonant three-photon ionization, so the ion current was observed only when the absorption of the first two photons led to the excitation of a resonant state or, in stepwise processes, when the absorption of the first and second photons was a resonant absorption. In the former case we used a single laser, and in the latter case two lasers. One of the two lasers was tuned to the frequency of the $6s^2(^1S_0)\rightarrow 6s6p(^1P_1)$ resonant transition (18060 cm^{-1}), while the other was scanned over frequency.

Making use of the possibility of exciting the same discrete state by means of a two-photon process (non resonant) and by a two-step process (resonant) (see Fig. 1) we can determine the absolute cross section of the two-photon excitation, $\sigma^{(2)}$. In two-photon excitation, the number of ions per unit volume which are formed during a pulse is found from the relation

$$n(2)=n_0 \cdot W(2)\,\frac{\tau^2}{2}\,\sigma \cdot N_\omega$$

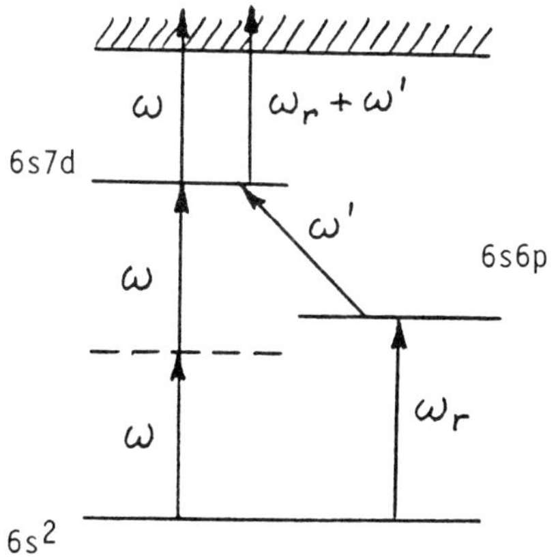

Fig. 1

where n_0 is the density of atoms in the ground state, N is the photon flux density, σ is the cross section for ionization of the excited state and $W^{(2)}$ is the probability for the two-photon excitation of the resonant state [2].

It is convenient at this point to introduce the cross section $\sigma^{(2)}$ for the two-photon excitation of the resonant state by means of the expression [2]

$$W^{(2)}=\sigma^{(2)}\cdot N_\omega \cdot N_\omega(\omega)$$

where $N_\omega(w)$ is the spectral density of the light. In the case ot the two-step process, for the pulse energy cited above, both transitions are in saturation and the number of ions produced during a pulse is given by

$$n_r=k\cdot n_0\cdot\sigma\cdot(N_\omega+N_r)$$

where N_r is the photon flux density at the resonant frequency ot the transition $6s^2(^1S_0)\rightarrow 6s6p(^1P_1)$ and k=1/3 reflects the ratio of statistical weights with the laser polarization taken into account.

In the experiment we measured the ratio of the number of ions produced during a pulse in these two cases: $\alpha=n^{(2)}/n_r$. Knowing this ratio, we can find the cross section for the two-photon excitation of the resonant state:

$$\sigma^{(2)}=\frac{2\cdot\alpha\cdot k\cdot(N_\omega+N_r)}{N_\omega^2\cdot N_\omega(\omega)}$$

In the experiment we verified that the dependence of the ion signal on the intensity of the laser light corresponds to the relations given above. The intensity of the laser light was measured by a calorimetric method. The intensity of the ion signal was determined by the amplitude. We found the following values for the excitation cross sections:

$\sigma^{(2)} = 7\cdot 10^{-30}$ cm^4 for the state $6p^2(^3P_2)$, $E = 35616$ cm^{-1},

$\sigma^{(2)} = 3\cdot 10^{-29}$ cm^4 for the state $6s7d(^3D_2)$, $E = 35762$ cm^{-1}.

An estimate of the errors shows that these values have been determined within a facotr of three.

For excitation of the 6s7d state it is easy to estimate the cross section $\sigma^{(2)}$ theoretically in the lowest order of perturbation theory by retaining in the sum over intermediate states only a single term which corresponds to the following Feynman diagram:

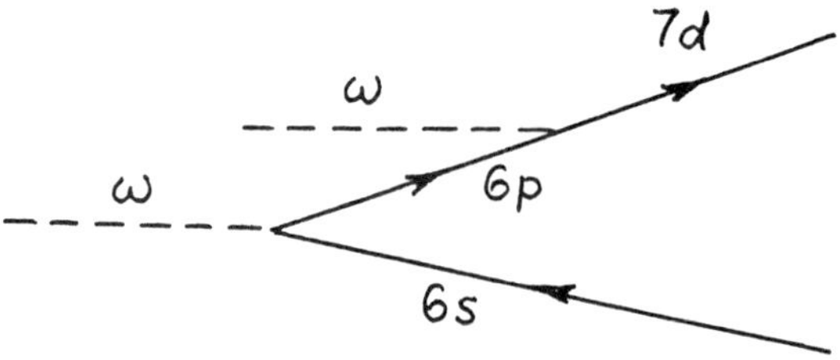

(*)

The result is $\sigma^{(2)} = 7.5\cdot 10^{-32}$ cm^4 which does not give us the correct cross section even by an order of magnitude. It is not difficult to see that the diagram of the first order in Coulomb interaction

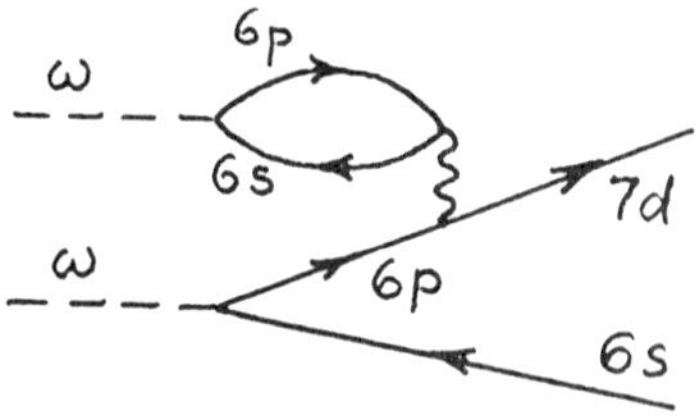

contains an additional small energy denominator and its contribution to the cross section exceeds that of the diagram (*) by two orders of magnitude if the Coulomb matrix element, respresented by the wavy line in the diagram, is taken to be 0.01 (in atomic units). The excitation of the $6p^2$ state is necessarily described by a diagram with a Coulomb interaction,

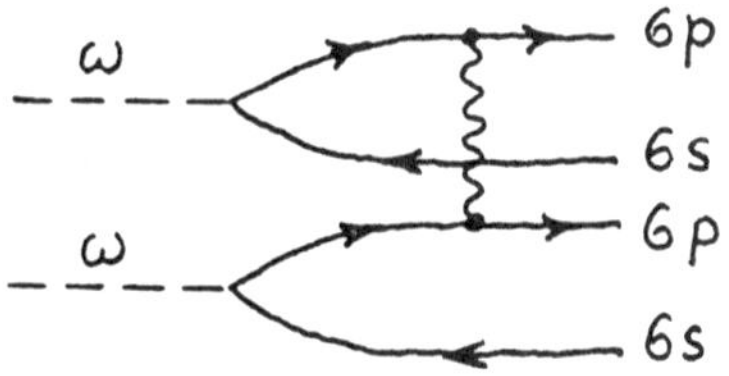

which contains a small energy denominator, so the cross section for the excitation of the two levels is of the same order of magnitude.

Two-photon excitation spectrum was used for calibration purposes in our investigations of autoionization resonances of configuration 6p7p with two lasers [3] (see Fig. 2). The first laser (ω_1) was fixed at the resonance transition $6s^2 \rightarrow 6s7p(^1P_1)$ and the second one (ω_2) was tuned. When the first laser was switched off, the two-photon resonances shown on Fig. 3 were observed, while when both lasers were on, the two-photon resonances usually disappeared (see Fig. 4). To excite simultaneously both autoionization resonances and two-photon resonances, as it is shown on Fig. 3, one needs to fulfill the condition $\Omega \leq W^{(2)}$ where Ω is the Rabi frequency for the transition $6s \rightarrow 7p$. Usually we were working under condition $\Omega > W^{(2)}$ when two-photon resonances are not seen. But when we decreased the output of the first laser, the two-photon resonances became clearly visible, as it is shown on Fig. 3.

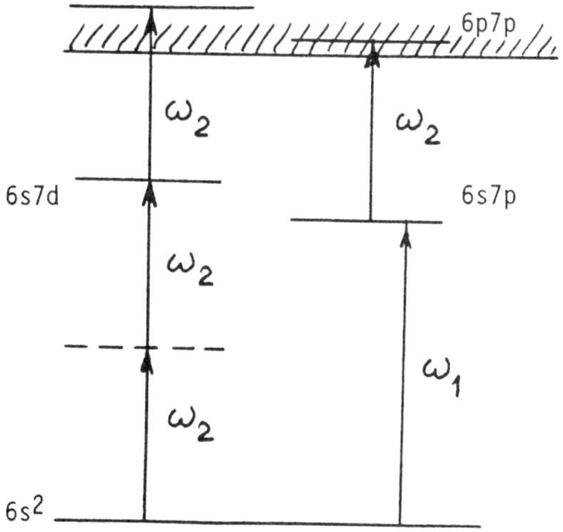

Fig. 2

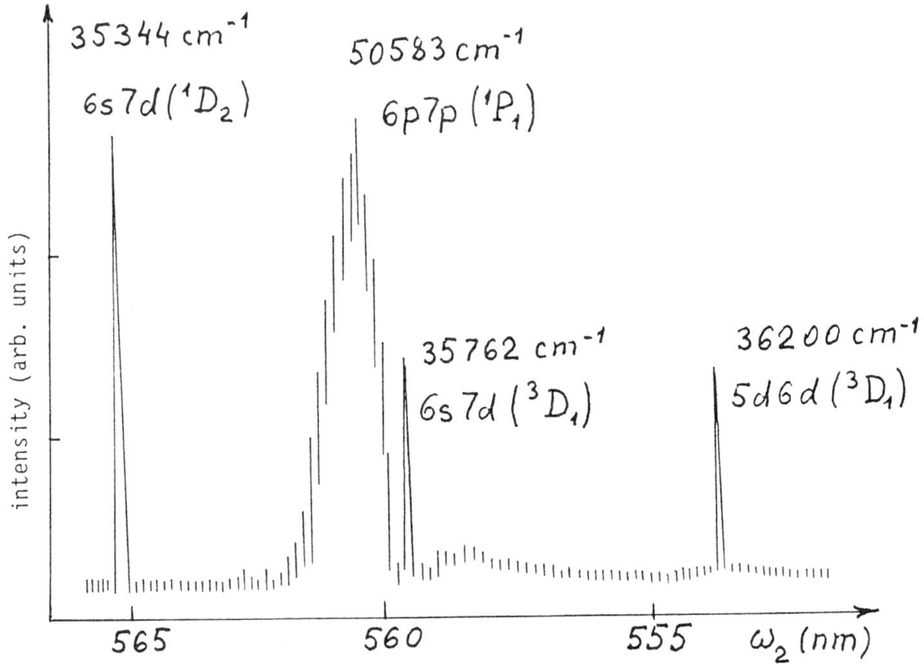

Fig. 3

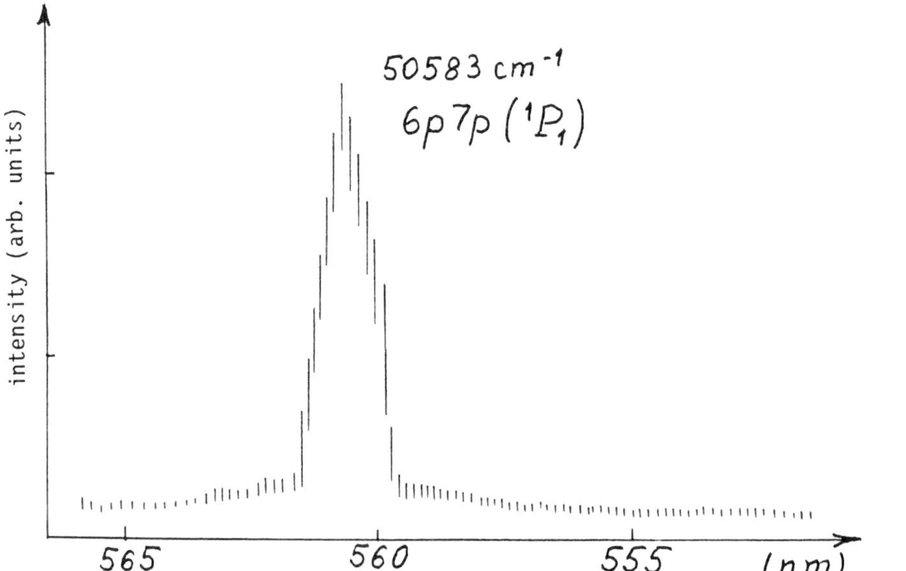

Fig. 4

[1] A.Yu. Elizarov and N.A. Cherepkov, Sov. Tech. Phys. Lett. 14, 93 (1988)

[2] P. Lambropoulos, Adv. At. Mol. Phys. 12, 87 (1976)

[3] A.Yu. Elizarov and N.A. Cherepkov, Sov. Phys. JETP 69, 695 (1989)

Inst. Phys. Conf. Ser. No 114: Section 2
Paper presented at RIS 90, Varese, Italy, 16–21 September 1990

Effect of tuning of laser axial modes on multistep ionization

K.Dasgupta*, K.G.Manohar, B.M.Suri, P.N.Bajaj, R.K.Talukdar, P.K.Chakraborty and P.R.K.Rao

M.D.R.S., Physics Group, Bhabha Atomic Research Centre
Bombay - 400 085, INDIA.
* Present Address : Institut für Physik, Universität Mainz, 6500 Mainz 1, Fed. Rep. Germany.

ABSTRACT : The effect of tuning of axial modes, of one of two multi-axial-mode dye lasers, on a two-color three-photon resonance ionization process involving narrow atomic transitions in uranium has been studied. Our results show that non-linear sum-frequency absorption, along with transition broadenings in two-photon excitation and three -photon ionization, can allow efficient ionization even when the axial modes are detuned from the transition.

Laser pumped narrowband pulsed dye lasers are frequently used in RIS. The spectra of these dye lasers with bandwidths of ≈ 0.03 - 0.1cm^{-1} usually consist of a number of sharp axial modes (AM) within the spectral envelope. The separation of the AM's are usually large compared to their widths. Atomic transitions, on the other hand, are often quite **narrow** compared to the separation of the AM's. It is therefore expected, *prima facie*, that an AM of each laser should be tuned precisely to the corresponding atomic transition, for an efficient ionization process.

A different situation exists when the atomic transitions are **inhomogeneously broadened**, (such that a no. of AM's are encompassed) e.g., by Doppler broadening in a poorly collimated beam and/or by hyperfine splitting. The AM's of the laser in the first step are then expected to burn narrow holes in the absorption spectrum. The AM's of the laser for the subsequent excitation should again be precisely positioned. Therefore, it is generally believed that the laser spectra should have a good overlap with the **inhomogeneously broadened** absorption spectra for an efficient photo-ionization (PI) process.

The conditions during our earlier experimental studies (Suri et al 1987, Bajaj et al 1988, Dasgupta et al 1988, Manohar et al 1989) on single-color and two-color three step PI of uranium, consists of a typical example of the former case. These studies were carried out with two copper vapor laser pumped home-made dye lasers (D1 and D2, Bandwidth ≈ 0.06-0.08cm^{-1}; AM width <0.005cm^{-1}; AM separation ≈ 0.02cm^{-1}). The atomic beam of uranium generated in the oven assembly of a commercial quadrupole mass spectrometer had a doppler broadened absorption width ≈ 0.002cm^{-1} (Bhowmick et al 1988). As seen in Fig.1, the AM's of the laser nearest to the transition may be detuned by as much as 0.01cm^{-1}. The AM positions of our dye lasers were seen to have a good short term stability. Long term drifts were < 0.01cm^{-1} in 20 secs. Effect of such mode drifts would cause changes in the excitation which has been reported recently by Broglia et al (1986).

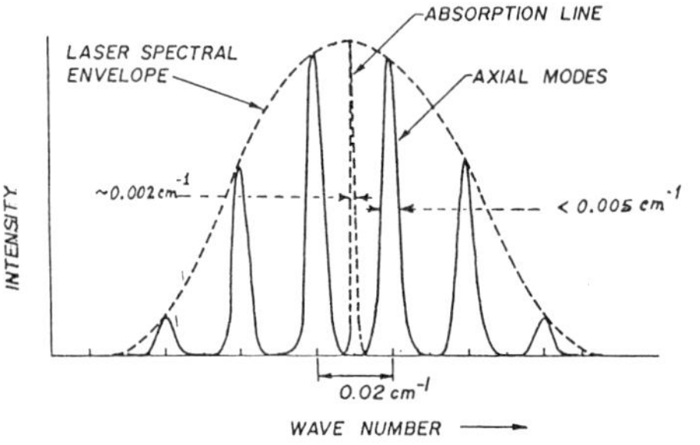

Fig.1 : Detuning of axial modes from **narrow** absorption line (schematic).

In this work we have studied the effect of detuning of the AM's of one dye laser (D1), not only on the excitation of the first step but also on the ionization signal. The two lasers were tuned (D1- 591.5nm, D2 ≈ 572 nm (Suri et al 1987)) to produce PI of UI from the ground state by a two-step three-photon process. The photoions were mass filtered and detected by a channeltron detector. The high repetition rate of the lasers allowed a phase sensitive detection, with the atomic beam chopped at ≈ 532Hz (Suri et al 1987). A piezoelectric translator was used for tuning the axial modes of D1 and this was monitored by recording the transmission through a Fabry-Perot Interferometer (FSR =0.17cm^{-1}, Finesse > 30). During tuning of the AM's, the envelope was held fixed by a wavelength stabilization system built in our lab (Manohar et al 1986) which doesnot resolve the AM's.

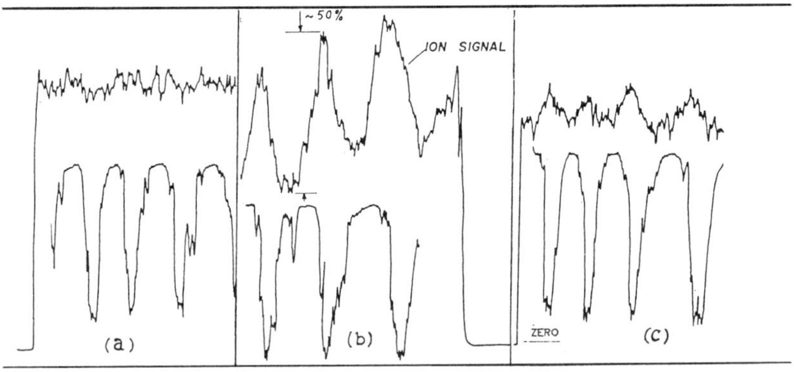

Fig.2 : Effect of axial mode tuning on two-color ion signal. Upper trace - **Ion signal**. Lower trace - **FPI transmission of laser D1.** Laser powers : D1 ≈ 10KW/cm^2 ; D2 ≈ 150KW/cm^2. (a)Simultaneous pulses; (b)D2 pulses delayed by 75ns; (c)Simultaneous but D2 pulses attenuated to 12%.

Fig.2 shows a typical result. With simultaneous laser pulses from D1 and D2, no significant changes are seen in the ion signal. When the D2 pulses are delayed by 75 ns, strong modulations (≈50%) were observed becuase of AM tuning. Here, D2 pulses essentially probe the excitation produced by the D1 laser which, as expected, depends on the AM detuning. With simultaneous laser pulses but with much lower D2 power, the modulations again started showing up. Thus, the ionization process was relatively independent of the detuning of the AM's although the excitation to the first step was not so.

To enquire into the reasons for this observation we note that for simultaneous laser pulses, the AM's of D2 ride on those of D1 creating a no. of Sum Frequencies (**SF**) $W_{nm} = w_{1n} + w_{2m}$, where $w_{1n} = w_{10} + ns_1$, $(n = 0, \pm 1, \pm 2)$ are the AM frequencies of D1 with a separation of s_1. Initially suppose that the AM's w_{10} and w_{20}, at the centre of the respective envelopes, are in exact resonance with the respective transitions at energies E_1 and E_2. Then the other **SF**'s nearest to the final transition at $E\ (=E_1 + E_2)$ are $W_{1-1} (=E/\hbar + s_1 - s_2)$ and W_{-11}. If $s_1 = s_2$, these **SF**'s are also exactly nonlinear two-photon (**2P**) resonant, but detuned with respect to the individual single-photon (1P) transitions at E_1 and E_2 by $\pm\ s_1$. If now the AM's of D1 are tuned by ε, $w_{10} = E_1/\hbar + \varepsilon$, then non-linear **2P** transitions will occur at $\varepsilon = s_1, s_2, s_1 - s_2$, becoming exactly 1P resonant at $\varepsilon = s_1$.

The efect of the **SF**'s are seen (fig.3) in the single-color 3P ionization signal obtained by tuning D1 to 16900.52cm^{-1} (near the strong 1P absorption at 16900.39cm^{-1}). The ionization takes place via a 2P absorption to an odd level at 33801.0cm^{-1} ($=2w_{10}$). Additional maxima in the PI signal appear when two photons are taken from adjacent AM's at $w_{10} \pm s_1/2$, with the **SF** exactly resonant with the 2P transition.

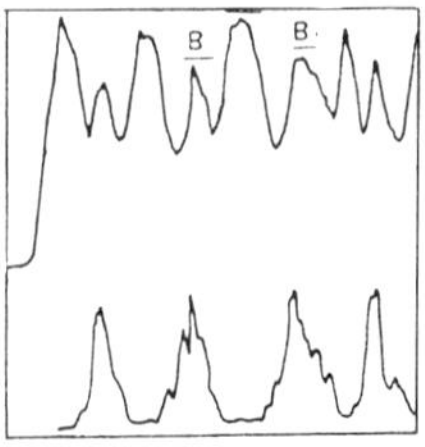

Fig.3 : Effect of tuning of AM's on single-color three-photon ionization via **2P** absorption.

Going back to the first experiment, since $s_1 \approx s_2$, it is very unlikely that, as the AM's are tuned, one or more of the **SF**'s will always be **2P** resonant. Hence some significant broadening must be present when the pulses are simultaneous to explain our observations. The cumulative effect of broadening of 1P processes from all sources (power broadening, AM fluctuations) as seen in fig.2b, cannot explain by itself the absence of modulation in the ion signal (fig. 2a).

Broadening of nonlinear **2P** transitions (Levenson 1982) will of course contribute. In addition, broadening due to population decay into the ionization channel will also occur. The joint effect of the last two processes can be seen by recording the population of the second excited state (at energy E) with a third dye laser, delayed considerably with respect to D1 and D2 pulses, the latter two being simultaneous. This third laser was tuned to a suitable autoionization transition from the second excited state such that no ionization was produced from the first excited state. As the AM's of D1 were tuned neither the two-color (D1 and D2) nor the three-color ionization signals showed any noticeable modulation. Thus, in this case, the excitation to the second excited level (at energy E) was also insensitive to the AM tuning.

A similar situation was studied by Chang and Schlier (1980) with two single-axial mode pulsed dye lasers, where the laser wavelengths did not allow any three-photon ionization. The **2P** excitation rate was shown to remain nearly same when each laser was detuned by ≈ 1Ghz from the intermediate level in opposite directions. However, the spectral intensities were high compared to our experiments, causing significant broadening of the individual **1P** transitions and allowing a two-step character to be acquired by the excitation process. It is quite possible in our case, that if D1 or D2 photons match with a strong autoionizing transition from the second excited state, (uranium has a rich autoionization spectrum) ionization broadening may also play a significant role in explaining our observations. Unlike as in the excitation processes, all the axial modes can contribute to this broadening as the autoionization transitions are relatively wide. For example, for an autoionization cross-section $\sigma \approx$ 1E-16cm^2 and D2 photon intensitiy ≈ 1E+24 - 1E+25 / (cm^2-sec), broadening due to ionization of the second excited step may be ≈ 100 - 1000Mhz.

Although further work needs to be done to understand the contributions from the different sources of broadening, two significant conclusions can be drawn from our observations :

(1) Not only the AM's which are in the immediate vicinity of the narrow atomic transitions, but also the more distant ones may contribute to the ionization process, increasing the photon efficiency of the process.

(2) If the transitions are inhomogeneously broadened, use of multi-axial mode dye lasers with closely spaced AM's spanning the absorption spectrum, may allow efficient ionization all over the absorption profile instead of just burning holes at the axial mode positions.

REFERENCES :

Bajaj P N, Manohar K G, Suri B M, Dasgupta K, Talukdar R, Chakraborty P K, Rao P R K, 1988 *Appl. Phys.* **B47** 55.

Bhowmick G K, Talukdar R, Bajaj P N, Kartha V B, Chakraborty P K, 1988 *Chem. Phys. Lett.* **144** 467.

Broglia M, Zampetti P, Benetti P, 1986 *Appl. Phys.* **B39** 73.

Chang C S and Schlier R, 1980 *Phys. Rev.* **A21** 872.

Dasgupta K, Manohar K G, Bajaj P N, Suri B M, Talukdar R, Chakraborty P K, Rao P R K, 1988 *J.Opt.Soc.Am.* **B5** 1257.

Levenson M D, 1982 *Introduction to Nonlinear Laser Spectroscopy*, Academic Press, New York.

Manohar K G, Dasgupta K, Suri B M, Bhawalkar D D, 1986 *Rev. Sci. Instrum.* **58** 920.

Manohar K G, Bajaj P N, Suri B M, Talukdar R, Dasgupta K, Chakraborty P K, Rao P R K, 1989 *Appl. Phys.* **B48** 525.

Suri B M, Dasgupta K, Bajaj P N, Manohar K G, Talukdar R, Chakraborty P K, Rao P R K, 1987 *J.Opt.Soc.Am.* **B4** 1835.

Inst. Phys. Conf. Ser. No 114: Section 2
Paper presented at RIS 90, Varese, Italy, 16–21 September 1990

The resonance ionization spectrum of the Ba atoms and Ba_2 molecules

Liu Daijing, Peng Weixian, Wu Donghong, Li Yuping, Wei Zhenqian, Jiang Zhankui

Department of Physics, Jilin University, Changchun
The People's Republic of China, 130023

Abstract. The resonance ionization spectrum of the Ba atoms and molecules, which occurred in a furnace containing Ba vapour, has been produced with laser spectroscopy technique in the wavelength region from 579.5nm to 614.8nm. Areas of discrete atomic lines and banded structure in the Ba vapour have been reported. The hybrid transition of Ba atom and molecule and Botlzmann distribution law were used to analyze the transition mechanism of atomic resonance lines. A preliminary calculation was made for some band heads of molecular spectrum.

In the present paper we report investigations on some transitions starting from several thermally populated low-lying states and one hybrid resonance transition in the 579.5nm- 614.8nm region.

The Ba vapour was produced inside a heat-pipe oven. The heat-pipe was made of stainless steel tubing 40cm long and 3.2cm inner diameter and was operated up to the temperature of about 900°C. Argon was used as buffer gas . The Ba vapour in the heat-pipe was irradiated by a laser beam of a CW ring dye laser (R_6G) pumped by Ar^+ ion laser. The linewidth of the laser is 2GHz, and there are no F-P systems inside the cavity of the ring dye laser. The scanning element of the dye laser consists of three pieces of birefringent filter. The laser resonance ionization spectra were recorded by means of a lock-in amplifier and a chart recorder. A F-P having 33GHz free spectral range gave the interference spectra used as frequency marks.

The heat-pipe containing Ba vapour was prepared for reducing some low-melting-point impurities before measuring. Ar gas was introduced with a pressure of 90 Torr as buffer gas. At the operating temperature of 885 °C, the vapour pressure was about 380 mttor. The laser beam was focused into the oven center with maximum power of 20mw. The thermionic detecter operated at static bias voltage of 2.6V. Scanning laser wavelength, we obtained spectra in which several high intensity atomic resonance lines overlie on the obvious band structure of molecular spectrum.

We measured the variation in the line intensity as a function pf laser power intensity. Four curves for lines numbered 1, 2, 4 and

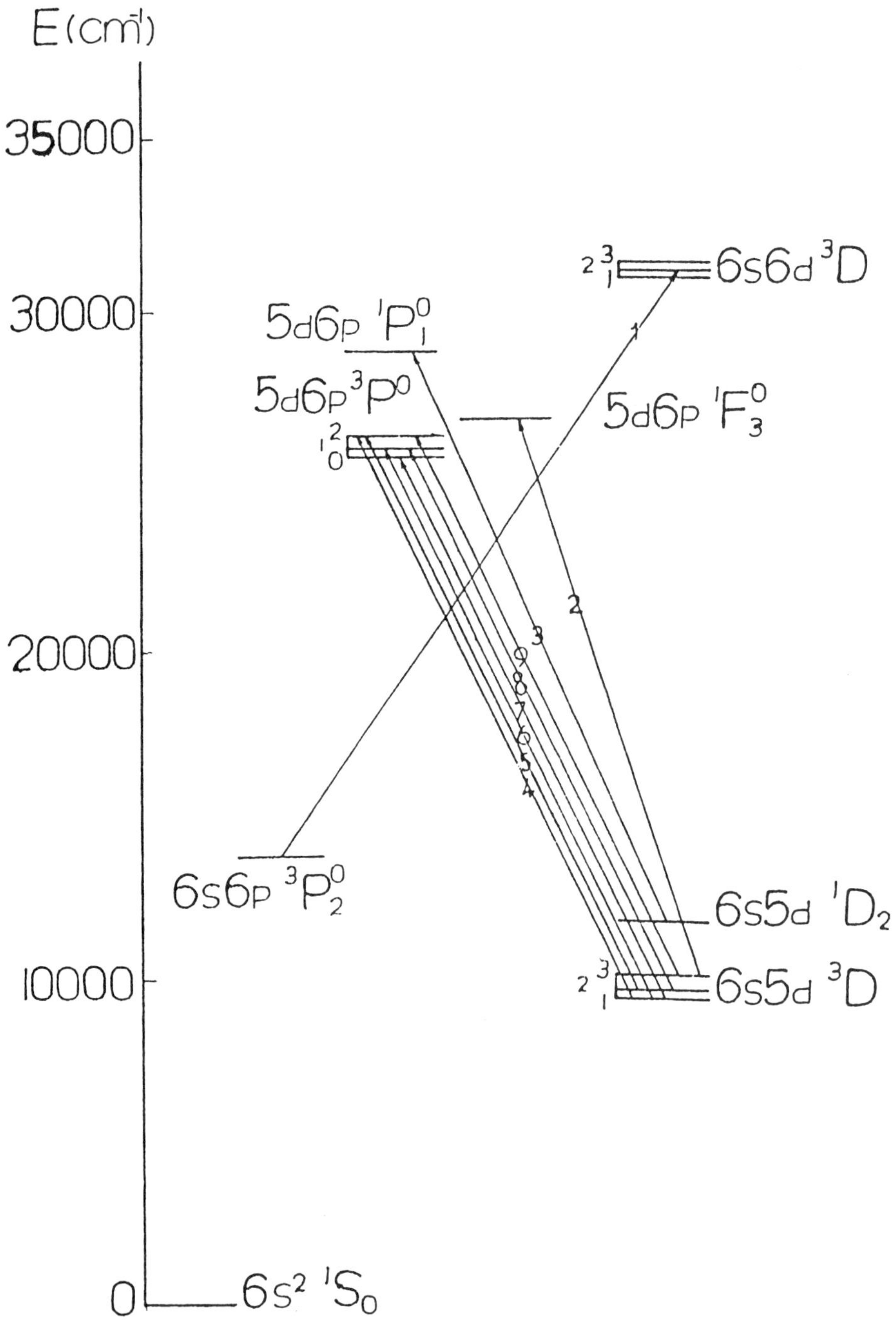

Fig. 1 Transition scheme

No.	Wavenumber(cm^{-1})	Wavelength(nm)	Resonance Levels
1	17235.93	580.02	$6s6p\,^3P_2 \rightarrow 6s6d\,^3D_2$
2	17219.74	580.56	$6s5d\,^3D_3 \rightarrow 5d6p\,^1F_3^o$
3	17158.88	582.62	$6s5d\,^1D_2 \rightarrow 5d6p\,^1P_1^o$
4	16922.56	590.76	$6s5d\,^3D_1 \rightarrow 5d6p\,^3P_2$
5	16741.03	597.17	$6s5d\,^3D_2 \rightarrow 5d6p\,^3P_2$
6	16670.16	599.71	$6s5d\,^3D_1 \rightarrow 5d6p\,^3P_1$
7	16608.17	601.95	$6s5d\,^3D_1 \rightarrow 5d6p\,^3P_0$
8	16488.62	606.31	$6s5d\,^3D_2 \rightarrow 5d6p\,^3P_1^o$
9	16360.00	611.09	$6s5d\,^3D_3 \rightarrow 5d6p\,^3P_2^o$

Tab. 1 Designation of transitions

5 listed in Tab.1 have been obtained.

We found that atomic line 1 is corresponding to transition $6s6p\,^3P_2 \rightarrow 6s6d\,^3D_2$, line 2, 4, 5 arose by transitions starting from $6s6d\,^3D$. See Fig 1.

The transitions corresponding to atomic lines 2, 4, and 5 are one-photon process.

We confirm that the atomic line 1 arose from a two-photon transition process which is called Hybrid Two-photon Transition. The wavelength of the absorbed photon is 580.0nm, the energy corresponding to this wavelength is just coincident with the energy difference between the higher level $6s6d\,^3D_2$ and the lower level $6s6d\,^3P_2$.

The hybrid transition involved both atomic and molecular transition. At the beginning, the Ba_2 molecules in their ground state were prompted to an excited molecule, relaxed to a low-lying state and dissociated into Ba ($6s6p\,^3P_2$) + Ba ($6s^2\,^1S_0$). Finally, the Ba ($6s6p\,^3P_2$) reached the $6s6p\,^3D_2$ state absorbing another photon.

Acknowledgment

The authors gratefully acknowledge the support to our work by The Natural Science Foundation of China.

Alignment and dynamics during resonant excitations

L W Green, G A McRae and J H Rowat

AECL Research, Chalk River Laboratories, Chalk River, Ontario, Canada
K0J 1J0

ABSTRACT: A discussion is presented of alignment and dynamics during resonant excitations and their effect on relative ionization rates of isotopes of the same element. The case of double resonant excitation by pulsed, broadband, linearly polarized light is described and variation of laser intensities, pulse durations and delays, and relative polarization are predicted to change isotopic ionization rates by up to 100%.

1. INTRODUCTION

Recent application of resonance ionization mass spectroscopy to isotopic abundance measurements have shown that observed isotope ratios may be different by as much as 37% from expectation, despite the use of lasers with frequency bandwidths much larger than isotope shifts (Fairbank et al, 1988, Bekov et al, 1981). Lambropoulos and Lyras (1988 and references therein) have completed density matrix calculations for 1+1 and 2+1 ionizations, and have shown that the relative rates of laser induced excitation, spontaneous decay, spin-orbit coupling, hyperfine coupling and ionization play critical roles in determining the overall photoionization rate. In this brief we qualitatively discuss a model for alignment and dynamic effects in two-color, time resolved, resonance ionization using linearly polarized broadband lasers.

2. ALIGNMENT AND POPULATION DISTRIBUTIONS

In the absence of an external field, the electric field of the linearly polarized laser (E_L) causes weak AC Stark splitting of the ground (initial) state and quantized alignment of the electronic angular momentum J. In the vector model, the alignment of J is represented by an array of vectors (Figures 1 and 2) that precess about E_L at a rate equal to the AC Stark splitting. The relative line strengths of the $\Delta M_J=0$ transitions for $\Delta J=0$ and $\Delta J=1$ electronic transitions vary as M_J^2 and $J^2 - M_J^2$, respectively, and are represented by the length of the vertical lines in the examples shown in Figures 1 and 2.

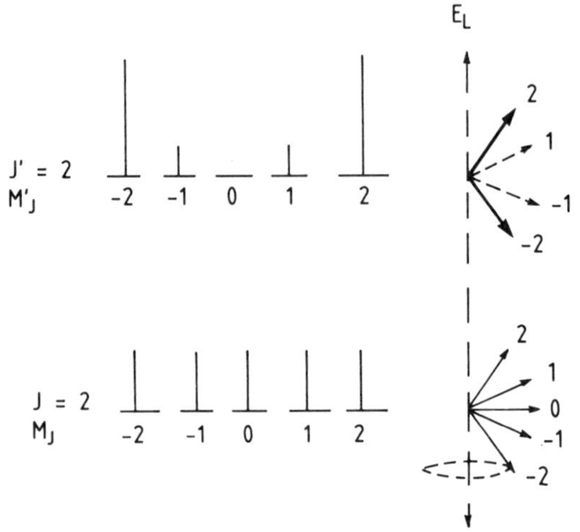

Figure 1. Alignment and population distribution in a $\Delta J = 0$ transition for broadband π light and J(initial) = 2. Relative populations are represented by the length of the vertical bars, and darkness of the vectors. Each vector represents a cone of precession about the electric field axis of the laser, E_L, as shown for the M_J = -2 component.

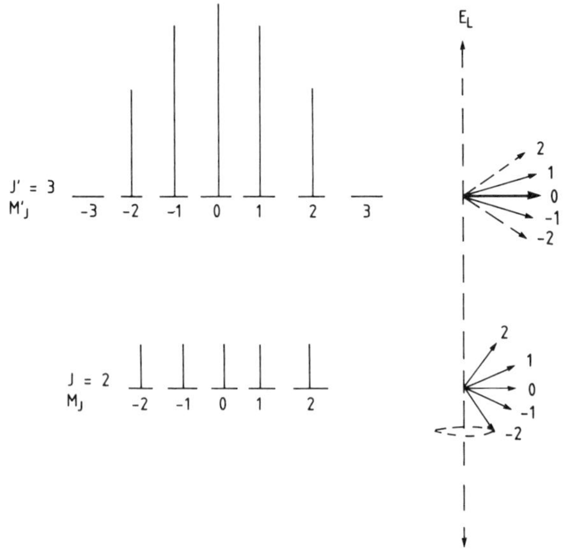

Figure 2. Alignment and population distribution in a $\Delta J = +1$ transition for broadband π light and J(initial) = 2. Otherwise as described in Figure 1.

The transitions can be described by the optical Bloch equations. When the Rabi frequencies, which are proportional to the laser electric field amplitude and the transition dipole moments, are small compared to the total rate of deexcitation (damping), then the Bloch equations reduce to the familiar rate equations. In this instance the population in the upper levels of a multi-level system increase monotonically to steady state with laser fluence and follow the distributions mentioned earlier. For intense laser fields the rate equations are no longer appropriate. When the Rabi rates are large relative to the damping rate the population of each M_J component oscillates at a rate proportional to its transition strength. The time average populations of upper and lower levels, and also of the various M_J components, become equal (saturated), provided there are many oscillations per laser pulse. For intermediate laser intensities, the stronger M_J - M_J' transitions may be in oscillation during the pulse, whereas the weaker ones may not. In this situation, the anisotropy caused by the different relative strengths of M_J transitions can be enhanced or diminished.

3. SECOND RESONANT TRANSITION TO A THIRD LEVEL

Unless saturated, the first resonant transition produces a spatially aligned, anisotropic, distribution for J that can be probed with a second resonant transition. When the electric field of the second laser is parallel to the first, a consideration of the line strengths presented in the figures suggests that if the change in J is the same for the second transition then a maximum overall line strength is expected. A minimum is expected if the change in J is different. For transitions involving $J=0$ and 1, it is possible to suppress completely the second excitation by choice of ΔJ. If J is large the population becomes diluted over a large number of sublevels and the effects become smaller (quantum dilution).

When the electric field axes of the two resonant lasers are orthogonal, ΔJ dependent intensities are also observed. For example, if the first transition is $\Delta J=1$, Figure 2, then the transition favours those M_J sublevels with least projection on the laser axis. If the second laser electric field is parallel to the first and if $\Delta J=0$ for the second transition then the sublevels with most projection on the second laser axis are favoured, and a relatively low combined transition rate is expected. As the relative polarization is rotated from parallel to perpendicular, the projection of the most populated components of the first excited state on the second laser axis increases. Correspondingly, the transition rate increases and reaches a maximum at perpendicular polarization.

Predictions of the magnitude of the changes in transition rate as the polarization or ΔJ is changed are possible, but require, where appropriate, estimates of the degree of alteration of the M_J populations due to Rabi flopping.

4. HYPERFINE INTERACTIONS

The coupling of the nuclear spin I with J to form the total angular momentum vector F, disrupts the J spatial alignment and causes quantum dilution to yield more uniform population distributions. For short pulse laser excitations, it is possible that the hyperfine interaction is too weak to be observed, and the transition follows ΔJ line strengths. However, if the time delay between laser pulses is long relative to the coupling rate, hyperfine coupling will be established before onset of the second laser pulse. If the second transition is also fast then again J and M_J are the good quantum numbers, but because the intermediate level alignment has been modified by the hyperfine interaction, the dependence of the overall transition rate on ΔJ or polarization will be greatly reduced from that expected without hyperfine interaction. Also, the overall transition rate will be higher if the ΔJ transitions and polarizations are poorly matched, and lower if they are well matched. This is particularly important for even-odd isotope measurements. For instance, isotope ratios have been found to be functions of the relative laser polarization for uranium 235 ($I=7/2$) and 238 ($I=0$) in resonance ionization measurements (Green et al, 1990).

5. DYNAMICS

The observed transition rates in any experiment will be determined by the relative rates of the above mentioned processes. Resonant excitations and Rabi flopping, which occur during the laser pulse, are in competition with spontaneous decay and hyperfine interactions, which occur at all times. A demonstration of the interplay of some of these parameters in two-color excitation of uranium is reported elsewhere in these proceedings (Green et al, 1990).

Ionization rates and external field effects also have significant influences on the dynamics and population variations in resonant transitions. A good example is the Hanle effect, observed in the presence of weak applied magnetic fields.

In applications of resonance ionization to isotopic abundance measurements, equal ionization rates for all isotopes are desirable. The alignment and dynamical effects described above may be reduced by saturating transitions or by polarization scrambling. Saturation of all M_J components of all isotopes may be difficult to achieve because of variations in the laser intensity over the laser frequency bandwidth, especially if the mode structure is resolved. Saturation, at high intensities, is further complicated by AC Stark broadening of the transition and power broadening of the laser.

6. REFERENCES

Bekov G I, Zherikhin A N, Letokhov V S, Mishin V I and Fedoseev V N *1981 JETP Leff* **33** 451

Fairbank W M, Spaar M T, Parks J E and Hutchinson J M R *1988 Inst. Phys. Conf.* **Ser. No. 94** 293

Green L W, Smyth M H C, Rochefort P A and McRae G A *elsewhere in this volume*

Lambropoulos P and Lyras A *1988 Inst. Phys. Conf.* **Ser. No. 94** 1

Inst. Phys. Conf. Ser. No 114: Section 3
Paper presented at RIS 90, Varese, Italy, 16–21 September 1990

Pulsed laser generation of intense atomic cluster beams

Paolo Milani and Walt A. de Heer

Institut de Physique Expérimentale, EPFL, 1015 Lausanne, Switzerland

ABSTRACT: Pulsed laser vaporization sources (PLVS) are widely used to produce atomic and cluster beams. We describe an improved PLVS for production of intense, stable and cold beams of neutral and ionized atoms and clusters. The very low temperatures attainable in the beam makes the particles suitable for Stern-Gerlach deflection experiments and photoionization spectroscopy.

1. INTRODUCTION

Among molecular beam techniques the generation of metal atomic and cluster beams is an art still in its infancy. Several methods have been developed for producing metal cluster beams of low vapor pressure materials. The introduction of laser vaporization sources allowed the generation of beams of virtually any element of the periodic table (Smalley 1983).

The principle of operation is very simple (Figure 1): an intense pulse of laser radiation is focussed onto a metal target, vaporizing a small amount in a flow of inert carrier gas produced by a pulsed valve. The inert gas quenches the vapor and then the mixture is expanded into a vacuum. It is thus possible to obtain pulsed beams of atoms and clusters cooled to low temperature by the thermalization with the carrier gas and the supersonic expansion.

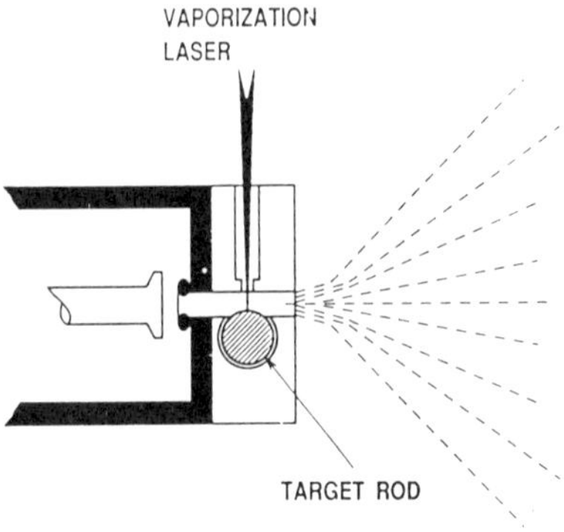

Fig. 1. Schematic diagram of the principle of operation of a standard laser vaporization source developed by Smalley (1983)

The heating is restricted to a small portion of the target so that no heat shielding or cooling is required. Despite its versatility and simplicity this source can be improved to attain higher beam intensities, better stability, lower temperatures and speeds. We present an improved PLVS which can produce intense stable and cold atomic and cluster beams (Milani 1990).

2. IMPROVED SOURCE DESCRIPTION AND OPERATION

A horizontal cross section of the source is shown in Figure 2. The source body is machined from a rectangular stainless steel block. Focused laser light enters the source via a channel which is also used to inject the carrier gas onto the target. A plug with a smaller diameter hole is inserted at the entrance of the light channel to reduce the gas losses. The channel terminates in a horizontal cylindrical cavity capped with an adjustable teflon plug on the rear and a teflon nozzle on the front. As shown in Figure 2 the target rod is in a thin walled vertically mounted stainless steel tube which intersects the vaporization cavity. Coaxially with the light channel a hole is drilled through this tube. to let the laser light pass through the source, so that the laser can be accurately aligned. The metal rod is pressed against the tube in order to minimize the gas leaking past the rod. This is very important for stable operation of the source as well. The source chamber is pumped by a 2500 l/sec diffusion pump fitted with a freon cooled baffle. Typical pressures in this chamber during operation at 10 Hz are about 10^{-4} Torr. The cluster beam is detected by means of a time of flight mass spectrometer located 2.4 meters downstream from the source. While the standard source we introduced a cavity to decouple the vaporization to the expansion process.

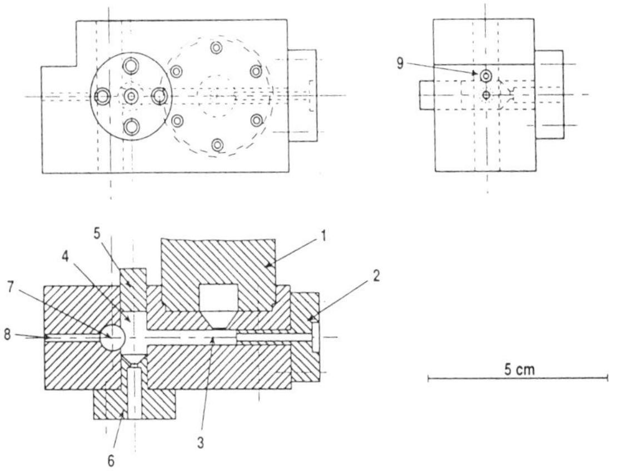

Fig. 2. Horizontal cross section and side view of the new source. (1) Pulsed valve head, (2) metallic plug to reduce gas losses, (3) light channel, (4) vaporization cavity, (5) adjustable teflon plug, (6) Teflon nozzle, (7) rod housing, (8) alignment tube coaxial to the light channel, (9) housing for a spring and a 2.5 mm steel ball to press the rod against the tube.

The dimensions of the cavity are chosen in order to optimize the cooling of the plasma plume and minimize the losses of vaporized material to the walls. The geometry of the nozzle does not dramatically affect the cluster production showing that the clustering is realized inside the cavity. The source can operate also with the nozzle cooled at liquid nitrogen temperature: this is accomplished by isolating the nozzle from the body of the source by a teflon ring.

3. PERFORMANCE

A mass spectrum of aluminum clusters generated by a single laser shot and ionized by an ArF excimer laser is shown in Figure 3. We estimate that each peak is produced by about 100 ions. The signal is stable for about 10^6 shots at 10 Hz. Neutral Al clusters for different mass ranges are shown in Figure 4. Note that some mass peaks are absent or very weak in the region between Al_3 to Al_{13} (more precisely Al_3, Al_4, Al_5, Al_6, Al_8, Al_9, Al_{13}). This is caused by the high ionization potential of these clusters (Cox 1988). Since these clusters are observed with the standard source using the same ionizing light wavelength (192 nm) (Cox 1988), this indicates that the clusters produced with our source are quite cold because high internal temperature will give rise to thermal tails in the photoionization efficiencies and therefore the clusters may be ionized at 192 nm. Velocity measurements have been performed by measuring the initial energies of the clusters entering the detector (de Heer 1989).

We find that the measured cluster velocities depend primarily on the the time delay between the vaporization and detection. In Figure 5 we show the results for two measurements taken varying the delay time Dt between the laser pulse and the triggering of the detector.

Comparing the flight times determined from the measured velocities with Dt, the dwell time of particles in the source can be determined and are found to be typically on the order of 100 μsec. This may be compared with dwell times in the conventional sources which are estimated to be about 10 μsec.

The most probable velocities are also found to be smaller than in a conventional source (Cox 1985).

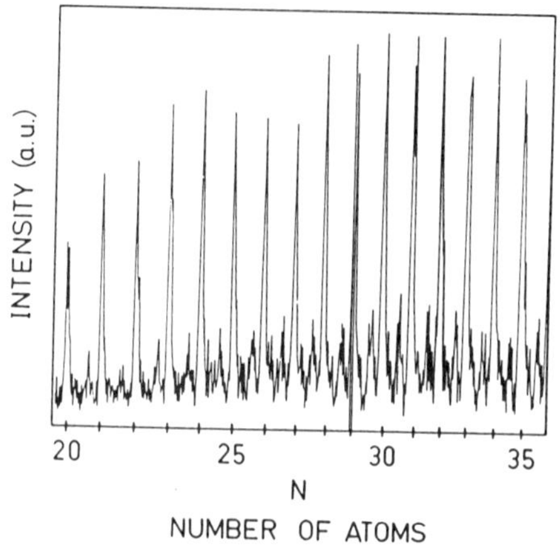

Fig. 3. Time of flight mass spectrum of aluminum neutral clusters from a single laser shot, ionized with an ArF excimer laser (fluence of $0.1 mJ/cm^2$) and detected 2.4 meters form the source. N is the number of atoms per cluster.

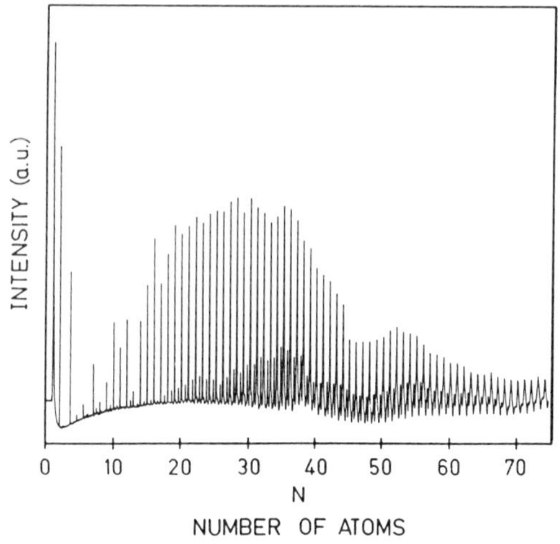

Fig. 4. Time of flight mass spectrum of aluminum neutral clusters ionized by the ArF laser. The signal is accumulated over 500 shots at 10 Hz. The beam is collimated with two 0.8 mm slits. the detector is 2.4 m away from the source. The first saturated peak on the left is the Al atom.

Since the clusters undergo on the order of 10^5-10^6 collisions with the carrier gas in the nozzle tube, they will be at least as cold as the nozzle and are then cooled further by the adiabatic expansion processes.

Using the adiabatic expansion temperatures as an upper limit we find for iron clusters (de Heer 1990) that absolute minimum (translational) temperature are 15 K and 4 K for nozzle temperatures of 300 K and 77 K respectively. Due to the velocity slip, however, the particles have (rotational) temperatures which are greater. We compared the operation of the new source with a standard one taken from literature and constructed in our laboratory. The latter was definitively inferior in stability and it was not possible to produce high intensity of large clusters. We conclude that the introduction of a cavity before the expansion channel make the source capable of producing intense and cold beams suited for cluster studies in general and photoionization spectroscopies in particular.

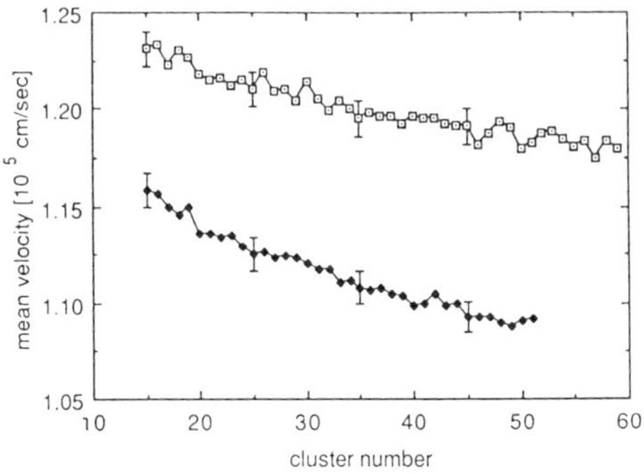

Fig. 5. Mean cluster velocities as a function of mass. The curves correspond with two different time delays between the vaporization laser pulse and the trigger of the detector (see text).

BIBLIOGRAPHY

Smalley R E 1983 Laser Chem. **2** 167 and references therein
Milani P and de Heer W A 1990 Rev. Sci. Instr. **61** 1835
Cox D M, Trevor D J, Whetten R L and Kaldor A 1988 J. Phys. Chem. **92** 421
de Heer W A, Milani P and Chatelain A 1989 Phys. Rev. Lett. **63** 2834
Cox D M, Trevor D J, Whetten R L, Rohlfing E A and Kaldor A 1985 Phys. Rev. B. **32** 7290
de Heer W A, Milani P and Chatelain A 1990 Phys. Rev. Lett. **65** 488

Isotopically resolved imaging in supersonic atomic beams by Doppler-tuned resonance ionization

X.W. Qian, R. Sturgeon, P.A. Hackett and D.M. Rayner

Steacie Institute for Molecular Science and Institute for Environmental Chemistry
National Research Council Canada, Ottawa, Ontario, K1A 0R6, Canada

ABSTRACT: A technique for the direct measurement of isotopic ratios by spatially resolved Doppler tuned laser spectroscopy of supersonic atomic beams is discussed. The development of a supersonic atomic beam source based on electrothermal atomization for the application of the technique as an analytical method is described.

1. INTRODUCTION

Supersonic atomic beams are ideal environments for atomic spectroscopy with significant advantages in terms of reduced Doppler widths, concentration of populations in a few low-lying states and suitability for resonance ionization time-of-flight mass spectroscopy. In addition, there is the possibility, explored here, that isotope ratios can be measured directly by exploiting the Doppler shifts associated with the velocity component perpendicular to the direction of propagation of the atomic beam. To realize these advantages, in analytical applications, requires an atomization source which can handle routine samples and which can produce a supersonic beam. Graphite furnace technology provides an attractive basis for such a source. Electrothermal atomization in a graphite furnace is a simple and effective technique. As the basis for atomic absorption spectroscopy (AAS) it is currently the premier technique for trace element analysis. Used as an effusive source, it has already been combined with resonance ionization spectroscopy, as witnessed by the proceedings of the last symposium in this series (Bushaw and Gerke 1989, Bekov and Letokhov 1989). A technique which enables isotope dilution protocols to be applied to the wide range of samples which can be handled in a graphite furnace can be expected to be useful in analytical chemistry, especially in the certification of chemical standards and in isotope tracer studies.

2. DOPPLER-TUNED ISOTOPE IMAGING

Reduction of the Doppler width has been recognized as an advantage of spectroscopy in atomic beams for some time. By crossing the interrogating laser beam and the atomic beam at right angles the velocity component in the direction of the laser beam, and hence the Doppler width observed on scanning the laser, is significantly reduced. It is usual in this type of spectroscopy to reduce the Doppler width further by the use of collection optics which restrict the field of view to the centre of the atomic beam (Levy 1980). Resolution down to 20 MHz can be obtained in this manner (Hackett *et al.* 1988).

One may also take advantage of the distribution of off-axis velocity components to obtain high-resolution spectra, resolved in space, using a fixed frequency laser. The concept is depicted in Figure 1. At a distance l downstream from the nozzle the resonance condition for a fixed laser frequency ν_0 and an isotope having a transition frequency ν is

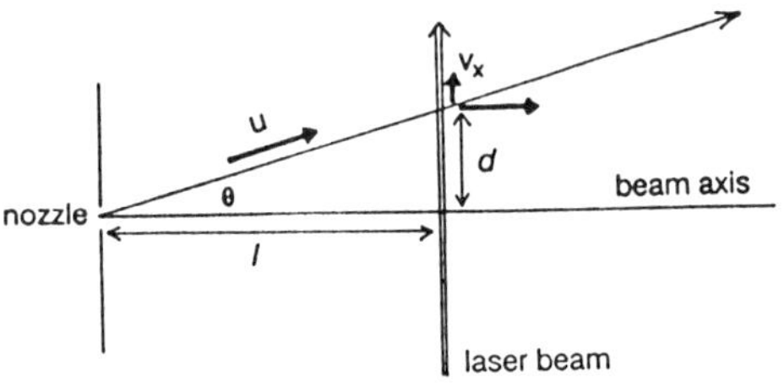

$$\nu = \nu_0 (1 + v_x/c)$$

$$= \nu_0 (1 + u \sin\theta/l)$$

Figure 1: Doppler-tuned isotope imaging

where c is the speed of light, v_x is the residual off-axis velocity component, u is the flow velocity and θ is the angle subtended by the streamline on the beam axis. Through the relationship $\sin\theta = d/\sqrt{(d^2 + l^2)}$ the resonance condition relates ν to the displacement from the atomic beam, d. Isotopes having different ν will absorb at, and therefore be detected as originating from, different displacements proportional to $\nu - \nu_0$. For laser induced fluorescence (LIF) detection the spectrum can be recorded using an imaging array detector. For RIS, the ionization step can be driven non-selectively by a second broad-band laser or field ionization and the ions then imaged on an imaging ion detector. We have demonstrated this technique by the observation of spatially resolved LIF from the isotopes of zirconium (90, 92 and 94; isotope shift ~90 MHz/amu) in a Zr atomic beam generated by the laser ablation of metallic Zr in a pulsed He expansion. The adoption of a similar technique to study the hyperfine spectrum of molecular iodine has recently been reported (Levinger *et al.* 1990).

Of concern for isotope ratio measurements are the spectral range and resolution of the method. The spectral range which can be covered at a single probe laser frequency is a function of the geometry shown in Figure 1. For a fixed image length the frequency range covered can be adjusted by varying the distance of the laser beam from the nozzle, l. For $u = 2 \times 10^3$ m s^{-1}, the terminal velocity of a He expansion, and an object length of 10 mm the frequency range, $\Delta\nu$, is 800 MHz at $l = 50$ mm, sufficient to resolve the isotopic structure of atoms with the smallest combined mass and volume effects on their isotopic shifts. At $l = 10$ mm $\Delta\nu$ is 4 GHz and this can be increased by using an Ar expansion with its lower terminal beam velocity. Assuming a sufficiently narrow laser bandwidth the resolution is limited by the finite size of the nozzle and/or the width of the atom beam velocity distribution. Levinger *et al.* (1990) achieved a resolution of 3.5 MHz in their I_2 experiments. It is clear that the beam must be supersonic in order to satisfy both the geometrical requirements for the spectral range and the narrow u distribution requirement for resolution.

Some advantages of this concept for isotope ratio measurements are apparent. The interrogating laser need only be tuned close to the atomic frequency. It does not have to be rigorously locked to the peak of the absorption. If the frequency does drift during a run it will result only in a broadening of the observed lines and all isotopes will be affected equally. The complete measurement is made in a single run removing completely errors associated with tuning the laser for the next isotope and repeating the experimental conditions exactly. Realization of the method as an analytical technique requires the development of a suitable atomic beam source.

3. ELECTROTHERMAL SUPERSONIC ATOM BEAM SOURCE

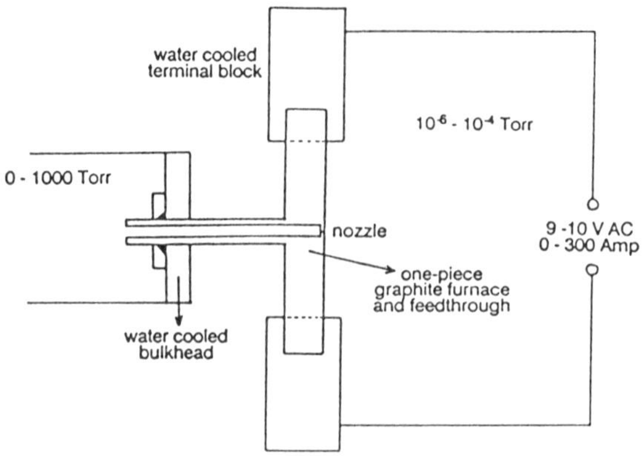

Figure 2: Electrothermal atomic beam source.

The atomic beam sources we are developing for this work are based on graphite furnace technology. Two approaches are under consideration. One involves a quasi-continuous beam using a heated graphite nozzle, the other a pulsed beam source with a heated graphite channel placed on the vacuum side of a pulsed valve. Here we concentrate on the first approach. The quasi-continuous source is shown schematically in section in Figure 2. It uses a gull-wing furnace piece of pyrolytically impregnated graphite closed at one end with the other end of the tube extended to form a gas inlet. The extended tube passes through a water cooled bulkhead where a high temperature o-ring makes a vacuum seal. The inside of the tube forms the high pressure reservoir for the nozzle and can be evacuated or filled with He or Ar backing gas. Samples are also introduced through this tube and deposited in the heating zone between the gull-wings using a micro-syringe. The heating current is passed between the wings of the piece using water cooled vacuum feedthroughs. A small hole is either laser (30-100 μm) or mechanically (150 - 300 μm) drilled in the closed end to form the nozzle. At room temperature a medium-sized diffusion pump (in our case a 6" Edwards Diffstack) is sufficient to maintain a continuous supersonic jet with nozzle diameters of ~100 μm (Amirav *et al.* 1980). However the throughput of the nozzle is dependent on the number density on the high pressure side and this decreases as the temperature is raised. To maintain the throughput required to produce a satisfactory supersonic beam at 2500 C and more means that the nozzle must be of a diameter which the pumping system cannot support at room temperature. To increase the pumping speed would require a prohibitively large pumping system. Instead we have chosen to keep the backing pressure at 10^{-3} Torr until the nozzle has heated to a preset temperature and to raise it quickly to the operating pressure just prior to atomization taking place using a control valve. This keeps the pumping needs modest but gives a continuous source whilst the atoms are emitted.

Initial tests of the source have been conducted using sodium as a test atom and LIF detection with a relatively broadband pulsed dye laser (Lumonics HD 300) 10 mm downstream from the nozzle. For any LIF experiments using a thermal source discrimination against the intense black-body radiation produced at high temperatures is an important concern. As well as careful design of baffles and the optical system we used a gated photomultiplyer (Hamamatsu C1392) for these pulsed experiments. For anticipated experiments using high resolution cw lasers lock-in techniques will be required. Ionization experiments will require pulsed field techniques similar to those used with effusive sources (Bekov 1989) to discriminate the thermal ion background. Figure 3 shows the LIF signal obtained with the laser tuned to the Na D_1 ($^2S_{1/2}$ - $^2P_{1/2}$) transition at 589.592 nm for the first furnace cycle following loading the source with an aqueous NaCl sample (70 ng). Na atoms are detected as the furnace reaches ~ 1100 C, as measured by an optical pyrometer. If the experiment is repeated without reloading or with a fresh sample but the laser tuned off-resonance no LIF is detected above the background, confirming that the signal is due to Na atoms. This initial experiment does little to characterize the beam directly however we can test whether the transitory gas load in the chamber results in the shock front moving too close to the source, essentially closing down the beam, to be responsible for the reduction in signal in the later stages of the cycle. For the 300 μm hole nozzle used in these experiments we see no difference in the Na LIF profile between backing pressures of 100 and 10 Torr. This would not be the case if the chamber pressure is going too high. Apparently the Na profile falls off due to complete atomization and not a disturbance of the gas dynamics. Considering the size of nozzle and backing pressures used and extrapolating from test measurements made on the system at room temperature using NO LIF, we are confident that the graphite furnace source is producing a supersonic atomic beam.

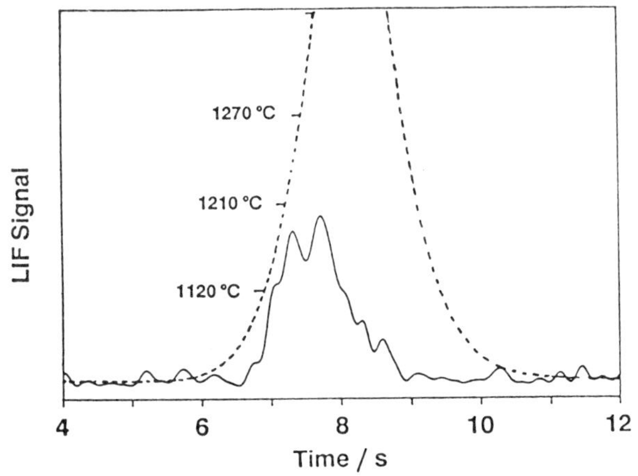

Figure 3: Na atom evolution from the electrothermal atomic beam source. The dashed line is the pyrometer signal.

4. CONCLUSIONS

The electrothermal atomic beam source described here shows promise as a technique for the direct, single furnace-cycle measurement of isotope ratios. Further experiments with higher resolution lasers will allow direct characterization of the atomic beam produced by this source and will test the practicality of the Doppler-tuned isotope imaging technique for isotope dilution experiments. The addition of RIS detection is expected to lead to a significant improvement in sensitivity. In addition the use of cw lasers can increase the detection duty cycle by up to 10^5. If these improvements can be achieved the sensitivity will approach or better that of non-isotopically selective graphite furnace AAS and will surpass by a factor of at least 10^3 that of inductively coupled plasma mass spectroscopy (ICP-MS), a technique currently gaining popularity for isotope dilution measurements.

REFERENCES

Amirav A, Even U and Jortner J 1980 *Chem. Phys.* **51** 31

Bekov G I and Letokhov V S 1989 *RIS 88, Inst. Phys. Conf. Ser. No. 94* 331 (Bristol: IOP Publishing)

Bushaw B A and Gerke G K 1989 *RIS 88, Inst. Phys. Conf. Ser. No. 94* 277 (Bristol: IOP Publishing)

Hackett P A, Morrison H D, Bourne O L, Simard B and Rayner D M 1988 *J. Opt. Soc. Am. B* **5** 2409

Levy D H 1980 *Ann. Rev. Phys. Chem.* **31** 197

Levinger A, Katzenellenbogen N and Prior Y 1990 *Opt. Lett.* **15** 625

A laser ion source for ion implantation applications

B.M. Suri[1], F. Ames, H.-J. Kluge, F. Scheerer
Institut für Physik, Universität Mainz, D - 6500 Mainz
N. Trautmann
Institut für Kernchemie, Universität Mainz, D - 6500 Mainz

[1]Permanent address: Bhabha Atomic Research Centre, Bombay-400085, India

ABSTRACT: Ion beams are nowadays being extensively used in applications involving, accelerators, ion implantation, nuclear physics research etc. We propose the use of a Laser Ion Source (LIS) based on resonance ionization for the generation of intense and pure ion beams for ion implantation and related applications. We discuss the potential and limitations of LIS and present results of pilot experiments on the generation of a gadolinium ion beam. Gadolinium photoion currents of $I > 3\ \mu A$ were observed.

Multi-step resonant photoionization of atoms trapped in a hot cavity can generate efficiently ion beams from atomic vapor. Such a laser ion source (LIS) operated with high-repetition rate dye lasers can produce ion beams of a large number of elements. An interesting possibility is that it might be possible to produce an ion beam of different elements simultaneously or sequentially from a single cavity. A conceptual diagram of such a source is shown in figure 1. The lifetime of such a hot cavity can be quite long and the rate of production of ions in the cavity is quite predictable. Since isotopic and isobaric selectivity is a built-in feature of resonance ionization spectroscopy (RIS) large-sized mass separators can be avoided.

Theoretically ion beam generation in a laser ion source has a high efficiency. Practically the maximum photoion current will depend on the ion extraction optics employed. The extracted ion current from a LIS cavity is given by (Ames et al 1990)

$$I_{ex} = \epsilon I_{in} = \frac{P_I}{P_I + P_D} \cdot I_{in} \qquad (1)$$

where ϵ is the efficiency of the LIS and I_{in} is the rate of injection of atoms into the cavity, P_I is the probability of photoionization and P_D is the probability for diffusion of atoms out of cavity. P_I is given by

$$P_I = (V_{int}/V_{tot})\nu_{rep}\ \tau_{flight}\ \epsilon_{photo}\ \epsilon_{pop}\ \epsilon_{ex} \qquad (2)$$

where V_{int} and V_{tot} are interaction volume and total volume of the cavity and τ_{flight} is the mean time of flight of atoms between collisions with the walls of the cavity. The repetition rate of the lasers is

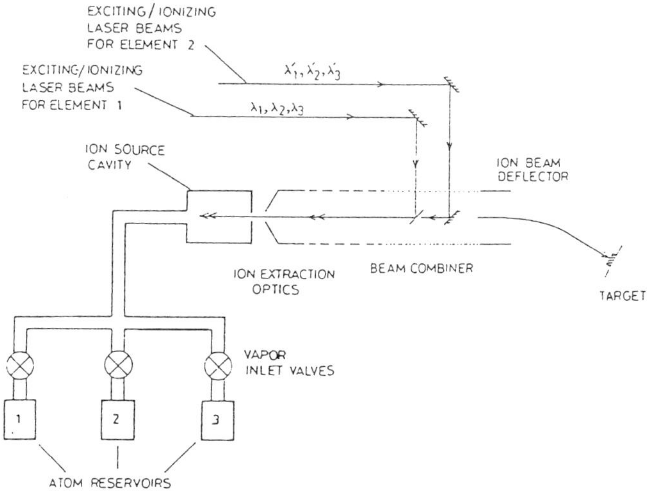

Fig. 1. A conceptual diagram of a proposed multi-element laser ion source.

given by ν_{rep}, ϵ_{pop} is the probability of ground state population, ϵ_{photo} that of photoionization with resonant laser light and ϵ_{ex} is the efficiency of ion extraction. It is assumed that surface ionization can be neglected. Presence of high charge densities may cause non-negligible internal fields. The maximum current density j extractable from a hole in the cavity is space charge limited and given by (Kaufman 1990)

$$j = k_0 \frac{q}{m} \frac{V^{3/2}}{d^2} \tag{3}$$

where q and m are the charge and the mass of the ion, V is the extraction voltage and d the interelectrode gap and k_0 is a constant depending on the geometry. Typically the gadolinium photoion current density extractable with our conditions (V = 10 kV, d = 1 cm) is of the order of 1 mA/cm^2. There is, however, a possibility of increasing the maximum extracted current by use of a cavity with an array of holes as it is often employed in ion implantation applications (Kaufman 1990).

The selectivity of the ion beam extracted is affected by charge exchange processes with neutral atoms inside the cavity as well as non-selective surface ionization of impurities present in the cavity. Except for alkali elements, charge exchange predominates in such a source for ion implantation. The fraction of selectively generated photoions lost by charge exchange processes is given by

$$F = \sigma n_0 l \tag{4}$$

where σ is the charge exchange cross section, n_0 the neutral number density and l is the length of the interaction region. This expression dictates the maximum number density usable in the cavity. Typically the cross section for resonant charge exchange is 10^{-14} cm^2. This resonant process is important in respect to isotopic selectivity if required. In case of impurities, the cross section for non-resonant charge exchange will be smaller than that given above.

Table 1: Choice of the size of the hole in the laser ion source cavity. Highest efficiency can be obtained in practice by saturation of the exciting/ionizing transitions and efficient ion extraction.

Hole radius [mm]	Range in cavity for electric field E>10V/cm (with extraction voltage 20 kV) [mm]	Energy density for 600 nm photons for 1 watt dye laser power (CVL pumped) [$\mu J/cm^2$]	Cross section of exciting/ionizing transitions that can be saturated [cm^2]
0.5	4.5	2.6×10^4	1.2×10^{-17}
1.0	9	0.6×10^4	4.7×10^{-17}
2.0	18	1.6×10^3	1.8×10^{-16}
5.0	45	2.6×10^2	1.2×10^{-15}

The efficiency is defined as the ratio of photoions produced and extracted to the number of particles leaving the cavity as neutral atoms as well as photo or surface ionized ions. This efficiency is independent of the size of extraction hole (Ames et al 1990). Thus the hole size can be optimized with respect to laser power available for all excitation steps, pumping speed and penetration of the electric field into the cavity. The laser power density is proportional to $1/r^2$, the penetration of an electric field in first approximation is proportional to r^3 and the pumping speed is proportional to r^2, where r is the radius of the hole. The penetration of the electric field is given by the expression for the potential ϑ on the other side of a conducting plane with a hole in it,

$$\vartheta(z) = \frac{Er^3}{3\pi z^2} \tag{5}$$

where E is the electric field far from the plane of the hole and z is the distance from the hole. Table 1 allows to decide on the choice of the hole size with respect to various parameters like excitation/ ionization cross sections of the transitions employed for photoionization.

Linear accelerators and synchrotrons require bunched ions for injection. LIS offers such a possibility as the production of ions is intrinsically pulsed. Quite often it is required to use a specific rare isotope in accelerators or storage rings, which is achieved by use of pre-enriched, expensive samples. Using LIS it is possible to generate an isotopically pure ion beam from natural samples. Also it might be possible to generate polarized ion beams with the help of appropriately polarized exciting/ionizing lasers, as they are required in some nuclear physics applications.

We have done some pilot experiments on gadolinium employing the laser ion source used earlier for trace analysis of technetium with suitable adaptions (Ames et al 1990). The main changes were, compared to the Tc experiments, incorporation of a Faraday cup before the slit of the mass spectrometer and an ion deflector which focuses the ion beam of the LIS either into the Faraday cup or onto the slit of the Mattauch-Herzog mass spectrometer. Hot cavities of two types were used. One was a tungsten cylinder with end plates on both sides and a hole in the front plate.

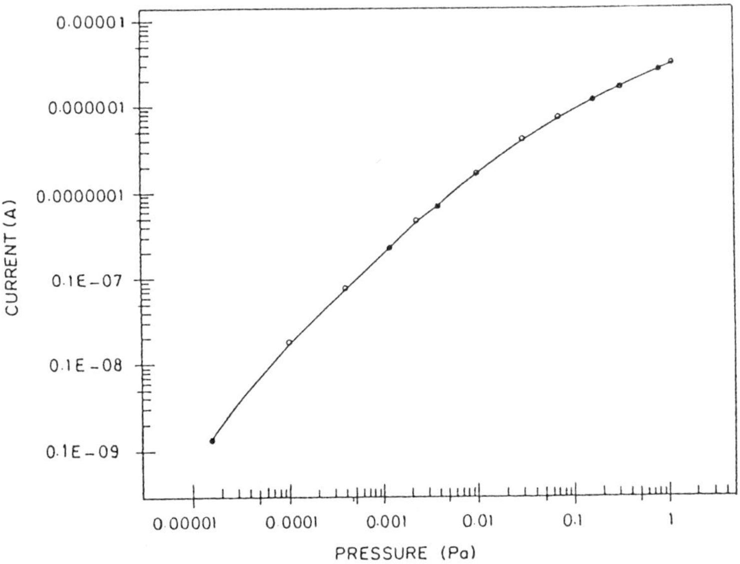

Fig. 2. Gd photoion current as a function of the Gd vapour pressure (inferred from measured temperature of the LIS cavity).

The length and inner diameter of the cylinder was 1 cm each and the hole size was 2r = 2.5 mm. The other cavity had the same shape but the cylinder was made of an insulator material (alumina) whereas the end plates were made of tantalum. Both cavities were heated by electron bombardment. Two copper vapour lasers (of 40 watts average power each) were employed to pump three dye lasers. The excitation/ionization wavelengths used were 561.8 nm, 635.2 nm, and 613.4 nm. The three dye laser beams were combined using quartz fibres and focused to a spot size of about 3 mm into the LIS cavity.

Figure 2 shows the current of the gadolinium ion beam obtained as a function of the gadolinium vapour pressure as calculated from the measured temperature of the LIS cavity. A maximum ion current of I_{Gd} = 3.2 μA was observed at the highest possible temperature of the cavity at 1850 K. Above this temperature no experiments could be performed because the sample put into the cavity was evaporated too quickly.

Ion beams with an intensity of about 10^{14} ions/sec seem feasible with the present setup which was originally designed for trace analysis experiments, i.e., low ion beam intensity. The presently achievable ion current is already quite attractive for a number of applications. Further experiments are required to check the purity of the gadolinium ion beam.

References

Ames F, Brumm T, Jaeger K, Kluge H-J, Suri B M, Rimke H, Trautmann N, Kirchner R 1990 Appl. Phy. **B51** 200

Kaufman H R 1990 Rev. Sci. Instr. **61** 342

Inst. Phys. Conf. Ser. No 114: Section 3
Paper presented at RIS 90, Varese, Italy, 16–21 September 1990

A laser ion source for on-line isotope separation

V.N. Fedoseev, Y.A. Kudryavtsev, V.S. Letokhov and V.I. Mishin,
Institute of Spectroscopy, USSR Academy of Sciences, 142092 Troitzk,
Moscow Region, USSR,
H. Ravn and S. Sundell, CERN CH–1211 Geneva, Switzerland,
H.-J. Kluge and F. Scheerer, Institut für Physik, Joh. Gutenberg Universität,
D-6500 Mainz, Fed. Rep. of Germany

ABSTRACT: A laser ion source has been developed for efficient production of isobarically pure ion beams at the on-line mass seperator ISOLDE at CERN. In first off-line tests with radioactive ^{169}Yb, an efficiency of about 15% was achieved. An elemental selectivity between 10 and 10^4 was observed. The maxium value could be obtained at the off–line separator with TaC as construction material. A first test at the on-line separator ISOLDE–3 was performed recently wit Yb isotopes. The lasers produce a pulsed ion beam of about 10 μs pulse length. In order to suppress the continuous background due to surface ionization a pulsed deflector was used so that the selectivity was improved by a factor of 10.

1. INTRODUCTION

An important problem of conventional ion sources at on-line isotope separators is their chemical selectivity. An especially illustrative example is given by the Ta foil target with a surface ionizer, where the ions of nearly all rare earth elements are produced simultaneously. As a result the mass separated ion beams are heavily contaminated by isobars. Sometimes the contamination is larger by some orders of magnitude than the yield of the isotope under investigation. In such a case, the selectivity but also the efficiency of a conventional ion source used for on-line mass separation can be improved applying resonance ionization spectroscopy. Moreover a pulsed ion beam is obtained which might be useful in some applications.

Such a laser ion source (LIS) has been set up at the on-line isotope separator ISOLDE–3/CERN/Geneva (Ravn and Allardyce 1987). It consists of a hot cavity where the atoms of the wanted isotopes are photo ionized. A laser system for RIS was set up at the ISOLDE mass separator which serves for analysis of the ions produced. The laser system consists of two copper vapour lasers, working as oscillator and amplifier, and three dye lasers pumped by the green lines of the pump lasers. Copper vapour lasers have the advantage of a high repetition rate (ν_{rep} = 10 kHz), which is essential for high efficiency. The atoms are excited and ionized via an autoionizing state by three-step

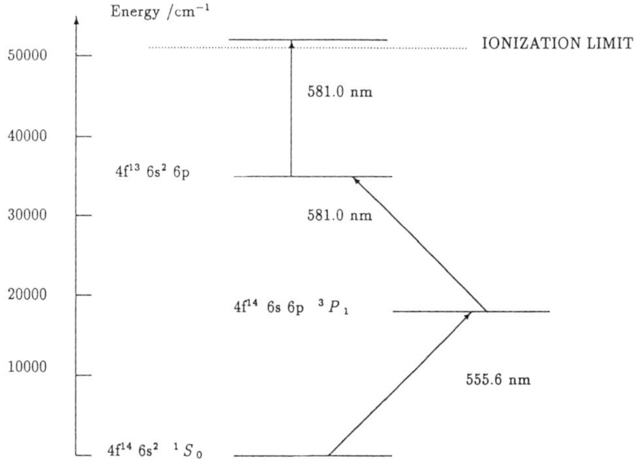

Figure 1: Level scheme for two-color, three-step resonant ionization of ytterbium.

resonant ionization as shown in fig.1. Generally three different wavelengths and hence three dye lasers are required for effective RIS of rare earth elements. In case of Yb an excitation scheme can be used, where the wavelengths for the second and third step are identical within the laser linewidth ($\Delta\nu_L = 24$ GHz), so that only two dye lasers are required.

2. EFFICIENCY

The efficiency for photo ionization of the ion source ϵ is determined by the ratio of photo ionization relative to the total number of atoms or ions leaving the cavity. The efficiency can be calulated as a function of the rate of photo ionization R_I, the rate of surface ionization R_S, and the rate of diffusion as neutral atom through the orifice R_D as

$$\epsilon = \frac{R_I}{R_I + R_S + R_D}.$$

An efficiency near to unity can be obtained, since the atoms are trapped in the cavity for a long time compared to the time between two laser pulses. Neglecting surface ionization one obtains for a tube of lenght L

$$\epsilon = \frac{\nu_{rep}\,\epsilon_{ion}}{\nu_{rep}\,\epsilon_{ion} + \frac{v}{4L}}$$

where ν_{rep} is the repetition rate of the lasers, and ϵ_{ion} is the probability for resonant ionization of the atoms in the volume of interaction with the laser beams during one laser shot. The velocity of the atoms in the tube is given by v. This equation shows that a high repetition rate of the lasers used for resonant ionization and a long cavity are essential for high efficiency.

3.EXPERIMENTAL RESULTS

Such a laser ion source has been tested at the off-line mass separator at CERN/Geneva and recently at the on–line mass separator ISOLDE–3/CERN in respect to efficiency and selectivity. The selectivity S is defined as the ratio of the current of photo plus surface ions to the one of surface ions. As shown in off-line experiments with Tm a very high selectivity can be reached at low temperatures and with TaC as construction material for the ionizer tube which has a low workfunction.
The data shown in fig. 2 were obtained with an special arrangement to provide a constant flux of Tm–atoms out of a reservoir into the ionizer tube. The efficiency of thermal ionization is nearly independent of the material used for construction of the cavity. A much lower photo ion current is observed for low temperatures, although, there is a constant flux of Tm–atoms into the cavity. This can be explained by a low extraction efficiency of the produced photo ions and sticking of the atoms to the wall. With TaC an efficient extraction of the photo ions can be obtained even at lower temperatures, so that a high selectivity (10^4 at T=1650 K) is reached. On the other hand, a short release time of the ion source is essential in case of the application of a LIS to ionization of short-lived isotopes, since the decay losses depend on the sticking time of the atoms on the walls of the ionizer tube. Here, a compromise between high efficiency and high selectivity may have to be found.

At the on–line separator ISOLDE–3 the laser beams were focused over a distance of about 20 m through a bore in the 90° magnet into the ionizer tube, where the investigated atoms are ionized. Inside the ionizer tube the laser beams were focused to a diameter of about 1 mm. Different ionizer tubes (W or Ta) were used with inner diameters between 1 mm and 3 mm and lengths between 30 mm and 40 mm. The foci of the laser beams could be controlled with the help of a theodolite which allowed to

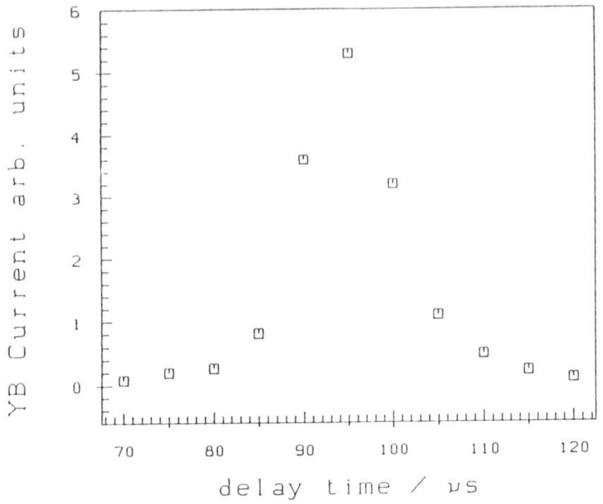

Figure 2: Currents of thulium ions obtained by resonance ionization (laser) and by surface ionization (surface) as a function of temperature of the ionizer tube. Different construction materials were used for the ionizer tube.

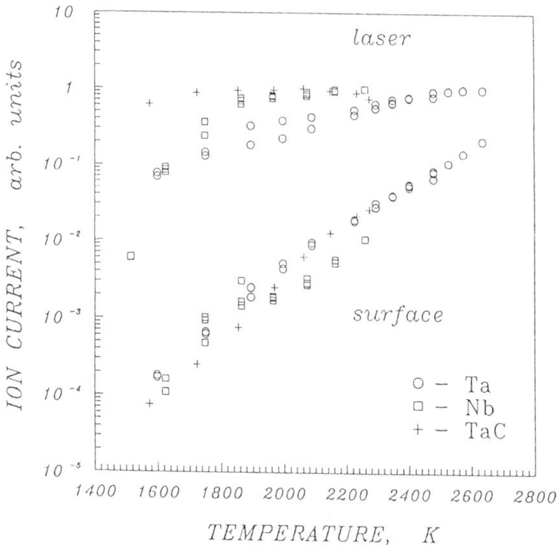

Figure 3: The pulse shape of a stable ^{174}Yb ion beam is shown. It was measured by the aid of a pulsed deflector which lets the ions pass through a slit during a time gate of $5\mu s$. The pulse shape was measured by changing the delay of this gate relative to the laser pulse. The delay of the ion pulse relative to the laser pulse is determinated by the time of flight through the mass separator to the deflector of about 64 μs and the time of extraction out of the cavity.

observe the the hot cavity as well as the laser beams. The cavity was heated by a dc current of about 100 A causing a voltage drop of approximately 1 V over the lenght of the tube. The potential difference serves for extraction of the ions. The pulsed laser beams produce a pulsed photo ion current. By use of a pulsed deflector up-stream the ion beam, a gated detection of these ions enables to suppress the continuous background of surface ions. In this way the selectivity was increased by a factor of 10 to S = 90 at a temperature of about 2000° C.

By varying the delay of the time gate relative to the laser pulse one can measure the pulse shape of the extracted ion beam. The pulse form observed is shown in fig. 3. The width of the pulse ($\approx 10\mu$s) is not determined by the laser pulse width (20 ns) but by the time it takes to extract the ions out of the cavity. This time is rather long because of the low voltage drop across the ionizer tube. The higher the voltage drop the smaller is the width of the photo ion pulse.

This work was supported by the Bundesministerium für Forschung und Technologie under grant number 06Hz188–I.

REFERENCES

Ravn H.L. and Allardyce B.W. (1987) Treatise on Heavy–Ion Science 8, ed. D.A.Bromley (New York: Plenum Press) p. 363

Continuous wave field ionization laser spectroscopy

R. Hergenröder and K. Niemax
Institut für Spektrochemie und Angewandte Spektroskopie (ISAS)
Dortmund, Federal Republic of Germany

ABSTRACT: A new technique of laser ionization spectroscopy is introduced. It is based on Doppler-free excitation of atoms to Rydberg states by cw lasers in a electric field free region of an atomic beam and the field ionization of the long- lived atoms in a region spatially separated from the excitation region.

1. INTRODUCTION

Field ionization laser spectroscopy (FILS) in an atomic beam has been proven to be a very sensitive analytical technique (Bekov and Letokhov 1989). It is based on stepwise excitation of atoms to a Rydberg state by pulsed lasers and subsequent ionization of the highly excited atoms by the application of an electric field pulse. The ions are separated from the beam by the applied field and detected by an ion detector. The absolute detection limits of this technique could be further improved if it would be possible to use cw lasers for excitation. With cw lasers the duty cycle for excitation would be orders of magnitude better than with pulsed lasers. Furthermore, cw excitation with single mode lasers would allow Doppler-free, isotope selective spectroscopy in the collimated atomic beam and a very efficient population in the excited states with moderate powers, because the laser line widths are in the order of the line widths of the atomic transitions. Pulsed lasers have line widths which are typically three or more orders of magnitude larger than of cw lasers. Therefore, the peak power of a pulsed laser has to be at least three orders of magnitude larger than of a cw laser to obtain similar population densities as with a cw laser in the excited states. On the other hand, the higher powers of pulsed lasers induce a very strong increase of the ion background due to off-resonant photoionization of matrix atoms. Note that the increase is at least 10^6 and 10^9 for off-resonant two-photon and three-photon ionization, respectively.

2. CONTINUOUS WAVE FIELD IONIZATION LASER SPECTROSCOPY

The basic idea of continuous wave field ionization laser spectroscopy (cw-FILS) is simple. The region of laser excitation to Rydberg states has to be separated spatially from the region of field ionization. Because Rydberg atoms have long lifetimes (μs-ms) they are able to stay excited in thermal atomic beams until they reach the electric field region. In our experiment described below, Rydberg atoms in states with principal quantum numbers above 30 (lifetimes $\geq$ 100 μs) are field ionized. With an average velocity of 1000 m/s these atoms can fly 10 cm and further before they recombine.

3. EXPERIMENT

Fig. 1 shows the experimental arrangement. A beam of lithium atoms was formed by electrothermal evaporation into vacuum. The lithium atoms were excited by three-step transition (2S-2P-3S-nP) to high-lying nP levels ($n \geq 15$). The first and the third step (671.0 nm and 630 nm, respectively) were induced by commercial single mode cw dye ring lasers,

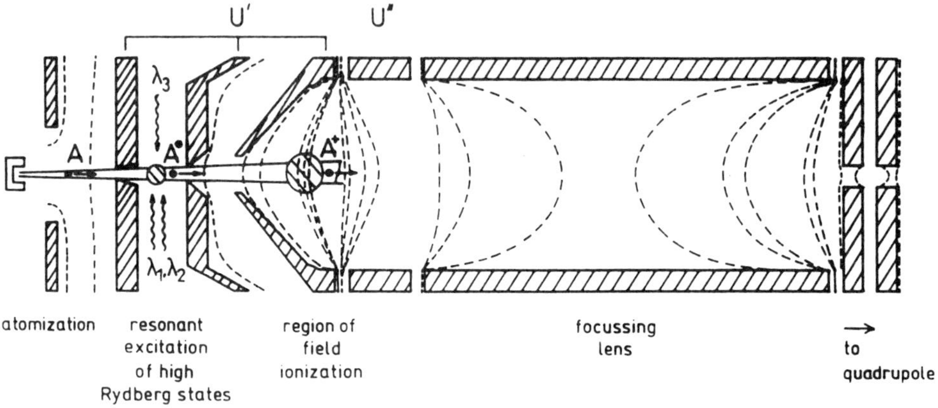

Fig. 1 Experimental arrangement.

while a single mode semiconductor laser was applied for the second step (812.9 nm). The line widths of the dye lasers and the diode laser were about 1 MHz and 20 MHz, respectively. The powers applied were about 100mW (dye lasers) and 10 mW (diode laser). The copropagating beams of the first lasers overlapped with the counter-propagating beam of the third laser in the atomic beam. The excitation region was free of electric fields. The excited atoms passed an exit hole of the Faraday cage and entered the region of the focusing lens of a quadrupole mass spectrometer. Here they experienced an electric field which increased almost linearly between the first two segments of the ion optics. At a critical electric field E_C the atoms ionized. The ions were focussed into the mass quadrupole and detected by a multiplier. The isotope selectivity of the Doppler-free laser excitation was high enough that the mass quadrupole filter could be operated in the integral mass mode. Hyperfine components of ^7Li were measured.

4. RESULTS

The critical electric field E_C for field ionization of Li nP Rydberg atoms can be determined. In Fig. 2 we show the potential and the electric field in dependence on the distance from the exit hole of the Faraday cage. Conditions as during the measurement of 34P Rydberg atoms. The data were calculated with a computer program (Dahl and Delmore 1987). Now the potential of the quadrupole was varied. At a potential U_C of +150 V the ion signal disappeared. This value correspondended to the potential at about 100 mm. The ions were created in this position. The electric field, the critical field E_C, was about 260 V/cm at 100 mm (see Fig. 2). In Fig. 3 we have plotted the experimental data of E_C in dependence on the effective principal quantum number n^*. The data agree very well with the critical fields calculated from the empirical formula given by Letokhov (1987) $E_C=3.5x10^8/n^{*4}$ V/cm (full curve), but they are slightly larger than the theoretical data from the classical approximation

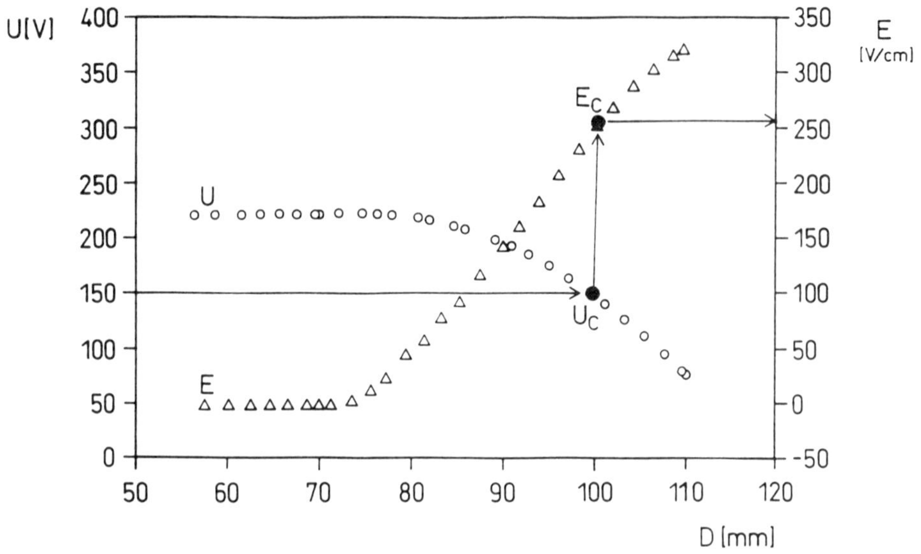

Fig. 2 Potential and electric field in dependence on the distance of the exit hole of the Faraday cage. From Hergenröder and Niemax (1990).

formula E_C=3.13x10^8/n^{*4} V/cm which does not take into account the field ionization by electron tunneling (see, e.g., Letokhov 1987).
In Fig. 4 the ratio of the field ionization to the photoionization signal from the 3S level

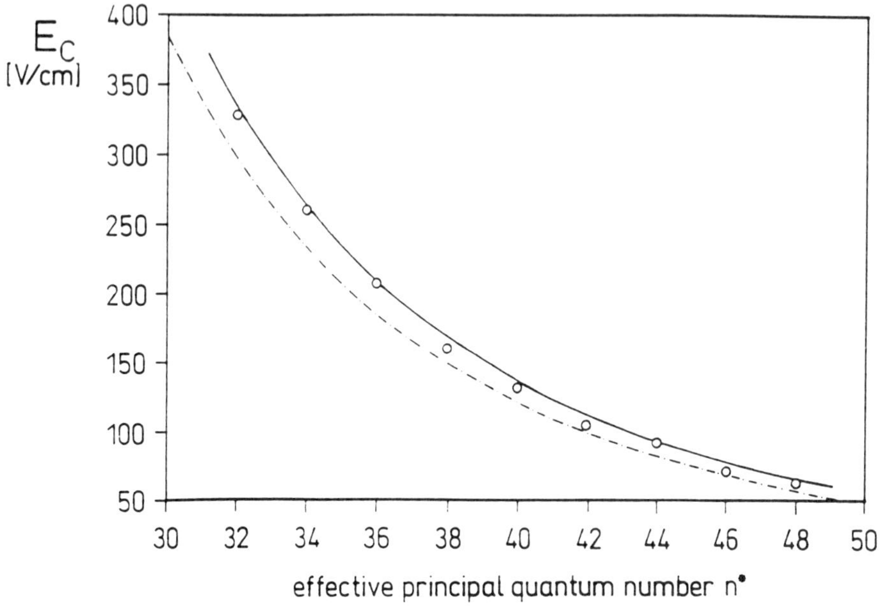

Fig. 3 Critical electric field E_c for field ionization of Li nP Rydberg atoms in dependence on the effective quantum number n*. From Hergenröder and Niemax (1990).

S_{FI}/S_{PI} is plotted in dependence on the effective principal quantum number of the final nP level. At about n=32 we have measured a signal which was about 1000 times larger than the photoionization signal. Because there were small residual electric fields in our Faraday cage the transition to the very high nP levels were significantly broadened (Stark effect). The n-dependence of the broadening is the reason why the measured ratio did not show a n^{*-3}-dependence which was expected (dashed curve in Fig. 4). Because of the broadening the laser radiation field could interact only with a decreasing fraction of atoms as n increased. Field ionization of atoms with 20 < n < 30 which also arrived in the region of field ionization could not be achieved because we could not generate electric fields in our apparatus which were high enough for ionization.

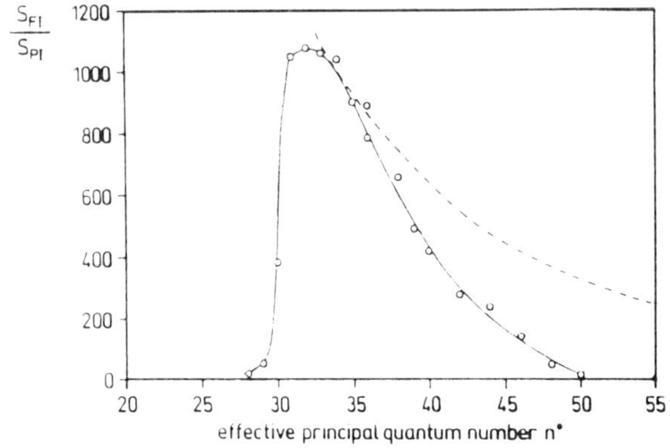

Fig. 4 Ratio of the field ionization signal S_{FI} to the photoionization S_{PI} from the 3S level in dependence on the effective principal quantum number of the final nP level.

5. CONCLUSION

The technique of cw-FILS presents an optimum duty cycle for excitation of atoms. The narrow band widths of cw lasers allow Doppler-free, isotope selective detection of elements and efficient excitation of analytes in a collimated atomic beam. The possibility to get high population densities with low laser intensities diminishes the ion background due to off-resonant multiphoton ionization.

Bekov GI, Letokhov VS 1989 in Resonance Ionization Spectroscopy 1989 (Eds TB Lucartorto and JE Parks, Institute of Physics) 331

Dahl D A and Delmore J E 1987 The Simion PC/PS2 User's manual. Informal Report EGG-CS-7233 Idaho National Engineering Laboratory

Hergenröder R, Niemax K, 1990 submitted for publication

Letokhov VS Laser Photoionization Spectroscopy (Academic Press, Orlando 1987)

Inst. Phys. Conf. Ser. No 114: Section 3
Paper presented at RIS 90, Varese, Italy, 16–21 September 1990

High-sensitivity cavity-enhanced detection of surface laser-ablation

H.A. Schuessler, E.C. Benck, S.H. Chen and Z. Rong

Department of Physics, Texas A & M University, College Station, Texas 77843

ABSTRACT: The time evolution of minute laser ablation plumes produced by microjoule laser pulses from different surfaces was investigated using cavity-enhanced detection. The vaporization amplitude, temporal structure and velocity of the ablated materials were measured. The limits in sensitivity of the novel technique are discussed.

1. INTRODUCTION

Light absorption in thin films and the subsequent conversion to thermal energy has been the basis for a large class of photothermal spectroscopies. In photoacoustic (Tam 1986) and surface-photoacoustic spectroscopy (Raja 1990) an acoustic wave is detected with high sensitivity. In photothermal interferometry (Stone 1973) and photothermal deflection spectroscopy (Jackson 1981) the change in the index of refraction due to the small absorption serves as the signal. We report orders-of-magnitude improvement by cavity enhanced detection of photo-vaporization signals where also the refractive index is changed, and in addition absorption and scattering of a probe beam increase the damping of the cavity. Cavity-enhanced photo-vaporization spectroscopy is therefore a sensitive method for monitoring laser-surface interactions. In particular, it can detect the material ejected from a solid or liquid film surface, when a laser is focussed onto it. The tiny plume is monitored inside an optical cavity by a fixed frequency cw probe laser. Since the probe beam is resonant with the optical cavity, the change in optical path, the absorption, and the scattering by the plume are amplified by the quality factor of the high finesse optical cavity. Such a technique could possibly be used to monitor and via feedback control laser ablation which is the first step in laser microprobe mass spectroscopy and analytical resonance-ionization techniques (Beekman 1984).

2. THEORETICAL CONSIDERATIONS

In the following the interesting case of minute single pass losses due to a tiny laser plume is considered. The signal is then dominated by the changes in refractive index while absorption and scattering losses are negligible. For this case the output intensity I_A of the optical cavity is given by the well known Airy formula

$$I_A = I_0(1-R)^2/[(1-R^2)^2 + 4R^2 \sin^2 \delta/2] \qquad (1)$$

where I_0 is the incident intensity, R is the reflectivity of the cavity mirrors and δ is the phase difference of the interfering beams. To derive a sensitivity limit also the expression of the slope of the transmission fringes is needed

$$dI_A/d\delta = -2I_0R^2\ (1-R)^2 \sin \delta/[(1-R^2)^2 + 4R^2 \sin^2 \delta/2]^2 \qquad (2)$$

The sensitivity limit η for small optical path changes $\Delta_f|x|$ and the corresponding small phase changes $\Delta_f|\delta|$ is related to the noise $\Delta_f|I_A|$ of the transmitted beam intensity according to

$$\eta = \Delta_f|x| = \lambda|\Delta_f\delta|/2\pi = \lambda\Delta_f|I_A|/2\pi(dI_A/d\delta)_{z0} \qquad (3)$$

Here $z_0 = \sin^2\delta_0/2$ is the optimal tuning parameter and occurs at the side of the transmission fringe where $dI_A/d\delta$ has its maximum value ($d^2I_A/(d\delta)^2 = 0$). For R close to 1 it holds (Yu 1989)

$$z_0 = [(2A+3B) - \sqrt{(2A+3B)^2 - 8AB}]/4B \qquad (4)$$

where $A = 2R^2(1-R^2)^2$ and $B = 8R^4$. Inserting Eqs. (4) and (2) into Eq. (3) yields

$$\eta = (\lambda/2\pi)\ (\Delta_f|I_A|/I_{z0})\ [(I-R^2)^2 + 4R^2z_0]/[4R^2\sqrt{z_0(1-z_0)}] \qquad (5)$$

where I_{z0} is the output beam intensity at the side of the transmission fringe. The noise $\Delta_f|I_A|$ of the transmitted beam intensity is minimized in a suitably chosen detection circuit where the difference between the equal intensity probe and locking beams is measured. The remaining noise originates from a non ideal adjustment of the transmission fringes, in which the pointing and intensity fluctuations of the HeNe laser, the error signal of the locked cavity, and the ambient noise due to convection and mechanical vibrations do not completely cancel.

3. EXPERIMENTAL ARRANGEMENT

Fig. 1 depicts the experimental setup. The vaporization is produced by a tunable pulsed dye laser which is focussed onto a sample located inside a spherical etalon. The pulsed dye laser produces pulse energies in the range of 0.35 to 35 μJ around λ = 640 nm and with a pulse width of 10 nsec. When focussed to a diameter of 20 μm, this corresponds to laser fluxes in the 10^7 to 10^9 W/cm^2 range. The material removed from the surface is monitored by a weak HeNe-laser probe beam, which is at resonance with the optical cavity. A second HeNe-laser beam is used to lock the cavity to the side of a transmission fringe and therefore to the HeNe-laser frequency. For experimental simplicity the present experiments were carried out in air, but the technique works equally well also in ultra-high vacuum.

4. RESULTS

Over 25 different materials were examined to investigate the ablation signals. The most interesting features were obtained from metals and opaque films. Three features were studied in detail and are the vaporization amplitude, the temporal shape of the plume, and the velocity of the ablated material.

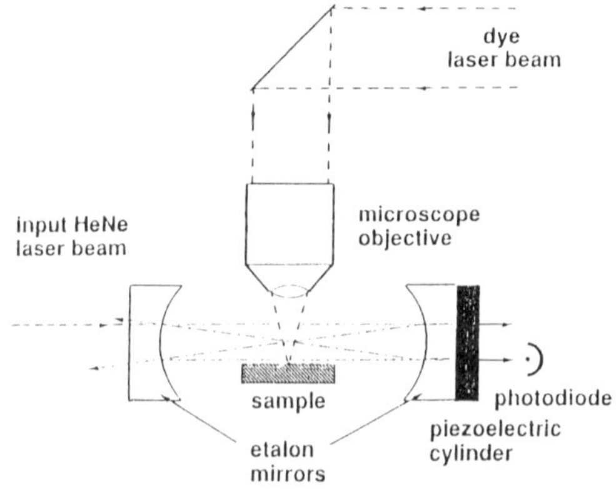

Fig. 1: Detail of the optical cavity.

The strongest signals were obtained from aluminium surfaces which had a peak 80 % attenuation of the probe beam. The weakest signal occurred for tungsten which produced a 1.5 % peak attenuation. Fig. 2 shows the time evaluation of typical photo-vaporization signals. The most common temporal structure observed consists of a single asymmetric peak. For some low melting point samples, a second broad peak, centered approximately 2 msec after the laser pulse, occurred in the averaged data. As the laser intensity was reduced, the second peak disappeared leaving only the normal single peak behavior. This second feature arises due to the localized melting of the surface which becomes deep enough to develop turbulent convection or even boiling. It can also be produced by a second laser ablation product, which has a different effective temperature and activation energy than the first. The material leaving the surface has a velocity made up of a hydrodynamic velocity of the dense vaporization cloud exploding into the ambient air, superimposed on a Maxwellian velocity distribution characterized by an effective temperature. Since the evaporated particles collide strongly with each other, the effective temperature differs from the temperature of the surface as thermodynamic equilibrium is not established. In addition the analytical form of the time distribution of the vaporization signals observed in the present experiment must have a viscosity term to describe the formation of the laser ablation plume in air.

The maximum velocity of the particles leaving the surface was investigated by measuring the time delay between the dye laser pulse and the start of the photo-vaporization signal as a function of the probe beam's height above the surface. In this way the maximum velocity of for instance laser ablation from lead was found to be 9 m/sec and from aluminium 2 m/sec.

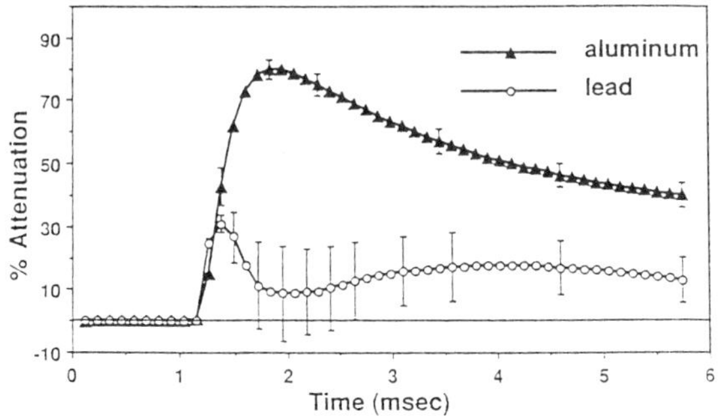

Fig. 2: Typical photo-vaporization signals for a laser energy of 35μJ per pulse.

The surface loading with iron of micron-size silicate dust particles is of interest in pulmonary research and was measured to demonstrate the sensitivity of cavity-enhanced detection of surface laser ablation. Although this method is not specific to iron, it was shown that the photo-vaporization signal increases with the iron content of silicate dust.

5. CONCLUSIONS

We have demonstrated that cavity-enhanced detection of minute laser plumes is ultrasensitive and can be used to monitor laser-surface vaporizations too small to be seen by the naked eye. We are in the process of developing a theory to describe the observations.

This work is supported by the Texas Advanced Research Program and the Center for Energy and Mineral Resources of Texas A & M University.

REFERENCES

Beekman D W, Tallcott T A 1984 Resonance Ionization Spectroscopy (Techno House, Bristol) p 143
Jackson W B, Amer W M, Boccara A C, and Fournier D 1981 Appl. Opt. 20 1333
Raja M Y A, Reicher D W, Brueck S R J, McNeil J R and Oates D E 1990 Opt. Lett. 15 66
Stone J 1973 Appl. Opt. 12 1828
Tam A C 1986 Rev. Mod. Phys. 58 381
Yu Q, Chen S H, Rong Z, Xu Y, and Schuessler H A 1989 Chin. J. Lasers 16 87

Kinetics of diffusive ionisation of the potassium Rydberg atoms

V. Švedas, Z. Kuprionis, V. Gontis, and B. Kaulakys
The Institute of Physics, Lithuanian Academy of Sciences, 232600 Vilnius, Lithuania

ABSTRACT: The kinetics of diffusive (multi-step) ionisation of potassium atoms excited by a laser to the 34P state have been studied in a vapour cell. The amount of ions J vs. the delay time T of ion-extracting electric field pulse was measured. The amount J was expressed as an integral of the diffusive ionisation rate w(t) and was calculated numerically. By the comparison of that integral with experimental amount J, the characteristic diffusion time τ' dependence of the vapour density value was determined.

1. INTRODUCTION

Collisional diffusive ionisation is one of the possible ways to ionise laser prepared atomic Rydberg states. Diffusive ionisation is the main mechanism of ionisation of Rydberg atoms with principal quantum numbers $n \geq 25\text{-}30$ by the parent vapour (Herrmann et al. 1986, Švedas 1988) and very probably by the buffer gases, too (Niemax 1983). A theoretical model of stationary diffusive ionisation has been suggested and the yield of this process has been calculated (Kaulakys and Švedas 1987). For the RIS aims, it is important not only to know the ionisation yield but the kinetics of that ionisation. In the present paper the kinetics of potassium 34P state ionisation, i.e. the diffusive ionisation rate vs. time, has been investigated.

2. EXPERIMENTAL METHOD AND SET-UP

Potassium atoms were excited from the ground 4S state by the second harmonic of the dye laser pumped by the second harmonic of the Q-switched YAG:Nd laser. The exciting λ = 286 nm pulse had the duration of about 10 ns and spectral width of 0.002 nm. The glass cell containing potassium vapour had a temperature of $\theta \simeq 520°$ K and has been described (Švedas 1988).

The combination of the stationary electric field E_s = 12 V/cm and the pulsed one E_p = -9 V/cm in the opposite direction were created in the excitation region of the cell. The appearing ions drifted the distance b = 0.6 cm in the field region with the drift-speed v_d. The field pulse with the duration Δ = 9 μs was delayed by the time T in respect to laser pulse. The amount of ions extracted by this field combination to the registering thermoionic diode is given by the following (Švedas et al. 1989):

$$J(T) = N_R^o \, V \int_{T - b/v_d}^{T + \Delta - b/cv_d} w(t)dt. \tag{1}$$

Here w(t) is the diffusive ionisation rate, N_R^0 the density of primary excited Rydberg atoms, V is the excitation volume ($V \simeq 3cm^3$) and $c = |E_p|/|E_s|$.

The relationship (1) is an exact one for a point (very small) ionic source. The laser spot diameter in the excitation region (~0.3 cm) is not negligible with respect to the drift distance. In this situation another integration over the drift axis z would be done with excited atom distribution $N_R^0(z)$ and with coordinate z instead of constant drift distance b. But it was very difficult to determine $N_R^0(z)$ because the intensity distribution was irreproducible inside our laser spot. For this reason we used relation (1) supposing all ions to be created in the laser spot centre. This simplification would not significantly influence the results, first, due to symmetry of the laser spot, and second, because the time of the drift through the spot region ($0.3/v_d \sim 1\ \mu s$) was small compared to the extracting field pulse duration time, Δ.

The amount of ions J(T) as a function of delay time T was measured in this work. The ionic response of thermoionic diode was digitised and stored in a computer's memory. The energies of the exciting laser pulse were attenuated to the values about 10^{-6} - 10^{-7} J, measured by the photomultiplier and stored in the same way.

3. APPROXIMATIONS FOR THE IONISATION RATE w(t)

The essence of our method is the repeated variation of the characteristic parameter τ of the function w(t) until integral (1) will coincide with the experimental J(T). Karpus (1988) obtained the ionisation rate analytically in the case when an unlimited number of Rydberg states are present (maximum principal quantum number $n_m = \infty$). This rate

$$w(t) = (\tau/t)^5 \exp(-\tau/t)/6\tau, \quad \tau = 3.9 \cdot 10^{-10}\, M/k\theta n_0^{*}\sigma N\sqrt{m} \qquad (2)$$

is plotted on Figure 1. In expression (2) the characteristic diffusion time τ is given in sec, M and m are the atom's and electron's mass in g, $k\theta$ is the thermal energy in J, σ is the elastic electron-vapour atom scattering cross-section in cm^2, N is the vapour density in cm^{-3}, and n_0^* is the effective principal quantum number of the primary excited state.

In our experiment the drift fields E_s and E_p ionise the atomic states with principal quantum numbers $n \geq 80$ ($n_m = 80$), and the rate (2) of Karpus (1988) is not suitable for us. For this reason the population balance equations were solved numerically to obtain diffusive ionisation rates as a sum of all transitions from the lower levels to the levels with $n \geq 81$ (Švedas et al. 1989). The results of those calculations are given in Figure 1 and are approximated by the following relation:

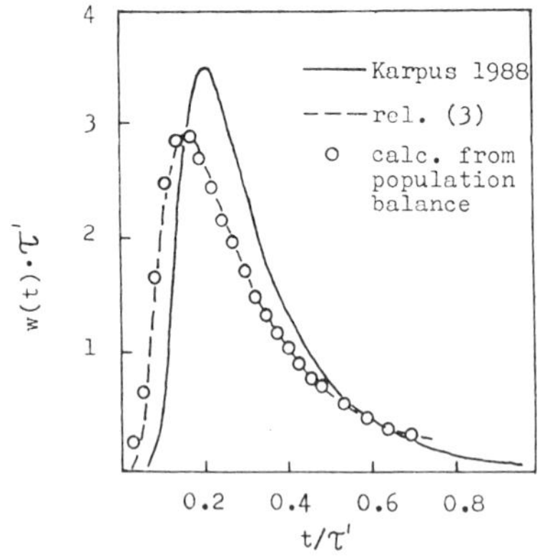

Figure 1. Dependence of Diffusive Ionisation Rate on Time

$$w(t) = \frac{0.28}{\tau'}\left(\frac{t}{\tau'} + 0.04\right)^{-4} \exp\left(-\frac{0.8}{t/\tau' + 0.04}\right), \quad \tau' = \tau(n_m - n_o^{*})/n_m \qquad (3)$$

Thus the "cutting" of the Rydberg states manifold at n_m reduces the characteristic time τ to τ' according to relation (3).

4. RESULTS

First of all, the drift time b/v_d needed for the integration (1) and its dependence on vapour density were determined. For that case potassium atoms were photoionised from the ground state by the help of the same laser. Then the ionisation rate is a Dirac delta function $\delta(t)$ and the integral (1) is a step function with the step width b/v_d. As was mentioned above the laser spot in the excitation region is not a point, therefore experimental J is not a neat step function, and we determined b/v_d according to half of step height in the falling edge of J. So it was obtained that the drift time increases from 1.5 μs to 3.5 μs as N increases from $5 \cdot 10^{13}$ cm^{-3} to $2 \cdot 10^{14}$ cm^{-3}.

The experimental ionic signal (or amount of the ions) versus delay time T is given in the Figure 2. The density of excited atoms in our experiment was about $N_R^0 \sim 10^8$ cm^{-3}. To exclude the influence of possible excited atoms pair collisions on kinetics, we have increased N_R^0 several times by increasing the energy of an exciting pulse. No dependence of kinetics on N_R^0 was found.

After the excitation, the small amount of Rydberg atoms diffuses out of the excitation and drift regions and is quenched on the internal surfaces of the cell. The surviving probability of excited atoms p(t) was expressed according to analysis of Kaulakys and Ciziunas (1985) in the paper of Švedas et al. (1989). This probability was inserted into integral (1) in the numerical calculations of the J(T). The correction due to this improvement is small because of $p(t) \simeq 1$ in our conditions. Concretely for Figure 2, p = 1 for $t \leq 12$ μs and p decreases to 0.92 at t = 30 μs.

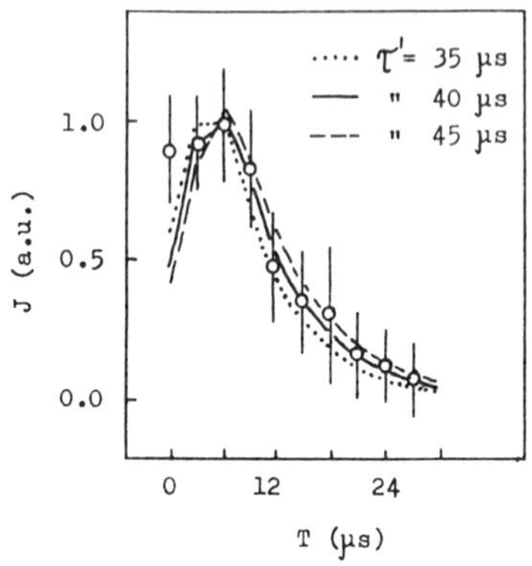

Figure 2. Amount of Registered Ions J vs Extracting Pulse Delay Time T. $N = 7.6 \cdot 10^{13}$ cm^{-3}.

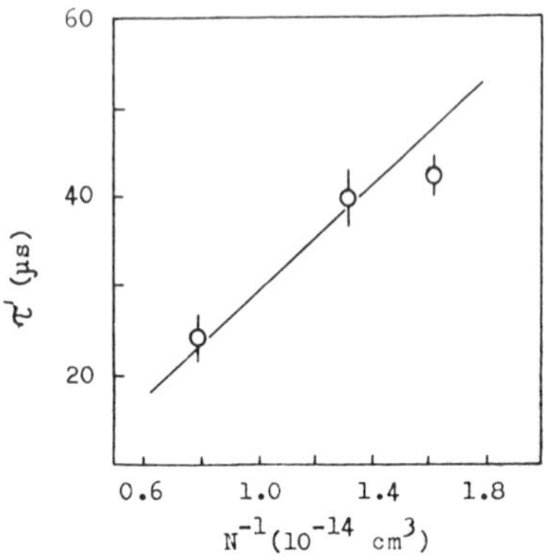

Figure 3. Dependence of Characteristic Diffusion Time τ' on Density of Vapour Atoms N.

The J(T) dependence calculated according to integral equation (1) and rate equation (3) with various characteristic time τ' values are plotted on Figure 2. The time τ' = 40 μs is a best fit for the experimental dependence. For T = 0 μs, the experimental J value exceeds the calculated one for all τ' values varied. Probably this discrepancy is caused by potassium dimer K_2 photoionisation by the λ = 286 nm radiation. The additional dimer photoionisation with the rate $\delta(t)$ gives an atom photoionisation with the additional step function disappearing after the time b/v_d (b/v_d = 1.8 μs at N = 7.6 $\cdot 10^{13}$ cm^{-3}) and the experimental J not being disturbed.

The different τ' values versus N^{-1} are plotted on Figure 3. The straight line is drawn according to experimental τ'. This line gives a characteristic time dependence on the vapour density,

$$\tau' (\mu s) \simeq 29/N\ (10^{14}\ cm^{-3}) \qquad (4)$$

From relations (2) and (3) for τ and τ', and relation (4), one gets the elastic electron-potassium atom scattering cross-section value $\sigma \simeq 8 \cdot 10^{-13}$ cm^2.

The diffusive ionisation rate of the potassium 34P state and its dependence on time and density of vapour atoms can be obtained by the insertion of relation (4) into (3). According to relations (2) and (3) for τ and τ', the expression of the diffusive ionisation rate (3) can be extended to the other Rydberg states and to the other atomic vapour or buffer gases.

REFERENCES

Herrmann, P. P., Hoffnagle, J., Schlumpf, N., Taqqu, D., Telegdi, V. L., and Weis, A., 1986 J. Phys. B: At. Mol. Phys. 19 1675

Karpus, V., 1988 J. Phys. B: At. Mol. Opt. Phys. 21 2311

Kaulakys, B., and Čiziunas, A., 1985 Diffusive flows and diffusive ionisation mechanism of Rydberg atoms (Preprint of the Institute of Physics Lithuanian Academy of Sciences, Vilnius 1985, in Russian).

Kaulakys, B., and Švedas, V., 1987 J. Phys. B: At. Mol. Phys. 20 L565

Niemax, K. 1983 Appl. Phys. B: Photophys. Laser Chem. 32 59

Švedas, V., 1988 J. Phys. B: At. Mol. Opt. Phys. 21 301

Švedas, V., Kuprionis, Z., Gontis, V., and Kaulakys, B., 1989, The investigations of kinetics of the collisional ionisation and redistribution of excitation over the atomic Rydberg levels (Deposed in Lit NIINTI at October 12, 1989, N2432-Li, in Russian)

Inst. Phys. Conf. Ser. No 114: Section 4
Paper presented at RIS 90, Varese, Italy, 16–21 September 1990

Resonance ionization spectra as a reflection of excited state dynamics

Philip M. Johnson
Department of Chemistry, State University of New York, Stony Brook, NY 11794

ABSTRACT: The MPI and multiphoton photoelectron spectra of CO_2 contain many unusual features which we have not come to expect in the multiphoton spectroscopy of molecules. These include:

1) Above threshold absorption at relatively low laser powers.
2) Sharp structure in the MPI spectra due to autoionizing resonances.
3) Long vibrational progressions in photoelectron spectra, due to ionization of dissociating molecular levels.

Pyrazine (1,4-diazine) is a classic molecule for the study of singlet-triplet coupling, being an intermediate case with respect to radiationless transitions. Two color threshold photoelectron spectra of S_1 are able to distinguish between singlet and triplet state components. Lifetimes of these indicate that a fraction of the molecules decouple themselves from the singlet.

1. INTRODUCTION

Resonance ionization processes are a reflection of the dynamic as well as the static properties of atomic and molecular species. Due to the sequential or quasi-sequential nature of photon absorption in resonant multiphoton events, the lifetimes of the intermediate states play an essential role in the overall cross-sections if they are short enough to be competitive with subsequent photon interactions. In molecules this is particularly important because there are many dissociative and other radiationless pathways which can contribute to a competitive channel. Many expected resonances are missing in MPI spectra because of the dissipative nature of numerous excited states. In theory, and sometimes in practice, the missing resonances can be regained by increasing the laser power so that the ionization process can compete with the dissipative channels. Under those conditions it should be possible to obtain information about the nature of the dynamics of the intermediate state from the multiphoton ionization process. This will involve looking at not only the ionization cross-section but also other observables such as the kinetic energy of the ejected electrons and possibly the distribution of fragment ions produced in the ionization event. The latter may not be indicative of the dynamics of neutral intermediates, however, but may be due to photofragmentation of the ion subsequent to the multiphoton ionization event.

Some channels which compete with ionization from an intermediate state lead to species which also can be ionized. Under those conditions the overall ionization cross-section may not be significantly altered by the competing process, but other characteristics of the ionization, such as the final state of the ion, or the symmetry of the outgoing electron wave, may be changed. Examples of such changes are intersystem crossing in larger molecules and collisionally induced m_J switching in smaller systems.

Whether the ionization amplitude is affected or not, the time scales of the dynamic events which alter the ionization path can vary over a large range from the femtoseconds of dissociation to the microseconds of some radiationless transitions in large molecules. When the competing channel has a time scale shorter than the laser pulse length, the kinetics of the ionization are intimately tied into the precise nature of the laser pulse. For time scales longer than the laser pulse, pump-probe ionization schemes in which one laser prepares a state while another does the ionization provide a particularly simple method for investigating the dynamics of the intermediate state. Here we will discuss examples from each of these regimes.

2. CO_2: A COMPETITION BETWEEN IONIZATION AND DISSOCIATION

Whereas the ground state of CO_2 is quite stable, almost all of its excited electronic states dissociate or predissociate to produce CO and O. The molecule therefore provides a very good object of study if one wishes to investigate the influence of rapid dissipative processes on multiphoton ionization. Indeed the excited state dissociation competes so effectively with ionization that only a few resonances are seen in the multiphoton ionization spectrum, and these require a photon flux considerably greater than is normally used for MPI spectra. The first information one gets from this is which are the more stable excited states, since those are the ones that appear. Secondly, one can examine the REMPI photoelectron spectra to see what effect an unstable intermediate will have on the ionization channel.

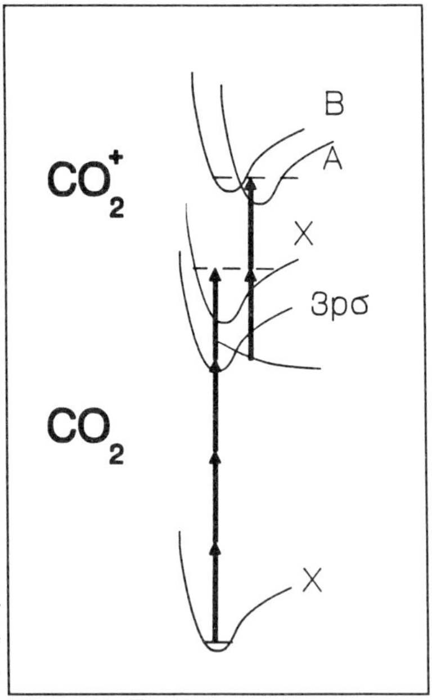

Figure 1 Excitation scheme

The (3+1) MPI spectra of CO_2 was studied in the range 275 to 338 nm[1] in a process shown schematically in figure 1. A new Rydberg state, $3p\pi_u\ ^1\Delta_u$ was observed along with several others of the $3p\pi_u$ complex. The strongest absorption band of the one photon spectra, $3p\pi_u\ ^1\Sigma_u$, was not observed in the MPI spectra apparently because the predissociation rate of this state is higher than the ionization rate even at the relatively high laser intensities (7-18 mJ/pulse) used in this study. The nf Rydberg series was also observed in three photons, accompanied by some very sharp structure to the blue of the 4f and 5f resonances, as shown in figure 2. Extensive simulation of the expected three photon spectra, including l-uncoupling and the Renner-Teller effect of the ionic core, were unable to account for much of this sharp structure.

Upon further analysis it was found that the sharp structure appearing with the three photon resonances of the 4f and 5f Rydberg states was due to four photon resonances with autoionization states in the Rydberg series converging to the second and third ionization limits.[2] Apparently the predissociation of the three photon resonances diminishes their intensity so that the four photon resonances are of comparable magnitude, and the f states act as near-resonances for the four photon absorption. Accidentally the energies of the photons used to excited the f levels are nearly those needed to cause the B-X transition of the ionic core. Therefore the autoionization states seen in the spectra are f Rydbergs converging to the A and B states of the ion. Ionic A and B states are quite strongly mixed and it is likely that the B component provides the oscillator strength while the A component gives the structure and the narrow linewidth because the principle quantum numbers are relatively high for A resonances.

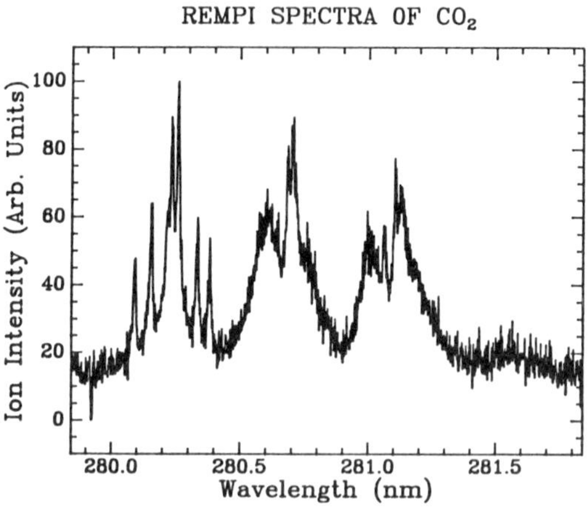

Figure 2 Structure near the 5f resonance

From the photoelectron spectra obtained when resonant with the various long-lived three photon levels, we see that the ionization is not a simple process proceeding through well defined bound levels. Since Rydberg states have the same geometric structure as the ion, one would expect that the vibrational structure in the photoelectron spectra would consist primarily of the 0-0 transition and this is what is seen for some of the resonances. For others, one sees long progressions in either the symmetric stretching mode of the molecule or in two quanta of the asymmetric stretch, depending upon the symmetry of the states involved in the transition. In figure 3 the structure associated with being resonant with the 3pσ state while terminating in the ionic X state shows the mentioned progressions in both the symmetric stretch and the asymmetric stretch. We believe that this is a manifestation of the fact that the molecule is ionizing while in the process of dissociating, as depicted on the right of figure 1, or at least relaxing to a valence state which would subsequently dissociate.

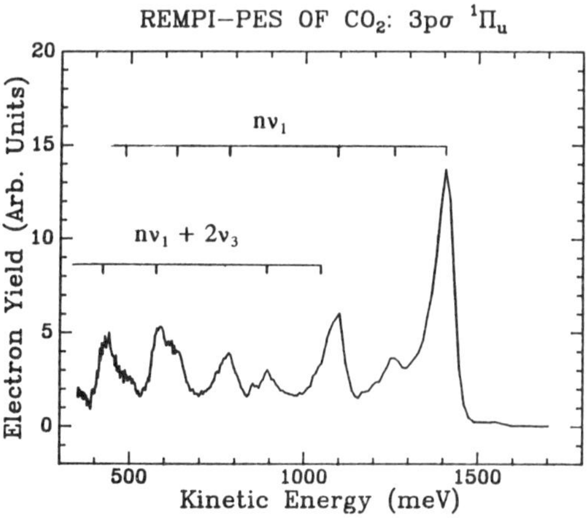

Figure 3 PES with long progressions

The rapid ionization event thus provides a "snapshot" of continuum vibrational function of the molecule as it is falling apart. Carbon dioxide's dissociation mode involves primarily the asymmetric stretch, but there is also a significant mixture of the symmetric stretch in the early part of the reaction coordinate.

The dynamics of the ionization step are also illuminated by the photoelectron spectra. Figures 4 shows spectra when resonant with a 4f level (upper) and when terminating on an autoionizing resonance (lower). In both cases, the low energy peaks represent additional photon absorption occurring in the continuum because of long-lived resonances there. The additional photon absorption allows the ionization to terminate on the C state of the ion. We call that resonantly enhanced above threshold absorption or REATA to distinguish it from the more common above threshold ionization (ATI), which is based on the pondermotive effect. ATI is not expected to occur under our experimental conditions because the laser power is too low to produce a significant pondermotive force. What is of interest in these spectra is that the resonance with the long-lived Rydberg autoionization state produces a much enhanced REATA. The high principal quantum number of the f electron (about 18), coupled with the high angular momentum, creates an orbital which has poor interaction with the excited core. Therefore enough time exists for an additional core excitation, which would take place on a picosecond time scale. When the photon energy is resonant with the f states at the three photon level, at the four photon level it hits states converging to B which are much more diffuse, and presumably have shorter lifetimes. REATA is therefore diminished.

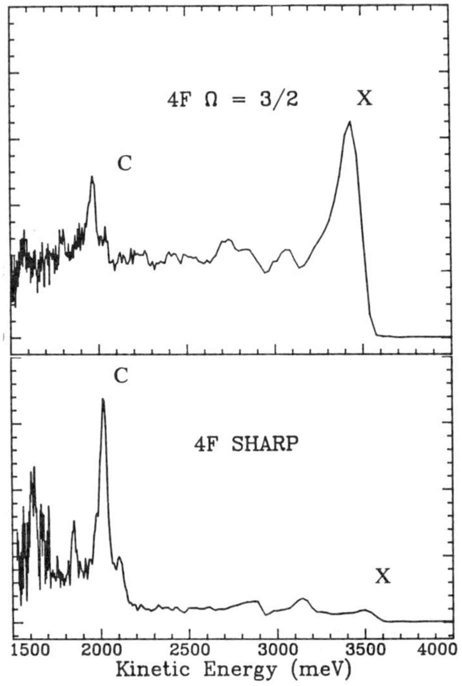

Figure 4 PES spectra showing the enhanced REATA when on an autoionizing resonance.

3. FIELD IONIZATION SPECTROSCOPY TO EXAMINE THE DYNAMICS OF PYRAZINE

The radiationless transitions of aromatic molecules provide a case where excited state dynamics are slow enough to be followed by nanosecond lasers in a pump-probe experiment. This can be done by measuring the total ion yield as a function of time delay between the two lasers[3], but more subtle techniques involve conventional photoelectron spectroscopy[4] field ionization photoelectron detection.

If one tunes a laser to be resonant just below a transition from a bound state of a molecule to a state of its ion, it is possible to populate very high Rydberg states which have long lifetimes. It is also possible to ionize these states with a small pulsed electric field, but only if there is a negligibly small energy gap to the resonant ionic state. Exploiting this, it is possible to map out the vibronic states of an ion by scanning a laser through the ionization continuum, detecting these almost-resonances by means of electrons produced by a delayed electric pulse (<1 V/cm). Any fast electrons produced by normal photoionization are thrown away. This technique has been popularly known as zero kinetic energy photoelectron spectroscopy[5]. The resolution obtained in this type of spectrum greatly exceeds that in ordinary photoelectron spectroscopy and problems created by background contaminants and contact potentials are eliminated, provided that one can tune a laser through the appropriate wavelength region.

We have applied this to pyrazine, a molecule in which the mixing between the S_1 and T_1 states is strong and the exact nature of the mixing of current interest. High resolution spectroscopy has shown that there is substructure inside each S_1 0-0 rotational line which is representative of transitions to mixed singlet-triplet states called "molecular eigenstates."[6] There is a disagreement whether there is also a broad background absorption which represents states with shorter lifetimes. Since ionization is not dependent upon spin selection rules in the usual sense, it is possible to probe either the singlet or triplet component of a mixed state in contrast to most spectroscopic methods. In addition to spectral information about the resonant and ionic states we obtain dynamic information about the resonant state by varying the delay between the pump and probe lasers.

The singlet component of the field ionization spectrum, shown at the left of figure 5 with almost laser limited resolution and excellent signal to noise, enables an extensive vibronic analysis of the pyrazine ion. Pumping different vibrations in S_1 produces even more vibrational information in the spectrum. The lifetime of the intermediate states when looking at the singlet components is the same as when measuring the total ion yield. A triplet component appears at higher photon energy because the different Franck-Condon factors in the hot triplet favor ionization to ionic states approximately at the energy of the S_1-T_1 gap. In this continuous part of the spectrum intermediate state lifetimes are shorter than when measuring ion production and monotonically decrease with increasing ionization photon energy, as shown in figure 6. Since it is unlikely that triplets which retain their coupling to the singlet would have other than the representative composite lifetime measured for the singlet component, this indicates that there is vibrational relaxation (IVR) in at least some of the states crossing over to the triplet.

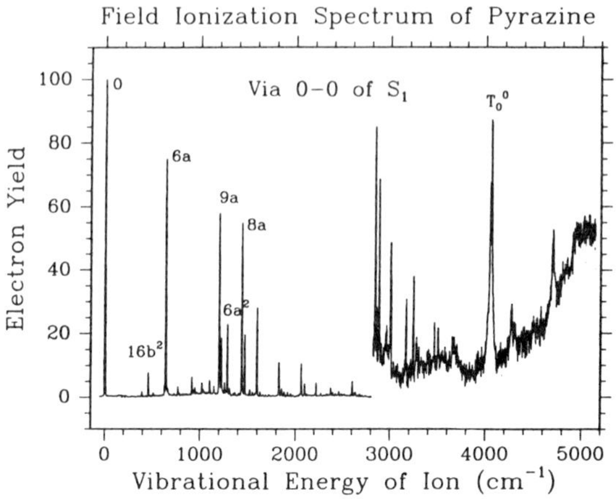

Figure 5

The IVR produces a statistical distribution of vibrational states with a recurrence time to the singlet longer than competing dissipative processes such as intersystem crossing back to S_0. The lifetimes of these relaxed states are shorter because promoting modes for intersystem crossing are populated to a greater extent.

4. CONCLUSIONS

The multiphoton ionization process contains within it much information about the dynamics of resonant intermediate states. However, it is usually necessary to discriminate the energies of the outgoing electrons to extract much of this information. The photoelectron or field ionization spectra are able to reflect changes in the intermediate over time scales ranging from picoseconds to microseconds, even using nanosecond lasers. Thus we are able to follow processes ranging from photodissociation to internal energy rearrangement.

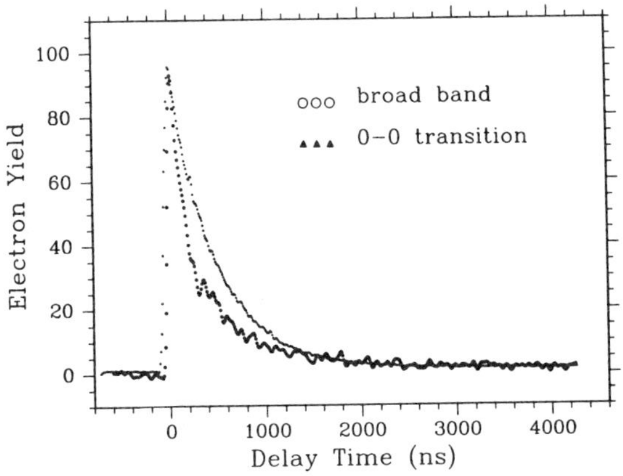

Figure 6 Intermediate state lifetimes at different probe resonances using field ionization detection.

5. ACKNOWLEDGEMENTS

The work reported here has been carried out by Ming Wu and David Taylor (CO_2), and Susan Hillenbrand and Lang-Chi Zhu (pyrazine). It has been supported by a grant from the Department of Energy, ED-FG02-86ER13590.

6. REFERENCES

1. Ming Wu and P. M. Johnson, J. Chem. Phys. **91**,7399 (1989).

2. P. M. Johnson and M. Wu, J. Chem. Phys., To be published.

3. J. Knee and P. Johnson, J. Phys. Chem. **89**, 948 (1985).

4. S. Hillenbrand, L. Zhu, and P. Johnson, J. Chem. Phys. **92**, 870 (1990).

5. K. Muller-Dethlefs, M. Sander and E. Schlag, Z. Naturforsch. **39a**, 1089 (1984).

6. B. van der Meer, H. Jonkman, J. Kommandeur, W. Meerts, and W. Majewski, Chem. Phys. Lett. **92**, 565 (1982).

Inst. Phys. Conf. Ser. No 114: Section 4
Paper presented at RIS 90, Varese, Italy, 16–21 September 1990

Autoionization in naphthalene trimer clusters

John Wessel

Chemistry and Physics Laboratory, The Aerospace Corporation, Los Angeles, CA 90009

ABSTRACT: Direct spectroscopic evidence is presented for biexcitonic 2-photon ionization of small molecular clusters. The process resembles resonant multiphoton autoionization of 2-electron atoms.

1. INTRODUCTION

Whetten and coworkers[1] concluded that large benzene clusters ionize by an exciton annihilation mechanism, involving sequential excitation of two or more molecules in the cluster. They described the ionization process in terms of diffusion and annihilation involving a pair of excitonic states, as depicted in Figure 1b. This accounted for the anomalously high efficiency observed for cluster photoionization. Previously, biexcitonic annihilation was identified in bulk molecular crystals[2] and indirect evidence was obtained for biexcitonic autoionization in gallium arsenide[3,4] Whereas biexcitons in semiconductor crystals are characterized by the large dimensions of Wannier excitons, tightly bound Frenkel excitons in molecular crystals are comparable to the size of small molecular clusters. We present direct spectroscopic evidence in support of a biexcitonic autoionization process for the naphthalene trimer cluster. The spectra suggest that the biexcitonic states of van der Waals clusters are attractive, but short lived. (A detailed description appears in the recent literature.[5])

The exciton annihilation mechanism shares characteristics of multiphoton excitation of autoionizing 2-electron atoms, such as strontium[6] and barium.[7] In barium, highly efficient ionization occurs when two $6s^2$ electrons are sequentially excited in the overall 3-photon process $6s^2$--$6p^1s^1$--$20s^16s^1$--$20s^16p^1$, leading to a final state above the lowest ionization potential. Configurational mixing with the lower energy ionization continua results in rapid autoionization, however the mixing is insufficient to destroy the 6s--6p character associated with the third photon resonance, although this resonance is shifted by configuration interaction and broadened relative to the ground state based resonance. This leads to a characteristic power dependent spectral signature. However, ionization excitation spectra for large molecular clusters and bulk crystals are dominated by the 1-photon ground state absorption spectra, because the majority of second photon interaction events occur at substantial distance from the original site.

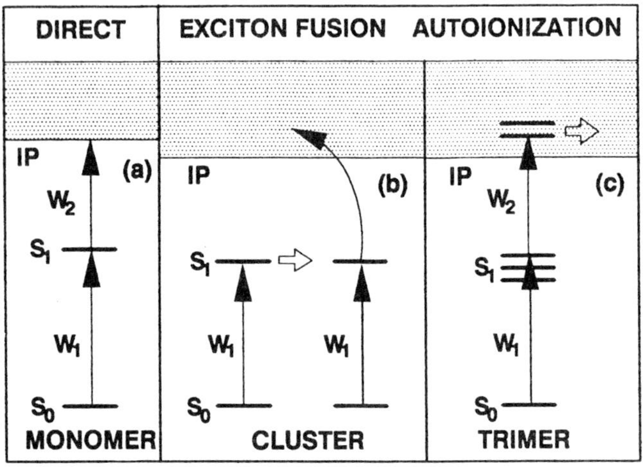

Fig. 1. 2-photon ionization processes include (a) resonant (1 + 1) ionization, (b) exciton annihilation, and (c) resonant 2-photon autoionization, where the multiplet of levels in S_1 represent exciton components and the discrete levels in the ionization continuum are doubly excited trimer states undergoing autoionization.

The electronic spectra of trimer clusters of benzene[8] and naphthalene[9] are characterized by exciton-like splittings in electronic transitions between the ground (S_0) and first excited electronic state (S_1) of the neutral clusters. Each parent monomer state contributes a cluster state energetically related to the parent monomer vibronic state. Resonant interaction (multipolar and possibly higher order effects) induce splittings (related to factor group splittings of molecular crystals) of from 1 to 20 cm^{-1} between related components. The 1-color spectrum of the isotopically pure protonated naphthalene trimer (HHH) measured in the region of the strong vibronically induced transition (denoted 8_0^1, corresponding to excitation of one quantum of vibrational mode 8 of b_{3g} symmetry in the S_1 state) is shown in Figure 2. Only one component is observed for this transition because the monomer transition moments are all parallel in the trimer structure (shown in the inset in Figure 2, as derived from a dipolar interaction analysis of the trimer spectrum). The trimer ionization potential (IP) is not yet established, however two photons in resonance with 8_0^1 are expected to exceed the trimer IP (which is below the monomer IP of 65,665 cm^{-1}).

2. RESULTS

Our experimental investigations are based on 2-color pump probe spectroscopy. Naphthalene clusters, cooled below 100 K rotational and vibrational temperature in a supersonic expansion are excited to the S_1 states by interaction with a photon of wavelength W_1. This wavelength is selected to coincide with a strong discrete cluster transition. A probe beam of wavelength W_2, time delayed by 10 ns (W_1 and W_2 pulse widths are 5 ns), is scanned across the S_1--S_0 spectrum and the region of the IP--S_1 threshold. The combined-beam ionization signals are analyzed for ionization in excess of the sum of

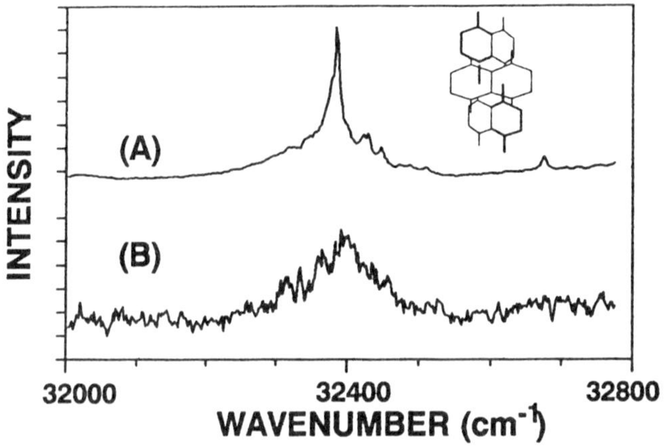

Fig. 2. Observed cluster ionization spectra for 8_0^1 of the isotopically pure protonated trimer (HHH) recorded (a) with W_1 only (1-color) and (b) the difference spectrum, where the nonoverlapping beam spectrum is subtracted from the overlapping beam spectrum.

the sum of independent W_1 and W_2 beam signals. This represents processes requiring a photon from each of the two sources. In Figure 2 the 1-color spectrum of 8_0^1 for the isotopically pure h-8 cluster (a), is compared to the $W_1 + W_2$ difference spectrum, (b). The difference spectrum is centered near the 1-color transition, and it is substantially broadened relative to the 1-color spectrum. Ionization is enhanced whenever both W_1 and W_2 beams are overlapped and tuned to the vicinity of S_1--S_0 vibronic transitions. This is distinctly different from typical ionization step spectra (IP--S_1) observed for the monomer. In this case the signal rises abruptly to a plateau when the total energy exceeds the ionization potential.

3. DISCUSSION

The new spectral characteristics suggests the ionization mechanism involves two ground state based S_1--S_0 transitions. 1-color and 2-color spectra were also measured for the isotopically mixed DHD trimer. The 1-color spectrum consists of three principal sharp components that are directly related to the monomer transition. (Only one conformer is observed in the DHD and HDH spectra.) In the 2-color experiment, all three sharp components are attenuated when W_1 is resonant with the H transition. The presence of W_1 clearly alters the W_2 spectrum, revealing broadening and a small red shift. A red shift ranging from 30 to 80 cm^{-1} is estimated based on intensity centroids of difference spectra of the origin band. The enhancement always occurs near the corresponding 1-color transition. The reduction in all three DHD components demonstrates the coupled nature of the trimer transitions.

The 2-color ionization spectra were modeled by incoherent population rate equations, with provision made for excitation by a W_1 beam at constant excitation rate, combined with a spectrally varying excitation rate for the W_2 beam. The predicted spectra closely simulate the observed 2-color spectra. At low W_1 and W_2 fluences, sharp spectra are predicted, and broadening occurs as W_2 increases. With increased W_1 fluence, the amplitude of the sharp transition decreases and rounded shoulders appear on the wings, extending approximately 30 cm^{-1} from line center. Experimental results fully demonstrate these predicted characteristics.

The above considerations support the resonant 2-photon autoionization mechanism for trimer ionization. The ground state based vibronic transitions are strong, with optical cross sections on the order of 10^{-13} cm^2, versus a cross section of 1.6 x 10^{-19} cm^2 for the monomer S_1 ionization transition (IP--S_1). Thus, the ionization step limits the overall rate of (1 + 1) ionization in the monomer, whereas large cross sections govern the transitions in the case of trimer autoionization. The broadening and shift of the combined W_1 + W_2 spectrum is also in accord with the model, implying an autoionization lifetime on the order of 100 fs, which is comparable to autoionization rates measured for Rydberg states of the monomer that are isoenergetic with cluster biexcitonic states. The red shift indicates excitonic interactions are attractive as in semiconductor biexcitonic states, however the lifetime determined molecular biexciton width is as large as the binding, in contrast to semiconductors, where binding exceeds linewidth. 1-color spectra obtained by this technique are governed by a convolution of spectra for the first and second photon, leading to a characteristic broad, intensity dependent pedestal. We conclude that small size homomolecular van der Waals clusters will ionize by the autoionization mechanism, provided the intermediate state is stable on the time scale of the excitation pulse.

*This work was supported by the Department of Energy, Office of Health and Environment, Grant No. DE-FG03-86ER60391.

---

1. K. E. Schriver, M. Y. Hahn, and R. L. Whetten, Phys. Rev. Lett. **59**, 1906 (1987).
2. S. Arnold, K. T. Hsieh, and M. Pope, Phys Stat. Solidi **B94**, 263 (1979).
3. A. Maruani and D. S. Chemla, Phys. Rev. **B23**, 841 (1981); D. S. Chemla and A. Maruani, Prog. Quantum Electron. **8**, 1 (1982).
4. Hanamura and H. Haug, Phys. Rep. **33C**, 210 (1977); V.B. Timofeev in Excitons, edited by E. I. Rashba and M. D. Sturge, (North Holland Press, Amsterdam, 1987) page 229; also M. Lampert, Phys. Rev. Lett. 1, 450 (1958).
5. J. E. Wessel, Phys. Rev. Lett. **64**, 2046 (1990).
6. W. E. Cooke, T. F. Gallagher, S. A. Edelstein, and R. M. Hill, Phys. Rev. Lett. **40**, 178 (1978); P. Lambroupoulos, X. Tang, P. Agostini, G. Petite, and A. L'Huillier, Phys. Rev. **A38**, 6165 (1988).
7. W. E. Cooke and T. F. Gallagher, Phys. Rev. Lett., **41**, 1648 (1978); L. D. Van Woerkom and W. E. Cooke, Phys. Rev. **A39**, 3326 (1989); J. G. Story, E. G. Yap, and W. E. Cook, Phys. Rev. **A39**, 5127 (1989).
8. K. O. Bornsen, S. H. Lin, H. L. Selzle, and E. W. Schlag, J. Chem. Phys. **90**, 1299 (1989).
9. J. A. Syage and J. E. Wessel, J. Chem. Phys. **89**, 5962 (1988); J. Wessel and J. A. Syage, J. Phys. Chem. **94**, 737 (1990).

Inst. Phys. Conf. Ser. No 114: Section 4
Paper presented at RIS 90, Varese, Italy, 16–21 September 1990

Multiphoton ionization mass spectrometry: Technique, results and prospects

G.R. Kinsel, J. Lindner, C. Köster, K. Zipperer, J. Grotemeyer*

Institut für Physikalische und Theoretische Chemie, TU München,
Lichtenbergstr. 4, D-8046 Garching, Germany

ABSTRACT : Laser radiation in the UV-regime can be used to generate cations from biological molecules by different experimental setups. Laser Desorption of neutral molecules followed by multphoton absorption leads to mass spectra which show adjustable fragmentation pattern depending on the laser intensity. The mass of the molecules investigated by this method is limited to about 6.000 Da at the moment. Fairly large molecules, masses up to 300.000 Da are sucessfully investigated by the matrix-assisted-simultaneous laser desorption and ionisation. Some results of these two methods are described and discussed.

1. INTRODUCTION

The adoption of lasers with different wavelength, such as in the infra-red (IR), visible (VIS) and ultraviolett (UV) regime, to mass spectrometry yielded in a large number of applications. Beside the use in ion activation (photodissociation) [1] lasers have been introduced throughout the years also in ion production. First applications were dealing with simultaneous desorption and ionization [2]. In recent years this technique has found large interest in the desorption of large bioorganic ions, such as intact proteins, enzymes and nucleotides, from a matrix [3].

An alternative method for ionization uses multiphoton absorption of already gas phase based molecules [4]. Through its features multiphoton ionization (MUPI) added some further opportunities to mass spectrometry. These are the possibilities of *soft ionization* with the formation of mainly molecular ions, *tuneable fragmentation* with an easy adjustment of the degree of fragmentation and the *wavelength dependence* of the mass spectrum. For these investigations it is necessary to separate the desorption from the ionization in space and time. To reach this requirement different techniques have successfully been examined [5], which incorporates laser desorption and liquid introduction of neutrals.

2. EXPERIMENTAL SETUP

The mass spectrometer system used for the study consists of three different parts: the laser desorption unit, the laser ionization source and the **Re**flectron-**T**ime-**o**f-**F**light (RETOF) mass spectrometer. For the investigations presented here either a home-build system or a Bruker TOF-1 have been used. In our experimental setup neutral molecules are transformed into the gas

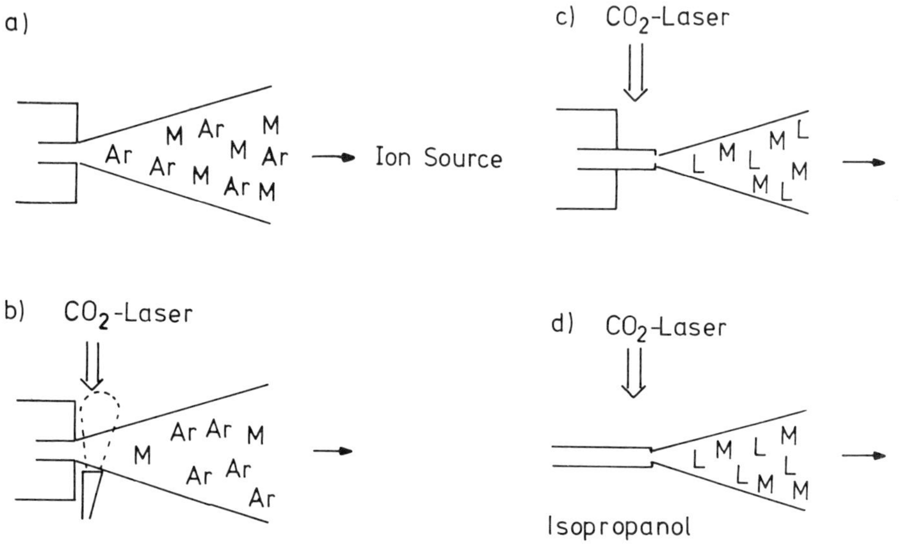

Fig. 1. Principles of the production of neutral molecules for multi-photon ionization.

phase separately from the ionization step. In the case of the matrix-assisted laser desorption-ionization the ion source of the mass spectrometer system has been rebuild in a way that the probe tip is mounted directly in the ion source. The laser beam is focussed onto the surface. Therefore simultaneous desorption and ionization takes place. The complete experimental setup is described in larger detail elsewhere [6].

3. PRODUCING NEUTRALS BY VARIOUS METHODS

To fulfil the needs of the mass spectrometer system in conjunction with the multiphoton ionization we have investigated different inlet methods, such as the laser desorption of neutrals, liquid introduction etc.. Some of the different principles are shown in figure 1.

Samples with sufficient volatility can be introduced by seeding them in the gas stream used for the supersonic jet before expansion (fig.1.a). For the desorption of molecules with low vapor pressures three different methods can be used. Figure 1.b shows the standard laser desorption procedure, while fig.1.c and fig.1.d. demonstrate possibilities of interfacing liquids and supercritical fluids.

To introduce involatile, thermally labile samples into the gas phase, different lasers can be used. Figure 2 shows a comparison of multiphoton ionization mass spectra of a pentapeptide, *methionine-enkephaline*, obtained through evaporation of neutrals by either a IR- or a UV-laser desorption. The results demonstrate clearly that it is possible to desorb neutrals by using UV-wavelength. Note also the equal intensities of the desorbed species as indicated by the axis shown in fig. 2. But it should also be noted that the appearance of most of the signals of smaller ion are also due to the desorption process and not linked to the ionization procedure. These signals

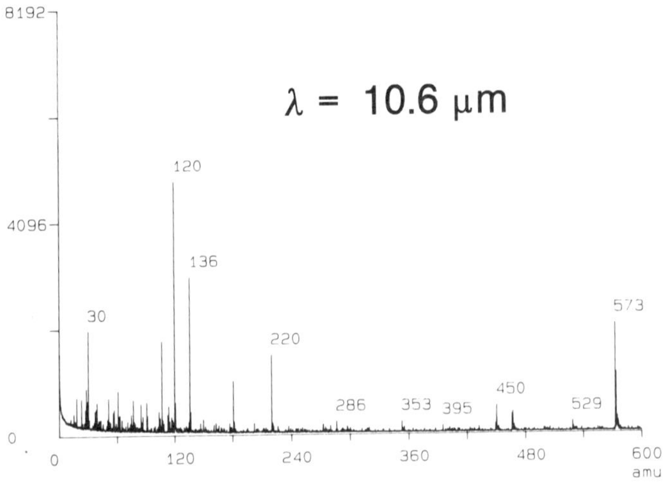

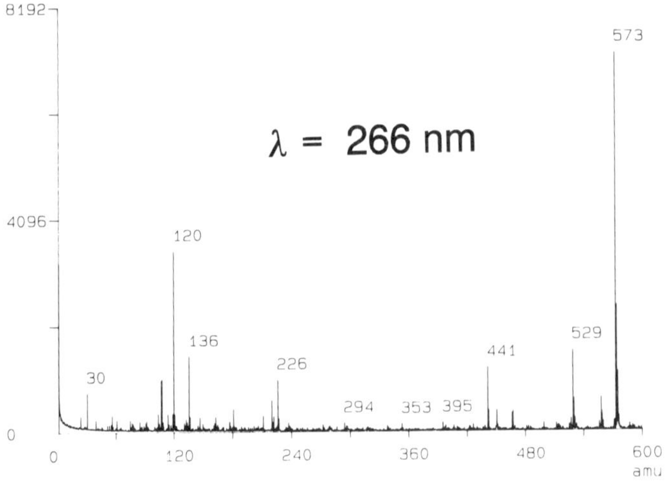

Fig. 2. Comparison of mass spectra from *methionine-enkephaline* obtained by laser desorption of neutrals with either 10,6μm or 266 nm. Note the equal absolute intensity of the signals.

stem from cleavage reactions of the neutrals.

To overcome these problems we have investigated the influence of matrices on the desorption process of molecules [7]. From the work of Hillenkamp et al.[8] it is known that matrices can have a drastic influence on both the desorption of ions and neutrals by the impact of an UV-laser beam. Knowledge about the desorption process is still rather limited. But a couple of experimental details may help to a better understanding of the process [9].

Figure 3 shows three multiphoton ionization mass spectra of a strong infrared absorbing sample, the fully protected mono-nucleotide *N-benzoyl-5'-O-dimethoxy-trityl-adenonsine-3'-(2-chloro-phenyl-cyanoethyl)-phosphate*.

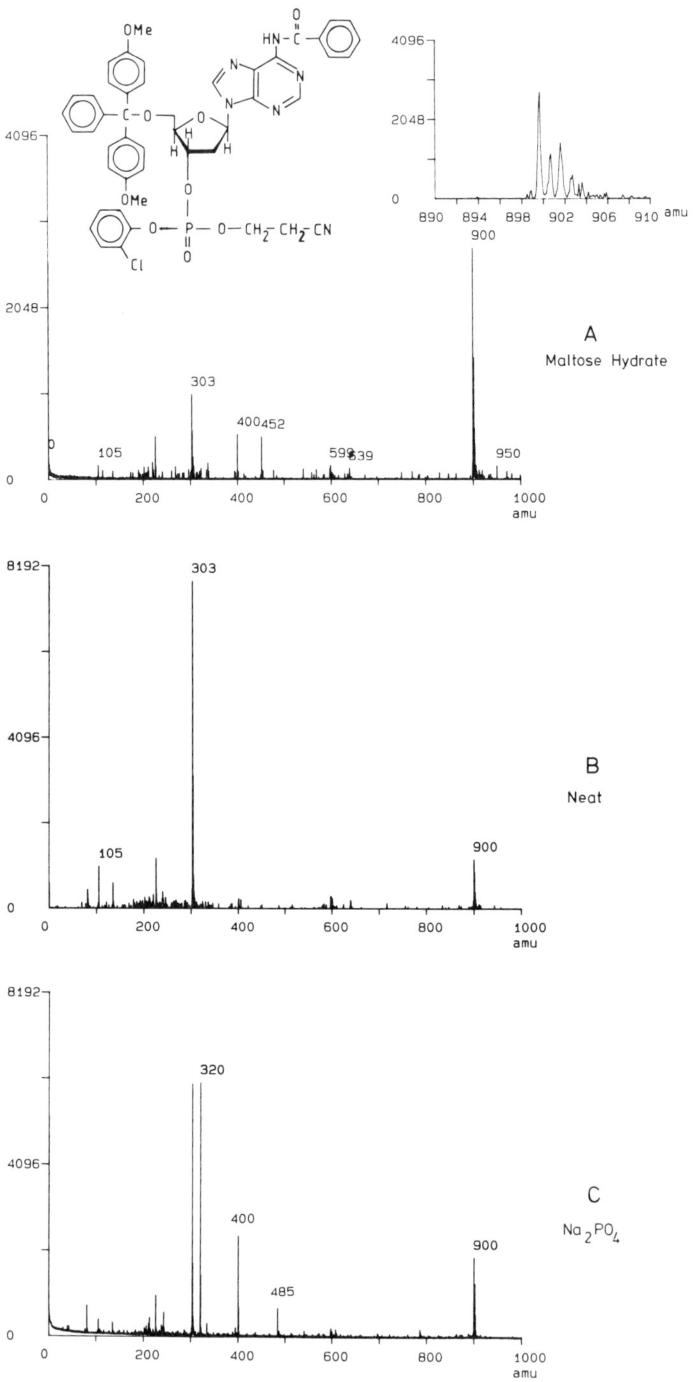

Fig. 3. MUPI mass spectra of the protected N-Bz-DMT-adenosine-phospate. a) with maltose hydrate as matrix, c) Na_2PO_3 as matrix added, b) neat.

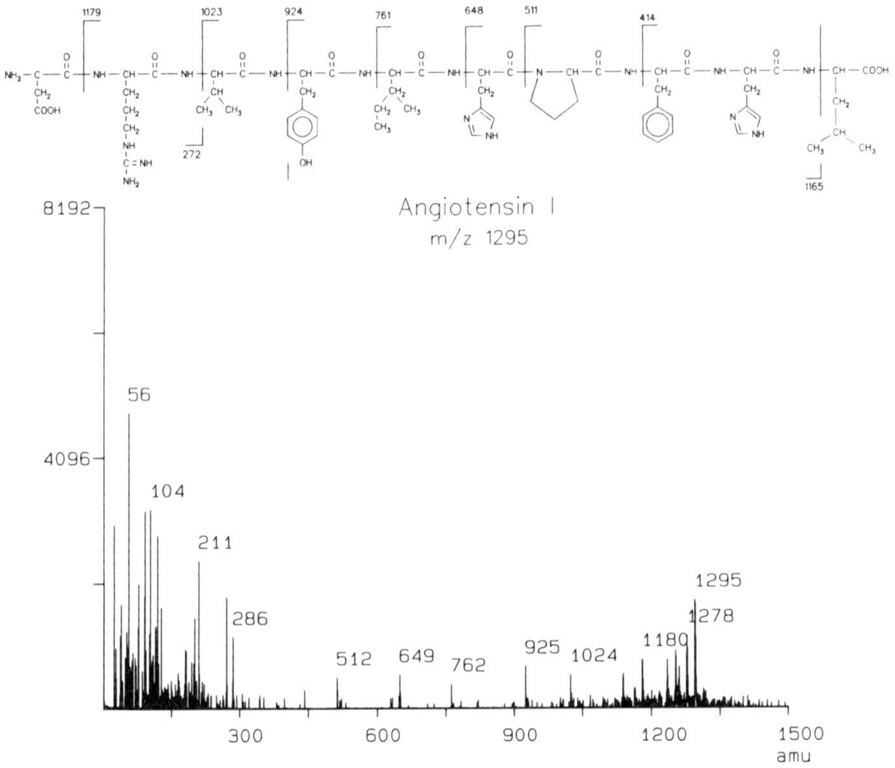

Fig. 4. Multiphoton ionization mass spectrum of *angiotensin I*.

Clearly the influence of the matrices is seen. If the sample is desorbed without any matrix material, a small molecular ion signal is detectable, while the observable fragment ions, m/z 303 especially, are mainly stemming from pyrolytic decomposition of the molecule. Mixing a stronger absorbing matrix with the sample yields in a change in the mass spectrum. At the same laser intensity, molecular ions are formed in higher yield, but still most of the signals, especially the new contributions, are due to a non-ionic fragmentation during the desorption process. Adding a semi-transparent matrix, such as the maltose hydrate, is finally leading to a multiphoton ionization mass spectrum, with the molecular ion as base peak. These results underline the importance of matrices in the desorption of neutral molecules from a surface.

4. MULTIPHOTON IONIZATION OF BIOMOLECULES

Details of the multiphoton ionization and/or fragmentation procedure have been published throughout the years in great detail [10]. During the last years MUPI-mass spectra have been presented for a large number of substances, such as drugs, peptides, nucleotides, aromatic compounds and others.

As example for the possibilities of multiphoton ionization applied to smaller biomolecules the mass spectrum of *Angiotensin I* is shown in figure 4. We and others have shown in our investigations that it is possible to sequence peptides in the gas phase. The ionization wavelength of the laser was tuned to 255 nm. The molecular ion is easily identified at 1295 mass units. As indicated in the figure Y'-fragments, according to the nomenclature of Fohlman and Roepstorff [11], can easily be detected throughout the mass spectrum. This behaviour found as well in the mass spectra of the *methionine-enkephaline* and other peptides allows the easy deduction of the primary structure of the peptide.

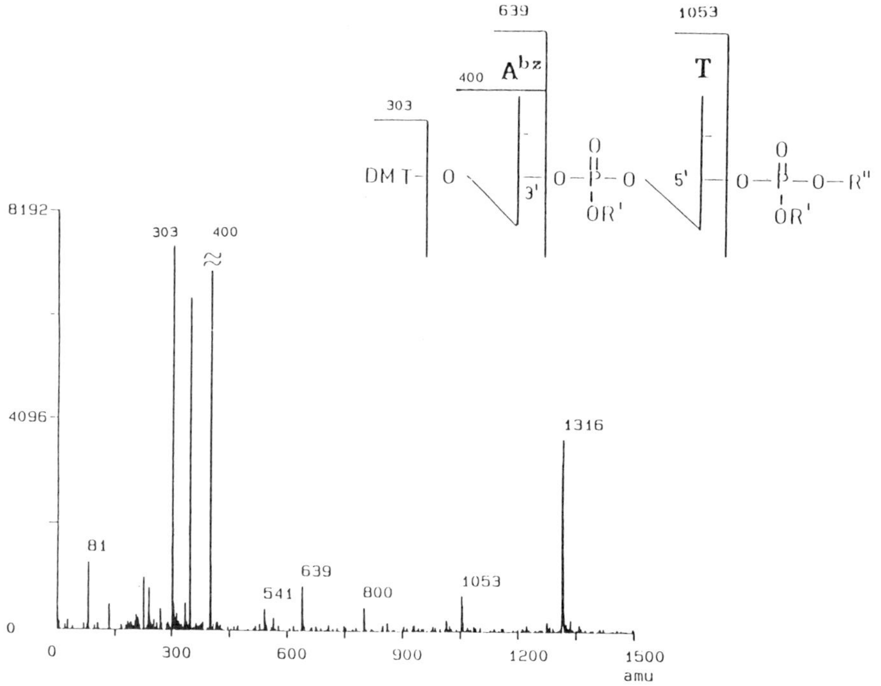

Fig. 5. MUPI-mass spectrum of the protected dinucleotide d(ApTp); Molecular weight 1314 Da.

Beside the investigations of peptides we also have obtained multiphoton ionization mass spectra of smaller protected nucleotides. As a result the mass spectrum of a fully protected deoxyribonucleoside 3'-phosphates d(ApTp) is shown in figure 5. The ionizing wavelength was tuned to the maximum of the molecular ion yield, which was found to be at 242.5 nm. Fig. 5 shows the partial hard MUPI-mass spectrum of d(ApTp) and the short hand structure. The molecular ion is formed with high intensity. As shown previously [12], a characteristic group of fragment ions includes the DMT-residue, the aromatic protective group of the $C_{5'}$-OH-group. Cleavage of the $C_{5'}$-O-bond leads to the formation of the DMT-fragment ion at m/z 303. The fragment ion at m/z 400 is assigned to the DMT-methylenfuran-ether, formed by cleavage of both the $C_{3'}$-O bond and the glycosidic linkage to the nucleic base. One of the most interesting fragment ions is formed by cleavage of the $C_{3'}$-O-bond of the first nucleoside with the charge remaining on the nucleoside. In this mass spectrum this sequence specific ion is found at mass 639. Thus the first nucleobase is unambigously determined as benzoyladenine.

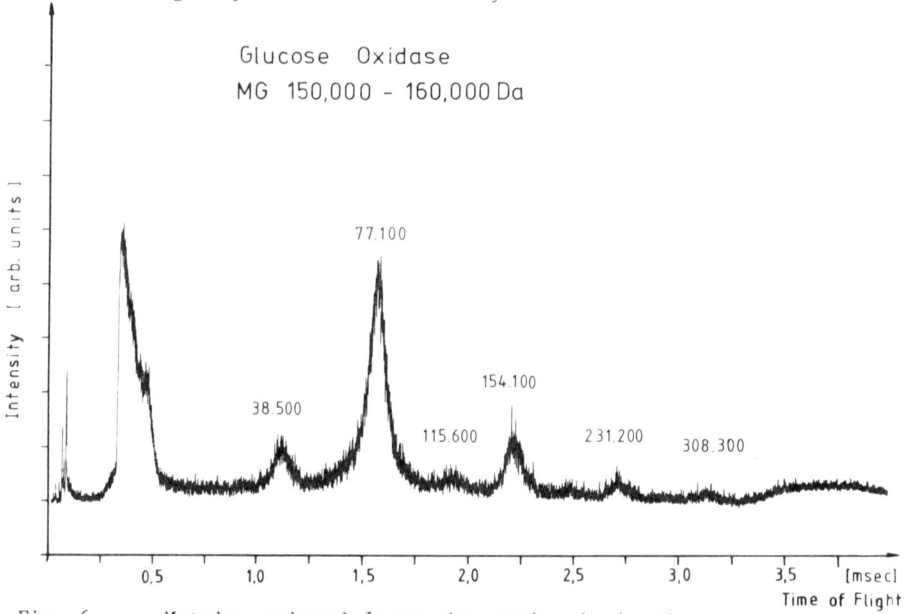

Fig. 6. Matrix-assisted laser desorption-ionization mass spectrum of *glucose oxidase*.

6. MATRIX-ASSISTED LASER DESORPTION-IONIZATION

During the last years Hillenkamp et al. as well as other groups [3] have demonstrated that it is possible to desorb, ionize and detect very large biomolecules in a mass spectrometer utilizing laser light. One essential of this technique is the use of a good matrix in which the sample is embeded as well as the exact ratio between sample and matrix. As shown in figure 6 it is possible with this technique to obtain mass spectra of intact enzyms like the *glucose oxidase* with a molecular weight of 154.100 Da. Even higher clusters of this substance up to mass 308.300 Da can be identified.

7. CONCLUSION

The combination of laser activation and mass spectrometry yield in clear and easy interpretable mass spectra of small and medium sized molecules. Although a large number of different molecules have been investigated, the question of a general application of this ionization method to all problems in analytical mass spectrometry remains open. The technique of soft ionization allows, in cases of resonant excitation, the nearly exclusive production of molecular ions, while simply increasing the photon density can lead to structure dependent signals. For larger biomolecules the technique of matrix assisted laser desorption-ionization seems to be appropiate.

7. ACKNOWLEDGEMENT

This work is supported by grants from the Deutsche Forschungsgemeinschaft (GR 917/1-2) and the Bundesministerium für Forschung und Technologie (13N 5307). The authors thank Prof. Dr. E.W. Schlag for his helpful discussions and his continuing interest in the multiphoton ionization mass spectrometry.

8. REFERENCES

1. For a recent overview see : Busch K.L., Glish G.L., McLuckey S.A., *Mass Spectrometry/Mass Spectrometry*, VCH Publisher Inc., Weinheim, New York, 1988.

2. Hillenkamp F., Unsöld E., Kaufmann R., Nitsche R.; 1975 *Appl.Phys.* **8** 341.

3. a) Karas M., Hillenkamp F.; 1989 *Anal. Chem.* **60** 2299.
b) Beavis R.C., Chait B.T.; 1989 *Rapid Commun.Mass Spectrom.* **3** 233.
c) Salehpour M., Perera I., Kjellberg J., Islamian M.A., Hakansson P., Sundqvist, B.U.R.; 1989 *Rapid Commun.Mass Spectrom.* **3** 259.

4. a) Lubman D.M.; 1988 *Mass Spectrom.Rev.* **7**, 535, 559.
b) Grotemeyer J., Schlag E.W.; 1988 *Angew.Chem.Int.Ed.Engl.* **27** 447.

5. Köster C., Dey M., Lindner J., Grotemeyer J.; 1990 *Proc. 2nd International Meeting "Spectroscopy Across The Spectrum"* ed. A.M.C. Davis, (The Royal Society of Chemistry, London) in press.

6. Boesl U., Grotemeyer J., Walter K. and Schlag E.W.; 1987 *Anal. Instrum.* **16** 151.

7. Beavis R.C., Lindner J., Grotemeyer J., Schlag E.W.; 1988 *Chem.Phys. Lett.* **146** 310.

8. a) Karas M., Bachmann D., Hillenkamp F.; 1985 *Anal.Chem.* **57** 2935;
b) Hillenkamp F., Karas M., Holtkamp D., Klüsener P.; 1986 *Int.J.Mass Spectrom.Ion Proc.* **69** 265;
c) Karas M., Bachmann D., Bahr U., Hillenkamp F.; 1987 *Int.J.Mass Spectrom.Ion Proc.* **78** 53.

9. Beavis R.C., Lindner J., Grotemeyer J. and Schlag E.W.; 1988 *Z.Naturforsch.* **43a** 1083.

10. see ref. [4]; Grotemeyer J., Schlag E.W.; 1989 *Acc.Chem.Res.* **22** 399.

11. Roepstorff P and Fohlman J.; 1984 *Biomed.Mass Spectrom.* **11** 601.

12. Lindner J., Grotemeyer J. and Schlag E.W., 1989 *Int.J. Mass Spectrom.Ion Proc.* in press.

Paper presented at RIS 90, Varese, Italy, 16–21 September 1990

Resonance ionization spectroscopy of molecular cluster using picosecond lasers

John C. Miller and D. Barton Smith*

Chemical Physics Section, Oak Ridge National Laboratory, Oak Ridge, Tennessee 37831-6125, USA

ABSTRACT: High peak-power picosecond laser pulses have been used for the first time to effect nonresonant or resonant multiphoton ionization (MPI) of clusters generated in a supersonic nozzle expansion. The resultant ions are subsequently detected and characterized by time-of-flight mass spectroscopy. Specifically, we present results involving MPI of the cluster series Ar_nNO, $(NO)_n$, and Ar_nI_2. Previous MPI studies of the latter two species using nanosecond lasers have not been successful in detecting the parent ion, presumably due to fast dissociation channels. It is proposed that the present technique is a new and rather general ionization source for cluster studies which is complementary to electron impact but in addition, may provide unique spectroscopic or dynamical information.

1. INTRODUCTION

In recent years the technique of mass-resolved multiphoton ionization (MPI) has been applied to the study of van der Waals (vdW) molecules and larger clusters. The advantages include high sensitivity and selectivity, as well as direct mass spectral identification of the cluster.

We have extensively studied the electronic spectra of ArNO associated with the $A^2\Sigma^+$, $B^2\Pi$, $C^2\Pi$, and $D^2\Sigma^+$ states of uncomplexed NO using (2+1) and (1+1) resonantly enhanced MPI. For several of these states we have also studied the Ne, Kr, Xe, and CH_4 analogues. Data on the ground neutral and ionic states of ArNO are known from other studies. Consequently, a rather complete picture of the spectroscopy, bonding, and photophysics of this weakly bound molecule has emerged (Miller and Cheng 1985, Miller 1987a, Miller 1989 and references therein).

Similar MPI studies on other important vdW molecules have been impeded by the presence of dissociative resonances or ion fragmentation. Very recently, we have begun to use a tunable picosecond laser which can overcome these problems in several ways (Smith and Miller, 1990). First, the high peak power allows nonresonant MPI to be observed, thus allowing characterization of a cluster beam without knowledge of the location of specific excited states. Secondly, when dissociative resonances are encountered, the short pulse (10-20 ps) and high power allow ionization to compete with bond breaking. Results from studies of Ar_nNO, $(NO)_n$, and Ar_nI_2 will be presented to illustrate these points. Previous MPI studies of the latter two species using nanosecond lasers have not been successful in detecting the parent ion, presumably due to fast dissociation channels.

In particular, the nitric oxide dimer is an important species for atmospheric chemistry. In the present study, results from a number of different visible and ultraviolet wavelengths are compared and interpreted with reference to the known electronic states of $(NO)_n$ and $(NO)_n^+$. When resonant with the dissociative B_1 state of $(NO)_2$ at $\sim$ 217 nm, cluster ions are observed with picosecond but not with nanosecond laser pulses. In contrast, nonresonant MPI at certain wavelengths results in parent cluster ions with either nanosecond or picosecond lasers. The use of such high-peak-power lasers for the characterization of cluster beams is discussed and compared with more conventional electron impact techniques.

2. APPARATUS

The pulsed supersonic nozzle (R. M. Jordan Co.), Wiley-McLaren-type time-of-flight (TOF) mass spectrometer, and vacuum system are similar to those described elsewhere (Miller and Compton 1986). The laser system consists of a Nd:YAG laser (Quantel, YG571C Nano-Pico) which delivers 1.2 J in a 10 ns pulse (Q-switched) or 75 mJ in a 20 ps pulse (mode-locked) at the 1064 nm fundamental wavelength. In either operating mode, the second, third, or fourth harmonics (532, 355, or 266 nm) are available. Other wavelengths may be obtained by Raman shifting. In addition, a short-cavity dye laser (Quantel, PTL-10) provides tunable, picosecond pulses in the red (580-710 nm) which may be frequency-doubled or mixed. The laser beam is typically focused by a 50 mm lens into the expanding gas. The TOF spectrum is observed for each shot with a digitizing oscilloscope (Tektronix 11402). The signal-to-noise ratio is improved by averaging spectra from $\sim$ 100 consecutive laser shots.

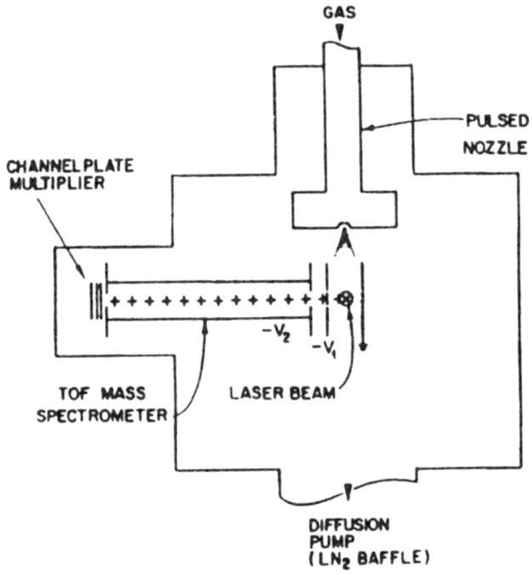

Fig. 1. Schematic diagram of the apparatus used for MPI studies of atomic and molecular clusters.

3. RESONANCE IONIZATION SPECTROSCOPY OF RARE GAS-NITRIC OXIDE MOLECULES

For some time we have been studying the spectroscopy and bonding of the van der Waals molecules formed between rare gas atoms and nitric oxide molecules (Miller and Cheng 1985, Miller 1987a, Miller 1989). These interesting species are only very weakly bound in the ground state ($\sim$ 87 cm^{-1} for ArNO); however, the ion formed upon removing an electron is much more strongly bound ($\sim$ 1000 cm^{-1}). For the excited states one might expect the valence states to be weakly bound, similar to the ground state. However for Rydberg states, which are more polarizable, the bonding should be stronger, more like that of the ion.

Using resonantly enhanced multiphoton ionization (MPI) coupled with mass-selective detection, we have spectroscopically characterized a number of excited states for NeNO, ArNO, KrNO, XeNO, and CH_4NO. Figure 2 shows schematically the known potential functions for ArNO where the internuclear distance coordinate is taken to be the weak van der Waals bond. The wall depths and vibrational levels are determined from the MPI spectra.

Because the strength of the van der Waals bond is related to the polarizability of the partners, changing the rare gas will increase or decrease the bond strength relative to ArNO depending upon the "size" of the rare gas. Table 1 of Miller, 1987b illustrates this trend for the constants relevant for the C-state of the rare gas·NO molecules. The monotonically increasing bond strength as the polarizability of the rare gas increases is shown by the increased "matrix shift" (ΔT_0) and vibrational force constant (K) derived from the spectra.

4. PICOSECOND MPI OF $(NO)_n$ CLUSTERS

Nitric oxide and $(NO)_n$ van der Waals molecules have been the subjects of numerous spectroscopic studies not only due to their intrinsic interest but because NO plays a prominent role in the chemistry of the upper atmosphere. Although NO dimers and rare gas-NO clusters are readily formed in a free jet expansion, nanosecond MPI experiments with these species, such as those described in the last section, have failed to detect the $(NO)_2^+$ parent ion. This failure to detect the dimer ion may be attributed to the presence of dissociative states in the $(NO)_2$ and $(NO)_2^+$ manifolds: the only known excited state of the neutral dimer dissociates very rapidly (Billingsly and Callear 1971), and the dimer ion is readily photodissociated by visible light (Smith and Lee 1978). The electronic states of NO and $(NO)_n$ are shown schematically in Figure 3 along with the multiphoton steps for the wavelengths used in the present study. We do not observe the dimer ion (or larger polymers) when we use visible light (532 nm or 585-605 nm). However, when either nanosecond or picosecond UV light (355 nm, 266 nm, or frequency-doubled 585-605 nm) is used, $(NO)_n^+(n = 1 - 5)$ ions are readily observed. Figures 4a and 4b show TOF

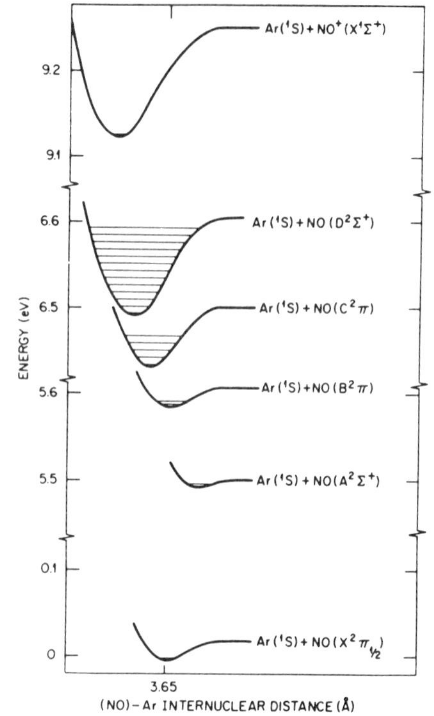

Fig. 2. Schematic representation of the potential energy curves for ArNO.

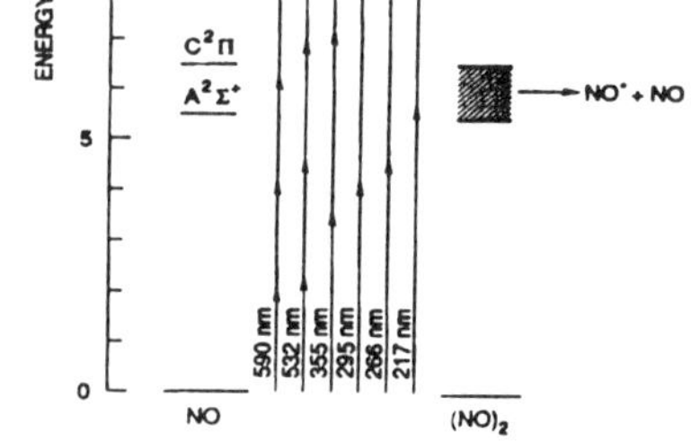

Fig. 3. Energy level diagram for NO and $(NO)_2$ showing the various MPI schemes discussed in the text.

spectra following MPI with 217 nm light (the second anti-Stokes Raman-shifted wavelength from the 266 nm output). In this case the two-photon ionization is resonant with the dissociative excited state in the neutral manifold. Only NO^+ ions are observed with nanosecond pulses, while $(NO)_n^+$ (n = 1 – 4) ions are observed with picosecond pulses. Clearly, the higher peak power available in the picosecond pulses enables MPI to compete effectively with the fast dissociation channel. More details may be found in a recent publication (Smith and Miller 1990).

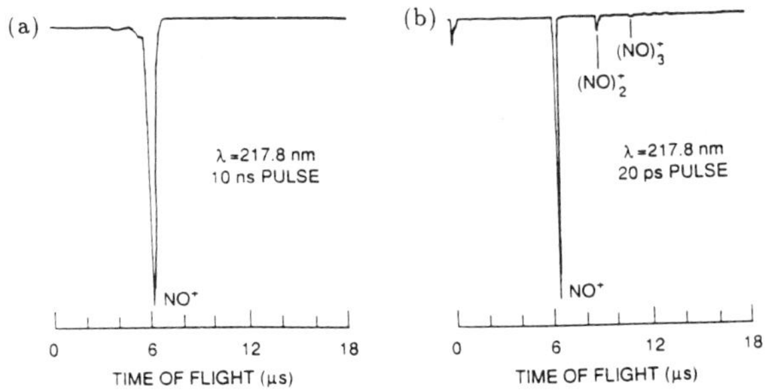

Fig. 4. MPI mass spectra of $(NO)_n^+$ with (a) nanosecond and (b) picosecond pulses.

5. RARE GAS·IODINE VAN DER WAALS MOLECULES

An iodine molecule complexed with a rare gas atom constitutes another example where nanosecond MPI experiments have failed to observe the van der Waals molecule (Miller 1987b). Figure 5 shows the TOF spectrum following MPI at 532 nm of species formed in an expansion of an I_2/Ar mixture (Smith and Miller 1989). The ion clusters observed are the Ar_nI^+ series rather than the $Ar_nI_2^+$ series. This observation of the daughter ion series is consistent with the very efficient one-photon photodissociation of uncomplexed I_2^+, but the fragmentation might also be occurring via so-called "half-collisions." The chemi-ionization reaction

$$Ar^* + I_2 \rightarrow ArI^+ + I + e^- \quad , \tag{1}$$

which has been previously observed (Henglein and Muccini 1960), occurs via a transition state $Ar^* \cdot I_2$. This same transition complex can be achieved by exciting the ArI_2 van der Waals molecule to an autoionizing state from which dissociative ionization occurs to ArI^++I. Likewise, intracluster ion-molecule reactions analogous to

$$I_2^+ + M \rightarrow IM^+ + I \quad \text{or} \quad I_2 + M^+ \rightarrow IM^+ + I \quad , \tag{2}$$

where M is a rare-gas atom, could also be responsible for our observation. Experiments to be performed will use intermediate resonances to produce localized excitation or ionization in either the rare gas or iodine part of the molecule.

6. CONCLUSIONS

In the present work we have used high-peak-power lasers (both nanosecond and picosecond) to effect multiphoton ionization of $(NO)_n$, $Ar_n(NO)$ and Ar_nI_2 clusters. When combined with results published elsewhere on Xe_n and Kr_n clusters using the same apparatus, the potential utility of this technique is apparent. The observation of other minor species such as Ar_n, $NO(H_2O)_n$ and $(NO)_n \cdot H_2O$ testifies to the sensitivity and selectivity which can be achieved, particularly if higher resolution or reflectron-type mass spectrometers were used.

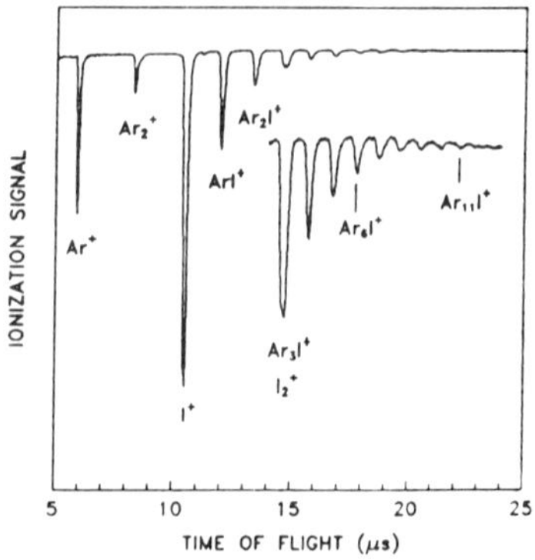

Fig. 5. Time-of-flight mass spectrum of Ar_nI^+ clusters following picosecond MPI with 532-nm light.

We suggest that laser mass spectrometry using high-power lasers may constitute a near 'universal' detector of atomic and molecular species. Unlike resonant MPI, no knowledge of intermediate states is required. Although more complex and more expensive than the traditional electron impact source, MPI has some unique advantages (and disadvantages) which are discussed below.

A major factor involved in the MPI technique is that atoms or molecules with a high ionization potential (E_i) are discriminated against relative to those with a low E_i as, typically, an additional photon is required for ionization. An nth order MPI process usually has a cross-section several orders of magnitude larger than that for n + 1 MPI. This need not always be a disadvantage, however. For instance, common expansion gases such as He, Ar, or N_2 all have high IPs and thus would appear much less prominently (if at all) in the MPI mass spectrum. The spectrum of the dopant and its clusters would then be less congested. Furthermore, space-charge problems are also reduced if the most abundant species are not ionized. Since clusters typically have a reduced E_i relative to monomers, judicious choice of the wavelength for MPI could allow discrimination against the monomer while all of the cluster ions would be efficiently ionized. For instance, the E_i of $(NO)_2$ is 8.736 eV compared to 9.264 eV for uncomplexed NO. The choice of a laser wavelength between 284.8 and 267.7 nm would allow two-photon detection of all $(NO)_n$ clusters while three photons would be required to ionize NO.

Compared to resonant MPI, the non-resonant analogue gives a more 'faithful' representation of the populations of species in the beam. (Of course, the ionization selectivity is also lost.) Furthermore, since precise tuning is not required, the higher power and better beam quality of the Nd:YAG harmonics may be used. The option of using different order MPI schemes by choosing different harmonics can be used to avoid resonances, as amply demonstrated in the present study. Finally, many of the adjunct techniques developed for resonant MPI can also be used in the non-resonant experiment. These might include, for instance, photoelectron or ion kinetic energy analysis (Compton and Miller 1989), two-color techniques (Compton and Miller 1985, Blazewicz and Miller 1988), or polarization dependencies (Miller and Compton 1986).

7. ACKNOWLEDGMENTS

This research was sponsored by the Exploratory Studies Program of Oak Ridge National Laboratory and the Office of Health and Environmental Research, U.S. Department of Energy under contract DE-AC05-84OR21400 with Martin Marietta Energy Systems, Inc. John C. Miller acknowledges use of a NATO International Collaboration Grant (0474/87) during the course of this work.

8. REFERENCES

Billingsly J and Callear A B 1971 *Trans. Faraday Soc.* **67** 589
Blazewicz P R and Miller J C 1988 *Phys. Rev. A.* **38** 2863
Compton R N and Miller J C 1985 *J. Opt. Soc. B.* **2** 355
Compton R N and Miller J C 1989 in *Laser Applications in Physical Chemistry* ed D K Evans (Marcel Dekker, Inc. New York) pp 221-306
Henglein A and Muccini G A 1960 *Z. Naturforsch Teil A* **15** 584
Miller J C 1987a *J. Chem. Phys.* **86** 3166
Miller J C 1987b *J. Phys. Chem.* **91** 2589
Miller J C 1989 *J. Chem. Phys.* **90** 4031
Miller J C and Cheng W C 1985 *J. Phys. Chem.* **89** 1647
Miller J C and Compton R N 1986 *J. Chem. Phys.* **84** 675
Smith D B and Miller J C 1989 *J. Chem. Phys.* **90** 5203
Smith D B and Miller J C 1990 *J. Chem. Soc. Faraday Trans.* **86** 2441
Smith G P and Lee L C 1978 *J. Chem. Phys.* **64** 5395

*Present address: Applied Technology Division, Oak Ridge National Laboratory, Oak Ridge, Tennessee 37831-7280, USA

Paper presented at RIS 90, Varese, Italy, 16–21 September 1990

Laser tandem mass spectrometry with two lasers in a reflection time-of-flight analyzer

C.Weickhardt, R.Weinkauf, K.Walter, U.Boesl

Inst. für Phys. und Theor. Chemie, Techn. Universität München
Lichtenbergstr. 4, D-8046 Garching, FRG

ABSTRACT: A new tandem-mass spectrometer is presented, where one laser is used for resonance enhanced multiphoton ionization and a second laser for secondary, mass selective dissociation in the second order space focus of a special ion source. The mass resolution of the primary, very short (10 cm) linear TOF is better than 700 (R50%). Pure secondary and pure metastable mass spectra of one preselected molecular or fragment ion are registered by a special tuning mode of a reflectron.

1. INTRODUCTION

The basic concept of tandem-mass spectrometry (MS/MS) includes the production of primary ions, their mass selective secondary excitation, and the mass selective registration of secondary fragment ions. The transfer of this principle to time-of-flight (TOF) mass spectrometry was not trivial, resulting in low mass resolution of hitherto existing instruments. On the other hand the use of pulsed laser techniques, which are perfectly adapted to TOF mass spectrometers, opens up a variety of interesting possibilities:

A first laser produces ions by resonance enhanced multiphoton ionization. This technique is molecule selective (no special purification of the probe) and allows a tunable degree of fragmentation. Ions can be produced with a very narrow distribution of internal energy, an important feature for experiments on unimolecular kinetics.

Secondary excitation can be achieved by a second laser. In contrary to collision induced dissociation, which is mostly used in conventional MS/MS experiments, the amount of energy transfered to the ion by photoexcitation is well defined. In addition more total energy can be placed in the ion by this method (due to MP excitation) than with conventional techniques.

2. EXPERIMENTAL

We combine a new linear TOF with a reflectron TOF (RETOF) to an MS/MS instrument. For the first stage we use a special ion source, which allows in its space focus a time-of-flight correction of second order concerning variations of the ion kinetic energy (valid for all masses). This results in a high mass resolution better than 700 (R50%) after only 10 cm drift region /1/. Figure 1 shows the signal of the benzene cation and its ^{13}C-isotope detected by microchannel plates in the space focus. The width of the signal is limited by the pulse length of

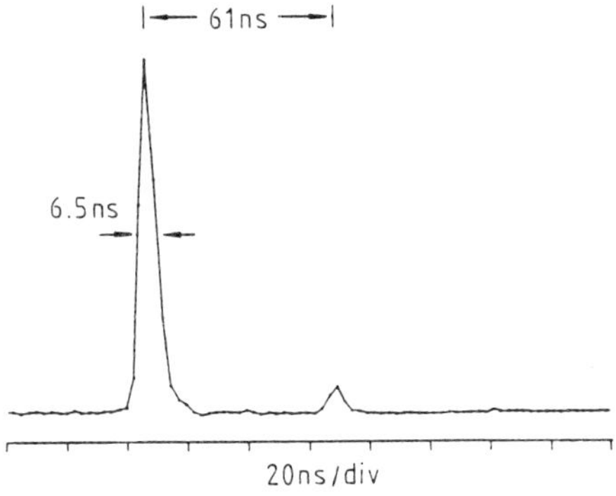

Fig.1 $C_6H_6^+$ and its ^{13}C-isotope detected in the space focus

the ionizing laser. The space focus is the ideal point for secondary excitation by laser 2 because ions of different masses are well separated and the well focused ion clouds allow a good focusing and thus high intensities of laser 2.

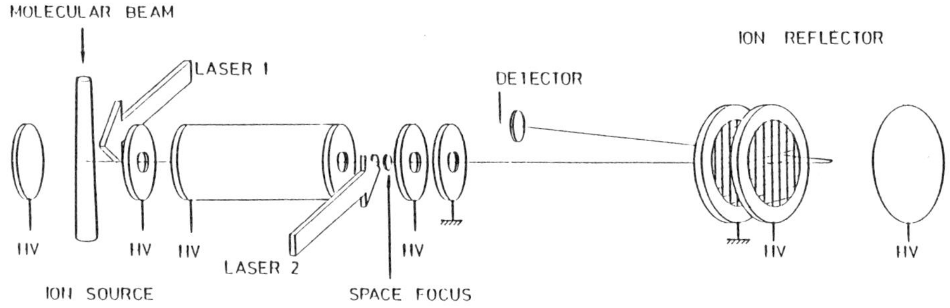

Fig.2 Experimental setup

Due to the wide range of kinetic energy of the secondary photofragments, secondary mass spectra can not be recorded at a single adjustment of the reflector. So we use a special tuning mode of the reflectron: At the decay the kinetic energy of the precursor ion is distributed among the fragments proportional to their mass so that E/m is the same for all fragments from one selected precursor. The penetration depth into the reflector field (and thus time-of-flight) for optimal correction intrinsically is the same for all ions. Therefore, if the reflector voltage is scanned, the time-of-flight of all masses of one precursor gets the same at their reflector voltage of optimal correction. This is exactly the time-of-flight of the precursor ion at its optimal correction; Its secondary mass spectrum can be obtained by setting the time window of a gated integrator on the signal of the precursor in the primary mass spectrum and recording the signal within this gate while scanning

the reflector voltage to zero. This technique allows the measurement of either the secondary or the metastable mass spectrum of one selected precursor with the excellent mass resolution typical for RETOF mass spectrometers. The disadvantage of this method is the long registration time due to scanning.
This is overcome by the Post Acceleration Method: The relative differences of the kinetic energies of the fragment ions can be reduced by adding a constant amount of kinetic energy to all ions after fragmentation (Postacceleration PA). E.g. the lightest and the heaviest fragment of benzene cations of 500 eV, C^+ and $C_6H_5^+$, have a relative difference in kinetic energy of about 73 %. After postacceleration with 2 kV it is reduced to about 9 %:

<u>E.g.:</u>

$C_6H_6^+$ (500 eV) —light→ C^+ (77 eV) $\xrightarrow[2\ kV]{PA}$ C^+ (2077 eV)

ΔE/E=73 % → ΔE/E=9 %

$C_6H_6^+$ (500 eV) —heavy→ $C_6H_5^+$ (494 eV) $\xrightarrow[2\ kV]{PA}$ $C_6H_5^+$ (2494 eV)

This energy difference can be corrected by the reflector at a single adjustment. Scanning is not necessary anymore. With this method it is possible to get the whole secondary mass spectrum at every lasershot. The secondary mass spectrum of $C_6H_6^+$ in Figure 3 was recorded by averaging over 400 lasershots in only 20 sec.

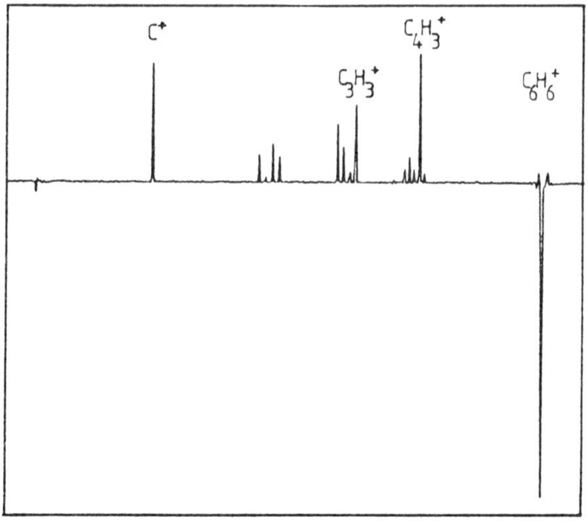

Fig.3 Secondary mass spectrum of $C_6H_6^+$

For using this technique the postacceleration voltage is added to the voltages of the electrodes of the ion source. Between the ion source and the space focus the ions pass a metallic tube, which is set at the postacceleration voltage. After fragmentation in a field free region the ions are postaccelerated between an additional pair of electrodes (see Figure 2). The reflector is adjusted for the average energy of the fragment ions.
Stable primary ions strike the reflector endplate and disappear.

To eliminate disturbing (e.g. metastable) primary ions we use a pulsed transversal electric field near the space focus, produced by a set of parallel wires, to deflect single masses or whole groups of primary ions, and difference measurement with shot to shot subtraction switching laser 2 off every other cycle.

3. RESULTS

In a first demonstration of our MS/MS-technique we investigated the decay of the benzene cation and several of its fragment ions after excitation with the harmonics of a Nd:YAG-Laser. After recording the secondary mass spectra of all members of the $C_3H_x^+$- and the $C_4H_x^+$- group we were able reveal the fragmentation tree of benzene, formed by the alternating absorption and dissociation processes from the benzene cation to the final carbon ion /2/.

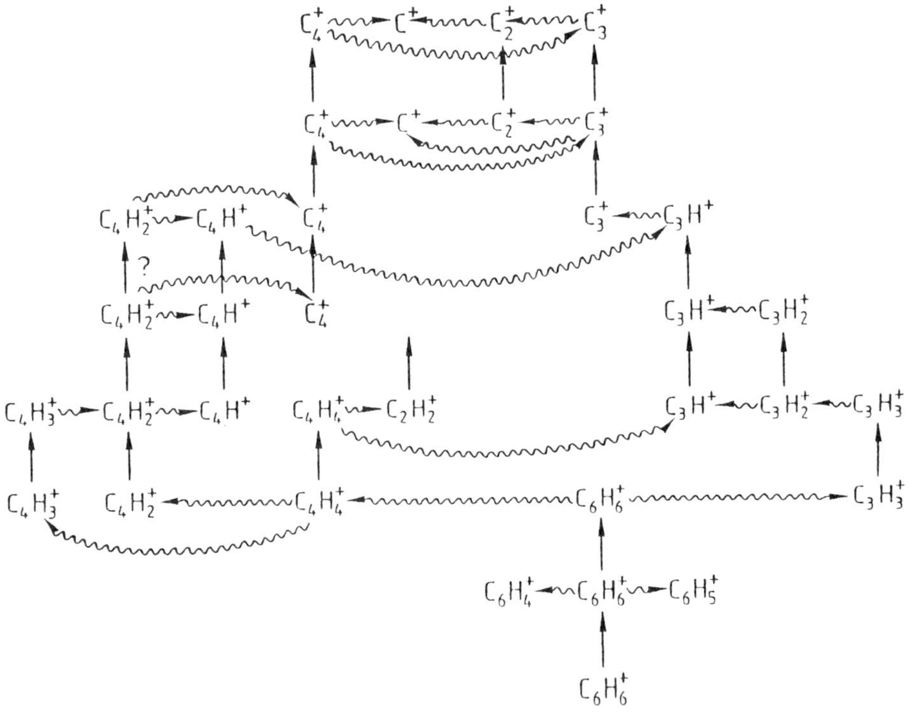

Fig.4 Fragmentation tree of the benzene cation for secondary excitation with 266 nm

/1/ R.Weinkauf, K.Walter, C.Weickhardt, U.Boesl, E.W.Schlag, Z. Naturforsch. 44a, 1219-1225 (1989)

/2/ U.Boesl, R.Weinkauf, K.Walter, C.Weickhardt, E.W.Schlag, J.Phys.Chem. to be published

Resonant two-photon ionization of aromatic molecules: Wavelength dependences and fragmentation data

A Marshall, A Clark, R Jennings, K W D Ledingham, C J McLean & R P Singhal

Department of Physics & Astronomy, University of Glasgow, Glasgow G12 8QQ, Scotland

ABSTRACT: Laser ionisation mass-spectrometry is used to study the wavelength dependence of resonant two-photon ionisation (R2PI) in the aromatic molecules, Aniline and Toluene. Molecular fragmentation patterns are recorded as are wavelength dependences of both parent and fragment ion yields. Similarities in the R2PI spectra of parent and fragment ions could lead to greater trace detection sensitivity if the total ion yield is monitored instead of only the parent ion.

1 INTRODUCTION

One of the attractive features of the multi-photon ionisation technique for analysis of molecular samples is the ability to produce parent ions by careful control of the laser fluence level (Grotemeyer *et al* (1986)). It has been found, however, that for some Amino acids (e.g. PTH-Lysine and PTH-Cysteic acid (Engelke *et al* (1987))), the parent ion yield is small compared with fragment ion production. Also, for some higher ionisation potential aromatic molecules (Apel and Nogar (1986)) no parent ion is observed in the laser ionisation mass spectrum. Wavelength dependent ionisation spectra of fragment ions have been recorded for Benzene (Zandee and Bernstein (1979(a)&(b))) and Acetylene (Robin (1980)) which are in close agreement with that of the parent ion. In this paper we examine two singly substituted aromatic molecules, Aniline ($C_6H_5NH_2$) and Toluene ($C_6H_5CH_3$), recording the relative abundances of parent and fragment ions and comparing the wavelength dependence of the parent and fragment ion yields.

2 EXPERIMENTAL

A Lumonics excimer-pumped dye laser system was frequency-doubled using an Inrad auto-tracking system to produce tunable laser radiation in the wavelength range 261-298nm using appropriate dyes. The bandwidth of the doubled laser output was less than 0.015nm. Maximum pulse energies were ~ 600μJ but typical energies depended on dye type, wavelength and dye condition. Flux densities were increased by focussing the beam to various degrees. Sample vapour was introduced into the cage of a VSW Ltd quadrupole mass spectrometer (resolution approximately 80 at mass 50) via a needle-valve controlled capillary tube. The quadrupole was located in a vacuum chamber evacuated by a turbo pump to a base pressure of ~2.10^{-6}mBar, the partial pressures of Aniline and Toluene used being~2.10^{-6} and ~$2\text{-}15.10^{-6}$ mBar respectively. Ion signals from the pre-amplifier were amplified before being sent to the ADC of a CAMAC LSI-11 data acqusition system together with a joulemeter signal which had been delayed for coincidence purposes. The ADC was triggered by a photodiode signal from the laser and the joulemeter was a Molectron J3-05 model. A diagram of the system is shown in Figure 1. Samples were used as supplied from manufacturers without further preparation.

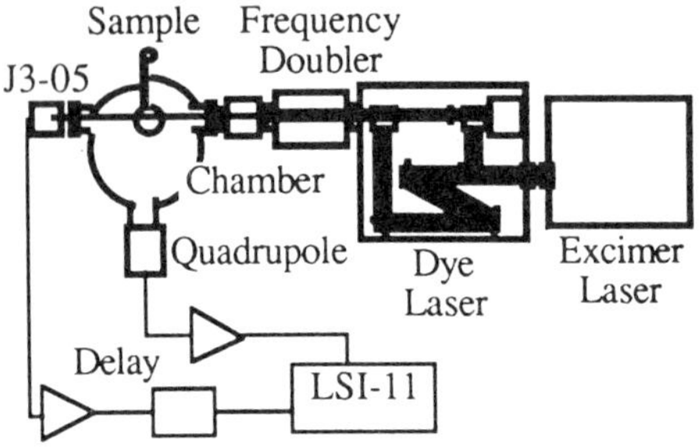

Figure 1 Experimental apparatus

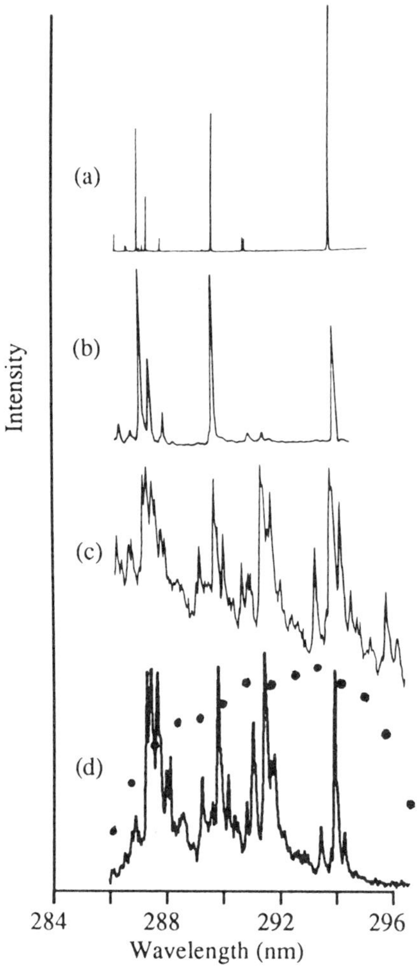

Figure 2 Aniline spectrum recorded by;
(a) Jet-cooled LIF (Mikami (1980)),
(b) Jet-cooled R4PI (Brophy & Rettner (1979)),
(c) Single-photon absorption (Brand et al (1966))
(d) R2PI ((•) denotes laser power profile)

3 RESULTS

The relatively large absorption cross-section of the $^1A_1$-$^1B_2$ transition in Aniline has led to the study of the spectroscopy of this molecule by a variety of different techniques. In this study we have extended the R2PI range of Brophy and Rettner (1979) to 285-294nm. Figure 2 shows the R2PI spectrum compared with jet-cooled laser induced fluorescence (Mikami *et al* (1980)) and 2-photon resonant, 2-photon ionisation (R4PI) spectra (Brophy and Rettner (1980)), and a single-photon absorption spectrum of a bulk sample (Brand *et al* (1966)). The good agreement with the single-photon spectrum of Brand *et al* (1966) supports the opinion that the cross-section of the second, ionising photon is approximately wavelength independent over this wavelength range.

Figure 3 shows the mass spectrum at 294.0nm and ~200μJ with medium focussing. There is good agreement with the spectrum of Rettner and Brophy (1981), and no variations in the relative intensities of the various peaks were observed when the wavelength was tuned to coincide with the peak ionisation yields at 287.2, 290.0 and 291.5nm. In fact, as Figure 4 shows, the wavelength dependence of the fragment ion yields matches very closely that of the parent ion over the wavelength range considered.

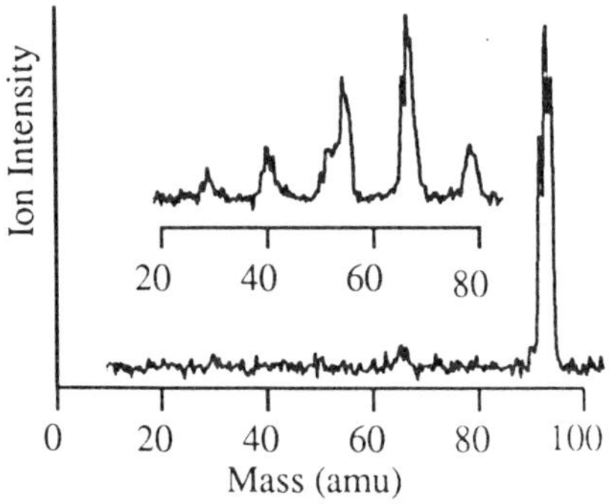

Figure 3 Mass spectrum of Aniline showing fragments at increased gain

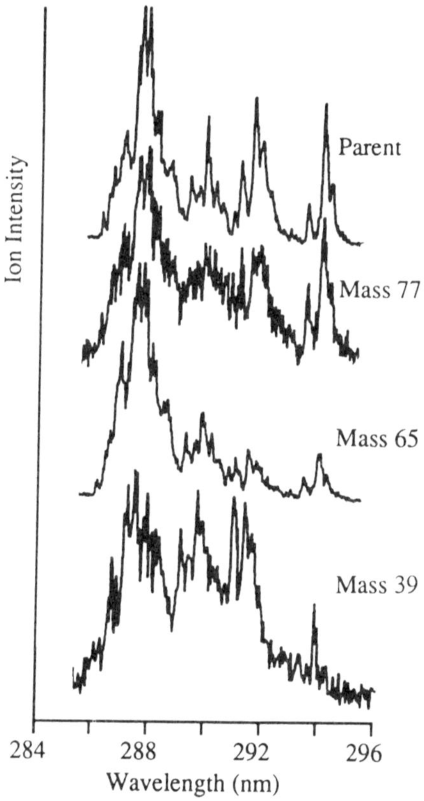

Figure 4 Wavelength dependence of the Aniline fragment ion yields compared with the parent ion

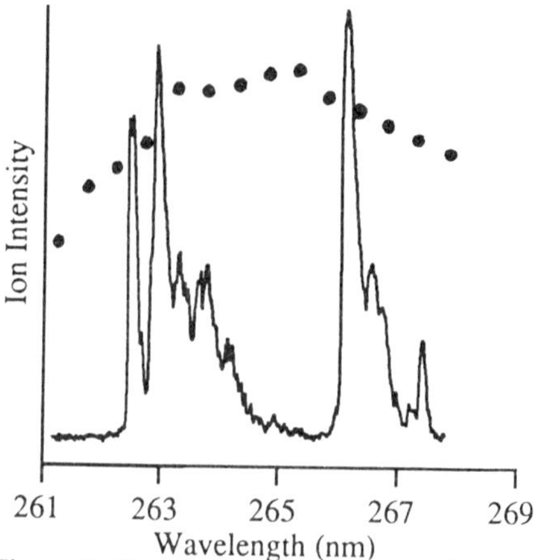

Figure 5 Wavelength dependence of Toluene parent ion showing the laser power profile (•)

Toluene has a higher ionisation potential than Aniline (8.82eV (Watanabe (1954)) compared with 7.76eV (Watanabe and Mottl (1957))), and a single-photon absorption band origin which occurs at a shorter wavelength than for Aniline. We have repeated the R2PI spectrum of Drysdale et al (1986) at higher resolution and approaching saturation fluences for some of the absorption transitions (Figure 5). Again, the R2PI spectrum agrees well with the single-photon absorption spectrum. Mass spectra (Figure 6) have also been recorded at a laser wavelength of 267.2nm, and approximately 350μJ/pulse. We note that at higher pressures the parent ion dominates, while at lower pressures there is more extensive fragmentation, with the 39 amu group of ions being particularly abundant. The R2PI spectra of the parent and some of the fragment ions is shown in Figure 7 and again there is good agreement-subject to the noise level.

4 CONCLUSION

The wavelength dependence of fragment-ion production suggests that the ionisation route is via the parent ion as proposed by Robin (1980). There is a possibility that for certain molecules which do not yield parent ions, the wavelength dependence of the total fragment yield will follow the single-photon absorption spectrum. With the possibility of tunability further into the UV using BBO crystals, detection of other singly-substituted aromatic molecules such as Nitrobenzene ($C_6H_5NO_2$) might be carried out more sensitively and selectively by monitoring certain fragments or even the wavelength dependence of the total ion signal.

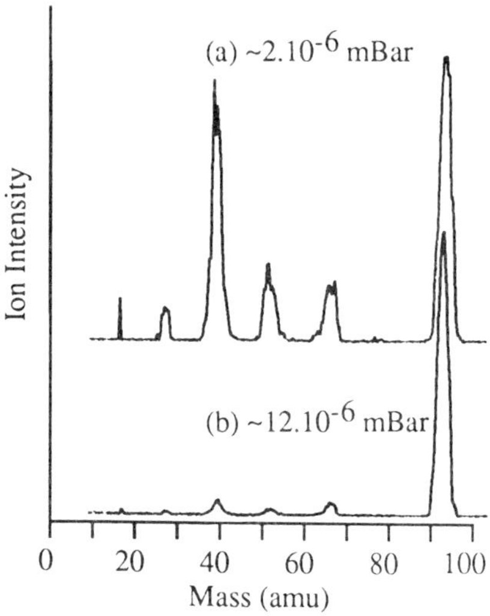

Figure 6 Mass spectrum of Toluene recorded at (a) lower pressure and (b) higher pressure

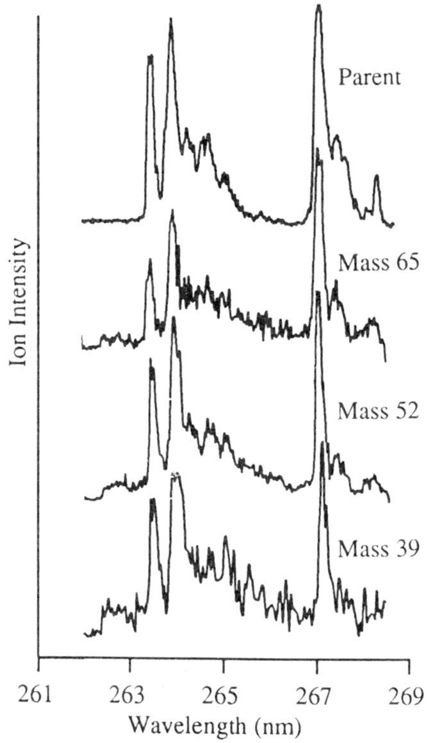

Figure 7 Wavelength dependence of Toluene fragment ions compared to the parent ion

5 ACKNOWLEDGEMENTS

This work has been carried out with the support of the Procurement executive, Ministry of Defence. AC is indebted to the SERC for postgraduate funding.

6 REFERENCES

E.C. Apel and N.S. Nogar, Int. J Mass Spec. and Ion Proc. **70** (1986) 243
Brand J C D, Williams D R, and Cook T J, J. Mol. Spec. **20** (1966) 359
Brophy J H and Rettner C T, Chem. Phys. Lett. **67** (1979) 351
Drysdale S L T, Ledingham K W D, Raine C, Smith K M, Stewart D T , Towrie M and Houston C M, Nucl. Inst. and Meth. **A252** (1986) 251
Engelke F, Hahn J H, Henke W and Zare R N, Anal. Chem. **59** (1987) 909
Grotemeyer J, Boesl U, Walter K and Schlag E W, Org. Mass Spec. **21** (1986) 645
Mikami N, Hiraya A, Fujiwara I and Ito M, Chem. Phys. Lett. **74** (1980) 531
Rettner C T and Brophy J H, Chem. Phys. **56** (1981) 53
Robin M B Appl. Opt. **19** (1980) 3941
Watanabe K, J. Chem. Phys. **22** (1954) 1564
Watanabe K and Mottl J, J. Chem. Phys. **26** (1957) 1773
Zandee L and Bernstein R B, J. Chem. Phys. **70** (1979(a)) 2574
Zandee L and Bernstein R B, J. Chem. Phys. **71** (1979(b)) 1359

Inst. Phys. Conf. Ser. No 114: Section 4
Paper presented at RIS 90, Varese, Italy, 16–21 September 1990

REMPI-MS: Application to the study of the gas phase dynamics of radical cations

Joseph J. BelBruno* and Gary Siuzdak

Department of Chemistry, Dartmouth College, Hanover, NH 03755 USA

ABSTRACT: Non-resonant and resonantly enhanced multiphoton ionization studies are presented for the isomeric propylamines, butylamines and *n*-pentylamine. Fragmentation patterns are described as a function of laser wavelength and fluence. These time-of-flight (TOF) mass spectra are compared with electron impact data and a discussion of the dynamics of the decomposition process is presented in terms of competitive ionic fragmentation pathways.

1. INTRODUCTION

Multiphoton ionization (MPI) and multiphoton ionization mass spectrometry (MPI-MS) have been employed in recent years to great advantage in the study of the mass fragmentation of fragile molecules as well as in the observation of photoinitiated chemical dynamics (Boesl, et al., 1978, Rockwood, et al, 1979). The technique has been especially valuable in those instances where electron impact mass spectrometry leads to extensive and somewhat indiscriminate fragmentation of the parent molecule.

We report here our efforts involving the use of MPI-MS in the study of the chemical dynamics of the isomerization of alkylamine radical cations (Siuzdak and BelBruno, 1990). In particular, the mass spectral fragmentation patterns of the propylamines, butylamines and *n*-pentylamine are presented and discussed in terms of the mechanism of parent radical cation fragmentation and rearrangement. A skeletal size threshold for isomerization and novel competitive kinetic pathways have been observed.

2. EXPERIMENTAL

The experiments described in this report were performed with a linear time-of-flight mass spectrometer of the standard Wiley-McLaren design. The sample was admitted to the source region via a molecular leak from a reservoir pressure of ~50-200 torr. The pressure in the source region was always less than ~8 x 10^{-5} torr and in the drift/detection region was maintained in the 10^{-6} torr range at all times. The resolution of the apparatus was 1 amu throughout the spectrum of the halocarbon calibration gases (m/z = 110).

The ionization potentials of all of the amines are clustered in a small range around 9 eV, so that depending upon the wavelength chosen for a particular experiment, 3 or 4 photon ionization of the parent molecule was expected. Resonant enhancement of the ionization process could occur through the n$\rightarrow$3s Rydberg transition, which has previously been reported only for the lower alkylamines.

3. RESULTS

The electron impact (EI) mass fragmentation patterns for the target molecules are well-known. For *n*-propylamine, the most intense feature occurs at m/z 30 and corresponds to α-cleavage of the parent radical cation to a fragment with the empirical formula CH_4N^+. The *iso*-propylamine EI fragmentation pattern follows analogously with the major feature at m/z 44 and due to the $C_2H_6N^+$ fragment. Small quantities of lower m/z fragments and parent ion are also detected. The electron impact spectra for the butylamines exhibit pure α-cleavage of the amine cation, such that the largest neutral fragment is produced. Finally, the *n*-pentylamine EI mass spectrum is dominated by α-cleavage to produce the m/z 30 fragment as observed in *n*-propylamine. Very small yields of parent ion and *larger* fragments are also observed and are attributed to molecular rearrangements prior to fragmentation or higher order bond cleavage.

The 355nm non-resonant MPI mass fragmentation patterns for *n*-propylamine at low laser intensities (~10^7 W cm^{-2}) indicate only fragmentation to m/z 30 as in the EI mass spectrum. As the laser intensity is increased to 10^9 W cm^{-2}, more extensive fragmentation of the molecule is observed. Interestingly, a larger quantity of the M^+-H ion is produced at the higher energy. The MPI fragmentation patterns for *iso*-propylamine are analogous to those for *n*-propylamine. That is, the MPI spectrum at low energy resembles that obtained from electron impact, the total ion yield increases dramatically with laser energy and the fragmentation becomes more extensive at higher laser energies. Again, molecular rearrangements prior to fragmentation appear to be minimal and the presence of hydrogen atom loss channels, which do not appear to be readily available to the electron impact produced parent ion, are observed.

At 355nm excitation, the major mass fragment for *tert*-butylamine occurs at m/z 58, corresponding to an apparent α-cleavage and loss of a CH_3• group. The *sec*-butylamine also appears to follow an α-cleavage fragmentation pathway with the major mechanism involving loss of C_2H_5• (m/z 44) and a minor contribution from CH_3• elimination (m/z 58). However, the remaining two isomers exhibit mass fragments which cannot entirely be attributed to α-cleavage pathways. For *iso*-butylamine, the expected fragment due to formation of $CH_2NH_2^+$ is observed at m/z 30, but additional fragments with significant intensities were recorded at m/z 44 and m/z 58. Similarly, *n*-butylamine ionization results in the expected m/z 30 product, but fragments with m/z 58 and m/z 44 are also observed. At higher levels of laser intensity, higher yields of smaller fragments are observed.

Comparison of the 355nm *n*-pentylamine spectra with those for EI indicates that MPI and electron impact ionization are nearly identical at low laser intensities, but that at higher laser intensities the *larger fragment ions are produced in much higher yields by MPI* than by EI. We note that H-atom loss channels again seem to be more prevalent with the MPI technique and that the spectrum has much the same appearance as a butylamine spectrum. Study of MPI fragmentation spectra, of all of the amines, for excitation at 532nm indicates a striking similarity to those recorded at 355nm. In this particular instance, the frequency dependent nature of the MPI fragmentation is lost.

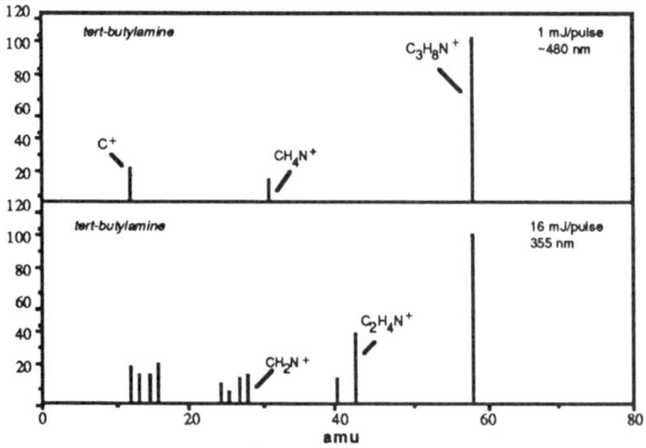

Figure 1. Comparison of MPI-MS for resonant (480nm) and non-resonant (355nm) excitation.

The mass spectra for resonant excitation at 480nm are different from the non-resonant results as shown in Figure 1. These spectra are obtained by excitation through low rovibrational states of the 3s Rydberg level. This result is a manifestation of the wavelength selectivity, i.e. the variable energy input, of MPI-MS, and one of its major advantages. The *tert*- and *sec*-butylamine mass spectra at this wavelength are intriguing, since they exhibit considerably more extensive fragmentation than for excitation with a greater laser energy at 355nm. Contrary to this observation is the indication that the remaining two isomers fragment along the expected reaction pathway, i.e., the additional peaks, attributed to some process other than α-cleavage, observed at 355nm are significantly reduced at this wavelength.

4. DISCUSSION

The MPI mass spectra, especially at the higher laser powers, exhibit considerably more extensive fragmentation than the EI spectra. This is typically observed in MPI-MS and is attributed to very efficient ladder-switching. Contrary to most MPI studies, the EI and MPI fragmentation patterns for the propylamines show many similarities. However, the butylamine and *n*-pentylamine results indicate that rapid (within the 6 ns laser pulse) isomerization mechanisms may effectively compete with direct fragmentation. *Isomerization is more favorable* than α-cleavage for the larger amines. One plausible explanation for the difference between the propyl- and larger amines is the fact that many fewer low frequency vibrational modes are available for the propylamines. The availability of such low frequency sinks for the excess excitation energy was found to be a critical factor in the opening of isomerization channels in the butylamine cations. It appears that the three carbon chain is simply below the threshold for the accessibility of isomerization pathways. The results involving the *n*-pentylamine mass spectrum are an interesting corroboration of this interpretation. The mass fragmentation pattern appears to indicate that aside from the expected α-cleavage, a significant number of the radical cations undergo γ-cleavage to produce the *n*-butylamine cation which is subject to both direct fragmentation as well as isomerization to the *sec*-butylamine cation (by a 1,2-methyl shift) with subsequent fragmentation. The greater probability of the M^+-H fragment ions for the amine cations at higher laser intensities is an intriguing and, initially, a counterintuitive result. However, this fragment lies ~1 ev above that resulting from C-C bond fission (in the parent ion). Therefore, absorption of an additional photon is required for creation of this product. Absorption of that photon is in competition with bond cleavage in the parent ion and the rates become comparable only at the higher laser intensities. This type of kinetic behavior in radical cations has not been previously reported.

We have carried out RRKM calculations for all of the butylamine radical cations and these calculations are consistent with the following mechanism. For example (see Figure 2), production of the *iso*-butylamine ion requires absorption of three photons at 355nm. The ion is created with an excess energy as large as 1.7eV, enough to energetically lead to α-cleavage and isomerization. However, the rate of both processes is extremely slow and neither would not be observed on the timescale of the experiment. Absorption of an additional photon is necessary to produce any observable α-cleavage product. The additional energy also leads to isomerization at a rate which is competitive with fragmentation. The multiphoton mass spectra confirm that isomerization is occurring. The calculations also indicate that although production of a M^+-H fragment is energetically possible after absorption of a single photon by the parent ion, the rate of production is sufficiently slow as to render the product unobserveable. Only after absorption of a second photon, *in the parent cation manifold*, is this process kinetically feasible. Again, the experiments yield results consistent with the calculations. At the higher laser intensity, up-pumping of the parent ion successfully competes with decomposition to CH_4N^+ and isomerization. The cross section for further absorption may be estimated (for an allowed transition) as $\sim 10^{-17}cm^2$. At a laser intensity of $10^9 W \cdot cm^{-2}$, the transition probability is estimated to be $\sim 2 \times 10^{10} s^{-1}$. This is comparable to the α-cleavage rate constant, but a factor of ~3 less than the isomerization rate. However, the amount of product resulting from absorption and subsequent fragmentation would easily be sufficient for detection. At 480nm, the α-cleavage and isomerization rates after absorption of a single photon by the parent ion are nearly identical. Production of M^+-H is not expected to be significant even after absorption of two photons by the cation. Again, the calculations are in accord with the experiments. Finally, irradiation at 532nm requires absorption of two photons by the parent cation for either α-cleavage or isomerization. Fragmentation of the N-H bond requires yet an additional photon. These calculated results are in agreement with the observation that MPI mass spectra are only observed at high laser intensities for this wavelength.

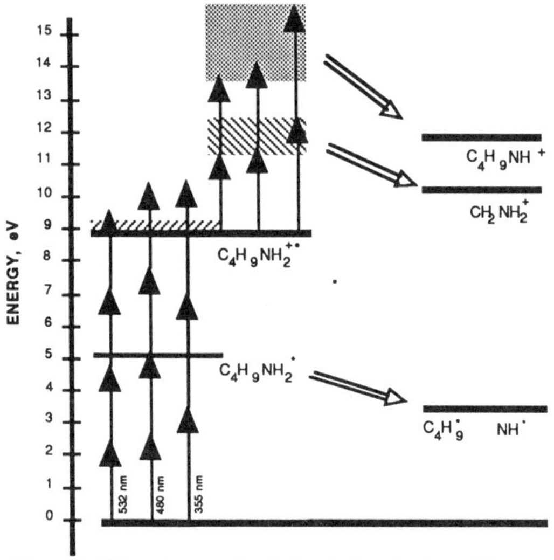

Figure 2. Relevant energy levels for the butylamines. Three possible fragmentation mechanisms are shown, The two uppermost shaded are correspond to one- and two-photon absorption by the radical cation.

One interesting aspect of this research is the lack of any wavelength specificity in the non-resonant excitation. Our butylamine research has indicated that the lowest electronic transition corresponds to an n→3s Rydberg excitation (X→A) with an origin at approximately 42,000cm^{-1}, depending upon the particular chain length of the alkylamine. The higher excited states have been relatively unexplored. Two photons at 355nm and three photons at 532nm would be required to attain an energy greater than the vibrationless ground state of the 3s level. Both of these excitation schemes would result in approximately 56,000cm^{-1} of energy input. Production of the molecular ion via either wavelength would result in an ion with excess energy and a high probability for fragmentation. One must also conclude from the similarity of the mass spectra, that the ions involved in ladder-switching mechanism readily absorb either wavelength.

The butylamine MPI spectra are wavelength dependent, implying more than one decomposition mechanism, and one which is not related to the electron impact pathway. The initial step in the MPI excitation mechanism at 480nm is absorption into the 3s Rydberg level. Other important differences for the three wavelengths used in the study are the order of the absorption process and the level of excitation in the lowest ionic state. The nearly equivalent and large levels of excess energy deposited in the amine at 355 and 532nm are expected to cause the pathways for these two wavelengths to be nearly identical. Previous studies have shown that the excitation from the Rydberg level to the ionic state occurs with little change in the vibrational quantum number, i.e., the geometry of the Rydberg state resembles that of the ion. Therefore, excitation with either 355nm or 532nm radiation will ultimately yield similar ions due to the approximate energy matching at the two and three photon level respectively. In contrast, excitation at ~480nm results in the formation of Rydberg molecules in the ground vibrational level and subsequent production of vibrationally cold ions. This difference could be cause of the variation in resonant and non-resonant mass spectra A second possible factor is that the *tert*-butylamine and *sec*-butylamine radical cations might possess larger absorption cross sections at 480nm than do the remaining isomers.

5. REFERENCES

Boesl U, Neusser H J, Schlag E W 1978 *Z. Naturforsch.* **33a** 1546
Rockwood S, Reilly J P, Hohla K, Kompa K L 1979 *Opt. Commun.* **28** 175
Siuzdak G and BelBruno J J 1990 *J. Phys. Chem.* **94** 4559
Siuzdak G and BelBruno J J 1990 *Appl. Phys. B* **50** 221

Inst. Phys. Conf. Ser. No 114: Section 4
Paper presented at RIS 90, Varese, Italy, 16–21 September 1990

Spectroscopy of molecular ions prepared by resonant multiphoton ionization: The OCS^+ ION

R.Weinkauf, U.Boesl

Inst. f. Phys. u. Theor. Chemie, Techn. Uni. Munic
Lichtenbergstr. 4, 8046 Garching, Germany

ABSTRACT: Resonance enhanced multiphoton ionization has been used for state selective preparation of molecular ions. This ion source has been combined with dissociation spectroscopy.

1. INTRODUCTION

Resonant multiphoton ionization is an excellent means for preparing molecular ions in very few, often even in only one vibrational level of the ionic ground state/1/. In many cases this level is the vibrationless ground state. This can be achieved by selective excitation of a vibronic intermediate state which resonantly enhances the multiphoton ionization. For low rotational excitation rotational cooling of the neutral parent molecules in a supersonic beam or selective excitation of rovibronic neutral intermediate states is used. The preparation of cold molecular ions is a necessary condition particularly for well resolved spectroscopy of larger organic molecular ions. With conventional ion sources this condition mostly cannot be realized. As ion spectroscopic technique we use predissociation spectroscopy (one photon dissociation) as in the case of OCS^+ presented here or resonant enhanced multiphoton dissociation. This methods allow observation of ionic states independent of fluorescence (as for fluorescence spectroscopy) or high ion yield (as for absorption spectroscopy). A time-of-flight mass analyzer (reflectron) is used for ion detection. The high mass resolution as well as some options available with reflectrons allow a good background suppression. There already exists a number of molecular ions to which state-selective preparation by MPI and dissociation spectroscopy was applied/2-4/. In most of these cases well resolved spectra have been observed for the first time, or have been extended down to the electronic origin or been entangled so much that interpretation was possible. For the example of OCS^+/4/ we demonstrate that simplified as well as supplementary ion spectra can be obtained by state selective ion preparation.

2. EXPERIMENTAL

We control the vibrational population of the ionic ground state by photoelectron spectroscopy. In a 50 cm long field free drift region the kinetic energy of the electrons is determined by time-of-flight analysis. For high energy electrons (>1eV) a retarding field is used to achieve long flight times. Our time-of-flight photoelectron

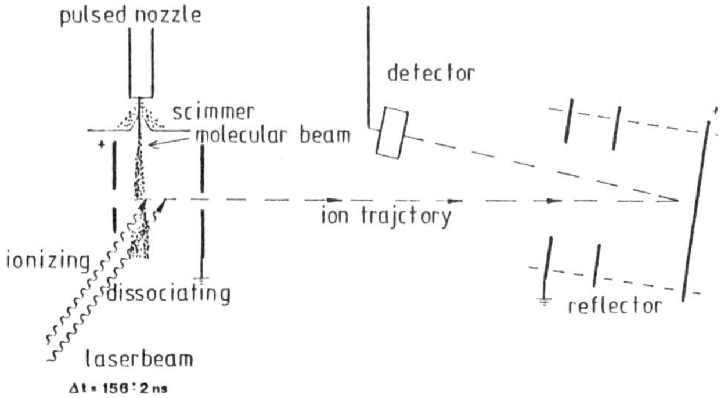

Fig.1 Experimental setup for dissociation spectroscopy of molecular ions.

spectrometer has a resolution of 10 to 15 meV. As mass selective detector for the ion spectroscopy a reflectron was used (fig.1). With the first laser, OCS^+ ions were state selectively produced involving a (2+1)-photon absorption. The necessarily high laser intensity resulted in a strong fragmentation (see mass spectrum of fig.2) interfering with the fragmentation of the second laser for dissociation spectroscopy. In the insert of fig.2 the solution of the problem that primary and secondary fragments are interfering is demonstrated. At the top of the insert only primary fragments (only ionizing laser on) are shown, in the middle also the dissociating laser (for ion spectroscopy) is switched on with a time delay of 156 ns resulting in a time-of-flight separation (actually in energy) of primary and secondary fragment ions. At the bottom of the insert (fig.2) the potential of the reflector endplate is decreased so that the primary ions with their higher kinetic energy strike it and cannot reach the ion detector anymore. This kind of separation only works for large mass differences of molecular and fragment ion, in more unfavourable cases other techniques involving reflectrons are available such as tandem mass sepectrometry /3/.

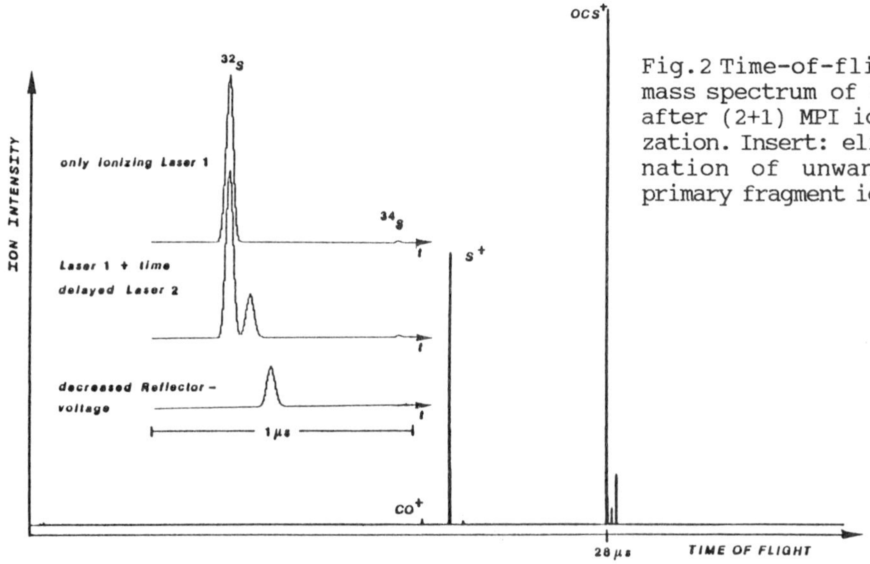

Fig.2 Time-of-flight mass spectrum of OCS^+ after (2+1) MPI ionization. Insert: elimination of unwanted primary fragment ions.

3. RESULTS

In fig.3 the excitation scheme for state selective preparation of OCS^+, the (2+1)-multiphoton ionization spectrum of the intermediate Rydberg states and the resulting photoelectron spectra (giving information about the ionic ground state population) are displayed. From fig.3 it can be clearly seen that the vibrationless D Rydberg state as an intermediate state induces the population of the vibrationless X(1/2) (1/2 spin orbit component) of the molecular ion. Involving vibrational excited D Rydberg states (ν_1 or ν_3) results in the population of the same vibrations in the 1/2 spin orbit component of the ionic X state. However, using the F Rydberg state as an intermediate sate leads to the population of the 3/2 spin orbit component. This allows to entangle the ion spectra of the X(1/2)-A(1/2) and of the X(3/2)-A(3/2) spectra of OCS^+. On the other hand, the preparation of one single vibronic level within the ion ground state led to a considerable simplification of the ion spectra which allowed for the first time an unambiguous assignement of the OCS^+ UV spectrum in the 30000 - 35000 cm^{-1} region.

In fig.4 two ion spectra due to the two selectively prepared spin orbit components of the vibrationless ground state of OCS^+ are displayed. For better comparison one spectrum is displayed with negative polarity. It can be clearly seen that in the upper spec-

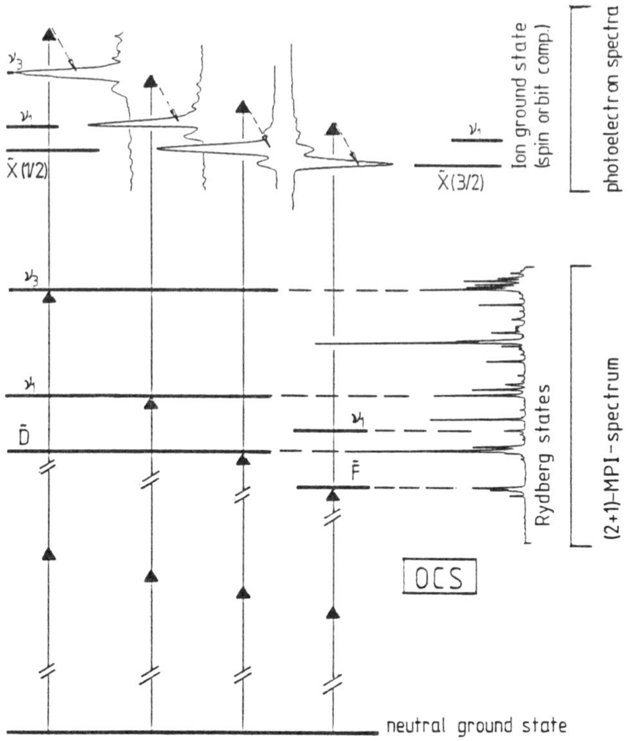

Fig.3 Selective prepartion of OCS^+ in different levels (vibrational levels; spin orbit components) of the ion ground state by MPI via different intermediate Rydberg states. The population of the ionic ground state has been checked by photoelectron spectroscopy.

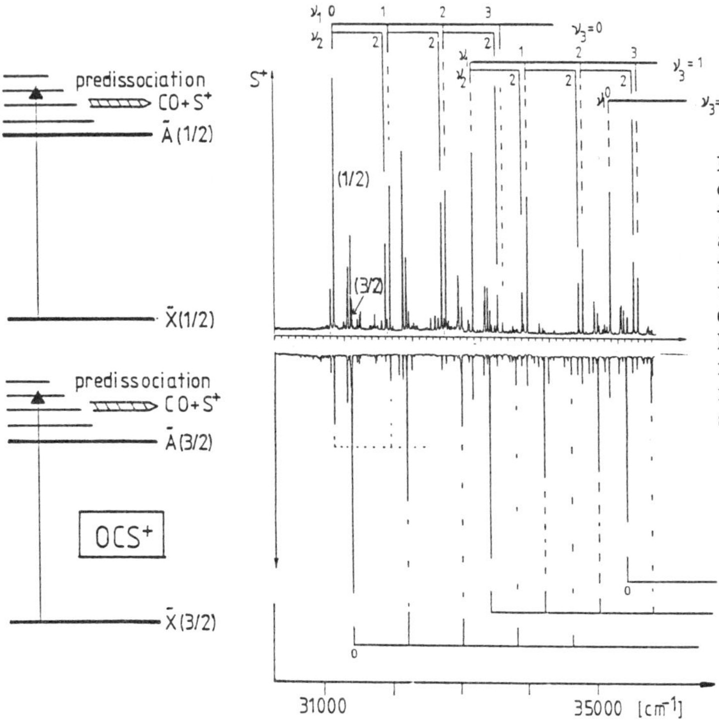

Fig.4 Ion spectra of (top) the X(1/2)-A(1/2) and (bottom) the X(1/2)-A(3/2) transition of the OCS^+. The ions have been prepared in the 1/2 resp. in the 3/2 spin orbit comp.

trum the (1/2)-transition dominates and only very small contribution of the (3/2)-transition appears and vice versa in the lower spectrum. Due to the fact that major contribution of vibrational hot bands now can be excluded and by comparison with ion spectra starting from the vibrationally populated ionic ground state a meaningful assignment (as insicated in fig.4) is possible. Of high interest is the excitation of the ν_2 vibration (bending mode) which is degenerate; its vibrational angular momentum can couple with the angular momentum of electronic states. Of course, the resolution of ion spectra is much better than of conventional photoelectron spectra and include more details about the population of the ionic ground state. Both photoelectron spectra from resonance enhanced multiphoton ionization and ion spectra deliver feed back information on the intermediate Rydberg states and may be used for an assignment of Rydberg states which is often very cumbersome.

References:

/1/ For a review see: K.Kimura, Int.Rev.Phys.Chem. **1987** 6 195
First OCS^+ work: B.Yang, M.H.Eslami, S.L.Anderson, J.Phys.Chem. **1988** 89 5527
/2/ A.M.Woodward, S.D.Colson, W.A.Chupka, J.Phys.Chem. **1986** 90 274; R.Weinkauf, K.Walter, U.Boesl, E.W.Schlag, Chem. Phys.Lett. **1987** 141 267; X.Ripoche, I.Dimicoli, J.LeCalvé, F.Biuzzi, R.Botter, Chem.Phys. **1988** 88 305; K.Walter, R.Weinkauf, U.Boesl, E.W.Schlag, Chem.Phys.Lett. **1989** 155 8
/3/ K.Walter, U.Boesl, E.W.Schlag, Chem.Phys.Lett. **1989** 162 261
/4/ R.Weinkauf, U.Boesl unpublished results

Observations on carbon clusters during trace elemental analysis of coal samples by RIMS

R P Singhal, S L T Drysdale, R Jennings, A P Land, K W D Ledingham, P T McCombes and M Towrie

Department of Physics and Astronomy, University of Glasgow, Glasgow G12 8QQ, Scotland.

ABSTRACT: Carbon clusters, C_n with n up to 214 have been observed using laser desorption from a graphite sample in a reflectron time of flight (TOF) mass spectrometer. The clusters produce unwanted background in resonance ionisation mass spectroscopy. The clusters, however, may be used to calibrate TOF mass spectrometers. A study of the variation of resolution with mass provides some insight into their formation.

1. INTRODUCTION

Resonance ionisation mass spectroscopy (RIMS) offers the possibilities of very high sensitivity and selectivity in trace impurities detection. In a typical RIMS arrangement the sample is first atomised by an ablation laser or an ion beam. The neutrals are subsequently resonantly ionised by a suitably tuned laser. However, the ablation produced ions in the vapour plume constitute an unwanted background signal and various methods are used to suppress their contribution. For example, in a reflectron type TOF-MS system commonly used for detecting ions, the ions produced at the target are more energetic than RIMS ions and are not reflected by the electrostatic mirror. However, none of these methods is able to eliminate totally the background ions and these may cause interference with the detection of the RIMS signal.

In a study involving the detection of trace metals in coal samples (Towrie *et al* 1990) it was necessary to investigate trace impurity levels in high purity graphite. In this communication we report the observation of carbon cluster ions with up to 214 atoms in the laser ablation of the graphite target and its implications for trace impurity analysis for targets with a high carbon content. Carbon clusters have previously been observed in the ablation of carbon targets Rohlfing *et al* 1984 and O'Keefe *et al* 1986) both in supersonic jet-cooled beams and in room temperature ablation. Carbon clusters C_n appear with $\Delta n = 1$ for $n < 31$, while beyond $n = 31$ only those with even number of carbon atoms are produced. Among the larger clusters, $n = 60$ is the most prominent. These observations are consistent with the model of even clusters having closed spheroidal shell structures (Zhang *et al* 1986).

2. EXPERIMENTAL DETAILS AND DISCUSSION

The ablation laser is a Quantel Nd-YAG 585. The frequency doubled output was used to irradiate a graphite target. The power was 10^8 Wcm-2 and the pulse duration was 10 ns. The laser spot size on the target was about 1 mm2. The carbon clusters observed are shown in fig.1 and they display all the prominent features observed previously (O'Keefe *et al* 1986 and Zhang *et al* 1986). Clusters of low n values are highly abundant. The intensity of observed clusters falls up to $n = 31$, beyond which a significant increase in their formation is measured. In addition, for $n < 31$, clusters for all values of n are obtained, but for $n > 31$ only those with even n values are observed. From a study of fragmentation patterns of these clusters (O'Brien *et al* 1988), it has been suggested that larger clusters ($n > 31$) have edgeless, spheroidal cage structures while the small ones have linear chain or ring structures. The clusters are clearly visible up to n = 214. It is interesting to note (fig.1) that between n = 32 and n = 214, the general envelope of cluster formation and survival probability has a characteristic shape with a maximum around n = 50 and nearly linear decrease beyond that. Superimposed on the smooth envelope are clusters with n = 60 and 70 which are more abundant, while n = 62 and 72 exhibit large depleted strengths. Clusters with n = 60 and 70 have been theoretically shown to be particularly stable and it seems reasonable to assume that in this experiment, the clusters are formed with a probability defined by the smooth envelope. Larger clusters fragment by losing a C_2 cluster (O'Brien *et al* 1988) with exceptionally stable ones like n = 60 and 70 appearing as the most abundant among clusters with n > 31.

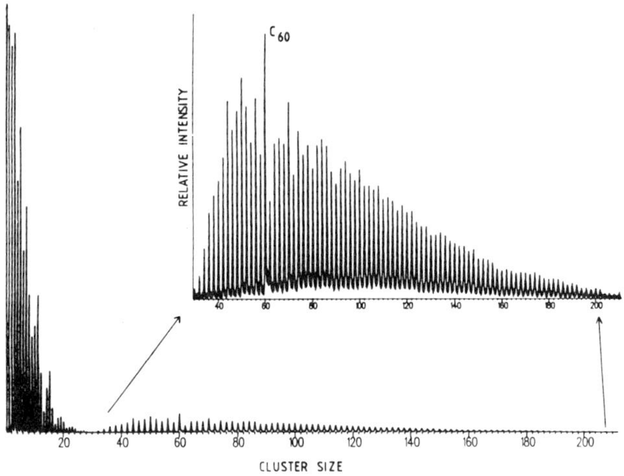

Fig. 1 Carbon clusters produced in the ablation of a graphite target by a 532nm laser beam.

Targets with a high carbon content, could have accompanying cluster production in an analysis run. These clusters would cause an unwanted background with long tails in the time of flight spectrum and reduce the signal to background and hence degrade the lowest detectable limits. Fig.2 shows a spectrum in the region of rubidium isotopes obtained by ionising Rb atoms by a combination of two photons having wavelengths 421.7 and 532 nm. However, also present in the spectrum are C_7 and C_8 carbon clusters produced during the ablation of the target sample (NBS coal sample SRM 1632a). The long tails of the ablation produced clusters limit the minimum detectable limit of Rb to 1 part per million (ppm). This situation is to be contrasted with that for a carbon-free target. In fig.3 a preliminary TOF spectrum for gold in copper matrix at 10 ppm is shown. A detection level of 0.3 ppm is already apparent and can be improved considerably by accumulating more statistics on the data points.

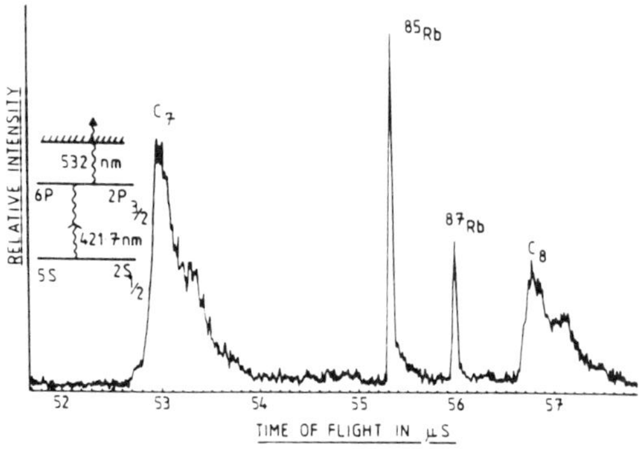

Fig. 2 Rb signal at a concentration of 15ppm in coal.

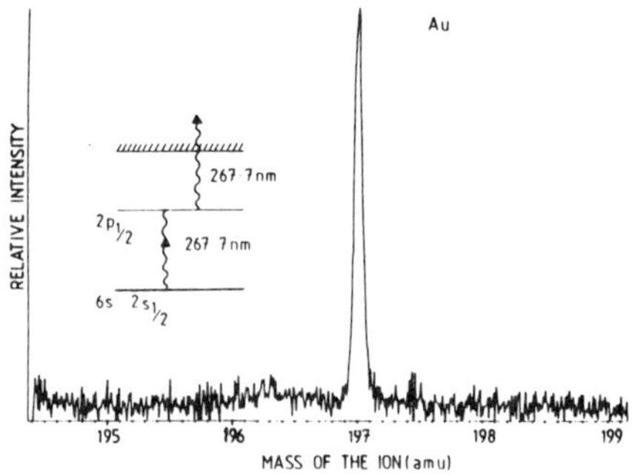

Fig. 3 Gold signal at 10ppm in a Cu matrix.

In TOF mass spectrometers, the mass of an ion is determined by its flight time as follows

$$\mathrm{sqrt}(m) = a * t + b \tag{1}$$

where a and b are constant for a given arrangement and depend on the physical dimensions of the spectrometer and the voltages on the various electrodes. For low masses upto a few hundred daltons, the calibration of the mass spectrometer is straightforward since one can use atomic masses as reference. However, for larger masses an extrapolation is required and the extracted masses may be less accurate. Carbon clusters can be used to define a mass scale and mass determination upto about 3000 amu may be performed with confidence. To illustrate this point, the spectrum shown in fig.1 was fitted to determine a and b in eqn.(1). Two other fits were made with clusters only to n = 10 and n = 20. On extrapolation to mass 2544 amu, n = 212, an error of 2.34% and 1.09% was calculated from the fits to truncated spectra. The fit to n = 214 is presented in fig.4. The clusters provide a convenient and accurate method of calibrating TOF mass spectrometers for large masses. The main advantage of this procedure is in the calibration of LIMA and LAMMA systems. In RIMS, ions are produced not at the target surface but generally one or two millimeters above it and the flight times are different. However, a correction may be calculated from the known voltages on the extract electrodes and the position of the ionising laser.

The present observation that the cluster masses and flight times are related according to the expected TOF-MS equation would suggest that all clusters in their respective abundances are formed at the target surface. On vaporisation of the target by the incident laser beam, graphite atoms and ions would interact to form the clusters. Any fragmentations and redistribution of their relative populations must happen in a time much shorter than a micro-second, since in-flight fragmentations would produce peaks in the time spectrum with inconsistent values of n, and the resolution of the TOF-MS would also be affected. These effects were not observed.

Fig.5 presents the resolution of the TOF-MS for carbon clusters. The resolution is constant at about 30 for n > 32. The full width at half maximum (FWHM) for these clusters was measured to be 450 ± 50 ns. For n < 32, there is some deviation from this value. Typically the smallest clusters have FWHM of 650 ns while in the range 16 < n < 32, the FWHM is 350 ns.

In a reflectron TOF-MS, the time of flight is mainly determined by two factors, the position and time of the ion formation. The initial velocity spread of the ions is compensated by the reflection of the ions at the electrostatic mirror. Thus, two ions formed with different initial velocities but formed simultaneously and at the same position will arrive at the detector at the same time. If it is assumed that the clusters are formed very close to the target surface then the main contribution to the width will arise from the interval over which they are formed. In fact the peak shape should reflect approximetely the ion formation probability as a function of time elapsed since the ablation

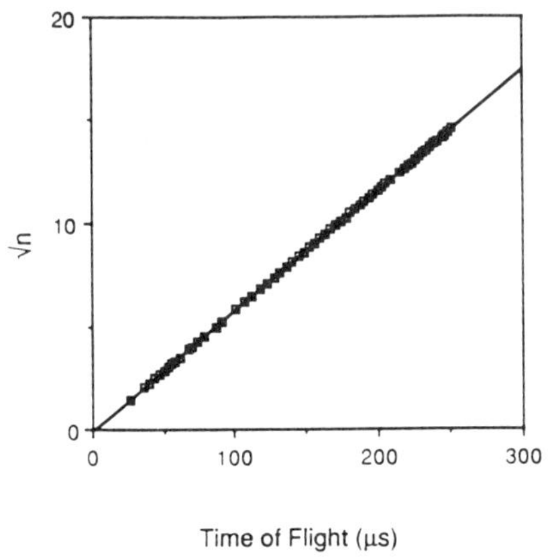

Fig. 4 A fit to the time of flight spectrum of carbon clusters up to n = 214. A straight line fit provides a good description of the data.

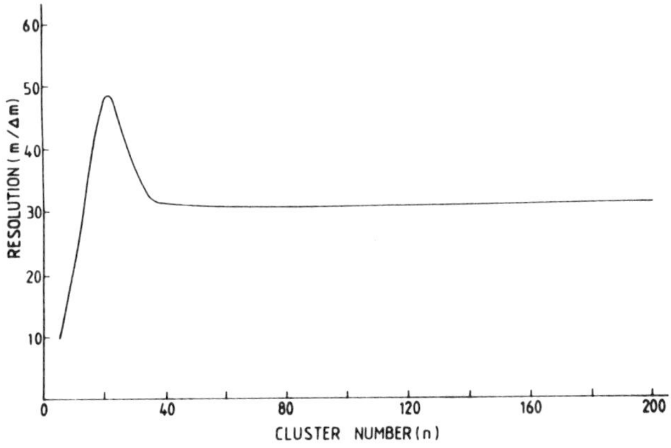

Fig. 5 The resolution of the ablated carbon clusters in the TOF-MS

pulse. Qualitatively, from the results of this experiment, one may construct a model as follow. The ablation laser pulse heats the target to a high temperature over a time scale of about 20 nanoseconds. The vapour pressure at the surface of the target could reach very large values of the order of many atmospheres (Pang and Yeung 1989) and cluster formation happens. Due to expansion the plume density decreases rapidly and affects cluster formation. Since, clusters with $16 < n < 31$ have the smallest widths, 350 ns, their formation must stop first. Hence, higher densities in the plume are required to form clusters in this range. The spheroidal clusters, $n > 32$ are formed up to 450 ns while the smallest clusters could be formed for lower plume pressures and have correspondingly larger widths.

CONCLUSIONS

For trace analysis in RIMS it is important to eliminate the ablation produced ions, especially in the analysis of carbon based samples e.g. biological material.

Carbon cluster ions may be used to obtain mass calibration of TOF-MS up to 2500 amu. Cluster ions of other compounds e.g. CsI could extend the range upto 13000 amu.

A reflectron type TOF-MS is well suited to study cluster formation mechanism. A detailed study of widths, shapes and relative intensities of the clusters is required to understand the dynamics of the ablated plume where cluster formation takes place.

REFERENCES

O'Brien S C, Heath J R, Curl R F and Smalley R E 1988 *J. Chem. Phys.*,**88**, 220
O'Keefe A, Ross M M and Baronavski A P 1986 *Chem. Phys. Letts,* **130**, 17
Pang Ho-ming and Yeung E S 1989 *Anal. Chem.*, **61**, 2546
Rohlfing E A, Cox D M and Kaldor A 1984 *J. Chem. Phys.*, **81**, 3322
Towrie M, Drysdale S L T, Jennings R, Land A P, Ledingham K W D, McCombes P T, Singhal R P, Smyth M H C and McLean C J 1990 *Inter. J. Mass Spectr.and Ion Proc.*, **96,** 309
Zhang Q, O'Brien S C, Heath J R, Tittle F K, Curl R F, Kroto H W and Smalley R E 1986, *Phys. Chem.*,**90**, 525

Inst. Phys. Conf. Ser. No 114: Section 4
Paper presented at RIS 90, Varese, Italy, 16–21 September 1990

REMPI on CF_3I in bulk and molecular beam

G. van den Hoek
Molecular and Laserphysics, University of Nijmegen, The Netherlands
S. Stolte
Chemistry Department, University of Amsterdam (V.U.), The Netherlands
D.W. Chandler
Sandia National Laboratories, Livermore, U.S.A.

CF_3I has been detected, in a molecular beam experiment, by resonance-enhanced multiphoton ionization (REMPI) technique. The two plus one process, with the electronic $\tilde{C}$-state as resonant intermediate level, leads to sufficient strong ion-signals of unfragmented CF_3I^+, so that determination of vibrational modes (and populations) is possible.

1. INTRODUCTION

REMPI results are presented, for CF_3I, without rotational resolution. These measurements were undertaken because CF_3I can easily be excited by an infrared multi-photon process using a powerful CO_2 laser. The intention has been to resolve the various vibrational modes in order to be able - in a later phase - to measure the state distribution of highly excited molecules.

CF_3I has never been investigated before, by the REMPI technique. We will present bulk and molecular beam spectra. In the bulk spectra, thermally populated vibrational states are visible but badly resolved. For efficient expansion, the beam spectra show hardly any population of vibrational modes of the electronic ground state. For moderate expansion, many hot bands have been observed.

2. EXPERIMENTAL

The apparatus used in our experiments is designed for the use of a molecular beam. The beam is crossed by a laser pulse from a dye laser, wich is pumped by a Nd-YAG laser (Quata Ray DCR-2). The dye-laser pulse is doubled by an angle tuned KDP crystal. The doubled wavelength has been tuned, for the $\tilde{C}$-band, from 336 nm to 350 nm using the dye pyridine 1. The repetition rate has been 20 Hertz. The bandwidth of the dye-laser was 0.2 cm^{-1}, with a pulse duration of 5 ns. In figure 1, the layout is shown schematically. The molecular beam is pulsed (0.5 ms pulse duration FWHM). The valve is formed by a modified fuel injector (Bosch). The skimmer diameter is 5 mm, the skimmer - nozzle distance about 4 cm. The beam is formed from a 10% CF_3I

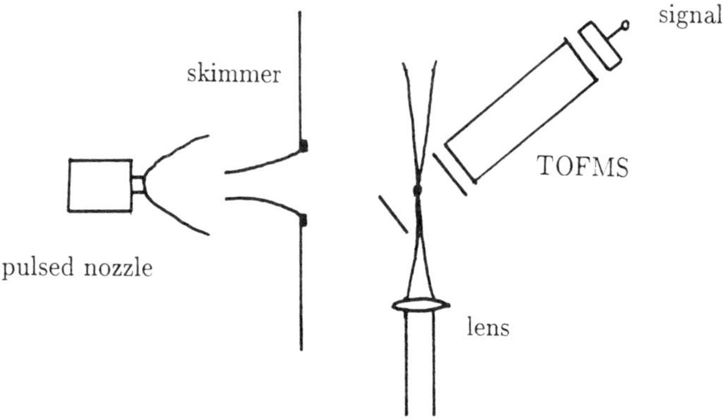

Figure 1: Experimental set-up

in He mixture source (T_0 = 300 K, P_0 = 1 bar).

The laser-beam interaction zone is 0.3 m downstream from the nozzle. The background pressure before the arrival of the molecular pulse amounts to 10^{-6} mbar. The detection chamber, where interaction occurs is pumped by an oil diffusion pump with water baffle (200 l/s). This detection chamber is equipped with a time of flight mass spectrometer (TOFMS), $m/\Delta m = 70$. An electric field of 3 kV/ 2 cm was used for accelerating the ions, typical flight times amounts to 5 μs, for CF_3I (M=196 a.m.u.). Ions are detected by two cascaded multichannel plates (MCP's, army refused items). The different isotopes of carbon could not be distinguished in the mass spectrum. Our intention has been to laser-ionize CF_3I without fragmentation.

The ionization spectrum reveals the distribution of the CF_3I molecules over various vibrational states (rotational resolution has not been attempted). To this end the dye laser has been calibrated by means of a hollow cathode lamp (HCL), which has been transversed by part of the laser pulse. Its response was monitored simultaneously with the ion yield as of function of the frequency. Between the narrow lines of the HCL, we interpolated using fringes of an etalon consisting of a microscope slide. The relative accuracy of the wavelength determination was better then 1 wave number. The frequency-scan extended over 1000 wave numbers (actually 2000 cm^{-1} in energy, since the doubled radiation frequency caused a two photon transition).

3. RESULTS

CF_3I possesses an electronically excited (partially) non-dissociating state at 57497 cm^{-1} (the $\tilde{C}$-state). REMPI has been realized by a 2 + 1 proces, with a photon energy thus of about 28748.5 cm^{-1}, i.e. λ = 347.8 nm for the doubled dye laser. For preliminary search, we also used the fourth harmonic of the Nd-YAG laser directly (266 nm) and

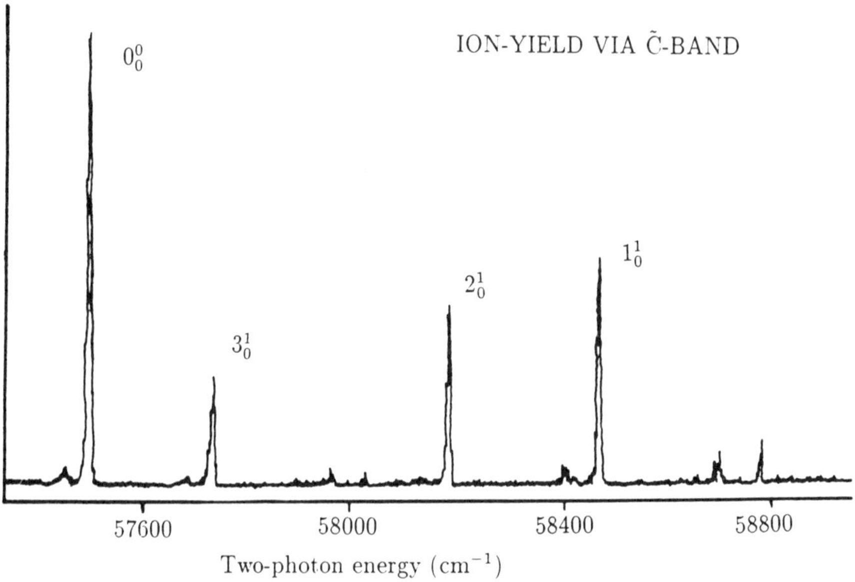

Figure 2: REMPI spectrum of CF_3I, m_l^u denotes: vibration mode m, l quanta in the ground state and u quanta in the C̃-state.

Vibrational energies C̃-band			
mode	ν_1	ν_2	ν_3
cm^{-1}	966	679	231

Table 1: Measured values

observed strong dissociation by probing the iodine by a second laser tuned to a resonance of I or I*. CF_3I (266nm) $\Longrightarrow$ CF_3 + I(I*), I (304.66nm) or I* (304.03 nm) $\Longrightarrow$ $I^+ + e^-$. This channel has not been further investigated because during the first step information on the initial state of CF_3I got lost (excitation of CF^3).

In figure 2, the ionization yield versus the laser frequency is shown. The different peaks belong to various vibration excitations of the C̃-state, the utmost left due to the 0_0^0 transition . These vibrational modes (and some combination bands) have been assigned. The vibration frequencies are also shown in the table. Note that this spectrum has been obtained at the Sandia laboratories, Livermore, U.S.A. in a precursor experiment. The precise frequency values were determined in Nijmegen.

Also other electronic excitations were tested; B̃-state, λ = 390 nm, no CF_3I ions observed; D̃-state, λ = 319.8 nm, strong CF_3 and I ions because of one photon resonance with the strong dissociating Ã-band (broad band contour from 30.000 to 45.000 cm^{-1}); Ẽ-state, λ = 283.2 nm, very weak CF_3I^+ signals, and again strong dissociation. For the C̃-state excitation, the rotational envelope shows an asymmetric steep rise on the blue side. From this we conclude that $B' < B''$: The rotational constant in the C̃-state is larger than in the groundstate.

Resonance CARS in Br_2 molecules and Br atoms

Ilse Aben, Pieternel Levelt, Wim Ubachs and Wim Hogervorst

Lasercentre Free University Amsterdam, The Netherlands

Abstract : Resonance-enhanced CARS processes were studied in molecular bromine. On the basis of the known spectroscopic constants of the two electronic states involved, the features in the spectra could be identified. CARS signals from Br-atoms produced from dissociation of Br_2 were obtained by tuning (ω_1 - ω_2) in resonance with the $^2P_{1/2}$ - $^2P_{3/2}$ spin-orbit splitting (3685 cm^{-1}). Different hyperfine components were resolved in the spectrum.

In this study (resonance-enhanced) CARS features were observed using a two-color set-up with a fixed-frequency pump-wave ω_1 (18788 cm^{-1}) and a tunable Stokes-wave ω_2 (13800 - 18300 cm^{-1}). A generated wave at frequency $\omega_3 = 2\omega_1 - \omega_2$ was detected. Strong overtones are observed in resonance-enhanced CARS spectra, similar to the appearance of overtones in Resonance Raman (Rousseau et al 1976). All distinct lines in the resonance-enhanced CARS spectra of Br_2, shown in the case of Δv=7 in fig.1, are produced by coincidental enhancement of the signal.

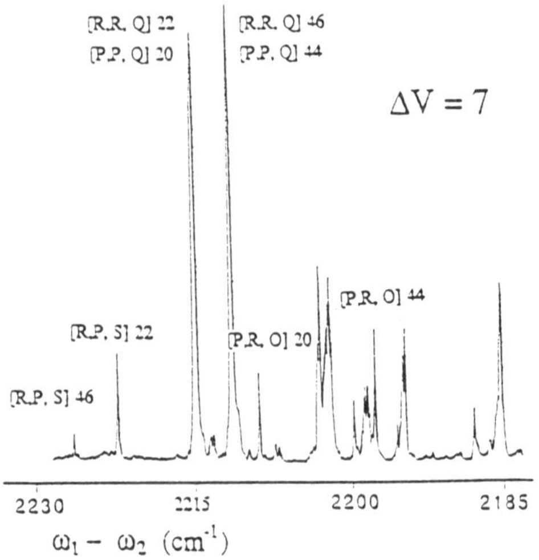

Fig. 1. Resonance-enhanced Δv = 7 CARS spectrum of Br_2.

The intensity of the CARS signal is proportional to the square of the tensor elements of the third order susceptibility tensor $\chi^{(3)}$ which can be written as (Oudar et al 1981):

$$\chi^{(3)}(\omega_3) \propto N \sum_{0,n}\sum_{n'''}\sum_{v} \rho_{00}^{(0)} \frac{\mu_{0n'''}\,\mu_{n'''v}\,\mu_{vn}\,\mu_{n0}}{(\omega_{n0} - \omega_1 - i\Gamma_{n0})(\omega_{v0} - \omega_1 + \omega_2 - i\Gamma_{v0})(\omega_{n'''0} - \omega_3 - i\Gamma_{n'''0})}$$

N = number density of molecules,

ω_{ij} = transition frequency from state j to state i,

μ_{ij} = transition moment between states i and j,

$\rho_{ii}^{(0)}$ = initial population of state i,

Γ_{ij} = relaxation parameter.

In this expression the summations over |0>, |v>, |n> and |n'''> refer to all possible rovibronic states of Br_2. From the denominator of $\chi^{(3)}$ it follows that enhancement of the CARS signal is achieved for states |0> for which the fixed frequency ω_1 is in near resonance with an allowed transition in the molecule. As the frequency ω_1 is in near resonance with just a few absorption lines in Br_2, this introduces a certain selectivity in the overall CARS process leading to distinct lines in the spectra as shown in fig. 1. The Stokes frequency ω_2 is tunable and a resonance on $(\omega_1 - \omega_2)$, corresponding to the second frequency factor in the denominator of $\chi^{(3)}$, may always be achieved. The generated anti-Stokes wave at ω_3 is in resonance with a continuum state, related to the repulsive $^1\Pi_{1u}$ state and the dissociative continuum above the $B^3\Pi_{u0}{}^+$ state. This continuum effectively enhances the signal but again does not introduce any extra coincidental enhancement of the signal. So the distinct features in the resonance-enhanced CARS spectra of Br_2 can all be explained by the coincidental enhancement of the signal at the fixed frequency ω_1.

The assignment of the lines used in fig. 1 indicates the four-photon sequence of the total CARS process. P- and R- (ΔJ = -1 or + 1) denote one-photon discrete ω_1- and ω_2-resonances, whereas O-, Q- and S- are used for the continuum-enhanced two-photon resonances (ΔJ = -2, 0 or + 2). The assignments given in fig. 1 all concern the $^{79}Br_2$ molecule.

In the same set-up strong CARS signals from Br-atoms were observed. The atoms were produced from dissociation of Br_2 by the pump-wave ω_1. As $(\omega_1 - \omega_2)$ is tuned into resonance with the fine-structure splitting $^2P_{1/2}$ - $^2P_{3/2}$ of 3685 cm^{-1}, a generated wave at $\omega_3 = 2\omega_1 - \omega_2$ is observed. This signal is not electronically resonance-enhanced as there are no discrete energy levels in Br-atoms corresponding to frequencies ω_1 or ω_3. Different hyperfine components in the Br-atoms were probed shown in fig. 2 due to the use of an injection seeded Nd:YAG laser ($\Delta\omega_1$ = 0.005 cm^{-1}) and a narrow-band dye-laser ($\Delta\omega_2$ = 0.05 cm^{-1}).

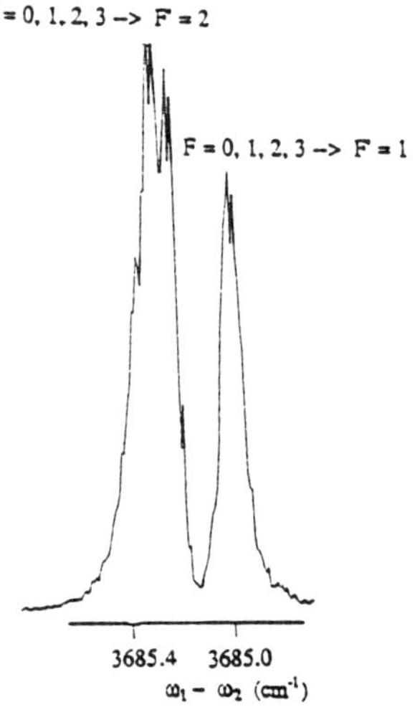

Fig. 2. CARS spectrum of Br-atoms. F denotes the hyperfine quantum number.

References :

Oudar J.-L. and Shen Y.R. 1980 *Phys. Rev.* **A22** 1141

Rousseau D.L. and Williams P.F. 1976 *J. Chem. Phys.* **64** 3519

Paper presented at RIS 90, Varese, Italy, 16–21 September 1990

Intermolecular vibronic spectroscopy of small van der Waals clusters: 1:1 and 1:2 complexes between a monosubstituted benzene molecule and rare gas atoms

M Schmidt, F Piuzzi, M Mons, J Le Calvé and I Dimicoli

C.E.A., CEN Saclay, Département de Physique Générale, SPER, Bât. 522, 91191 Gif sur Yvette cedex, FRANCE

ABSTRACT: Using resonant one- or two-color two-photon ionization and time-of-flight mass spectrometry techniques, van der Waals mode spectra of 1:1 and 1:2 complexes between a monosubstituted benzene molecule and rare gas atoms have been obtained and assigned. The effects of substitution on the potential energy surface and external mode structure are modelized using an atom-atom interaction sum model. Comparison of the spectra in the 1:1 and 1:2 complexes supports the symmetric 1,1 structure of the latter.

1. INTRODUCTION

The van der Waals (vdW) vibrational spectroscopy in the S_1 state of 1:1 and 1:2 complexes involving a monocyclic benzenic molecule and one or two rare gas atoms has been investigated in detail.

Using a supersonic expansion coupled with time-of-flight mass-spectrometry and one- or two-color laser ionization, well-resolved R2PI vibronic spectra of the 1:1 and 1:2 complexes of phenol, aniline, chlorobenzene or toluene with rare gas atoms have been experimentally obtained (Mons et al 1989 ; Schmidt et al 1990).

2. SPECTRA OF THE 1-1 COMPLEXES

In the 1:1 complexes (Mons et al 1989), of C_s symmetry, the stretching and symmetric (A') bending vdW modes could be observed, as well as the asymmetric (A") bending mode in its second harmonic (Figure 1). Correlations have been shown to exist between the red shift of the S_1-S_0 0_0^0 molecular transition or the stretching force constant of the vdW bond and the rare gas polarisability or the origin transition energy of the substituted benzene molecule (Figure 2).

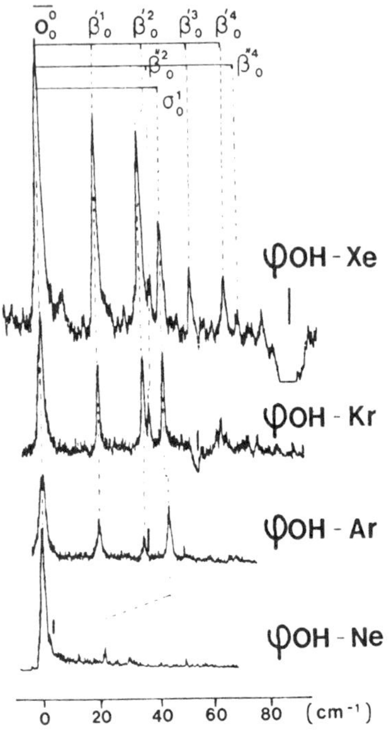

Fig. 1. R2PI vdW mode spectra of the φOH-Rg 1:1 complexes

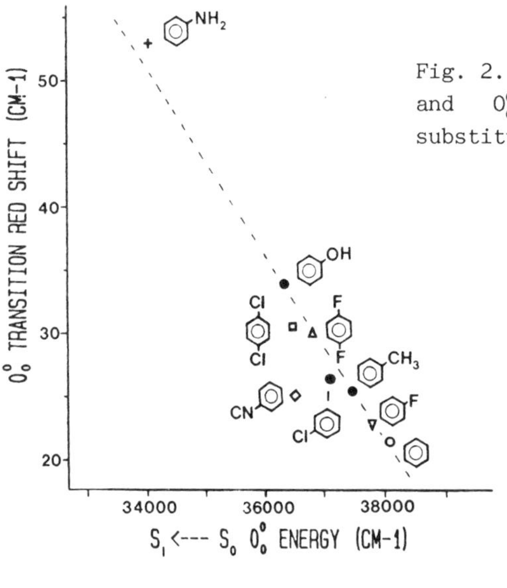

Fig. 2. Correlation between red shift and 0_0^0 energy for a series of substituted benzene-Ar complexes

3. CALCULATED POTENTIALS AND VDW MODE STRUCTURES

The potential structure of these complexes has also been theoretically studied using an atom-atom interaction sum model. In particular, the perturbations induced on the potential energy surface (Figure 3) and vdW mode structure, due to the presence of the substituent on the molecule, have been discussed relatively to the benzene-argon complex and the results of modelling compared with the experimental data (Mons et al 1990). The substituent generally induces four main effects on the PES : i) an increase of the well depth, ii) a shift of the Ar minimum position towards the substituent, iii) a tilt of the Ox potential gutter, iv) changes of the potential curvatures. A clear correlation is found between the calculated stretching (K_z) or bending (K_x) force constant and the potential well depth of the studied φX-Ar complexes. The dissymmetry effects due to the presence of the substituent on the vdW mode structure are shown to be small. The main effects concern the reduced mass and force constant changes (Mons et al 1990).

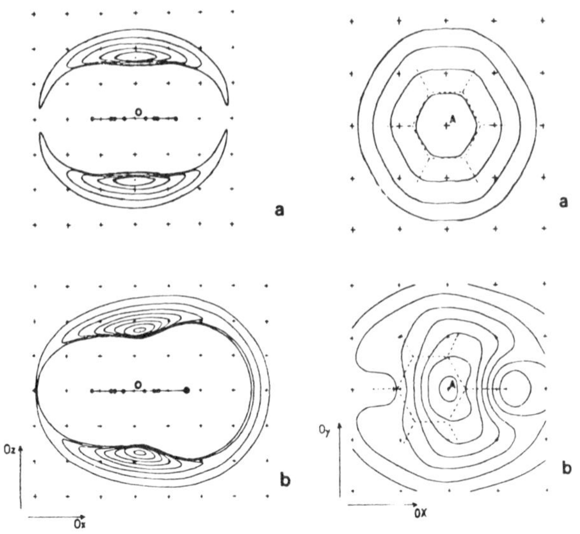

Fig. 3. PES contour maps of benzene-Ar and chlorobenzene-Ar in the Ox,Oz and AX,Ay planes

4. SPECTRA OF THE 1:2 COMPLEXES

The 1:2 complexes between phenol or aniline and two argon atoms have also been studied (Schmidt et al 1990). The vdW mode spectra support a symmetric (1,1) structure in each complex, the two totally symmetric stretching and bending modes being clearly assigned (Figure 4).

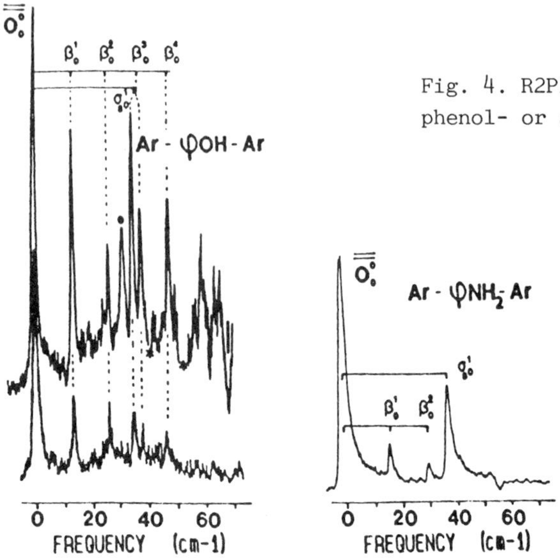

Fig. 4. R2PI vdW mode spectra of the phenol- or aniline-Ar_2 1:2 complexes

Comparison of the results in the 1:1 and 1:2 complexes confirms that the first rare gas atom solvation above the benzene plane has a negligible effect on the symmetric second solvation on the other side.

Mons M, Le Calvé J, Piuzzi F and Dimicoli I 1989 *J. Chem. Phys.* 92 2155
Mons M and Le Calvé J 1990 *Chem. Phys.* 146 195
Schmidt M, Mons M and Le Calvé J 1990 *Z. Phys. D Atoms, Molecules and Clusters* (in press)

Inst. Phys. Conf. Ser. No 114: Section 4
Paper presented at RIS 90, Varese, Italy, 16–21 September 1990

Two-colour two-photon threshold ionization spectroscopy of van der Waals complexes: Isomers of the first benzene-argon$_n$ complexes

M.SCHMIDT, F.PIUZZI, M.MONS, J.LE CALVE and I.DIMICOLI

CEA, CEN Saclay, Département de Physique Générale, SPER, Bât 522,
91191 GIF-SUR-YVETTE Cedex, FRANCE

ABSTRACT: The 2C-R2PI technique is used to identify different conformers of the first benzene-argon$_n$ van der Waals clusters.

INTRODUCTION

During the last few years numerous spectroscopic studies of van der Waals (vdW) complexes formed between mono- or polycyclic aromatic molecules and atomic or molecular polarizable species have revealed important information about their structural, energetic and dynamic properties. After the pioneering work of Leutwyler and Jortner (1987) it is now well established that the first successive solvations of polarizable species like rare gas atoms on the finite microsurface of large aromatic molecules give rise to increasing red-shifts $\Delta\nu$ of the $S_1 \leftarrow S_0$ origin transition and to decreasing ionization potentials. Moreover different complex conformations are expected, when two or more rare gas atoms are solvated. The existence of several isomers was brightly illustrated by three recent studies of phenanthrene- (Troxler et al 1989), carbazole- (Knochenmuss et al 1989) and perylene- (Shchuka et al 1989) complexes with argon. An interesting question concerns the possible existence of isomers in the case of monocyclic aromatic molecules, and in particular of benzene (Bz) itself.

Using two-color resonant two-photon ionization (2C-R2PI) technique in combination with a pulsed molecular jet and a mass-spectrometer (TOF) we present the first observation of different conformers of benzene-argon$_n$ (1:n) complexes.

RESULTS AND DISCUSSION

The one-color R2PI spectrum of the 1:2 complex clearly shows strong effects of larger cluster fragmentation processes, due to the high energy excess. Therefore only mass-selected two-color R2PI spectra are presented.

The $S_1 \leftarrow S_0$ 6^1_0- transition spectra of the first Bz-Ar_n complexes (n≥2) in figure 1 exhibit two nearly equidistant main bands (labelled A and B). The width of the more red-shifted band (A-band) relative to the monomer 6^1_0-transition increases with the number of solvated Ar atoms, while that of the B-band remains narrow, but nearly disappears for clusters higher than the 1:6 complex.

In the 1:2 complex the A-band position satisfies the red-shift additivity rule, when compared with the red-shifted 6^1_0- transition in the 1:1 complex, as for other 1:1 and 1:2 complexes of benzenic molecules with rare gas atoms (Dao et al 1985, Schmidt et al 1990). This indicates that (1) the Ar atoms occupy identical positions above and below the molecular plane and (2) a first complexation does not significantly affect the second solvation on the other side of the molecule. The A-band is thus attributed to the 6^1_0-transition of the symmetric (1.1) conformation with an Ar atom on each side of the molecular surface. Concerning the assignment of the B-band in the 1:2 complex several possibilities can be envisaged :

(1) a vdW band of the 1:2 (1,1) complex can be ruled out with intensity and frequency arguments.

(2) an eventual fragmentation of higher clusters in both S_1 and ionic states is excluded since no spectral coincidence exists between the successive B-bands (and first A-bands) of the 1:n complexes. In addition from pressure arguments the fragmentation of Bz_2- Ar_n cluster was also ruled out.

(3) finally the B-band could correspond to the 6^1_0- transition of a second isomer of the 1:2 complex. Indeed simple calculation models show that the most stable structures (namely the (1,1) symmetric species and the (2,0) complex with an Ar_2 dimer parallel to the benzene plane) have similar total binding energy values ($D_e \simeq$ 700 and 650 cm^{-1} respectively) and can thus simultaneously be present in the jet.

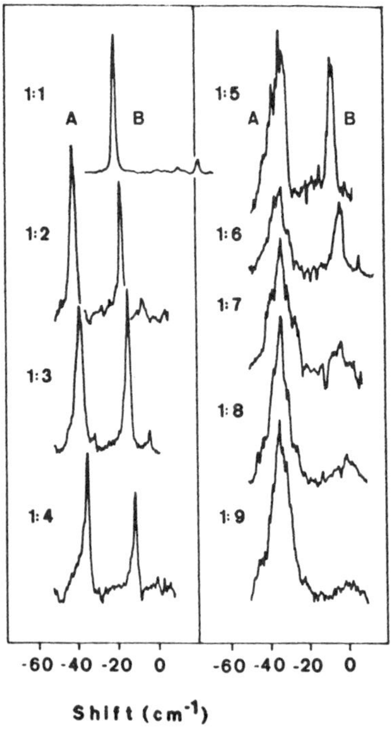

Figure 1 : Mass-selected 2C-R2PI spectra of the $S_1 \leftarrow S_0$ 6^1_0 transition of the first Bz-Ar_n clusters. The frequency shifts are relative to the monomer transition.

In order to check this assignment, 2C-R2PI ionization efficiency spectra of the monomer and the first 1:n complexes (n≤5) have been recorded. Fig.2(a) shows that the spectra of the monomer, the 1:1 and 1:2(A) complexes are very similar. The adiabatic threshold is observed to be red-shifted for the two complexes by 170 cm^{-1} and 335 cm^{-1}, in agreement with the red-shift additivity rule. A second threshold appearing at higher energy, (≈350 cm^{-1}) is assigned to the 6(3/2) Jahn-Teller split vibrational level in each cation (Long et al 1983). Fig. 2(b) shows the different ion yield spectra obtained by excitation of the A- and B-bands in the 1:2 complex. The spectrum of the 1:2(B) complex exhibits only one broad threshold blue-shifted by 143 cm^{-1} relative to the adiabatic threshold of Bz. This feature has to be attributed to the vertical ionization of another species than the 1:2 (1,1) complex. We assign it to the disymmetric (2,0) isomer of the 1:2 complex.

The 6^1_0- transition red-shifts of the 1:n clusters deserve some important remarks. The 1:2 (2,0) B-band position nearly corresponds to the 6^1_0- band of the 1:1 complex (Fig.1). This means that one or two Ar atoms on one side of the Bz plane result in very similar red-shifts. The assignment of the A- and B-bands in the 1:3 complex obviously follows, namely the A-band belongs to the (2,1) isomer and thus the B-band belongs to the only other conformer (3,0). Figure 3 presents schematically our proposed assignments of the A- and B-bands for all possible conformations of the first 1:n clusters. The A-band is generally attributed to conformers with Ar atoms distributed on both sides and the B-band to the unique conformation with all Ar atoms on

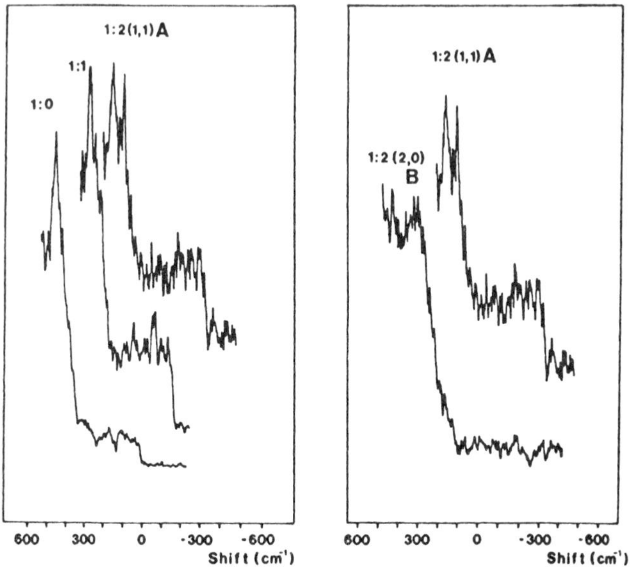

Figure 2 : Two-color ionization efficiency spectra of a) the monomer, the 1:1 and 1:2 (1,1) complexes ; b) the (1,1) A and (2,0) B isomers of the 1:2 complex.

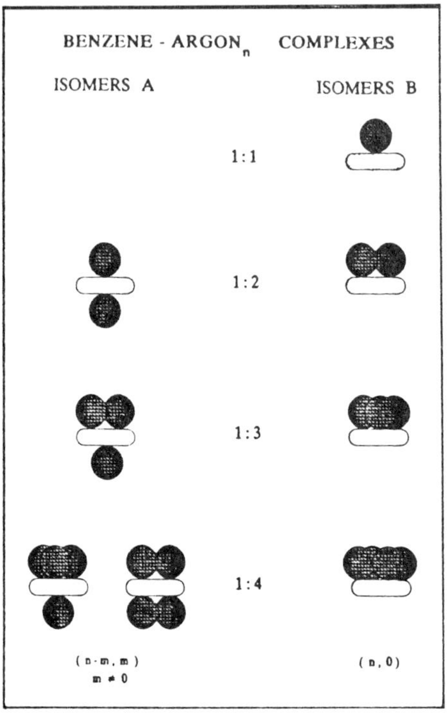

Figure 3 : Proposed assignments of the A and B bands in the spectra of the first Bz-Ar_n clusters.

the same side. This model obviously explains the increasing width of the A-band for higher clusters as arising from the 6^1_0- transitions of several isomers (n≥4) and on the other hand the narrow B-band as only due to the (n,0) conformers.

For the 1:7 (7,0) complex we suggest that the flower-shaped configuration of the Ar-atoms represents some stability limit for the (n,0) conformations, in agreement with the fact that the narrow B-band disappears for n≥7.

REFERENCES

Dao P D, Morgan S and Castleman A W 1985 *Chem. Phys. Lett.* **113** 219
Knochenmuss R and Leutwyler S 1990 *J. Chem. Phys.* **92** 4686
Leutwyler S and Jortner J 1987 *J. Phys. Chem.* **91** 5558
Long S R, Meek J T and Reilly J P 1983 *J. Chem. Phys.* **79** 3206
Schmidt M, Mons M and Le Calvé J 1990 *Z. Phys. D Atoms, Molecules and Clusters* (in press)
Shchuka M I, Motyka A L and Topp M R 1989 *Chem. Phys. Lett.* **164** 87
Troxler T, Knochenmuss R and Leutwyler S 1989 *Chem. Phys. Lett.* **159** 554

Inst. Phys. Conf. Ser. No 114: Section 5
Paper presented at RIS 90, Varese, Italy, 16–21 September 1990

Laser ionization of atoms through Rydberg states

E. Arimondo

Dipartimento di Fisica, Università di Pisa, Piazza Torricelli 2, 56100 Pisa, Italy

ABSTRACT: Experiments concerning ionization of Rydberg sodium atoms are described. Rydberg states are prepared by two-step laser excitation and ionized through either electric field ionization, or collisional processes, or absorption of radiation from i) blackbody, ii) dye laser radiation iii) CO_2 laser. Collective effects occurring in the collection phase of the laser produced electrons and ions will be also discussed.

1. INTRODUCTION

Rydberg states represent an important step in ionization through laser excitation. The specific properties of Rydberg states as their long lifetime,their low ionization energy make the ionization dynamics at the same time very efficient, but also quite complicate. The investigation of the ionization dynamics of Rydberg states is important for basic reasons and for the efficiency of a RIS scheme. The fundamental aspects of sodium Rydberg states dynamics has been investigated by us through a series of experiments in different experimental conditions. In this paper the properties of Rydberg states and their relevance to the ionization dynamics will be briefly recalled and illustrated on the basis of the experimental and theoretical results.

In the first part of this paper the properties of Rydberg states will be briefly recalled. Then the ionization channels for the Rydberg atoms will be discussed. A rate equation model for the ionization through the different channels will be described and numerical results presented. Finally experimental results concerning collision processes, CO_2 laser ionization, Stark effect, and collective phenomena in the charge collection phase will be presented.

2. PROPERTIES OF RYDBERG STATES

The Rydberg states correspond to highly excited atomic states, that, apart local perturbations, are similar to hydrogen atoms. Their energy E_{nl} is represented through a hydrogen-like formula for a state with quantum numbers n and l:

$$E_{nl} = - \frac{R_y}{(n^*)^2}$$

where R_y =13.6 eV is the Rydberg energy, and $n^* = n - \delta_l$ the effective quantum number is related to the quantum number n by the quantum defect δ_l. These Rydberg levels are characterized by a large orbital radius, large geometrical cross sectional area (that determines also the order of magnitude of most collisional cross sections), small ionization energy, small energy level separation, as discussed in the review by Gallagher (1988). For instance our experiments dealing with the n=20 Rydberg level show that its $1.4\ 10^5\ A^2$ cross sectional area and 0.03 eV ionization energy play an important role in the ionization processes. Features of Rydberg states depend very weakly on the core so that all Rydberg "atoms" have very similar properties.

3. IONIZATION

The ionization of Rydberg atoms may take place through an applied electric field, absorption of radiation and finally collisions, that represent an efficient ionization channel, but not a selective one, and will be discussed in the following Section.

3.1 Electric field

Rydberg states are very sensitive to an applied electric field, and Stark manifolds with linear Stark shifts are associated to each set of n levels, except the s and p states which have minimal Stark shift until they approach the Stark manifold, at which point they become part of the manifold. The energy levels of sodium Rydberg states in an electric field have been determined by Littman et al 1976. Because of the electric field, the Coulomb potential binding the electron is modified in an asymmetric way and high n states are not bounded by the potential whence ionize in presence of the electric field. The electric field E_c required for ionisation may be written as

$$E_c = \frac{1}{16n^4}\, 5.14\ 10^9 \mathrm{Vcm^{-1}}$$

Thus the ionizing field decreases as $1/n^4$ for high Rydberg states. In order to avoid a decrease in selectivity in the excitation process associated to the Stark shift of the Rydberg levels, the ionizing electric field may be applied after the laser excitation to Rydberg states. Time delay in a fraction of microsecond range is large enough to separate the laser excitation and the field ionization.

3.2 Radiation absorption

Ionization of Rydberg states may take place by absorption of an extra photon. The largest cross section for photon absorption occurs when the photon energy matches the separation between the Rydberg state and the ionization limit: for high Rydberg states this photon energy corresponds to the far infrared region. The cross section decreases with increasing photon energy, i.e. when for absorption of a photon with energy $\hbar\omega$ the released electron has increasing kinetic energy E_{kin} given by $E_{kin} = \hbar\omega - E_{nl}$ In Figure 1 we report the theoretical ionization cross section (in Mb) for ionization from the sodium 20d state versus the kinetic energy of the released electron, as derived from Manson, 1990. The upper line corresponds to the ionization cross section for hydrogen. The blackbody radiation peaked in the far infrared at room temperature provides an important source of ionization for the Rydberg states. Figure 2 reports the blackbody ionization rate for several sodium Rydberg states in presence of blackbody radiation at T=450K. For comparison the l-averaged spontaneous rate for n=30 results $3\times10^3\ s^{-1}$, so that a large ionization rate is produced by the blackbody absorption during the excited state lifetime. Absorption of visible photons, corresponding to kinetic energies in the 2-3 eV range for the 20d state, occurs with the smaller cross sections denoted by points in Figure 2. However the dye lasers used for the excitation to Rydberg states have a large photon flux, so that the ionization rate from this absorption channel may become comparable to the blackbody one.

4. APPARATUS AND SPECTRA

The experimental set-up for excitation and detection of sodium Rydberg states is composed by a diffusive atomic beam, entering a cooled, electrostatically shielded interaction cell. The cell cooling system allows to control the ionization rate associated to blackbody absorption. Laser beams, yellow and blue, for excitation to Rydberg states through 3s-->3p and 3p-->nd or ns two-step processes enter into the cell from a side. Infrared or far-infrared radiation for ionization of the Rydberg states may be also injected into the cell. Electric field plates inside the cell produce a highly uniform electric field for Stark splitting or ionization of the Rydberg states. The detection of the produced ions takes place through a mass selective system (time-of flight or quadrupole mass spectrometer) while electrons are collected on a plate.

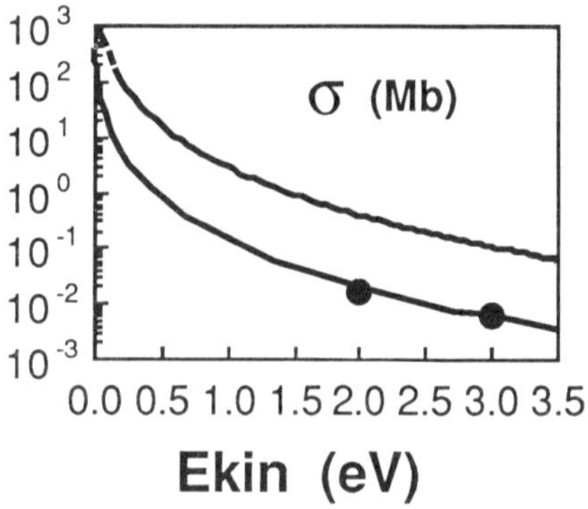

Fig. 1. Cross section for absorption of radiation from 20d state with release of an electron at kinetic energy E_{kin}. Lower line for sodium from Manson 1990; upper line for hydrogen as derived from Bethe and Salpeter 1957. Points for absorption of yellow and blue laser photons

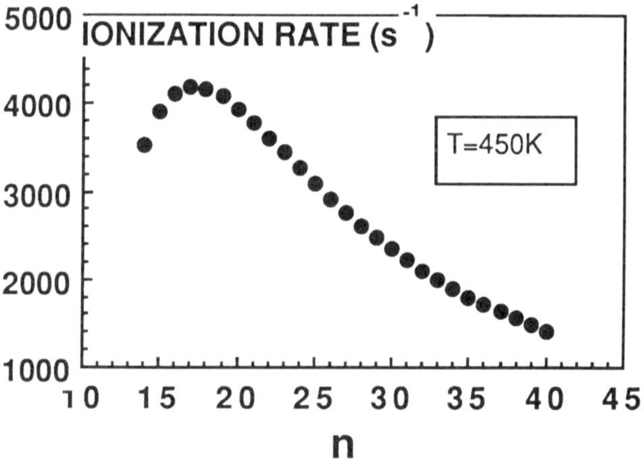

Fig. 2. Blackbody ionization rate for l-averaged Rydberg levels.

Figure 3 reports a typical spectrum for the Na ion production rate as a function of the blue laser wavelength from the 3p state to the Rydberg ns or np levels. The spectrum has been recorded for Na beam density $2x10^{11}$ cm^{-3}, yellow laser tuned 589.0 nm with intensity 0.25 MW/cm^2 and blue laser intensity 0.3 MW/cm^2.

5. RYDBERG IONIZATION MODELLING

A rate-equation model has been developed in order to describe the ionization of Rydberg states laser excited through a two-step process. Populations of i) ground state, ii) first excited state, iii) laser excited Rydberg state, iv) other Rydberg states coupled to the laser excited one by collisions or blackbody transfer, v) ions, vi) slow and vii) fast electrons created by different processes are included in the model. Atomic oscillator strength and collisional cross sections for ion production are derived from Spencer et al 1982 and Burkhardt et al 1986, while photoionization cross sections for the radiative ionization, as reported in Figure 2 for 20d, have been calculated by Manson 1990. The numerical solution of the rate-equation system has allowed us to make precise extimates of the ion production rates.

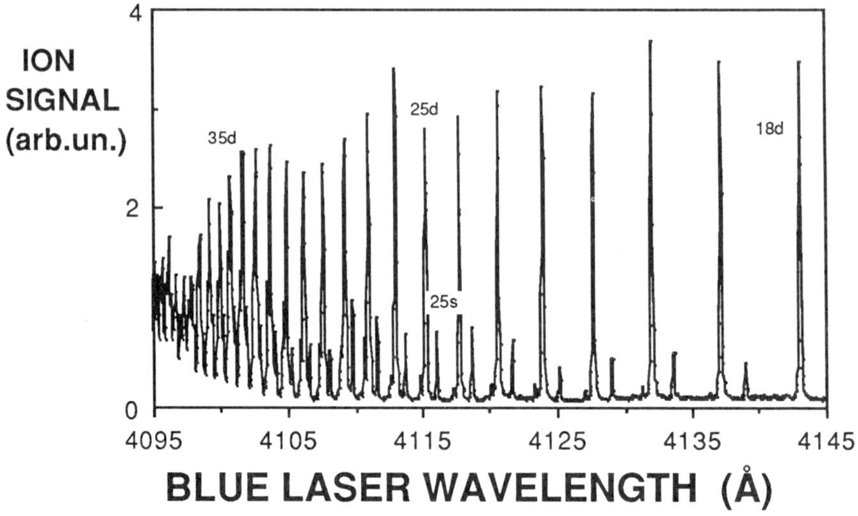

Fig. 3. Na ion spectrum as function of the blue laser frequency for 0.3 MW/cm^2 laser intensity

6. IONIZATION

We will analyze separately the ionization mechanisms associated to collisions, absorption by CO_2 radiation above threshold, and finally the collective effects modifying the charge collection.

6.1 Collisions

Several collision processes between Rydberg states have been observed and their cross section measured. For the case of sodium Rydberg states, the following ionization processes involving the Na** sodium atoms in Rydberg states and the Na* atoms in the first excited state contribute with a significan rate to the ionization:

Diffusive ionization:	$Na^{**} + Na \rightarrow Na^+ + Na + e^-$
Energy pooling:	$Na^{**} + Na^* \rightarrow Na^+ + Na + e^-$
Ion-pair formation:	$Na^{**} + Na \rightarrow Na^+ + Na^-$
	$Na^{**} + Na^{**} \rightarrow Na^+ + Na^-$
Associative Ionization:	$Na^{**} + Na \rightarrow Na_2^+ + e^-$

Review of cross-section for the different processes have been presented in Kopystynska and Moi 1982, Allegrini 1985, Klucharev and Vujnovic 1990.

The spectra obtained for the ion production versus the laser wavelength in the different collisional processe are not very different from that of Figure 3, because the main requirement in the process is the preparation of Rydberg states whence laser resonance through the ns and nd states. It may be interesting to report the values for the cross section of some processes, for instance a very exothermic one as the collision between two Rydberg states leading to ion pair formation. This process was investigated by Ciocca et al 1986, and by a comparison with the associative ionization the cross section for the ion pair formation resulted 10^4 A^2 at n=20. This value, while huge compared to cross sections for processes involving ordinary atoms is ten times smaller than the geometric cross section. That result illustrates the large importance of collisions in the dynamics of Rydberg levels.

6.2 CO_2 absorption

Absorption by radiation from a Rydberg level can be enhanced by the application of an additional laser field. In order to make use of the increase in the ionization cross-section near threshold reported in Figure 1, this additional laser has to be tuned at a frequency slightly larger than the energy splitting between the Rydberg level and the ionization limit. Experiments have been performed during the last year on Na Rydberg levels making use of a CO_2 laser operating with 5 Watts maximum output power. CO_2 laser is not continuously tunable, however its comb of output lines allows us to scan the threshold region. Attention has been concentrated on the 12s Rydberg level lying 0.094 eV below the ionization limit. Experimental observations have confirmed that an increase in ionization rate is obtained by tuning the laser above the threshold, with cross section in agreement with the theoretical values. However a large increase in ionization rate has been observed also for laser lines lying below the threshold value. This behaviour has been interpreted as originated by a Stark tuned resonance in the absorption between the 12s and an excited Rydberg level. In fact by operating in a regime of properly controlled and uniform electric field, the Stark tuning of different Rydberg levels into resonance with CO_2 laser radiation has been observed. An example of Stark tuning of the P32 laser line into resonance with transition from the 12s to the Stark components of the n=57 Rydberg manifold is presented in Figure 4. It has to be pointed that the resonant tuning occurs for applied electric field of few Volt/cm. The detection is based on the ionization of the n=57 state through a pulsed electric field.

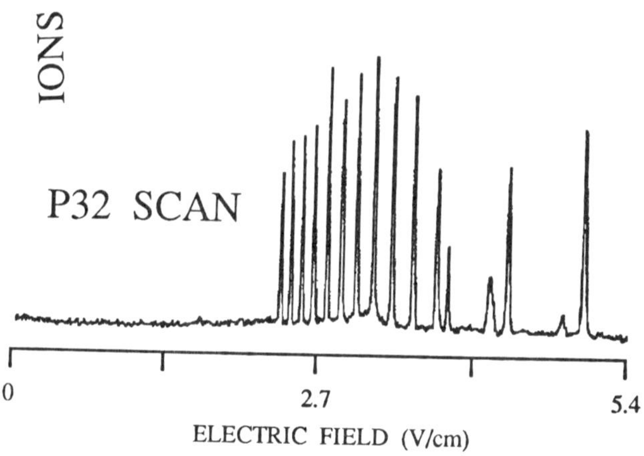

Fig. 4. Na^+ signal versus the applied electric field for excitation from n=12 to the n=57 Stark manifold with P32 CO_2 laser line in the top.

6.3 Collective effects

We have studied recently the phenomena occurring in the collection phase of laser ionization experiments when the number of created charges is increased. During the collection phase, the time-space behaviour of the produced charges may differ noticeably from that of non interacting charged particles. Trapping may occur when the electron kinetic energy is not sufficient to overcome the self generated electric field. Collective effects have been investigated in three-photon ionization experiments of sodium atoms where electrons are produced with an initial kinetic energy of about 1.5eV (Arimondo et al 1989), and in ionization experiment of Rydberg sodium atoms which produces both slow (at threshold) and fast ($\approx 2 \div 3$eV) electrons (Fuso et al, 1990). These investigations on charge collective effects in ionization experiments are at the border between atomic/molecular physics and plasma physics. A concentrated plasma with a low energy is created and the plasma evolution determines the charge collection. It results that in presence of an externally applied electric field the collective effects have a negligible role if $E>>4\pi e n_0 L$, with L the collection dimension and n_0 the initial charge density determined by the ion production mechanism. Furthermore a threshold condition exists, i.e. the charge density should be larger than a value dependent on the initial kinetic energy T_e of the ejected electrons: $n_0>n_{th}=kT_e/2\pi e^2d^2$ with d, the initial charge distribution width, that may differ from the laser spot size as a consequence of thermal motion(Giammanco and Arimondo,1990). For an initial kinetic energy kT_e =1eV and d=0.3 mm the density threshold value results $5x10^8$ charges/cm^3. Owing to the tiny volume irradiated by the laser, a total number of 10^3-10^4 charges produced by the ionization is large enough for collective effects to take place.

7. CONCLUSIONS

We have shown that the excitation through Rydberg levels could be an important path for ionization. The Rydberg level dynamics, even if complicated by the presence of competing mechanisms, may be modelled through rate-equations. Collision induced ionization processes, Stark tuning of transitions betweeen Rydberg levels and collective phenomena in the charge collections at charge densities in the 10^8- 10^9 cm^{-3} range have been observed.

8. ACKNOWLEDGMENTS

This paper is dedicated to the memory of Andrea Vellante who worked generously to set up experiments here described for his thesis at the University of Pisa. The present work originates from the collaboration of M. Allegrini, F. Fuso and F. Giammanco at the Dipartimento di Fisica dell'Università di Pisa, and C.E. Burkhardt, M. Ciocca, W. Garver, P. Schmidt , T. Hezel and J.J. Leventhal at the Physics Department, University of Missouri at St. Louis, Mo, USA. R. Dygdala from the Institut of Physics, University of Torun, Poland, has participated to the Pisa activity through a fellowship of the International Centre for Theoretical Physics in Trieste.S. Manson is thanked for the calculations of photoionization cross sections. This work has been supported by CNR-NSF Bilateral Research Program Grants No 89.00006.02 and INT-8719649.

6. REFERENCES

Allegrini M 1985 in *Fundamental Processes i n Atomic Collision Physics* ed H Kleinpoppen, J S Briggs and H O Lutz (New York Plenum)
Arimondo E, Buckhardt C, Qin L J, Giammanco F and Vellante A 1989 *Opt. Commun* **71** 52
Bethe H and Salpeter E E 1957 *Quantum Mechanics of One and Two Electron Atoms* (New York: Academic)
Burkhardt C E, Corey R L, Garver W P, Leventhal J J, Allegrini M and Moi L 1986 *Phys. Rev.* **A34** 80
Ciocca M, Allegrini M, Arimondo E, Burkhardt C E, Garver W P and Leventhal J J 1986 *Phys. Rev. Lett* . **56** 704
Fuso F, Allegrini M, Dygdala R, Giammanco F and Arimondo E, 1990 to be published
Gallagher T F 1988 *Rep. Prog. Phys.* **51** 143
Giammanco F and Arimondo E.,1990 *Europhys. Lett.* **11** 31
Klucharev A N and Vujnovic V 1990 *Phys. Rep.* **185** 55
Kopystynska A and Moi L1982 *Phys. Rep.* **92** 135
Littman M G, Zimmerman M L, Ducas T W, Freeman R R and Kleppner D 1976 *Phys. Rev. Lett.* **36** 788
Manson S T 1990 private communication
Spencer W P, Vaidyanathan A G, Kleppner D and Ducas T W 1982 *Phys. Rev.* **A26** 1490

Inst. Phys. Conf. Ser. No 114: Section 5
Paper presented at RIS 90, Varese, Italy, 16–21 September 1990

Large collective Lamb shifts in high pressure noble gases

M.G. Payne, W.R. Garrett,
Roger C. Hart[1], Irene Datskou,[1]
and James Wray[2]
Chemical Physics Section
Oak Ridge National Laboratory
Oak Ridge, Tn. 37831-6378

Abstract

When three-photon atomic excitation is produced by two laser beams of angular frequencies ω_{L1} and ω_{L2}, which are crossed at an angle θ , large co-operative pressure dependent shifts can be produced in the peak position of the excitation lineshape. By studying an experimental situation in xenon the theoretical predictions are shown to be in precise agreement with experiment for several angles and a wide range of pressures. The possibility of observing related shifts in nonlinear processes (such as backward stimulated hyper-Raman generation) is discussed.

Theoretical Results

In work[1–5] published between 1980 and 1986 it was shown that when three-photon resonant atomic excitation is attempted at elevated concentrations for a transition which is electric dipole allowed in both one- and three-photons, a third harmonic field is generated in the resonant medium which adiabatically follows the three-photon coupling due to the laser field. This field has an amplitude and phase such that a very complete destructive interference occurs between the coherent pumping of the transition by the three-photon laser excitation and by one-photon pumping due to the third harmonic field. As a result, the three-photon laser excitation probability can be shown to be strongly suppressed. This cancellation persists even when two (or three) different colors[4] from separate broad bandwidth lasers are used, or when any (2n+1)-photon (i.e., any odd-photon) resonant atomic excitation is attempted[2,5] for a transition that is electric dipole allowed in both one- and (2n+1)-photons. This cancellation effect differs from most nonlinear phenomena in being independent of the laser power density at low and intermediate power densities. The only requirement is that the power density should not be so high that saturation effects or large a.c. Stark shifts result.

In theoretical and experimental studies[1–5] of the odd-photon cancellation effect it was shown that the interference is avoided if counter propagating laser beams are used to carry out the excitation.[1] With counter propagating beams the magnitude of the signals is consistent with excitation occuring due to two photons being absorbed from one direction of propagation and one from the other, but with no excitation occuring due to the absorption of three parallel photons propagating in either direction. Ferrell, Payne, and Garrett[6] utilized three-photon excitation with counter propagating beams to study the selfbroadening of spectral lines. The use of counter propagating beams seemingly subverted the destructive interference in the excitation process and allowed excitation line profiles to be determined through multiphoton ionization measurements. However, large unexplained violet shifts were observed in the line position (about 70% of the full-width at half maximum of the line width expected from theory) for all of the transitions studied. Our experimentally observed shift was recently explained quantitatively by Friedberg, Hartmann, and Manassah[7,8] as being due to cooperative Lamb shifts[9] as they manifest themselves in three-photon excitation with counterpropagating laser beams of a single color.

In recent papers[10–12] we have considered the collective shift in a treatment which is similar to that of Ref. 8, but more general in that we treat cases involving two laser beams of different frequencies ω_{L1} and ω_{L2} crossing at an angle θ. Both $2\omega_{L1} + \omega_{L2}$

and $2\omega_{L1} - \omega_{L2}$ modes of excitation were treated. We have shown[10,12] that except for a small region within a few absorption lengths of the ends of the excitation volume, the total pressure-induced shift $\bar{\Delta}_P$ in an elementally pure atomic vapor or gas is made up of a contribution due to the sum of the Lorentz shift[9] and a conventional collision-induced shift and a third contribution due to the collective Lamb shift.[7] Specifically, we find (neglecting depletion of the ground state population)

$$\frac{d\rho_{11}}{dt} = R - \Gamma_I \rho_{11},$$

where,

$$R = \frac{2\Gamma_P |\Omega_{01}^{(3)}|^2}{\left[\delta_1 - \frac{\lambda_{mix}\kappa_{01}}{4\pi}\left(-1.11 + \frac{\lambda_{L1} \pm 2\lambda_{L2}}{(1-\cos(\theta))\lambda_{mix}}\right)\right]^2 + \Gamma_P^2}, \tag{1}$$

where ρ_{11} is the diagonal element of the density matrix corresponding to the population of the upper state, and

$$\lambda_{mix} = \frac{2\pi c}{2\omega_{L1} \pm \omega_{L2}},$$

$$\kappa_{01} = \frac{2\pi(2\omega_{L1} \pm \omega_{L2})N|D_{01}|^2}{\hbar c},$$

$$\Gamma_P = 2.33\frac{\lambda_{mix}\kappa_{01}}{4\pi}.$$

Here, λ_{L1} and λ_{L2} are the vacuum wavelengths of the two lasers, ω_{L1} and ω_{L2} are the laser angular frequencies, N is the gas concentration, θ is the angle between the propagation vectors for the two laser beams, and D_{01} is the electronic dipole matrix element between the ground state $|0>$ and the excited state $|1>$. Where $\pm$ appears, the choice of "+" corresponds to $2\omega_{L1} + \omega_{L2}$ excitation, while the choice "−" corresponds to three-photon excitation by $2\omega_{L1} - \omega_{L2}$.

The validity of Eq. (1) depends on strong absorption being present at the position of the shifted line. This in turn depends on θ not being too small, and λ_{L2} not being so large that $\kappa_{01}\Gamma_P L/(\bar{\Delta}_P^2) \leq 1$, where L is the length of the beam overlap region. Γ_P is the half-width of the absorption profile at half-maximum.[13] Once the shift becomes too large the excitation efficiency drops very rapidly. Note that when $\lambda_{L2} >> \lambda_{L1} > \lambda_{mix}$ the cooperative shift is much larger than the pressure broadened width, i.e., $\Delta_P >> \Gamma_p$. Moreover, in addition to its dependence on pressure and $|D_{0,1}|^2$ (or oscillator strength) the shift is also dependent on the relative magnitudes of the wavelengths used in carrying out the three-photon excitation and the angle between the two laser beams.

In reference (12) we have shown that for very large shifts – such that the far wing absorption β is no longer of the simple form $\beta \simeq 2\kappa_{01}\Gamma_P/(\delta_1^2)$ – Eq. (1) must be generalized to

$$R = \frac{2\Gamma_{eff}|\Omega_{01}^{(3)}|^2}{(\delta_1 - \Delta_{pm})^2 + \Gamma_{eff}^2}, \tag{3}$$

where

$$\Gamma_{eff} = \frac{\beta\Delta_{pm}^2}{2\kappa_{01}}, \tag{4}$$

and Δ_{pm} is the value of δ_1 at which phase matching occurs with the unfocused laser beams and β is the absorption coefficient at the detuning $\delta_1 \simeq \Delta_{pm}$. If $\beta \simeq 2\kappa_{01}\Gamma_P/(\delta_1^2)$, then $\Gamma_{eff} \simeq \Gamma_P$ and the width and amplitude of the peak in R is the same as in Eq. (1).

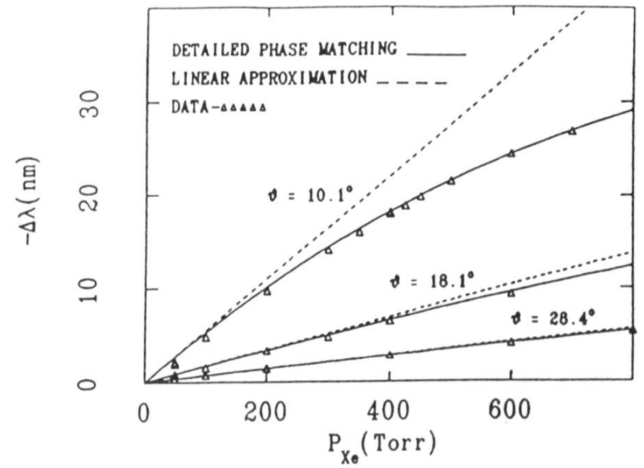

Figure 1: Pressure dependent shifts in the three-photon excitation of the $5p^6(J=0) \rightarrow 5d[1/2](J=1)$ transition in xenon for different angles between the beams at ω_{L1} and ω_{L2}.

All features of the cooperative shift phenomena predicted in References 10 and 12 for $2\omega_{L1} + \omega_{L2}$ excitation have been verified in a recent study by Garrett et. al.[11] In Figure 1 we compare[12] the measurements from Reference 11 to calculations based on theoretical estimates of the exact position of phase matching. This relationship between the position of phase matching and the collective shift only applies when the shift is very large compared with the pressure broadened width of the absorption line. In earlier studies[1–5] the prevailing view has been that the generation of the interfering four-wave mixing field could be ignored when counter propagating photons were used. We now know that this is not the case.[8,10–12] Indeed it is the interplay between the laser fields and FWM fields that produces the collective shifts and cancellation effects. At elevated concentrations, the generated field influences the three-photon excitation under all excitation geometries.

At the detuning where phase matching occurs, the FWM signal was shown[12] to provide far stronger pumping between states $|0>$ and $|1>$ than does three-photon pumping due to the laser. On the other hand, at detunings near the unperturbed resonance we have shown in Reference 12 that the FWM field provides a coupling nearly equal, but 180° out of phase, to that of the laser at all points in space. Consequently, three-photon cancellation occurs at and near the unperturbed resonance. If the absorption of photons at detuning δ_1 from the unshifted position is given accurately by the wing-absorption expression $\beta = 2\kappa_{01}\Gamma_P/\delta_1^2$, then when δ_1 corresponds to the point where "phase matching" occurs, an excitation peak appears. The peak has the same amplitude and pressure-broadened lineshape as a shifted atomic resonance with the same oscillator strength. The peak is due to the absorption of the FWM photons on the wing of the resonance line. In this limit the peak in the line occurs at the detuning where the four-wave mixing is phase matched. The lineshape is Lorentzian.

Collective Shifts in Stimulated Hyper-Raman Generation

Suppose, as shown in Figure 2, that a narrow bandwidth dye laser is tuned near (but not onto) a two-photon atomic resonance in a high pressure inert gas. Imagine that an additional weak electromagnetic wave at angular frequency ω_{HR} is introduced propagating either parallel or antiparallel to the laser beam. The latter frequency is chosen so that $2\omega_{L1} - \omega_{HR}$ resonantly pumps the three-photon resonance between level $|0>$ and level $|1>$. According to our previous discussion, if the field at ω_{HR} propagates parallel to the laser beam no excitation of level $|1>$ will occur. However, if this wave propagates in the direction opposite to the propagation direction of the laser, Eq. (1)

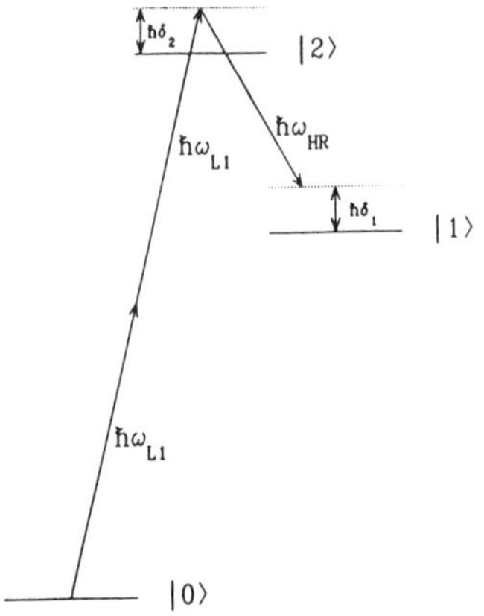

Figure 2: Energy level diagram defining the levels $|0>$, $|1>$, and $|2>$ and the detunings δ_1 and δ_2. The circumstance shown here is very favorable for the generation of stimulated hyper-Raman radiation at frequency ω_{HR}.

predicts that the transition is pumped at a rate given by

$$R = \frac{2\Gamma_P|\Omega_{01}^{(3)}|^2}{(\delta_1 - \bar{\Delta}_P)^2 + \Gamma_P^2}, \tag{4}$$

where $\Omega_{01}^{(3)}$ is given by

$$\Omega_{01}^{(3)} \simeq \frac{\Omega_{02}^{(2)}\Omega_{21}^{(1)}}{\delta_2}, \tag{5}$$

and $(\theta = \pi)$

$$\bar{\Delta}_P = \frac{\lambda_{mix}\kappa_{01}}{4\pi}\left[-1.11 + \frac{\lambda_{L1} - 2\lambda_{HR}}{2\lambda_{mix}}\right]. \tag{6}$$

In Eq. (5), $\Omega_{02}^{(2)}$ is one-half of the two-photon Rabi frequency between states $|0>$ and state $|2>$. The quantity $\Omega_{21}^{(1)}$ is one-half of the one-photon Rabi frequency between levels $|2>$ and $|1>$ due to the weak field at ω_{HR}. From energy conservation, every time an atom is promoted from $|0>$ to $|1>$ a stimulated emission must occur at the frequency ω_{HR}. Thus, for each excited atom produced an additional photon appears at ω_{HR} propagating in the direction of the original weak field. Consider the change in the intensity of the weak field which occurs between z and $z + dz$. We have

$$\begin{array}{l}\text{number of excited atoms produced} \\ \text{between z and z + dz per second} \\ \text{per unit area of the beam}\end{array} = NRdz. \tag{7}$$

$NRdz$ is also the number of photons at ω_{HR} generated per second per unit area between z and z + dz. Multiplying by $\hbar\omega_{HR}$ we obtain the change in the intensity, I_{HR}, of the beam at angular frequency ω_{HR} in propagating from z to z + dz. Thus,

$$\frac{dI_{HR}}{dz} = NR\hbar\omega_{HR}. \tag{8}$$

If we note that $\Omega_{21}^{(1)} = D_{21}E_{HR}/(2\hbar)$ we can write

$$\hbar\omega_{HR}|\Omega_{21}^{(1)}|^2 = \frac{2\pi\omega_{HR}|D_{21}|^2}{\hbar c} I_{HR}. \tag{9}$$

Letting $\kappa_{21} = 2\pi\omega_{HR}N|D_{21}|^2/(c\hbar)$ and using Eqs. (4) and (5) we find

$$\frac{dI_{HR}}{dz} = \frac{2\Gamma_P\kappa_{21}}{\delta_2^2[(\delta_1 - \bar{\Delta}_P)^2 + \Gamma_P^2]}|\Omega_{02}^{(2)}|^2 I_{HR}. \tag{10}$$

We have shown that a weak backward wave at ω_{HR} is amplified exponentially, with the gain peaked at $\delta_1 = \bar{\Delta}_P$. $\bar{\Delta}_P$ is given by Eq. (6). Since the maximum gain is at the shifted wavelength, the hyper-Raman wave starting from a hyper-Raman scattering event will amplify fastest at the shifted wavelength. Aside from the shift, the expression for the gain is identical to the result obtained using appropriate nonlinear susceptibilities and solving Maxwell's equations in a slowly varying amplitude and phase approximation. The fourth wave was included in deriving R and is responsible for the pressure dependent shift. In contrast, a weak field at ω_{HR} propagating in the forward direction produces no population in $|1>$, and therefore does not amplify. Thus, only a shifted backward stimulated hyper-Raman beam should be observed. As proven previously[14] there is no forward stimulated hyper-Raman beam.

By using focused laser beams in xenon it should be possible to achieve efficient stimulated hyper-Raman generation when the dye laser is tuned around the two-photon resonance between the ground state and the 6p[1/2](J=0) state. This resonance occurs at 249.6 nm, where more than 2mJ of tunable output is achievable with commercial lasers. Several lower states can serve as state $|1>$, with the 6s[3/2](J=1) state yielding the wavelength which is easiest to study. The wavelength generated will be 827.6 nm. The shift generated here with 500 Torr of xenon is $\bar{\Delta}_P = -4.2$ cm^{-1}, corresponding to the hyper-Raman wavelength shifting towards the blue.

ACKNOWLEDGMENT

Research sponsored by the Office of Health and Environmental Research, U.S. Department of Energy under contract DE-AC05840R21400 with Martin Marietta Energy Systems, Inc. [1] Graduate student from the University of Tennessee, Knoxville, Tennessee. [2] Student from Berea College, Berea, Kentucky.

REFERENCES

1. J.C. Miller, R.N. Compton, M.G. Payne, and W.R. Garrett, Phys. Rev. Lett. 45, 114 (1980). See also: J.C. Miller and R.N. Compton, Phys. Rev. A25, 2056 (1982).
2. M.G. Payne, W.R. Garrett, and H.C. Baker, Chem. Phys. Lett. 75, 468 (1980). See also: M.G. Payne and W.R. Garrett, Phys. Rev. A 28, 3409 (1983); M.G. Payne and W.R. Garrett, Phys. Rev. A26, 356 (1982).
3. D.J. Jackson and J.J. Wynne, Phys. Rev. Lett. 99, 543 (1982). See also J.J. Wynne, Phys. Rev. Lett. 52, 751(1984).
4. W.R. Garrett, Stuart D. Henderson, and M.G. Payne, Phys. Rev. A34, 3463(1986).
5. W.R. Garrett, Stuart D. Henderson, and M.G. Payne, Phys. Rev. A35, 5032 (1987).
6. Wanda R. Ferrell, M.G. Payne, and W.R. Garrett, Phys. Rev. A35, 5020 (1987).
7. R. Friedberg, S.R. Hartmann, and J.T. Manassah, Phys. Rev. A39, 93 (1989).
8. R. Friedberg, S.R. Hartmann, and J.T. Manassah, J. Phys. B: At. Mol. Opt. Phys. 22, 2211 (1989).
9. R. Friedberg, S.R. Hartmann, and J.T. Manassah, Phys. Rpts. 7C, 101 (1973).
10. M.G. Payne and W.R. Garrett, Phys. Rev. A42, 1434 (1990).
11. W.R. Garrett, Roger C. Hart, James E. Wray, Irene Datskou, and M.G. Payne, Phys. Rev. Lett. 64, 1717 (1990).
12. M.G. Payne, W.R. Garrett, R.C. Hart, and Irene Datskou, Phys. Rev. A (September 1, 1990 issue).
13. P.R. Berman and W.E. Lamb, Phys. Rev. 187, 221 (1969).
14. M.A. Moore, W.R. Garrett, and M.G. Payne, Optics Comm. 68, 710 (1988).

Resonant ionization spectroscopy of doubly-excited states including one-Rydberg electron

P. Camus, J.-M. Lecomte, C. R. Mahon, P. Pillet and **L. Pruvost**

Laboratoire Aimé Cotton*, CNRS II, batiment 505, Campus d'Orsay, 91405 Orsay cedex, France.
*This laboratory is associated to the Université Paris-Sud

Abstract

We report on autoionisation decay of di-excited states of barium atom including one Rydberg electron with different l angular momentum values. When l is low ($l \leq 2$) the great part of the large autoionisation linewidth is due to core penetration of the Rydberg electron. At the opposite, when l is large, the autoionisation decay is considerably reduced and mostly depends on the large range interaction between both excited electrons. In the first case, all the spectra caracteristics of the $7p_{j_1}nd$ series are well interpreted using an independant electron model based on the quantum defect theory. In the second case, because of the narrower autoionisation linewidths $6p_{j_1}nl$ and $7d_{5/2}nl$ series with $l = n-1, n-2, ...$ exhibit resolved $(j_1 - l)K$ splittings due to correlation effects between both excited electrons.

Introduction

To prepare di-excited barium atoms we use the isolated core excitation (ICE) method (*Cooke et al. 1978*), which is a two step sequential laser excitation of each electron. The first step consists to excite an initial $6snl$ Rydberg state from the $6s^2$ ground state. The second step consists to excite the $6s$ valence electron of this state through a resonant photoionisation to an autoionising Nl_1j_1nl di-excited state. A time-of-flight mass spectrometer is used for detecting produced ions after autoionisation.

For the low-l angular momentum states as $7p_{j_1}nd$ ($j_1 = 1/2$ or $3/2$, $10 \leq n \leq 50$), the core interaction between the outer electron and the rest of the atom is responsible for observed large autoionisation widths. For the high-l angular momentum states, $6p_{j_1}nl$ ($j_1 = 1/2$ or $3/2$, $10 \leq n \leq 15$ and $l = n-1, n-2, ...5$) and $7d_{5/2}nl$ ($11 \leq n \leq 14$ and $l = n-1, n-2, ...7$), narrower observed autoionisation widths, allow to clearly put in evidence the $(j_1 - l)K$ coupling effects between both excited electrons.

1. Experimental method

All the experiments have been realised with a barium atomic beam crossing perpendicularly the laser beams. In the case of low-l angular momentum states, the initial $6snd$ Rydberg state is obtained by using a two-step laser excitation which are the $6s^2 \rightarrow 6s6p\ ^1P_1$ (λ = 553.5 nm) and the $6s6p\ ^1P_1 \rightarrow 6snd\ ^1D_2$ ($\lambda \sim$ 415 nm) one-photon transitions. The resonant photoionisation of the $6snd$ state to $7p_{j_1}nd$ di-excited states, which is closed to the $6s \rightarrow 7p_{j_1}$ ionic Ba^+ transition uses a tunable $\sim$ 200 nm laser light, obtained by frequency mixing in a βBaB_2O_4 non linear crystal (*Pruvost et al. 1990*). The $7pnd$ states autoionisation products, which are mostly $7s$, $6d$ and $4f$ excited Ba^+ ions, are efficiently photoionised by the same $\sim$ 200 nm light. Then, the Ba^{++} ion signal recording allows spectroscopy of $7pnd$ autoionising states by scanning the $\sim$ 200 nm wavelength.

To prepare high-l angular momentum $6snl$ ($l = n-1, n-2, ...6$), we use the well adapted Stark switching method proposed by *Freeman et al.* in 1976, first developped by *Cooke et al.* in 1978 and more recently by *Jones et al.* in 1988. In presence of a high electric field ($\sim$1000 Vcm^{-1}) the well separated l-mixing $6sn\beta$ Stark levels can be selectively excited from the $6s6p\ ^1P_1$ state (see the before section) by an one-photon ($\lambda \sim$ 415 nm) excitation. Because of the long lifetime of the Stark levels ($\sim 1\mu$s for $n \sim 10$) the electric field can be decreased so slowly that the atom follows adiabatically the Stark diagram (see the dashed lines of the figure 1). So, at the zero electric field value we obtain a l well definite $6snl$ state which is non-degenerate because of very small quantum defects due to the core polarisation of the Ba^+ ion. Further, the excitation of the $6s$ valence electron of these high l angular momentum states has been studied in the vicinity of the one-photon $6s_{1/2} \rightarrow 6p_{j_1}$ ($\lambda \sim$ 493 or 430 nm) and the two-photon $6s_{1/2} \rightarrow 7d_{5/2}$ ($\lambda \sim$ 333 nm) Ba^+ transitions. The required laser lights are provided by a home-made dye laser pumped by an excimer XeCl* laser. In the case of the $6pnl$ series, the autoionisation products are low energy Ba^+ ions, which are directly detected because of their inefficient photoionisation in Ba^{++} by the laser light. This is not true for the $7d_{5/2}nl$ series, for which the more excited Ba^+ products are efficiently photoionised to Ba^{++} ions.

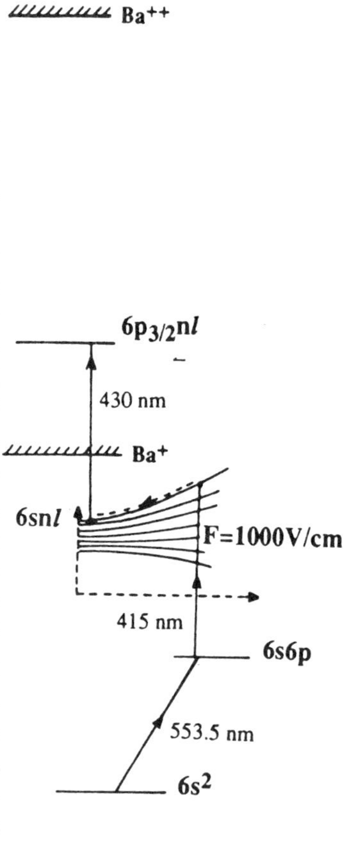

Figure 1

2. Low-l angular momentum states: $7pnd$ series

The very important autoionisation width, that we observe, is due to the core penetration of the Rydberg electron. This interaction induces a coupling between closed channels (when the outer electron is bounded) and opened channels caracterised by the kinetic energy of the free electron and the excitation of the Ba^+ ion. In the most cases, only two channels, one opened and one closed, are required (see the figure 2) for a parametrical fit of data. Only two parameters, a quantum defect δ' and a scaled autoionisation width γ' are then used to describe the coupling between a Rydberg series and an effective continuum (*Lecomte 1987*). Using the one-photon operator $T^{(1)}$ between the initial state $\mid 6snd\rangle$ and final state $\mid 7pnd\rangle$ wavefunctions, the excitation probability is given by the variation of the $\mid \langle 6snd \mid T^{(1)} \mid 7pnd\rangle \mid^2$ matrix element square. In an independant electron model (*Boulmer et al. 1988*) it is the product of a core excitation part, the spectral density g of the final Rydberg series and an overlapping integral square:

$$\mid \langle 6snd \mid T^{(1)} \mid 7pn'd\rangle \mid^2 = \mid \langle \chi_{6s} \mid T^{(1)} \mid \chi_{7p}\rangle \mid^2 .g. \mid \langle \phi_{n^*d} \mid \phi_{n'^*d}\rangle \mid^2$$

The spectral density g is given by lorentzian width series caracterised by the scaled autoionisation parameter γ'. The overlapping integral, which caracterises the excitation process, involves the initial and final radial wavefunctions of the outer electron (*Camus et al. 1985*) and depends on the difference of effective quantum numbers n^* and n'^* (*Bhatti et al. 1981*) of the outermost electron respectively of the $6snd$ and $7pnd$ states. The δ' and γ' fitted parameters are given in by *Pruvost et al. 1990*. Studying the $7p_{3/2}nd$ series we have observed an 20% enhancement of the scaled autoionisation width when n grows up above 16. As it is shown on figure 3 by the relevant energy position of the levels, the enhancement is due to the new opened channel $Ba^+_{7p_{1/2}}$. A selective analysis of the decay to this channel has been performed using an extra excitation of the $Ba^+_{7p_{1/2}}$ ions to Ba^+_{25d} Rydberg state, then microwave field ionised to Ba^{++} ion. This selective decay analysis method described in the above reference can be generalised to other studies.

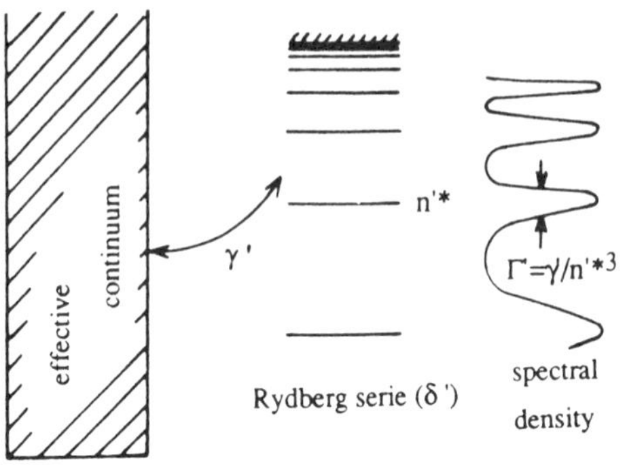

Figure 2

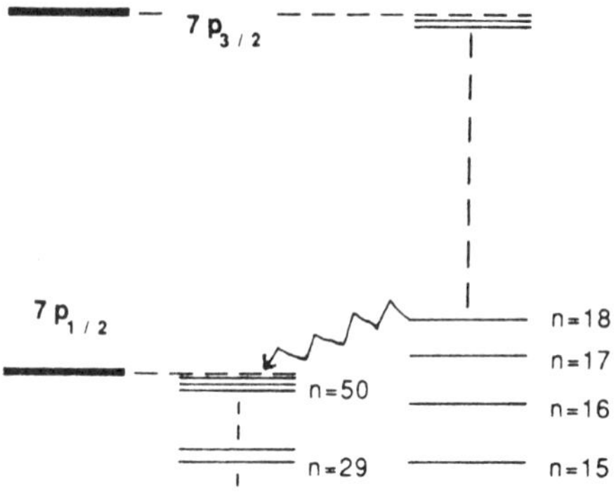

Figure 3

3. High-l angular momentum states: $6p_{3/2}nl$ and $7d_{5/2}nl$ series

As it is illustrated in the figure 4 for $n = 13$, the $6p_{3/2}nl$ states autoionisation widths are narrower than $7pnd$ states and rapidly decrease when l increases. Qualitatively, in our cases, the low autoionisation rate can be understood because of the high centrifugal barrier, the outer electron being submitted to come near the core. The low autoionisation width allows the observation of small effects due to the electrostatic interaction of both excited electrons as it is shown on the figure 4, for $l = 5$ for example. The observed splittings result of the $(j_1 - l)K$ coupling of the angular momentum of both electrons and cannot be interpreted using an independant electron model. Similar splittings are more clearly observed in the case of the $7d_{5/2}nl$ series (figure 5 with $n = 11$), by increasing the l value, because of the stronger correlation between both excited electrons. These new $(j_1 - l)K$ coupling effects have been conforted by accurate calculation of energy positions of the $6p_{3/2}nl$ levels using the perturbative theory.

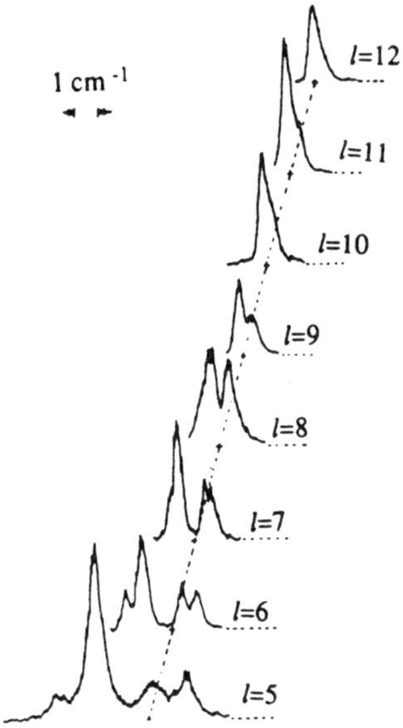

Figure 4

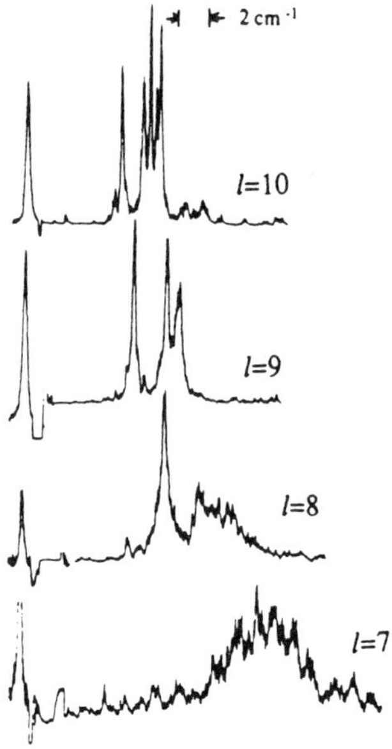

Figure 5

Conclusion

The two studies, we have presented here, show the evolution of autoionisation widths by increasing the angular momentum of the outermost electron in a doubly excited state. The correlation effects between both electrons, previouly observed for highly excited low angular momentum double-Rydberg states (*Camus et al. 1989*) are more clearly observable for high angular momentum di-excited states, because of their small autoionisation widths.

References

Bhatti S A Cromer C L Cooke W E 1981 *Phys Rev A* **24** 161
Boulmer J Camus P Lecomte J-M and Pillet P 1988 *J Opt Soc Am B* **5** 2199
Camus P Gallagher Lecomte J-M Pillet P and Pruvost L 1990 *Phys Rev Lett* **62** 2365
Camus P Pillet P Boulmer J 1985 *J Phys B: At Mol Phys* **18** L481
Cooke W E Gallagher T F Edelstein S A and Hill R M 1978 *Phys Rev Lett* **40** 178
Freeman R R Kleppner D 1976 *Phys Rev A* **14** 1614
Jones R R Gallagher T F 1988 *Phys Rev A* **38** 2846
Lecomte J-M 1987 *J Phys B: At Mol Phys* **20** 3645
Pruvost L Bolovinos A Camus P Lecomte J-M and Pillet P 1990 *J Phys B: At Mol Opt Phys* **23** L95

Inst. Phys. Conf. Ser. No 114: Section 5
Paper presented at RIS 90, Varese, Italy, 16–21 September 1990

RIS studies of weak perturbations in highly excited Rydberg series of heavy atoms

H. Maeda,* Y. Mizugai,** Y. Matsuo, A. Suzuki,* and M. Takami

RIKEN, The Institute of Physical and Chemical Research, Wako, Saitama 351-01, Japan

ABSTRACT: Feasibility of weak perturbations from low-lying nuclear excited states to electronic excited states of highly charged atomic ions is discussed. Preliminary experiments to study weak perturbations in highly excited Rydberg states of neutral atoms are reported as relevant experimental techniques to study perturbations from nuclear excited states.

Spectroscopic studies of highly excited Rydberg states provides us with crucial information on the electronic properties of highly excited atoms. This is particularly important for heavy atoms since their energy level structures near the ionization limits are still largely not well understood. However, such studies may also be useful for studying an exotic interaction, an interaction between nuclear and electronic excited states as very weak local perturbations in long Rydberg series of highly charged atomic ions.

Among various freedom of motions in atoms and molecules, internal motions of nuclei were almost always ignored in high resolution spectroscopy probably because the nuclear excitation energies are usually much larger than the energies of other motions. However, the lowest nuclear excited state known for primordial elements is only 75 eV above the ground state of ^{235}U. This is not far from the energy region where high resolution laser spectroscopy is applicable. There will be even lower excited states in unstable nuclei of heavy atoms.

These nuclear internal motions couple with electronic motions through electromagnetic interactions. A near resonant energy transfer from electronic to nuclear excited states through such interactions was first discussed by Morita (1973). Subsequently Otozai *et al.* (1978) demonstrated the existence of such energy transfer experimentally for ^{189}Os. This experiment was based on the internal excitation of a nuclear state with a virtual X-ray photon emitted when an outer shell electron filled a hole formed in the K-shell.

The above process occurs through weak Coulomb interactions between an electronically ground/nuclear excited state and an electronically excited/nuclear ground state. Such interactions, being off diagonal for the nuclear states, will cause perturbations to an ordinary electronic excited states from a nearby nuclear excited state. The perturbations also

---

*Faculty of Engineering, University of Tokyo, Tokyo, Japan
**Faculty of Science and Engineering, Sophia University, Tokyo, Japan

induce state mixing that will allow electronic dipole transitions between the nuclear ground and excited states.

However, such perturbations will be very difficult to find because the interactions are usually very weak. Therefore there should exist electronic bound states satisfying the interaction selection rules near the nuclear excited states so that one can find observable perturbations from nuclear excited states. This implies that the perturbations will be observed almost only in highly excited electronic states of highly charged ions to match the energy of nuclear excited states. Furthermore, since the energies of nuclear excited states are usually not accurately known, the perturbations will be best explored in the region where the density of electronic energy levels is high over a wide range of energy. One possible solution to satisfy the above requirements is to study weak local perturbations in long Rydberg series of highly charged ions.

Weak electronic perturbations to Rydberg series are frequently observed in heavy neutral atoms. Recently we observed one such example in highly excited Rydberg series of Lu. The ground state configuration of Lu I is $4f^{14}5d6s^2$ and that of Lu II is $4f^{14}6s^2$ $^1S_0$. Thus the Rydberg states of Lu I is hydrogen-like. Owing to the stability of 4f electrons, Lu atom has the simplest electronic energy level structure below the first ionization limit among Lanthanide elements. The electronic spectrum is rather well analyzed by Wyart(1978) providing reliable level assignment for the lower half of the first ionization limit.

The Lu atomic beam from a resistively heated oven was directed to an ion detector through three metal meshes. Two pulsed dye lasers pumped the Lu atoms to one of the $6s^2ns$ $^2S_{1/2}$, $6s^2nd$ $^2D_{3/2}$ and $^2D_{5/3}$, and $6s^2ng$ $^2G_{9/2}$ states between the first pair of meshes under field free condition. After a few mm free flight, the excited atoms entered into the second region where they were forced-ionized with 2 kV/cm electric field. This experimental configuration allowed us to detect the Rydberg atoms with life times longer than a few μs.

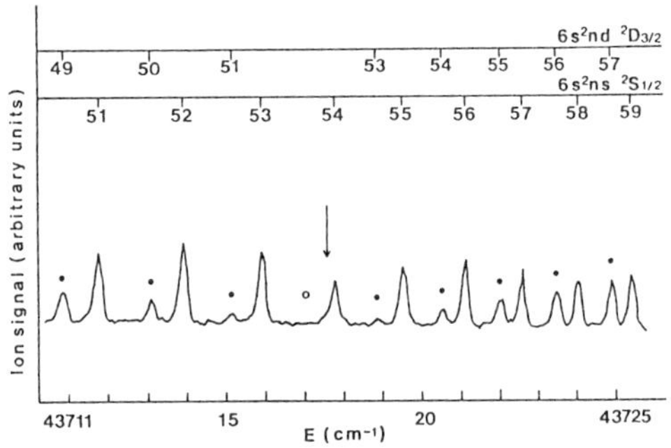

Fig.1. The perturbed $6s^2nd$ $^2D_{3/2}$ series.

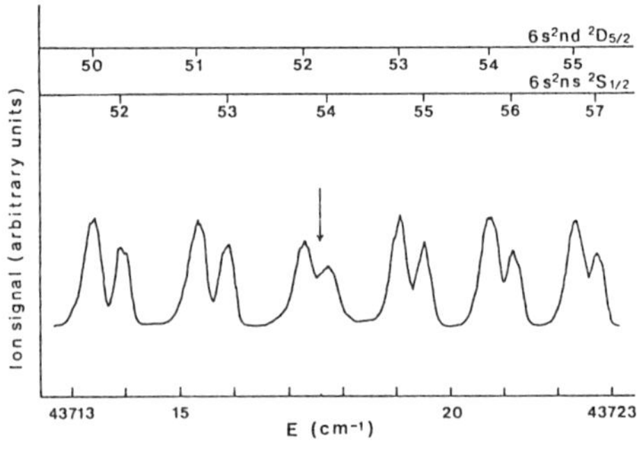

Fig.2. The perturbed nearby Rydberg series.

During this measurement, we found remarkable intensity dips among the $6s^2ns$ and $6s^2nd$ Rydberg series as shown in Figs. 1 and 2. The 52d $^2D_{3/2}$ line was entirely suppressed. The nearby lines of the same series showed decrease in intensity but frequency shift was not detected under the present resolution (0.1 cm^{-1}). When an electric field of 133 V/cm depressed the ionization limit below this region, a very strong ion signal appeared right at the center of the dip.

These observations clearly indicate the existence of a weak J = 3/2 perturber near the 52d $^2D_{3/2}$ level. The suppression of the ion signal occurred due to the radiative de-excitation of the perturbed levels since the radiative lifetime of the perturber is of the order of 10 ns: The perturbed Rydberg states lose their energy by spontaneous emission during the free flight time and are not ionized in the second region. Such depression of the ion signal has been observed also in the 6s24g Rydberg state of Ba by Sandner *et al.* (1986). However, the present perturber is much closer to the ionization limit and the observed interactions seems to be much weaker.

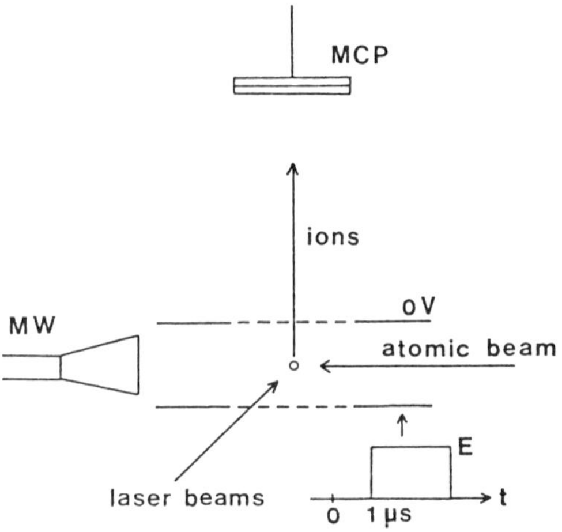

Fig.3. Experimental arrangement.

A particular interest in the present measurement is the weaker dips observed also in 54s $^2S_{1/2}$ and 52d $^2D_{5/2}$ since these J values are different from the most perturbed nd $^2D_{3/2}$ states. This breakdown of the J selection rule will be understood as a second-order nuclear quadrupole interaction since the major isotopic species of ^{175}Lu has extremely large nuclear quadrupole moment (5.68 b) (Miller *et al.* 1985).

For further study of such weak perturbations, microwave spectroscopy of Rydberg atoms will be useful since the energy level spacings are measured with much higher accuracy (MacAdam K B and Wing W H 1975). We performed a preliminary experiment on Yb I. Microwave spectroscopy of Rydberg atoms have been carried out extensively on alkali-metal atoms, and also on alkaline-earth atoms. However, to our knowledge no such study has been reported on rare-earth atoms. The Rydberg series of Yb has been well studied by Camus *et al.* (1980) and Aymar *et al.* (1984) by resonance ionization spectroscopy. The Yb atom (ground state: $4f^{14}6s^2\ ^1S_0$) was excited resonantly to the 6sns and 6snd series through the $4f^{14}6s6p\ ^1P_1$ intermediate state.

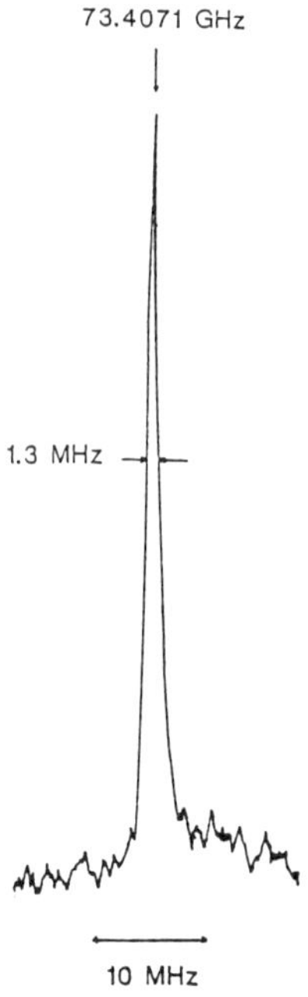

Fig.4. The 6s35p - 6s35s transition.

Weak microwave radiation further pumped up the excited atoms to odd states. The experimental scheme is shown in Fig.3. The strength of pulsed electric field was adjusted so that only the atoms pumped by microwave radiation were selectively ionized. The onset of pulsed field was delayed by 0.5-1 μs from the laser excitation.

So far the 6snp (J = 1) - 6sns $^1S_0$ microwave transitions have been observed for n = 35, 36, 39 - 50 in the frequency range of 23 - 74 GHz. A considerable singlet-triplet mixing has been reported between the 6sns $^1P_1$ and $^3P_1$ series (Aymar *et al.* 1984). The upper states of the observed microwave transitions are the upper component of the mixed state, denoted by P+ in the above literature. Figure 4 shows a typical trace of the 6s35p P_1+ - 6s35s $^1S_0$ transition at 73407.1 MHz. The observed line width, 0.65 MHz (HWHM), will be further reduced by improvement of the experimental conditions.

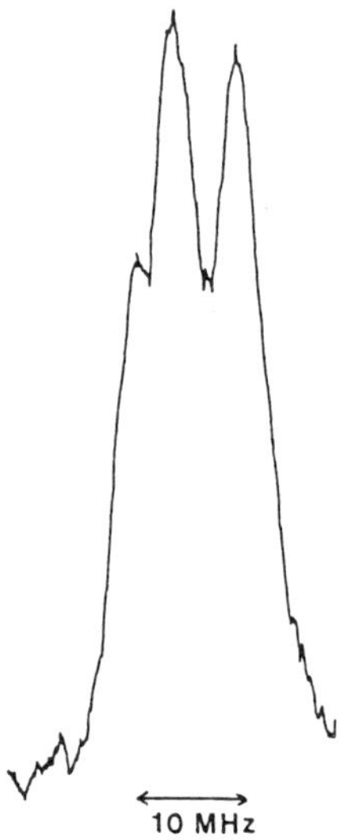

Fig.5. The perturbed 6s47p - 6s47s transition.

As the principal quantum number n increases, the measurement becomes more difficult since the Rydberg states become more sensitive to external perturbations like residual electric field or collisions with background gas. Figure 5 shows the 6s47p P_1+ - 6s47s $^1S_0$ transition observed at 27297 MHz. The line shows a triplet structure with larger line width. The 6snp P_1- series is known to be perturbed from $4f^{13}5d^26s$ J = 1 level around n = 47. Naturally this perturbation should influence the 6snp P_1+ levels, the upper levels of the microwave transitions observed in the present work. Although our poor experimental conditions do not allow definite conclusion, it is highly probable that the anomalous line structure is due to the perturbation since other lines near n = 47 also show anomalous line shapes. The perturbations will be complicated by the existence of five major isotopes as well as hyperfine structures associated with ^{171}Yb (I = 1/2) and ^{173}Yb (I = 5/2).

The perturbations from nuclear excited states will similarly be observed as anomalies in energy levels, life times, hyperfine structures, etc. They will be distinguished from the electronic perturbations by isotope-dependence of the effect.

References

Aymar M, Champeau R J, Delsart C and Robaux O 1984 *J. Phys. B: At. Mol. Phys.* 17 3645

Camus P, Debarre A and Morillon C 1980 *J. Phys. B: Atom Molec. Phys.* 13 1073

MacAdam K B and Wing W H 1975 *Phys. Rev.* A12 1464

Maeda H, Mizugai Y, Matsumoto Y, Suzuki A and Takami M 1989 *J Phys. B: At. Mol. Opt. Phys.* L511

Miller C M, Engleman R Jr and Keller R A 1985 *J. Opt. Soc. Am.* B2 1503

Morita M 1973 *Prog. Theor. Phys.* 49 1574

Otozai K, Arakawa R and Saito T 1978 *Nucl. Phys.* A297 97

Sandner W, Safinya K A and Gallagher T F 1986 *Phys. Rev.* A33 1008

Wyart J F 1978 *Phys. Scr.* 18 87

Monte-Carlo simulations of resonance ionization spectra

R Salomaa, T Kajava, H Lauranto, and X Zhan

Helsinki University of Technology, Department of Technical Physics,
Rakentajanaukio 2C, SF-02150 Espoo, Finland

ABSTRACT: Monte-Carlo simulations have been performed to study field fluctuation effects in resonance ionization spectroscopy. Considered topics include validity of rate equations, saturation and spectral properties, and statistics of the signal.

1. INTRODUCTION

The quality of laser beams is one factor that determines the performance limits of resonance ionization spectroscopy (RIS). Laser noise is partly due to quantum fluctuations, but often technical imperfections like uncontrolled multimode operation, mechanical vibrations etc., cause the dominant contributions. As regards the sensitivity of the method, laser noise has an important influence both on the saturation properties of the signal and on the background — on the latter one especially, when, e.g., amplitude fluctuations can enhance non-resonant multiphoton ionization. The RIS-signal is nonlinear and, therefore, it is difficult to predict its statistics. In such applications as isotope or isomer detection, which demand high spectral selectivity, phase and frequency fluctuations are expected to be particularly harmful.

Considerable theoretical effort has been devoted to studies on laser noise in nonlinear spectroscopy (see, e.g., Zoller 1984, Oraevsky 1988). Experimentally such investigations are most reliably done by mixing electronic noise into highly monochromatic laser light. The practical situation in RIS (Capelle *et al* 1987) as also, e.g., in CARS (Kröll *et al* 1987) is less ideal. In the present approach we resort to numerical simulation. Due to the large number of variables the simulations resemble real experiments and share in common, e.g., the difficulties in extraction of general conclusions.

2. NUMERICAL MODEL

We consider a two-step RIS scheme, which is assumed to bring forth the basic features of the system. The atoms are described with the density matrix ρ which obeys the equation of motion $d\rho/dt = -i\hbar^{-1}[H_0 + V, \rho] + \delta\rho/\delta t$. The commutator contains the unperturbed Hamiltonian and the atom-field (dipole) interaction $V = -\vec{\mu}\cdot\mathbf{E}$; the term $\delta\rho/\delta t$ includes reservoir phenomena such as spontaneous emission and collisions. The observed signal is assumed to be the total population escaping through the top level $|3\rangle$. Initially only the lowest level $|1\rangle$ is populated. The fields reside in the effective Rabi-frequencies $\alpha_p = \sum \alpha_p(k)\exp[i(\Omega_p - \Omega_k)t + i(\phi_k - \phi_p)]$ where the index p refers to the transition (1 to the lower step and 2 to the upper one), $\alpha_p(k) = \mu_p E_p(k)/2\hbar$ is proportional to the amplitude of the k th mode, Ω_k is the frequency of that mode and ϕ_k its phase; Ω_p is the center-mode frequency eliminated in the rotating-wave-approximation.

The behavior of ρ under single-mode excitation is well-known. The characteristic features include, i.a., *ac* Stark shifts and splitting of the resonances. When level degeneracies are accounted, the total signal is obtained as a superposition of the three-level subsystems, provided that circularly or linearly polarized light is used and no relaxational level mixing takes place. Calculations involving multimode fields are more intricate because of the rich variety of beat-phenomena. Numerical solutions of the multimode equations are, in principle, applicable, although problems may arise — in particular in the case of long pulses. Typically five modes at each transition is sufficient; already three modes are expected to introduce the basic properties of the system.

Laser noise is here simulated by introducing random variations into the field parameters. We have adopted a model where any of the parameters, α_i, ϕ_i, or the center frequency Ω_0 (mode spacings are kept fixed), may experience an abrupt jump to a new value uncorrelated with the previous value. We assume a Poissonian jump-time distribution $P(t) = \tau_{xc}^{-1}\exp(-t/\tau_{xc})$ where x refers to the random variable and τ_{xc} is the corresponding coherence time. The different random variables are fully uncorrelated. The new values of the variables are selected from a predefined distribution: top-hat for the phases, Gaussians for the center frequency and the mode amplitudes. If all τ_{xc} are longer than the pulse durations, it suffices to take the ensemble averages over the coherent responses. For shorter correlation times one has to sample realizations and thereafter calculate the statistics. Some preliminary results of the calculations were reported in the previous RIS-conference (Salomaa *et al.* 1988)

3. SATURATION BEHAVIOR

The best sensitivity of RIS is obtained when the transitions are saturated. Density matrix calculations predict that in the single-mode case the signal decreases at high intensities due to the *ac* Stark effect. This feature is preserved for multimode fields, too. In RIS models based on the rate equation approximation (REA) the atomic coherences and *ac* Stark effect are ignored, and, therefore, a different saturation behavior results. A question arises: how much laser field fluctuations improve the accuracy of REA?

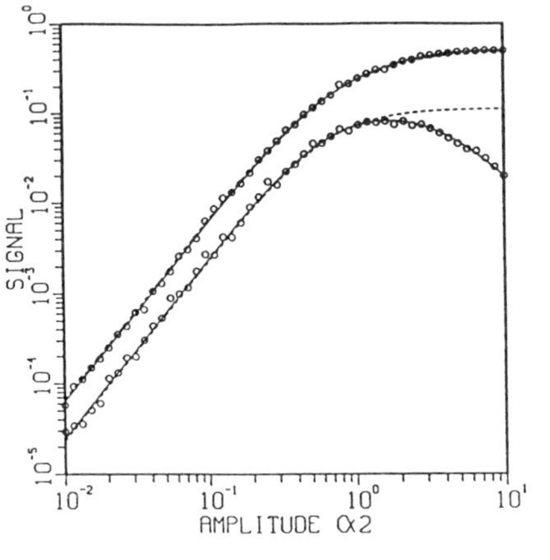

Fig. 1: Saturation curves

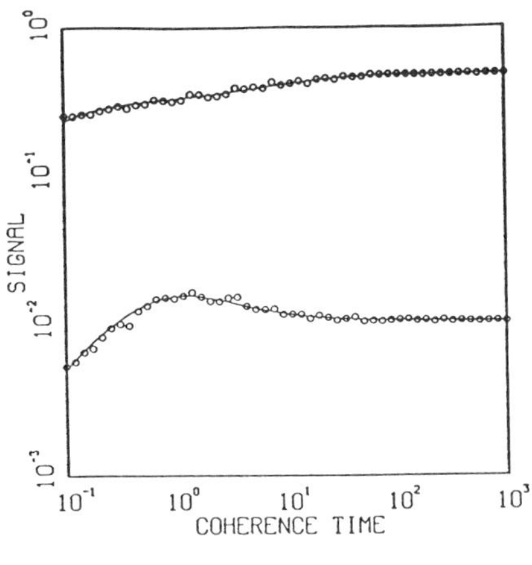

Fig. 2: Fluctuation dependence

In Figure 1 we show saturation curves calculated for resonant single-mode fields in the case of pure phase fluctuations with $\tau_{c1} = \tau_{c2} = 0.5$. In the lower curves the first step is weakly coupled, $\alpha_1 = 0.15$; in the upper curves $\alpha_1 = 1.50$. The population decay rates are $\gamma_1 = 0$, $\gamma_2 = 0.5$, $\gamma_3 = 0.5$, and the Gaussian laser pulses both have the length $\tau_p = 10$ (e^{-1} value). The averaging is done over 200 realizations only, which explains the small residual scatter in the points. For comparison we have also plotted the density matrix predictions (solid lines) and the REA results (dashed curves).

A formal derivation of the rate equations from the density matrix formalism yields the rate coefficient $R_{21} = 2\alpha_1^2\gamma_{21}/(\gamma_{21}^2 + \Delta_{21}^2)$, and similarly for R_{32}, where $\Delta_{21} = \omega_{21} - \Omega_1$ is the detuning and γ_{21} is the decay rate of the density matrix element ρ_{21}. The ensemble averages in the case of pure Poissonian phase fluctuations can be obtained analytically by simply making the replacements (Salomaa 1978) $\gamma_{21} \to \gamma_{21} + 1/\tau_{c1}$, $\gamma_{32} \to \gamma_{32} + 1/\tau_{c2}$, and $\gamma_{31} \to \gamma_{31} + 1/\tau_{c1} + 1/\tau_{c2}$ which formulas have been utilized when calculating the full curves of Figure 1. The agreement between simulations and analytical results is excellent; the surprisingly good accuracy of REA is explainable by the rather large effective value of $\gamma_{31} = 4.25$.

Similar saturation behavior is obtained, when α_1 is varied and α_2 is kept fixed. REA is accurate up to values $\alpha_1 \approx 1$, for the set of decay parameters considered in Figure 1. Interestingly, also the REA curves decrease in a pulsed regime for large values of α_1. The reason is that already the leading edge of $\alpha_1(t)$ strongly populates the intermediate level which thus has ample time to decay spontaneously before the maximum coupling to the top level takes place.

Figure 2 shows the dependence of the RIS signal on the coherence time for pure phase noise ($\alpha_1 = 1.5$, $\alpha_2 = 0.15$, lower curve, and $\alpha_1 = \alpha_2 = 1.5$, upper curve). As is evident, the simple recipe of taking into account phase fluctuations works. In the case of weak fields, fluctuations can, as a matter of fact, increase the signal. The underlying physical reason for this is that of the two different kind of processes contributing to the signal, the step-wise process disappears when $\gamma_1 \to 0$; only a finite spectral width of the fields makes this excitation sequence allowed (for details see Stenholm 1984). The effect is more pronounced when the lower transition is detuned: then the step-wise process, appearing at $\Delta_2 = 0$, is spectrally distinguishable from the two-photon process at $\Delta_2 = -\Delta_1$. According to steady state calculations the transition amplitude of the step-wise process is proportional to $(2\gamma_{21}/\gamma_2 - 1)$ which applied to the present situation yields $2/\gamma_2\tau_{1c}$.

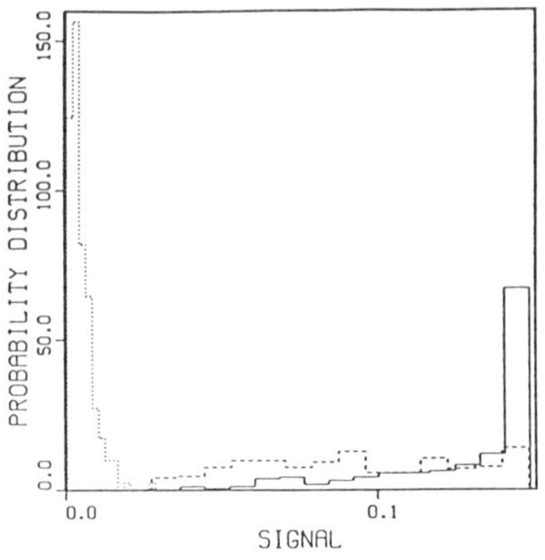

Fig. 3: Weak field distributions

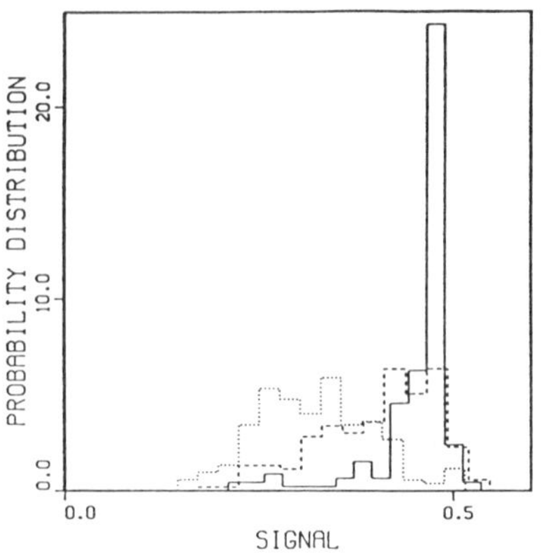

Fig. 4: Strong field distributions

The simulations, of course, provide also the distributions and thus the full information on the statistical properties of the signal. Figure 3 displays signal distributions for resonant single-mode fields with $\alpha_1 = \alpha_2 = 0.15$ and fixed coherence times; $\tau_{c1} = \tau_{c2} = 30$, 10, and 0.5 in the solid, dashed, and dotted curves, respectively. Figure 4 is the same, but for stronger fields, i.e., $\alpha_1 = \alpha_2 = 1.5$. As seen there are considerable differences in the weak and strong field behavior. For rapidly fluctuating fields the distribution is roughly exponential in the weak field limit. For stronger fields more complicated distributions are obtained: they neither fit well with a Poissonian nor with a Gaussian. The relative standard deviations of the cases shown in Figure 3 are 0.21, 0.36, and 0.75 in the order of decreasing coherence time. The corresponding values obtained from Figure 4 read 0.12, 0.19, and 0.23 which manifest the reduced effect of fluctuations in the presence of intense coherent fields.

4. SPECTRAL SELECTIVITY

Figure 5 gives examples of simulated resonance lineshapes for two different jump times, 100 and 0.5. We have assumed single-mode fields ($\alpha_1 = \alpha_2 = 1.5$, $\tau_p = 4$) and allowed, in addition to phase fluctuations, also laser frequency jumps within Gaussian distributions having a standard deviation $0.15\times2\pi$. For simplicity the mean jump times of the phases and frequencies are assumed all equal. In the average, the lower laser is in resonance, i.e., $\langle\Delta_1\rangle = 0$; the average detuning $\langle\Delta_2\rangle$ of the upper transition is varied. The line would be a Voigt profile if the central frequencies remained fixed during the pulse. Rather large line broadening in Figure 5 is evident, when the mean jump times are decreased from 100 to 0.5. For each detuning 50 realizations has been used in the averaging.

A basic question concerning the spectral selectivity of RIS is whether the signal is distinguishable from a background due to nearby perturbing transitions of unwanted species. Apparently the fluctuations fall off slower than the average signal. For instance, the mean value of the signal at $\langle\Delta_2\rangle = 0$ is 0.321 and its standard deviation 0.074 in the case $\tau_c = 0.5$; when the upper field is detuned by $\langle\Delta_2\rangle/2\pi = -1.5$ the signal drops to about 0.092, but the standard deviation only to 0.054. The inherent signal-to-noise ratio decreases towards the wings of the resonance line. To illustrate the effect of nearby perturbing transitions we show in Figure 6 the signal distributions at $\langle\Delta_2\rangle/2\pi = -1.5$ for different detunings of the lower transition and for $\tau_c = 0.5$. We have $\langle\Delta_1\rangle/2\pi = -1$, 0, and 1 in the dotted, dashed, and solid lines, respectively; the corresponding mean values (standard deviations) are 0.057 (0.044), 0.092 (0.054), and 0.101 (0.058). As is evident, the spread in the observed signal is rather insensitive to small detunings. Thus to get an accurate experimental value for the average of the signal, a large number of realizations must be sampled.

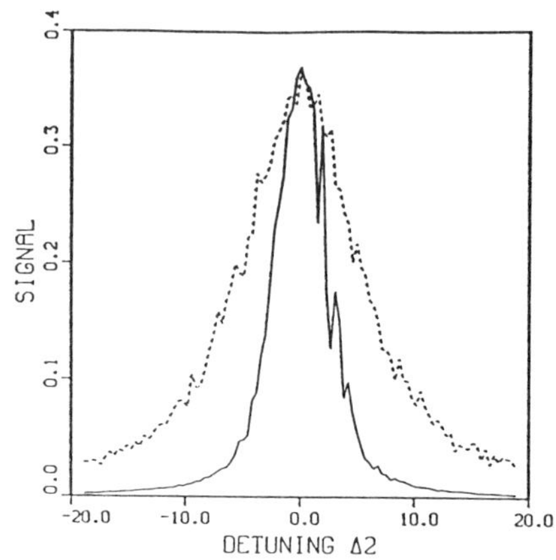

Fig. 5: Resonance lines

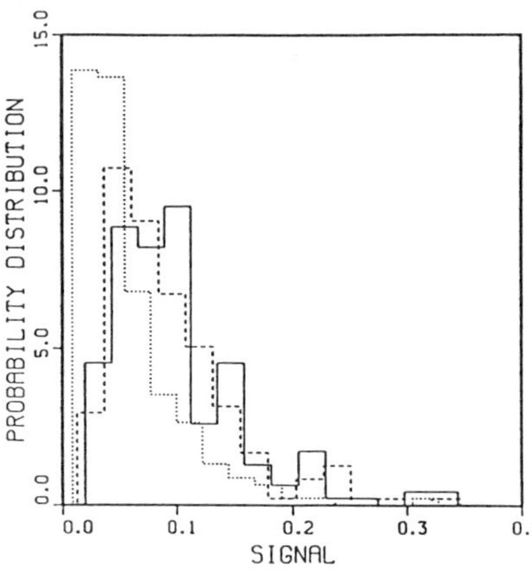

Fig. 6: Off-resonance distributions

REFERENCES:

Capelle G A *et al.* 1987 J. Opt. Soc. Am. B **4** 445.
Kröll S *et al.* 1987 Appl. Opt. **26** 1068.
Oraevsky A N 1988 J. Opt. Soc. Am. B **5** 933, and other papers in the same issue.
Salomaa R 1978 J. Phys. B: Atom. Molec. Phys. **11** 3745.
Salomaa R *et al.* 1988 *Resonance Ionization Spectroscopy 1988* ed T B Lucatorto and J E Parks (Bristol: Institute of Physics) p. 65.
Stenholm S 1984 *Foundations of Laser Spectroscopy* (New York: Wiley).
Zoller P 1984 *Multiphoton Processes* ed P Lambropoulos and S J Smith (Berlin: Springer) p. 68

Applications of diode lasers to resonance ionization mass spectrometry

J. P. Young, R. W. Shaw, and D. H. Smith

Analytical Chemistry Division, Oak Ridge National Laboratory, Oak Ridge, TN 37831-6142

ABSTRACT: Diode lasers have been utilized in resonance ionization mass spectrometry (RIMS). These devices have been used to excite the first transition of several 1+1+1 multiple photon processes for both lanthanum 138 and 139. Enhanced isotopic selectivity resulted; very-high resolution spectra were also obtained with the diode laser step to yield new atomic hyperfine structure constants. An isotopically selective RIMS instrument can be envisioned that uses diode lasers for optical excitation followed by non-optical ionization.

I. INTRODUCTION

Tunable semiconductor diode lasers are high intensity, coherent light sources that have found many technical and commercial applications. We have used these devices for RIMS. Diode lasers are very inexpensive, e.g., < $100 for the laser and $5,000 or so for commercial power supplies and temperature controllers to control wavelength. Diode lasers can operate in a single longitudinal mode with ample power, 3 mW, to excite bound-bound transitions; the FWHM of the emission is ca. 25 MHz (Lawrenz and Niemax 1988). Presently a wavelength limitation is that these semiconductor devices only emit in the red to near-infrared portions of the spectrum (670 nm to 1.5 μ) and do not cover this region completely. With the present commercial wavelengths, there are a number of opportunities for resonant absorption of red and near-infrared photons to excite atomic states of the lanthanides, actinides, and alkali metals. We have utilized a computer search program (Smith et al 1989) to find atomic transitions in lanthanides and uranium that are amenable to diode lasers in order to provide a catalog of transitions that would be useful for RIMS.

In this paper we report the application of diode lasers to the first transition of 1+1+1 multiple photon processes used in the study of lanthanum by RIMS. Lanthanum has two naturally occurring isotopes, 139 with a nuclear spin of 7/2 and 138 with a nuclear spin of 5. The natural isotopic composition of this element is 99.91% La-139 and 0.09% La-138.

2. EXPERIMENTAL

The experimental arrangement and the mass spectrometer used for these studies has been described (Shaw et al 1989). Light from either of two diode lasers was focused to approximately 1 mm dia. and allowed to intersect a focused beam, ~ 0.8 mm dia., from a copper vapor laser-dye laser (CVL/dye) in the source region of a magnetic sector mass spectrometer. These intersecting beams were between a thermal filament, that generated atoms, and the entrance slit of the mass analyzer. The pulsed (25 ns) dye laser was operated with Rhodamine 6G (R6G) solution at 6kHz. The spectral bandwidth was either 0.3 cm^{-1} or approximately 0.05 cm^{-1} with an intracavity etalon in the dye oscillator. The ASE of the dye laser was removed with a 3-66 Corning glass filter. The laser beams intersected at 90° but were sometimes collinear. Both natural abundance lanthanum metal and enriched La_2O_3 (14.3% La-138) were used as samples. Filament preparation and use has been described (Shaw et al 1989). In general, diode laser power in the sample region was approximately 100 μW and dye laser power varied from 20 to 200 mW average power. It was found that the attenuated power for both lasers was necessary for ultimate spectral resolution and isotope selectivity in the lanthanum studies.

3. RESULTS AND DISCUSSION

Diode lasers can indeed promote optical transitions that are part of a 1+1+1 multiple photon process leading to the ionization of lanthanum. As has been pointed out (Shaw et al 1989), several new peaks appeared in a plot of ion count vs. wavelength, over the R6G region of the CVL/dye laser system, if a diode laser was tuned to an allowed La transition. The new peaks represent a single-color 1+1 ionization route that was possible from this different quasi-initial state. In this dye region, six 1+1 single-color ionization routes were found from the $^4F^\circ_{3/2}$ level; two different diode laser transitions were found that would generate the $^4F^\circ_{3/2}$ level. These results are summarized in Figure 1. The plethora of resonant ionization routes leads to many studies. One can obtain information covering the relative population of the ground state compared to the first metastable state of lanthanum by measurements of the ion signal generated when one or the other diode laser transition is coupled with a common ionization route, e.g., 753.9 or 819.0 nm diode laser with 584.8 nm CVL/dye laser. Preliminary studies of relative ion count rates by these two routes have been carried out. The corrected count rate coming from the $^2D_{3/2}$ ground state is 45 times higher than that from the $^2D_{5/2}$ metastable state at a filament temperature of 1350°C and 25 times higher at 1500°C. Because of the hyperfine complexities of lanthanum, no further conclusions were drawn from the above result. It is obvious, however, that population information can be obtained and also obvious that it would be easier to obtain for non-hyperfine split levels.

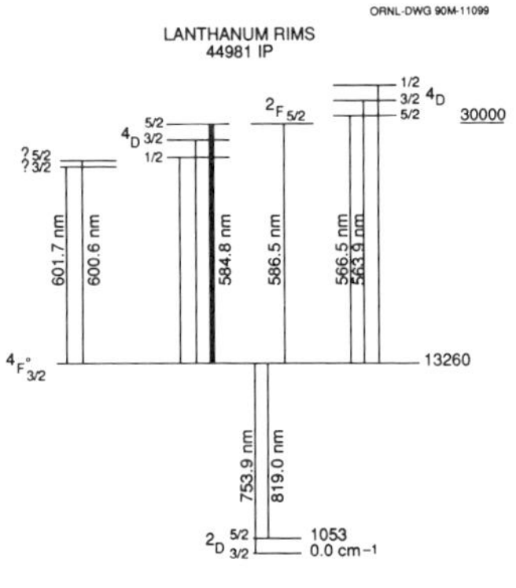

Fig. 1. Optical ionization routes for La

If the diode laser wavelength is fixed and the CVL/dye laser tuned, spectral results as described above (Shaw et al 1989) result; if one fixes the CVL/dye laser wavelength, say at 584.8 nm to promote the $^4F^\circ_{3/2}$ - $^4D_{5/2}$ ionization route, and tunes the diode laser wavelength, by varying the current, one gets spectral results as shown in Figure 2. The hyperfine spectrum of the $^2D_{3/2}$ - $^4F^\circ_{3/2}$ transition for La-139 as detected by RIMS is shown in the upper trace, and that for La-138 is shown in the lower trace. The enriched La_2O_3 sample was used for this study. It was also possible to observe the La-138 hyperfine structure in natural lanthanum (0.09% La-138); that

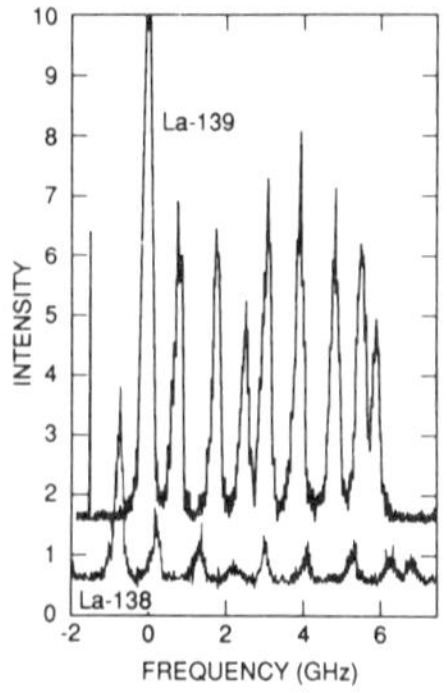

Fig. 2. Hyperfine structure of $^2D_{3/2}$ - $^4F^o_{3/2}$ for La138 and 139

result has been reported (Shaw et al, 1990). Although the expected Doppler width of transitions for these thermally excited atoms, generated at approximately 1350°C, is near 1 GHz, narrow absorption lines (FWHM 150 MHz) are observed. The reason for this is believed to be the geometry of our experimental setup; ions are generated at the intersection of two focused laser beams and are extracted by a narrow volume element of an ion lens.

The ability to obtain very-high resolution spectra opens many investigational areas. For isotopes with known nuclear spins one can determine level splittings and obtain atomic hyperfine structure constants. Such information for the lowest lying odd parity level of La-139, $^4F^o_{3/2}$, was obtained (Shaw et al, 1990). The splittings for both ^2D levels for La-139 were known (Ting 1957), and we calculated the atomic hyperfine splitting constants, A and B in two separate experiments using either diode laser transition. The final results for the 13260 cm^{-1} level are:

$$A = -347 \pm 2 \text{ MHz}$$
$$B = 60 \pm 8 \text{ MHz}$$

This study was carried out using relatively inexpensive diode lasers. Only one hyperfine transition for La-138 $^2D_{3/2}$ level has been reported (Childs and Goodman 1979), and from this the A and B constants have been estimated; we have made preliminary measurements of these constants for the $^2D_{3/2}$, $^2D_{5/2}$, and $^4F^o_{3/2}$ levels and will be able to obtain more precise measurements using the enriched La sample.

Looking again at the spectra shown in Figure 2, it is obvious that one can attain isotopically selective ion formation with diode-laser RIMS by selective tuning of the diode laser wavelength to a specific hyperfine transition of each individual isotope. Studies of isotope selectivity enhancement (ISE) by diode laser assisted RIMS using natural abundance lanthanum have been reported (Shaw et al 1990). With enriched lanthanum samples we have demonstrated ISE of 2000 for 138/139 ratios when the diode laser is tuned to the most intense La-138 transition. ISE is the ratio of a given isotope ratio to that obtained by thermal ionization mass spectrometry. Likewise, ISE of approximately 200 has been obtained for 139/138 ratios if the diode laser is tuned to the most intense La-139 transition. For these measurements the diode laser power was ~ 0.1 mW and that of the CVL/dye laser 20 mW or less; the linewidth of the CVL/dye laser beam was approximately 0.05 cm^{-1}. The ISE values reported are limited by a low and variable count for the unwanted isotope. There appears to be a pedestal effect, i.e., the hyperfine lines seem to be on a broadly tunable base. The reasons for this are not clear. There is a residual Doppler width. Another possibility is that the diode laser emission lies on top of a broad band emission. Counts from the unwanted isotope limit both the ISE and the precision of this measurement.

A partial summary of the possibilities that are known to exist for diode laser-assisted RIMS studies of gadolinium are given in Figure 3. It should be possible to measure ground state and metastable state populations with straightforward experiments. The influence of the method of atom generation on initial state populations can therefore be measured. Determination of optical excitation and de-excitation routes for these inter-related levels should be possible. ISE factors measured for gadolinium with diode laser-RIMS should be very large. Gadolinium has seven naturally occurring isotopes; two of them have nuclear spin > 0 and will exhibit hyperfine splitting. All of the lower transitions shown are accessible with commercially available diode lasers, all are allowed transitions. Ionization can be accomplished by a 1+1 optical process originating from the diode laser-excited level. Because of the lack of identified upper levels for gadolinium, exact wavelengths for these 1+1 processes are not known. It is likely that several ionization pathways can be found at a wavelength equal to or shorter than those given in the figure.

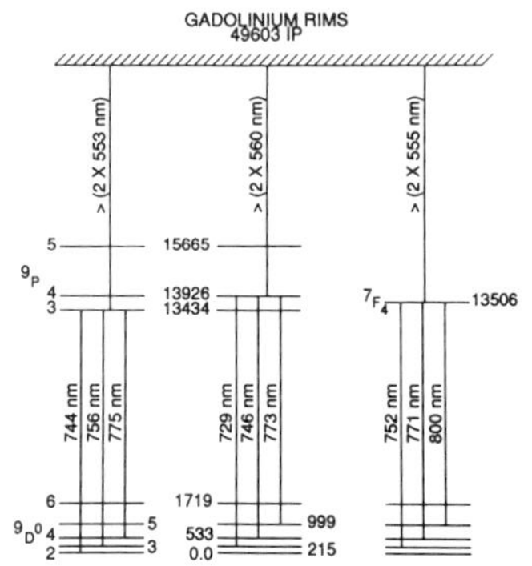

Fig. 3. Optical ionization routes for Gd

There are many applications of diode lasers to one transition of a multiple resonant photon ionization processes. Besides the routes shown in Figures 1 and 3, where diode lasers excite the first transition, there are some possibilities where diode lasers excite intermediate steps in a resonant ionization process. The ultimate technique would seem to be an all diode laser excitation scheme. Even though diode lasers probably do not process sufficient power to promote bound-continuum transitions, it may well be possible to generate upper Rydberg states that could be ionized using an electric field or other non-optical step. In the heavier alkali metals, routes for generation of upper Rydberg atoms by diode laser excitation are already known (Ohtsu and Tako 1988). In the case of the lanthanides and actinides it should be possible to generate an upper Rydberg level with three to four diode laser steps. A transportable analytical device with excellent isotopic selectivity can be envisioned that contains a set of tuned diode lasers for each isotope desired to be determined. No additional mass analyzer would be necessary.

4. ACKNOWLEDGEMENTS

Research is supported by U.S. Department of Energy, Office of Energy Research, under contract DE-AC05-84OR21400 with Martin Marietta Energy Systems, Inc.

5. REFERENCES

Childs W J and Goodman L S 1979 *Phys. Rev. A.* **20** 1922
Lawrenz J and Niemax K 1989 *Spectrochim. Acta* **44B** 155
Ohtsu M and Tako T 1988 *Progress In Optics* Vol. 25, "Coherence in Semiconductor Lasers", Elsevier, Amsterdam 262
Shaw R W, Young J P, Smith D H 1989 *Anal. Chem.* **61** 695
Shaw R W, Young J P, Smith D H, Bonanno A S, Dale J M 1990 *Phys. Rev. A* **41** 2566
Smith D H, McKown H S, Young J P, Shaw R W, Donohue D L 1988 *Appl. Spectrosc.* **42** 1057
Ting Y 1957 *Phys. Rev.* **108** 295

Inst. Phys. Conf. Ser. No 114: Section 5
Paper presented at RIS 90, Varese, Italy, 16–21 September 1990

Collective phenomena in laser-matter interactions

F.Giammanco-

Dipartimento di Fisica,Universita` di Pisa-Piazza Torricelli 2-56126 Pisa-Italy

ABSTRACT: The two-fluid description of the time-space behavior of charges in the presence of strong density gradients and relevant departures from the neutrality is extended to include energy conservation and Coulomb collisions. Collective interactions with an intense laser field are also discussed with special regard to the harmonic generation.

1. INTRODUCTION

The importance of collective effects depending on the electron-ion mutual interaction , in experiments involving production and collection of charged particles, has been highlighted and analyzed. It has been pointed out that the time-space behavior of charges, as produced in multiphoton ionization (MPI) and electron impact ionization (EII) (Giammanco 1989a, Giammanco *et al* 1990c) is strongly affected by the a macroscopic electric field, due to the different mobilities of electrons and ions.
The theoretical model (Giammanco 1989b) gives a quantitative description of the observed phenomena, as electron trapping, distortion and delay of ion signal that leads to anomalous dependences of produced yields on the laser power in MPI experiment (Giammanco 1989a), and dynamic electron and ion focusing in EII experiment (Giammanco *et al* 1989c). Collective interactions can play a relevant role also in phenomena, as generation of harmonics in gases ionized by multiphoton processes induced by an intense laser field (Li *et al* 1989, McPherson *et al* 1987, Reintjes 1984, Rosman *et al* 1988, Wildenauer 1987).
Whereas in the experiments of McPherson *et al* (1987) and Rosman *et al* (1988), harmonic generation occurs in a totally ionized medium, in the experiment of Li *et al* (1989) harmonics are still observed greatly reducing the ionization degree. Nevertheless, the kind of plateau from 7th to 27th harmonic in Ar or similar, but less extended, in Xe and Kr, that differentiates this experiment from the previous ones, is observed near or above the total ionization of the medium. In this work, the theoretical analysis of the time-space evolution of a Gaussian-shaped profile of charges, coupled by a self-generated electric field, is extended to include the energy conservation in the presence of energy exchange with the external laser field and Coulomb electron-ion collisions. Influence of Coulomb collisions is relevant in MPI produced plasmas, in the density range of the above-mentioned experiments (about 10 torr), because the electron temperature is quite low (of the order of 1 eV). It turns out that Coulomb collisions cannot be neglected in spite of the short laser pulse duration (picoseconds). It will be shown that Coulomb collisions play a fundamental role in the generation of odd harmonics, according to Ginzburg (1964), in the case of tenuous, cold plasmas as produced in multiphoton ionization. Moreover, density gradient effects and electrostatic oscillations are taken into account in a simple way. Finally, harmonic distribution as a function of the experimental parameters is discussed.

2. THEORETICAL MODEL

As previously pointed out (Giammanco 1989b), the fluid approximation constitutes a suitable approach to describe the time-space evolution of charges in spite of the fact that the velocity distribution function is, in general, quite far from the equilibrium Maxwell distribution. In fact, the fluid scheme, derived from the Boltzmann transport equation as momenta of different order averaged over the velocity distribution function, holds true providing that the averaged particle velocity is equal to zero. Thus, an isotropic velocity distribution, regardless of its shape, is sufficient to make the fluid approach valid. The basic equations describe the time-space evolution of the fluid mean quantities (velocity, density, and energy) referred to the laboratory frame. It turns out that they are functions of coordinates and of time. The fluid velocity represents the common component of collective motion, whereas the temperature is related to the mean kinetic energy of random motion of the particles. Since ions are at rest during laser pulses of some picoseconds, the analysis is restricted to the electrons, whose basic equations are given by

$$\frac{\partial \mathbf{v_e}}{\partial t} + (\mathbf{v_e}\nabla)\,\mathbf{v_e} = -\frac{e}{m}(\mathbf{E_o} + \mathbf{E_r}) - \frac{1}{n_e}\nabla\left[\frac{kT_e}{m}n_e\right] - \nu_{ei}\mathbf{v_e}$$

$$\frac{\partial n_e}{\partial t} + \nabla(n_e\mathbf{v_e}) = S_e$$

$$\frac{\partial U_e}{\partial t} + (\mathbf{v_e}\nabla)U_e - kT_e\frac{\partial \ln(n_e)}{\partial t} + (\mathbf{v_e}\nabla)kT_e = -\frac{e}{m}\,\mathbf{v_e}(\mathbf{E_o} + \mathbf{E_r}) - m\nu_{ei}\,v_e^2$$

$$\frac{\partial kT_e}{\partial t} + (\mathbf{v_e}\nabla)kT_e + \frac{2}{3}kT_e(\nabla\,\mathbf{v_e}) = 0$$

$\mathbf{E_0}$ represents the external laser field. S_e represents the source term responsible of the growth of ionization. U_e is the total energy of electrons including collective (fluid) motion and random motion (T_e). At t=0, $U_e=kT_e=E_e$, where E_e is the energy of ejected electrons given by $E_e=nh\nu - I$, where I is the ionization potential and n the order of MPI transition. The energy exchange between electrons and ions has been neglected because this term, that depends on the mass ratio m/M, is effective on a time-scale much greater than the laser pulse duration. $\mathbf{E_r}$ is the self-generated electric field given by Poisson equation. ν_{ei} is the electron-ion collision frequency, i.e. $\nu_{ei} = a\, n_i \ln \Lambda/U_e^{3/2}$, where $a \approx 10^{-6}$ (cgs units and U_e in eV) and $\ln \Lambda$ is the Coulomb logarithm (Ginzburg 1964). A quantitative description of the phenomena is simplified if the spatial profile of involved species is represented by a Gaussian-shaped profile. In the simplest case ($E_r=0$, $\nu_{ei}=0$ and $U_e \approx kT_e$), the motion is still represented by a Gaussian-shaped profile, whose width is a function of time and whose maximum translate according to the single-particle motion in an external field $E_0(t)$ (Giammanco 1989b). Thus, the one-dimensional time-space evolution of the density profile in presence of an uniform electric field E_0 along the x axis, is expressed as

$$n_e(x,t) = n_0 f_x(t)^{1/2} \exp[-f_x(t)(x-x_s)^2/d_x^2]$$

where d_x and n_0 represent the initial charge distribution width and density, respectively, while $x_s=\int v_s\, dt$, with $Dv_s = -eE_0(t)/m$, represents the electron single-particle motion in the external electric field, D being the total time derivative. The continuity equation provides the functional dependence of the fluid mean velocities on n_e and, thus, on the $f_x(t)$ functions, namely $v_e= v_s+0.5(x_s-x)[D\ln(f_x(t))]$. The functions $f_x(t)$ satisfy the initial conditions $f_x(t=0)=1$, and $Df_x(t)|_{t=0}=0$. A second order differential equation describing the $f_x(t)$ time evolution, is derived from the motion equation by replacing the $f_x(t)$ functional dependences for both mean velocity and n_e, and by equating the terms with the same power dependence on the spatial coordinates. Owing to the independence of motion along the axes, the three-dimensional motion is represented by a product of three Gaussian profiles with different initial widths, depending on the symmetry of the charges production. Thus, the time evolution of $f_k(t)$ (k=x,y and z) along any axes depends on the initial width d_k.

In general, the previous scheme does not hold if applied to the complete system. Nevertheless, if all terms are expressed in a form like G(t)+ F(t)**r**, where **r**=(x,y,z), the basic scheme of solution still holds true. Poisson equation is linearized around the coordinate of the single-particle motion for the electrons (ions are at rest, i.e. $f_i=1$), i.e. around the top of the Gaussian profiles, that, in the one-dimensional case, gives

$$E_r(x,t) = 4\pi e n_o \left[1 - f_x^{1/2}(x-x_s)\right] \quad .$$

The production term Se can be introduced in the previous scheme at a same degree of approximation . In fact, it is sufficient to replace n_o by the solution g(t) of the rate equation for the selected multiphoton transition. Since at t=0 grad T_e=0, solution for T_e is immediate, i.e. $T_e=T_o(f_x f_y f_z)^{1/3}$ which is equivalent to the adiabatic relation $T_e \propto n_e^{2/3}$ for a monoatomic gas. The analysis is restricted to a region whose dimensions are comparable with the width of the charge profile along any axis. Hence, in the collision frequency term a constant and uniform mean value is assumed for the ion density. By assuming a cylindrical geometry, a laser beam direction and polarization along z and x axes respectively, the complete system to be solved is

$$D^2 f_k = \frac{3}{2}\frac{(Df_k)^2}{f_k} - 2(f_k/\tau_k)^2 (f_r^2 f_z)^{1/3} + \left(\frac{2}{3}\right)\omega_{pe}^2 f_k[1 - f_r(f_z)^{1/2}] - \nu_{ei} Df_k$$

$$D^2 x_s = -\frac{e}{m}E_o - \frac{1}{3}\omega_{pe}^2 x_s - \nu_{ei} v_s$$

$$DU_e - kT_o(f_r^2 f_z)^{1/3} \, (D\ln f_r + 0.5 D\ln f_z) = -e\, v_s E_o - m\nu_{ei} v_s^2$$

where k=r,z. τ_k is the characteristic time constant of free expansion due to the radial and longitudinal density gradients, respectively given by $\tau_z^2 = \frac{md_z^2}{2nkT_e}$, $\tau_r^2 = \frac{md_r^2}{2nkT_e}$. d_z and d_r are longitudinal and radial laser widths related to the confocal parameter (Reintjes 1984) and n the order of the transition. ω_{pe} is the electron plasma density whose value depends on time through the function g(t). Nonlinear polarizability of medium can be carried out from the Fourier transform of the induced current, which in turns is proportional to the velocity v_s times $f_r f_z^{1/2}$. By the way, a simple inspection of the system shows that main contribution to odd harmonic generation comes from the collision frequency term. In the first approximation, U_e is modulated at a frequency twice the laser one, that leads to a third harmonic modulation of the velocity derivative. The simple scheme without collective interactions (i.e. with f=1 and Df=0) gives a rapid decrease of the harmonic field as the harmonic order increases. Moreover, it does not depend on the type of gas except for the saturation threshold of the ionization. On the contrary, collective interactions depend strongly on the density gradients, which are function of the transition order, and hence of the type of gas.

The phase matching, depending on the plasma refractive index, is modified by the electron motion. By assuming a Gaussian beam profile, the initial longitudinal width of charges (1/e of the maximum) is given by $d_z \approx .16\ b/\ n^{1/2}$, where b is the confocal parameter. Under approximation of weak field and of $d_z \ll b$ ($n \gg 1$ for noble gases, the phase matching factor is proportional to $(\sin\phi/\phi)^2$ where $\phi=(\Delta k_q+4/b)L_{eff}$ (Reintjes 1984).

Taking into account that $L_{eff}=d_z/f_z^{1/2}$, $\phi=\frac{0.08\ b\omega\rho_o^2}{\omega_L c n^{1/2}}(q-\frac{1}{q})f_r+\frac{0.64}{(nf_z)^{1/2}}$.

Figure 1 shows the harmonic field distribution in Ar at 5×10^{13} W/cm^2, 5×10^{17} at/cm^3, confocal parameter 0.4 cm and different values of pulse duration of Nd-Yag laser

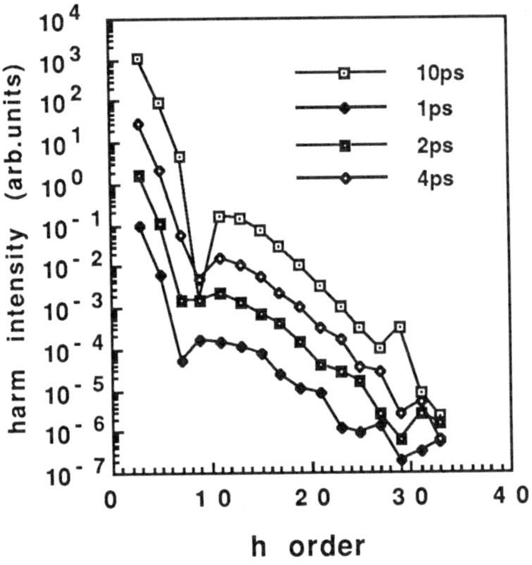

Fig. 1. Harmonic distribution in Ar.

From Figure 1, it is evident that pulse duration extends the range of produced harmonics. In fact, important increase of the higher order harmonics occurs whose distribution evolves towards a kind of plateau. The dependence of the laser intensity of each harmonic shows interesting features. Near the total ionization threshold, all harmonics follow the perturbative dependence on the laser field, i.e. I^q. As laser power increases, a subsequent saturation of the harmonics occurs. Harmonics in the plateau (from 9th to 19th) saturate too, but present a dip in a narrow region of laser intensity. Dip position scales towards higher laser intensity as the harmonic order increases. Dependence on laser intensity of last harmonics is far from the I^q dependence. For instance, 21th harmonic of Figure 1 follows slope ≈10, for 1 ps duration, and saturates as laser intensity increases. For 10 ps duration, in the same intensity range, slope becomes 21, whereas departure from the perturbative behavior occurs for higher order harmonics. Dependence on the gas density follows the quadratic dependence, if phase matching is neglected. Under conditions of Figure 1, phase matching factor is about 1 up to 33rd harmonic. In fact, radial (strongest) and longitudinal gradients induce decrease of f_r and increase of f_z, which reach minimum and maximum values at approximately the top of the laser pulse. Note that this behavior is peculiar of high order MPI processes and in this framework it is possible to explain differences due to the type of gas (different order of MPI, different ionization efficiency, initial electron kinetic energy) or to the laser wavelength. Calculations show that reduction of the laser wavelength, and thus of the order of the transition, implies a strong reduction of harmonic conversion efficiency.

REFERENCES:

Giammanco F 1989a *Phys.Rev. A* **40** 5160
Giammanco F 1989b *Phys.Rev. A* **40** 5171
Giammanco F ,Arena G,Bruzzese R and Spinelli N 1990c *Phys.Rev.A* **41** 2144
Ginzburg V L 1964 *The Propagation of Electromagnetic Waves in Plasmas* (New York: Pergamon Press) pp 428-52
Li X F, L'Huillier A,Ferray M,Lompré L A and Mainfray G 1989 *Phys.Rev.A* **39** 5751
McPherson A,Gibson G,Jara H,Johann U,Luk T S,McIntyre I A,Boyer K and Rhodes C K 1987 *J.Opt.Soc.Am. B* **4** 595
Reintjes J F 1984 *Nonlinear Optical Parametric Processes in Liquids and Gases* (New York: Academic Press)
Rosman R,Gibson G,Boyer K,Jara H,Luk T S,McIntyre I A,McPherson A,Solem J C Rhodes C K 1988 *J.Opt.Soc.Am. B* **5** 1237
Wildenauer J 1987 *J. Appl. Phys.***62** 41

Paper presented at RIS 90, Varese, Italy, 16–21 September 1990

Investigation of even-parity Ryderg levels using the resonant multi-photon ionization TOFMS method and their analyses by MQDT

Mingxing Jin, Dajun Ding, Hang Liu and Shoufu Pan
Institute of Atomic and Molecular Physics, Jilin University, Changchun 130023, P. R. China

ABSTRACT: Two-photon resonant ionization and TOF detection method have been employed in measuring energies of Sn even-parity Rydberg levels. Five bound and two autoionizing Rydberg series 5pnp, 5pnf with J values ranging from 0 to 3 have observed and identified. The J=0,2 bound Rydberg series have been analysed by MQDT. We have calculated the level positions and analysed the strong interaction among Rydberg series.

1. INTRODUCTION

Resonant multiphoton ionization spectroscopy has been successful in investigating highly excited states of atoms and molecules and detecting atoms and molecules (Chin and Lambropoulos 1984, Letokhov 1987). We have studied the even-parity states of atomic tin by two-photon resonant laser spectroscopy.

A considerable number of experiments have been performed in recent years employing multiphoton excitation of alkaline-earth and rare-gas atoms to perform laser spectroscopic studies of excited bound and autoionising levels. The tool used for the interpretation of these atomic data has been the mutichannel quantum defect theory (MQDT) (for a review, see Aymar 1984). The odd-parity bound Rydberg states of the carbon-group atoms have been studied by vacuum ultraviolet spectroscopy, with an analysis using MQDT (Brown et al 1977a, b, c). Ding et al (1989) measured the even-parity bound Rydberg series of lead atom via two-photon laser spectroscopy and analyzed the experimental results using MQDT. We report in this paper the results of our investigations on the multiphoton ionization spectra of tin. Even-parity highly excited np and nf Rydberg series of atomic tin converging to the ground state $^2P^o_{1/2}$ of Sn^+ have been investigated recently using two-photon resonant excitation starting from the ground state $5s^25p^2$ $(1/2,1/2)_0$ and the metastable $5s^25p^2$ $(1/2,3/2)_1$ level. Five Rydberg series np and nf with J values ranging from 0 to 3 have been observed. Two autoionizing Rydberg series have also been observed by two-photon resonance from the metastable $5s^25p^2$ $(1/2,3/2)_2$ level. The n values was as high as 46 in some case. Most of the Rydberg series observed belonged to J=0,2 Rydberg series. They are $5pnp(1/2,1/2)_0$, $5pnp(1/2,3/2)_2$, 5pnf1/2

$[5/2]_2$, 5pnp$(3/2,3/2)_2$ and 5pnf3/2$[5/2]_2$(Jin et al 1990).

2. EXPERIMENTAL RESULTS

To our previous an atomic beam apparatus (Ding 1989), we have added a photoionization time-of-flight (TOF) mass spectrometer apparatus. The plate valtages, timing and physical dimension of the spectrometer have long been optimized (Wiley 1955). Using dye LDS698, Quanta-Ray laser system provided the 335-360 nm tunable pulsed UV laser which had a linewidth of 0.2 cm^{-1}. The UV laser was focused into the vacuum chamber with a 25 cm focal-length spherical quartz lens to a power density of $10^7 \sim 10^8$ W/cm^2, orthogonal to the atomic beam of tin. In the interaction volume, the atom density was about 5×10^{10} atoms/cm^3. The photoions produced in multiphoton ionization process were accelerated by 3000 V electric field and drift through one-meter tube to a microchannel plate and an electron multiplier. The ion signal pulse were observed on the oscilloscope after a delay time about 25 μs (time of flight plus transit time in the microchannel plate detector and the electron multiplier). This signal is sent to a Boxcar averager with two M165 gated integrators, the gate width is about 100 ns and the signal is integrated over 100 laser shots. The detection is very efficient. Finally, the output was recorded on a multi-pen chart recorder. In this case, the photoionization spectra of atomic tin is not superposed to a background ion signal due to multiphoton ionization of the residual gases by the strong UV laser light. Therefore the photoionization spectra with TOF mass analysis results in a best signal-to-noise ratio. To our knowledge, this is the first observation of even-parity bound and autoionizating Rydberg series in atomic tin. The levels with different J can be identified experimentally by changing the polarization of the laser beam. The initial state $5s^25p^2(1/2,1/2)_0$ is with a zero total angular momentum. In this case, the J=0 resonance should not be present using a circularly polarised laser because of the condition m=2, and only the J=2 state can be populated. Otheriwise, both J=0 and J=2 states can be populated by a linear polarised laser beam. The energy levels are averages of measurments on seven different spectra recored under identical conditions and are reproducible to within ±0.15 cm^{-1}. We noted that the linewidth of autoionization resonance is larger than that of bound Rydberg series. We have not determined autoionization linewidth with high precision. We can, however, estimate approximate linewidths from the spectra directly. The two autoionizing Rydberg series have linewidths about $3\sim20$ cm^{-1}. The accuracy is estimated to be $\pm20\%$.

3. MQDT ANALYSES

Multichannel quantum defect theory (MQDT) has provided a powerful theoretical tool for analyzing spectra and collision processes of atoms and molecules (Fano and Rau 1986). The techniques of MQDT have used to analyse the even-parity J=0,2 bound Rydberg series of atomic tin quantitatively in this work. We adopted two-channel model and six-channel model for analysis J=0 and 2 even-parity Rydberg series of atomic tin respectively. The optimized values of MQDT parameters $U_{i\alpha}$ and μ_α are determined by minimizing the $(\sum_{n=1}^{N_m} F_n^2/N_m)^{1/2}$ to make the plot of $F(\nu_1, \nu_2)=0$ pass as close as possible to the points that represent the experimental energies, where $F_n=F(\nu_{1,n}, \nu_{2,n})$ and N_m is the number of energy levels that considered in the fitting procedure. We have introduced a linear variation with excitation energy for the eigen quantum defects. Using the parameters we calculate the level position with J=0,2 and even-parity. Comparison of these calculated values with measured values shows an excellent agreement. For the values with n>10 the rms error between them is about 0. 23cm^{-1}. MQDT does not well with the lowest states, the deviations between them are larger because these levels are affected by the polarization of the core by the outer electron and we had omitted them in our fitting procedures for MQDT parameters. The MQDT analyses focused on high excited levels provided a very good interpretation for the measured levels. The calculation was extended to some highlying levels which were unmeasured. The strong perturbations among Rydberg series can be illusrated graphically, namely, by a method of Lu-Fano plot (Lu 1970). The Lu-Fano plots have been shown if Fig. 1 and Fig. 2 for even-parity J=0,2 levels of tin. The effective quantum number ν_i is no longer a smooth function of energy in the perturbed Rydberg series, which corresponds to a variation of n-ν_1. This variation is caused

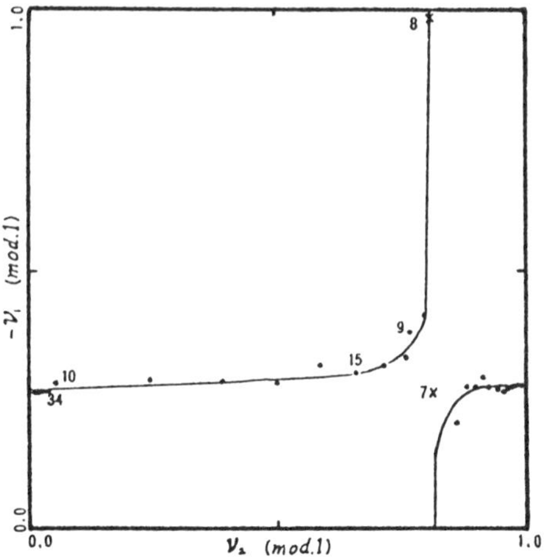

Fig. 1 Lu-Fano plot for the even-parity J=0 bound energy levels 5pnp (1/2,1/2)$_0$ in tin. ●: observed levels, ×: 5pnp(3/2,3/2)$_0$.

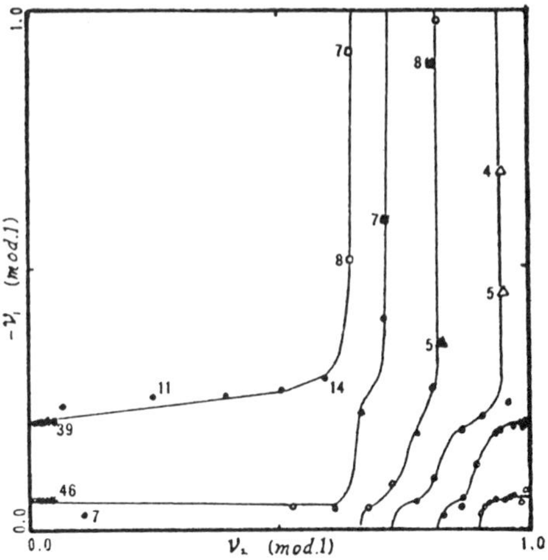

Fig. 2 Lu-Fano plot for the even-parity J=2 bound energy levels 5pnp (1/2,3/2)$_2$ and 5pnf 1/2 [5/2]$_2$ in tin. ●: 5pnp (1/2,3/2)$_2$. ○: 5pnf 1/2[5/2]$_2$, △pnf 3/2[5/2]$_2$. □: 5pnp (3/2,3/2)$_2$. ▲: 5pnf 3/2[3/2]$_2$. ■: 5pnp (3/2,1/2)$_2$.

by a perturbing levels, which are the low-lying Rydberg levels converging to a higher ionization limit. Therefore the curve bends obviously in the perturbed Rydberg series. In this case the wave function for a given state can be expressed as a superposition of i-channels (Lee 1973)

$$\Psi_n = \sum_i \Phi_i p_i^{(n)} z_i^{(n)}$$

The mixing coefficients $z_i^{(n)}$ for given state in i-channel can be calculated from the MQDT parameters μ_α and $U_{i\alpha}$. A quantitative analysis of the perturbation is performed with the wave function and the mixing coefficients $z_i^{(n)}$ for each level is calculated. Our data provide a further characterization of discrete levels through mixing coefficients z_i, which represent the mixing of i channels in the given level. Large channel mixing occurs nears the perturbation and the perturbation levels mainly perturb the levels located near the perturbers. The p-f interference effects are quite large as characterized by the elements $U_{i\alpha}$ with $i=1,2,3$, $\alpha=4,5,6$. The p-f mixing is about 50% which is one order of magnitude stronger than in rare gas atoms. Some levels tentatively labelled according to large mixing of channels and in fact they cannot be assigned by a single channel.

4. CONCLUSION

Resonant multiphoton ionization spectroscopy and TOF mass spectrometry have been shown to be useful tools for the study of excited bound and autoionising levels of atoms. The emphasis has been on the measured and evaluated properties of spectra for Sn even-parity Rydberg series. The MQDT analysis is successful in interpreting these spectra of strongly perturbed Rydberg series and give a good correspondence between the experiment and the calculation.

This research was supported by Fok Ying Tong Education Foundation, P. R. China

Aymar M 1984 Phys. Rep. **110** 163.

Brown C M, Tilford S G and Ginter M L 1977a J. Opt. Soc. Am. **67** 1240.
-1977b J. Opt. Soc . Am. **67** 607. -1977c J. Opt. Soc. Am. **67** 584.

Chin S L and Lambropoulos P ed. 1984 Multiphoton Ionization of Atoms, (London: Academic Press).

Ding D J, Jin M X, Liu H, and Liu X W 1989 J. Phys. **B22** 1979.

Fano U and Rau A R P 1986 Atomic Collision and Spectra, (London: Academic Press Inc.).

Jin M X, Ding D J, Liu H, and Pan S F 1990 J. Phys. **B.** to be published.

Lee C M and Lu K T 1973 Phys. Rev. **A8** 1241.

Letokhov V S 1987 Laser Photoionization Spectroscopy, (Orlando: Academic Press, Inc.).

Lu K T and Fano U 1970 Phys. Rev. **A2** 81.

Moore C E 1971 Atomic Energy levels, Vol. 3 (NSRDS-NBS35), P. 74.

Wiley W C and Mclaren I H 1955 Rev. Sci. Instrum **26** 324.

Resonance ionization mass-spectroscopy of neptunium

P Sattelberger+, R Deißenberger+, G Herrmann+, J Riegel+, H Rimke+, N Trautmann+, F Ames*, H-J Kluge*

+Institut für Kernchemie, Universität Mainz, D-6500 Mainz
*Institut für Physik, Universität Mainz, D-6500 Mainz

ABSTRACT: For ecological studies of ^{237}Np resonance ionization mass-spectroscopy has been explored. The experimental set-up consists of three tunable dye lasers pumped by two pulsed copper vapour lasers and a time-of-flight spectrometer. Several excitation schemes have been studied. For two schemes autoionizing levels could be observed. The hyperfine structure splitting of the resonance lines of the first and second excitation step have been investigated. The available laser powers were sufficient to saturate the first two excitation steps but not the ionization step. Without having optimum conditions, a detection limit of $\leq 10^{10}$ atoms could be reached.

1. INTRODUCTION

For studies of the ecological behaviour of ^{237}Np a sensitive, element- and isotope selective detection method is required. The conventional methods used so far are α-spectroscopy with a detection limit of $4 \cdot 10^{10}$ atoms at a counting time of 1000 min and inductively coupled plasma spectroscopy which is even more sensitive than α-spectroscopy. As an alternative, trace amounts of neptunium can be determined by laser resonance ionization mass-spectroscopy.

2. EXPERIMENTAL SET-UP

The experimental set-up is shown in figure 1. It consists of three pulsed dye lasers pumped by two copper vapour lasers which have an average output power of 30 W and a pulse repetition rate of 6.5 kHz. The laser beams are deflected with three quartz fibres and a lens system into the interaction zone of the time-of-flight spectrometer. Here, an atomic beam of neptunium is produced by evaporating it from a rhenium filament, on which it is electrodeposited and covered with a thin rhenium layer in order to reduce neptunium oxide to the metal. The filament is heated by a dc current to temperatures up to 2300 K. The laser light irradiates the atoms and ionization occurs in a three-step resonant photoionization process. The ions are accelerated to 2.9 keV and after passing a drift tube of 2 m they are detected by a microchannel-plate detector. With pulsed extraction voltage, a mass resolution up to $\Delta M/M$ =2700 can be reached. A more detailed description of the apparatus is given by W. Ruster et al (1989).

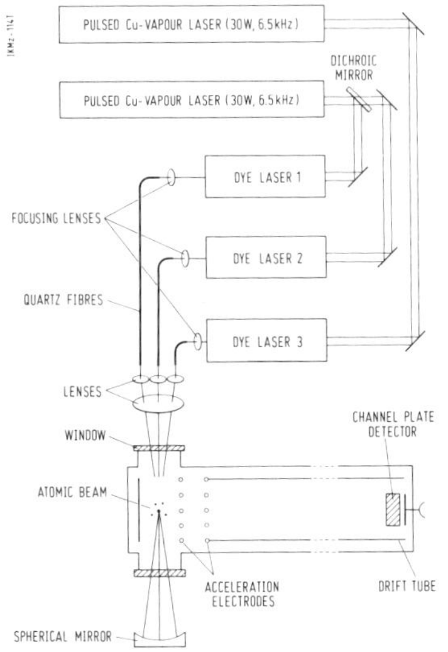

Fig. 1. Experimental set-up for resonance ionization spectroscopy

3. RESULTS

Several excitation schemes (figure 2) have been investigated for three-step photoionization. With scheme (a), no autoionizing levels could be found in the scanned wavelength range between λ_3=591.8-557.4 nm.

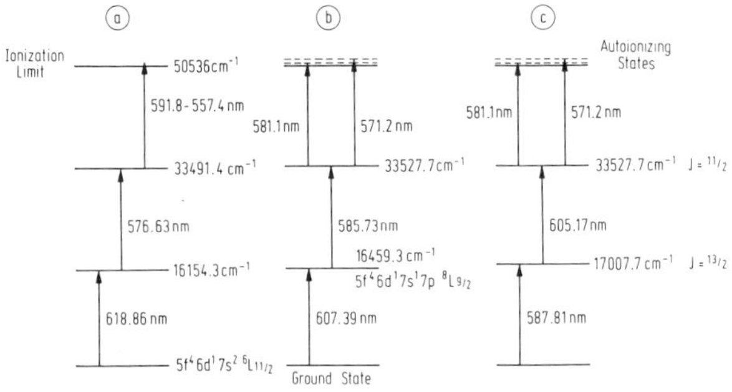

Fig. 2. Excitation schemes for the resonant ionization of ^{237}Np

For schemes (b) and (c), leading to the same second excited level, autoionizing levels were observed. The ion count-rate as a function of λ_3 with λ_1=607.39 nm and λ_2=585,73 nm fixed at resonance is shown in figure 3. The strongest resonances at λ_3=571.2 and λ_3=581.1 nm in the scanned wavelength range yield an enhancement in the ionization rate of about one order of magnitude compared to non-resonant ionization. ^{237}Np with a nuclear spin of 5/2 and an angular momentum of J=11/2 shows large hyperfine structure splitting of the resonance lines, as illustrated in figure 4.

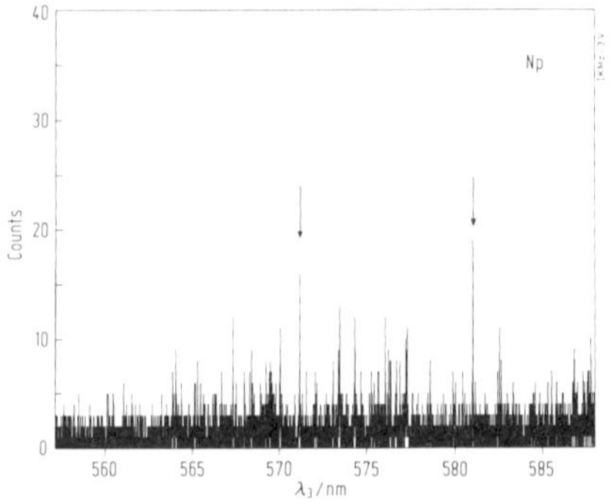

Fig. 3. Autoionizing levels in ^{237}Np with λ_1 and λ_2 in resonance at 607.39 nm and 585,73 nm

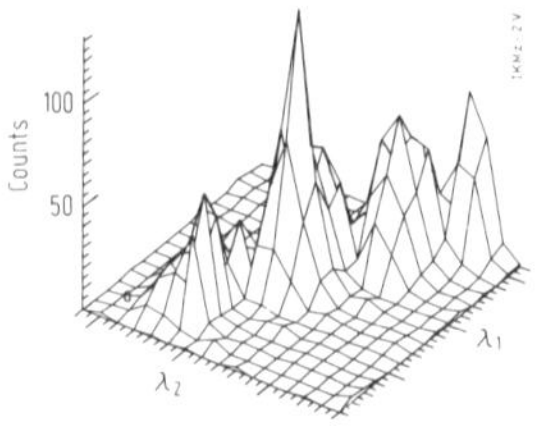

Fig. 4. Ion count rate as a function of λ_1= 607.39 nm and λ_2=585.73 nm

With excitation scheme (b) of figure 2 the total splitting is 66 GHz in λ_1 and 50 GHz in λ_2 at a laser bandwidth of 3 GHz in λ_1 and 5 GHz in λ_2. The hyperfine structure splitting of the ground state and the first excited state together with the strongest transitions between the sublevels of the ground state and the first excited state are shown in figure 5. Due to the opposite signs of the A and B-factors of the hyperfine structure of the ground state and the first excited state (Fred 1977), the splitting is large and even with broadband laserlight it is not possible to excite all atoms of the ground state simultaneously, resulting in a decrease of the overall detection efficiency. With excitation scheme (c) of figure 2, where the ground state and the first excited state have the same signs of A and B-factors, the hyperfine structure splitting of the resonances has been measured to be 26 GHz in λ_1 and 33 GHz in λ_2.

For an efficient ionization of the evaporated neptunium atoms all three steps should be saturated. The available laser powers were sufficient to saturate the first two excitation steps but not the ionization step. Figure 6 shows the dependence of the ion count-rate from the laser power with excitation scheme (b) of figure 2. For the first two excitation steps, the following cross sections have been determined: for λ_1=607.39 nm, $\sigma_{sat}=1.3\cdot10^{-14}$ cm^2, for λ_1=587.81 nm, $\sigma_{sat}=2.2\cdot10^{-15}$ cm^2, for λ_2=605.17 nm, $\sigma_{sat}=7.9\cdot10^{-15}$ cm^2, and for λ_2=585.73 nm, $\sigma_{sat}=1.4\cdot10^{-14}$ cm^2. The overall detection efficiency of the resonance ionization mass spectrometry has been measured with samples containing $\approx 10^{13}$ atoms of ^{237}Np. The number of evaporated atoms was determined by α-spectroscopy before and after heating the filament.

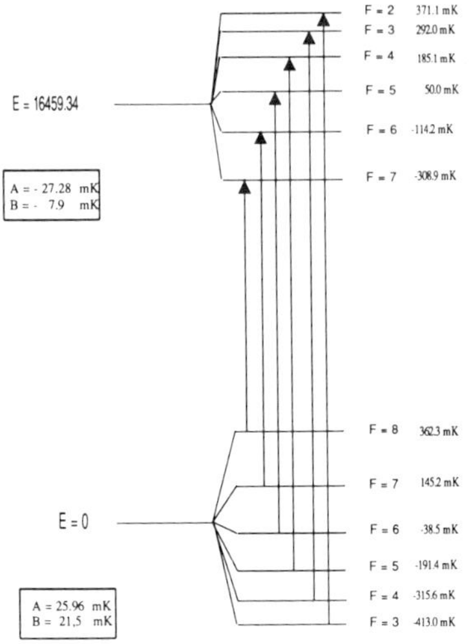

Fig. 5. Hyperfine structure of the ground state and the first excited state of excitation scheme (b)

From the ratio of the counted ions and the evaporated atoms a value for the efficiency of $\epsilon=1\cdot10^{-9}$ was received yielding a detection limit of 10^{10} atoms. However, this value was obtained without having optimum conditions in the time-of-flight spectrometer and without reaching saturation in the ionization step.

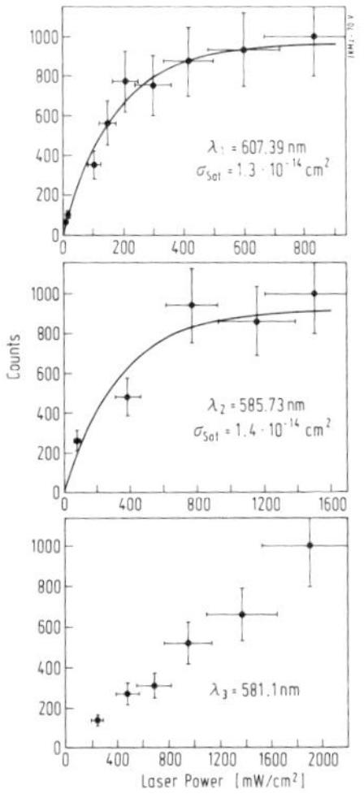

Fig. 6. Dependence of the ion count-rate from the laser power, measured with excitation scheme (b) of fig. 2

REFERENCES

Fred M, Tomkins F S, Blaise J E, Camus P and Verges J 1977 J. Opt. Soc. Am. **67** 7

Ruster W, Ames F, Kluge H-J, Otten E-W, Rehklau D, Scheerer F, Herrmann G, Mühleck C, Riegel J, Rimke H, Sattelberger P and Trautmann N 1989 Nucl. Instr. Meth. **A281** 547

Paper presented at RIS 90, Varese, Italy, 16–21 September 1990

Polarization and isotope shift effects in uranium isotope ratio measurements by resonance ionization mass spectrometry

L W Green, M H C Smyth, P A Rochefort and G A McRae

AECL Research, Chalk River Laboratories, Chalk River, Ontario, Canada
K0J 1J0

ABSTRACT: Many strongly resonant transitions and their corresponding isotope shifts were identified by two-colour, three-photon, pulsed photoionization spectroscopy and time-of-flight mass spectrometry. Steep wavelength sensitivity of the 235/238 ratio limited the precision of ratio measurements to 4% (1 std. dev.), and was caused by the inability to obtain uniform excitation of the isotopically shifted atomic lines within the bandwidth of the laser, despite the use of the lasers in the broadband mode. In addition, changes in the relative polarization of the lasers caused up to 20% bias in the observed 235/238 ratio, due to the preparation of aligned states in each resonant transition.

1. INTRODUCTION

Resonance ionization mass spectrometry has considerable appeal for isotopic ratio measurements because of its potential for high elemental selectivity and sensitivity. Alternatively, the inherently high photoselectivity is, for many elements, sufficient for isotopic spectral shifts to be resolved, and this forms the basis of laser isotope separation processes being developed in many countries. Fundamental to isotopic measurement and separation applications, are studies of the spectroscopy and photoexcitation dynamics of the vaporized atoms or molecules. This paper reports on studies of electronic states and isotope shifts in atomic uranium, and on the effect of the relative polarization of linearly polarized laser beams in two-colour excitations.

2. EXPERIMENTAL

Uranium vapours were generated in the ion source of a 5 m time-of-flight mass spectrometer by resistive heating of a rhenium ribbon, to which a 0.7 mg piece of U metal had been spot-welded. Temperatures between 1300 and 1700°C were used; the optimal temperature for minimization of thermal ion interference was 1360°C. An Oxford CU40 40 W copper vapour laser (CVL), fitted with unstable resonator optics, was used to pump two tunable dye lasers: a Lumonics HD300 and a Lumonics EPD330. The laser bandwidths were 0.06 and 0.12 cm^{-1}, respectively, but could be increased to 1 cm^{-1} by removal of the prism beam expanders. Only the oscillator cells were pumped, and the pumping efficiency was 4% for a typical pump energy of 0.7 mJ/pulse. A pulse repetition rate of 6.5 kHz was used, and dye laser pulse durations were

20 ns with the arrival of the second pulse at the ion source optically delayed by 40 ns from the arrival of the first. Rhodamine 6G was used in both lasers for two-colour studies in the 566 to 593 nm region. Small portions of the beams were deflected into a neon hollow cathode lamp and an iodine cell for wavelength calibration. Two $\lambda/2$ birefrigent plates were used to rotate the relative plane of polarization of the two linearly polarized laser beams.

Photoions were mass resolved in a linear, 5 m time-of-flight mass spectrometer ($M/\Delta M$ = 1200) and detected with a Galileo microchannel plate detector. The ion signal, neon optical galvanic signal, and iodine fluorescence signals were simultaneously monitored by Stanford Research boxcar averagers, the outputs of which were digitized and stored in computer memory, to allow subsequent display and calibration of the spectra. A gated ion counting detection system was used (Green and Sopchyshyn) for both isotope ratio measurements and polarization studies.

3. RESULTS AND DISCUSSION

A typical, two-colour U spectrum is shown in Figure 1; all of the peaks correspond to two-colour transitions, except that at 33801.22 cm^{-1}; which corresponds to a one-colour doubly resonant transition. Where comparisons are possible, the assigned energy levels are in good agreement with previously reported values: Blaise and Radziemski (1976) and Miron et al. (1979). Isotope shifts were measured for ^{235}U relative to ^{238}U, and ranged between 0.3 and 0.7 cm^{-1} for the states identified in Figure 1.

Two important features of the photoexcitation process are illustrated by a slow scan of the high resolution laser over the low to intermediate state ^{235}U transition at 567 nm (Figure 2). First, there is no evidence of saturation of the transition, likely due to the relatively low pulse energy of the CVL pumped

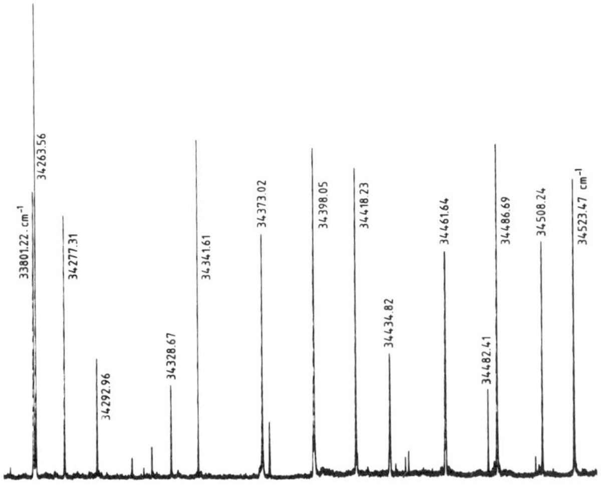

Figure 1. Two-colour spectrum of U from 580 to 592 nm for λ_1 = 576.0 nm: $^5L_6{}^0(g) \rightarrow {}^7L_6$ (17361.895 cm^{-1}). Filament temperature, 1650°C; pulse repetition rate, 6.5 kHz; λ_2 scan rate, 2 cm^{-1}/s.

dye laser. Second, there is no evidence of the hyperfine structure of ^{235}U, which has been observed with a CW pumped dye laser (Bohm et al. 1978). The spectrum has the approximate shape and width expected from a convolution of the Lorentzian shape of the laser with the Doppler broadened (0.035 cm^{-1}) Gaussian shape of the atomic line. The lack of hyperfine features was attributed to the slow rate of the hyperfine interaction (60-150 MHz) relative to the pulse duration, in the pulsed laser experiments.

For most isotopic ratio measurements, the prism beam expander was removed from the dye laser cavity to obtain a more uniform power output of the laser over a bandwidth sufficiently large, e.g. 0.5 cm^{-1}, to span most isotope shifts. However, even in this broadband mode, steep wavelength sensitivity of the isotopic ratio was observed (Figure 3), which resulted in the standard deviation of 235/238 ratios being much higher than expected; e.g. 4% compared to 0.4% expected from Poisson statistics for a particular data set. This wavelength sensitivity is similar to that observed by Miller et al. (1988) for Lu isotopes, except that a dip at the ^{235}U line centre was not observed. A slow scan, for which a low resolution laser ($\Delta\omega$ = 1 cm^{-1}) was used, over the first transition showed a peaked shape very similar to that in Figure 2, except that $\Delta\omega$ was 1.1 cm^{-1} (FWHM). Attempts to reduce the wavelength sensitivity of the 235/238 ratio by increasing the laser power, to achieve near saturation of the transitions, or by using a non-resonant second step, to avoid the higher isotope

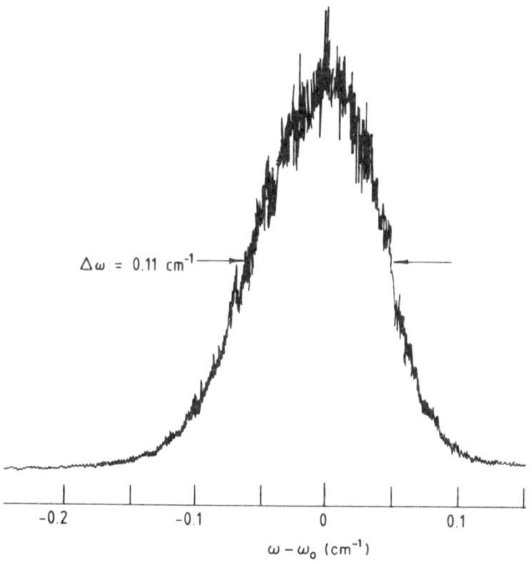

Fig. 2. Scan over $^5K_5{}^0$ (620.3 cm^{-1}) → $^3K_6$ (18253.93 cm^{-1}) transition of ^{235}U. Scan rate, 0.005 cm^{-1}/s; $\Delta\omega(\lambda_1)$, 0.07 cm^{-1}; λ_2, 589.3 nm (transition to $^5L_6{}^0$ 35223.8 cm^{-1} state). Other parameters as in Fig. 1.

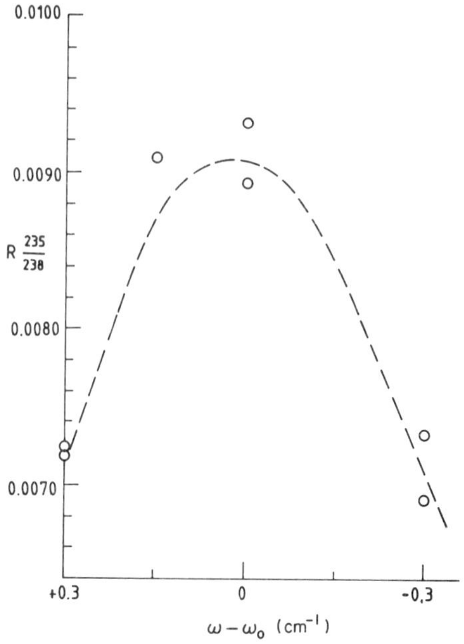

Fig. 3. Sensitivity of 235/238 Ratio (R) to tuning of λ_1, for the same transition as in Fig. 2. ω_o is the ^{235}U line centre.

shifts, were not successful. The moderate precision obtainable, e.g. 4% (1 std. dev.), although not useful for most nuclear applications, is sufficient for many applications such as environmental monitoring.

The relative photoionization rates for ^{235}U and ^{238}U were also dependent on the relative polarizations of the linearly polarized laser beams. For the transition described in Figure 2, the 235/238 ion ratio was 20% larger for parallel alignment of the polarization axes, compared to perpendicular. A systematic set of experiments showed that the effect was entirely due to laser-laser orientation, rather than being an artifact of ion source electric or magnetic fields. These experiments also showed that the change in the isotope ratio was almost entirely due to a corresponding change in the 238 signal, as the 235 signal was essentially unaffected.

The variation in ^{238}U photoionization rate was attributed to the preparation of an aligned, excited state by the first transition, which was sampled by the second laser. The transition line strengths of the M_J components varied as $(J+1)^2$ - M_J^2 for the first transition (Sobelman 1979) and those substates with least projection on the laser field axis were preferentially populated. The second transition line strengths varied as $M'_{J'}{}^2$ and those substates with most projection on the second laser field axis preferentially absorbed. As the relative polarization was varied, the match of the line strength distribution of the second transition with the anisotropic population distribution of the prepared first excited state went through a maximum at perpendicular, and a minimum at parallel, relative polarization. The resultant variation in the ^{238}U photoionization rate caused corresponding minima and maxima, respectively, in the 235/238 ratio.

The lack of a laser polarization effect on the 235 isotope was attributed to hyperfine coupling. Although, as mentioned earlier, the hyperfine interaction was too slow to cause observable features in Figure 2, nevertheless, it was sufficiently rapid to convert the aligned substates of the first excited level to a mixed distribution of substates, before onset of the second laser pulse.

4. REFERENCES

Blaise J and Radziemski L J *1976 J. Opt. Soc. Am.* 7 644

Bohm H-D V, Michealis W and Weitkamp C *1978 Opt. Comm.* **26** 177

Green L W and Sopchyshyn F C *1989 Int. J. Mass Spectrom. Ion Proc.* **89** 81

Miller C M, Feary B L, Palmer B A, Nogar N S *1988 Inst. Phys. Conf.* **Ser. No. 94** 297

Miron J, David R, Erez G, Lavi S and Levin L A *1979 J. Opt. Soc. Am.* **69** 256

Sobelman I I *1979 Atomic Spectra and Radiative Transitions* (New York: Springer-Verlag) p 177

High Rydberg states and their structure of ruthenium by RIS

X Y Xu, J Z Tang, W Huang, W Z Zhao and D Y Chen

Department of Modern Applied Physics,
Tsinghua University, Beijing, PRC

ABSTRACT: We report the first work in measuring and analyzing the level structure of high Rydberg states of a platinum group atom. The Rydberg energy levels in the Ru atom for the $4d^7(^4F)ns$ (n = 28–50) and $4d^7(^4F)nd$ (n = 24–64) series were measured by RIS. The energy levels and relevant quantum defects as well as the relative line strengths of doublet are calculated and compared with the experimental values. Also, the first ionization potential of Ru is obtained.

1. INTRODUCTION

The ground configuration of the Ru atom is (Kr, $4d^7(^4F)5s$ ($^5F_5$)). Its inner electronic shell is open. The structure of its Rydberg energy levels might be complicated. To our knowledge there are no experimental data for the high–lying excited states of Ru atom. Thus research into the Rydberg energy levels of Ru atom would greatly enrich our knowledge about Rydberg states. The hypersensitive analysis of elements of the platinum group is also of wide application in the exploitation of marine resources, cosmochemistry and geology.

2. EXPERIMENTAL METHOD AND RESULTS

The experimental set–up includes several parts: electro–thermal atomization, laser resonance excitation, electric field ionization and ion signal collection.

The Ru atoms in the ground state were excited to Rydberg states in two–steps:

$$4d^7(^4F)5s(^5F_5) \xrightarrow{\lambda_1} 4d^7(^4F)5p(^5G_5^o) \xrightarrow{\lambda_2} \begin{cases} 4d^7(^4F)ns \\ 4d^7(^4F)nd \end{cases} \quad . \qquad (1)$$

The two laser beams were produced respectively by two FL3002E dye lasers pumped by an excimer laser (EMG202). λ_1 was 3301.60Å. λ_2 was tunable in the range from 3465 Å to 3440Å. In order to show the line structures clearly, we tried to improve the resolution power. For the measurement of relative line strengths the pulse energy of the second laser beam should also not be too high. After excitation, a electric field pulse of 2000V was applied and then the excited atoms were ionized. In order to reduce the Stark splitting caused by the rise of electric field pulse, the pulse was delayed about 200ns–1μs relative to excitation.

An electron multichannel plate was used as the ion detector. In the experiment the resonance line corresponding to each Rydberg level of Ru can be obtained by fixing λ_1 and tuning λ_2 continuously. Some parts of the experimental spectrum are shown in Fig.1.

The measured ns (n=28–50) series has a singlet structure and nd (n=24–32) series has a doublet structure. When n is greater than 32 in the nd series, the doublet structure can not be seen in the experiment beacuse of the limit of energy resolution. The relative line strength of doublet in the nd series was measured so that the corresponding branching ratio was obtained. They varied from 1.7(1) to 2.1(1).

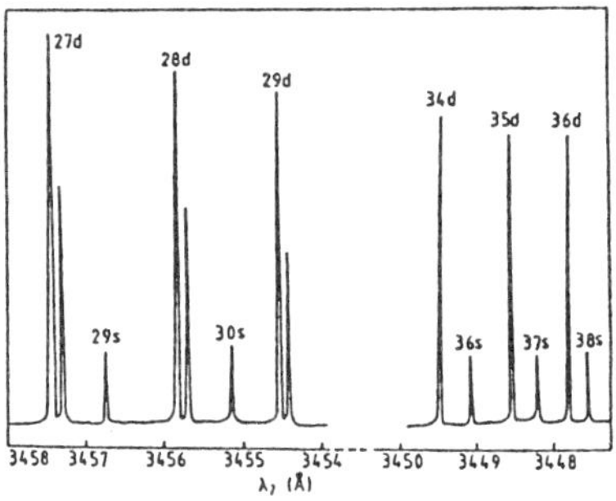

Fig.1. Two–step laser excitation–electric field ionization spectroscopy of the Ru atom.

Two lines of the Ru atom which are produced by the transitions from the first excited state to intermediate states were chosen as the calibration wavelengths (Moore 1971). After correction of the refractive index of air we obtained the calibrated experimental wavelengths λ_2, expt (cm^{-1}). From them we can get the measured values of the Rydberg levels of ns and nd series of the Ru atom.

3. THEORETICAL ANALYSIS

3.1 The calculations of the quantum defects and the ionization potential

The relativistic Dirac–Slater self–consistent–field method (Grant 1970, Lee 1986) is used to calculate the energy levels and quantum defects of the high Rydberg states of the Ru atom. The calculated values of quantum defects for s and d series of Ru atom are:

$$\mu_s = 3.5 \quad (n = 28 - 50)$$
$$\mu_d = 1.9 \quad (n = 33 - 56). \qquad (2)$$

Using 23 lines of the s series (n = 28–50) from experiment, we obtain the limit of 59365.88 (42) cm^{-1} for the ns series through nonlinear least–squares fitting. Using 24 lines of the d series (n = 33–56) from experiment we obtain the limit of 59365.96(42) cm^{-1}. Finally we take the first ionization potential of the Ru atom as:

$$I.P. = 59365.9(4) \quad cm^{-1} \tag{3}$$

Callender et al. (1988) used double–resonance field–ionization spectroscopy to determine the first ionization potential of Ru atom. It was 59366.4(3) cm^{-1}. Our result is a significant improvement over theirs.

The experimental quantum defects for each energy level of ns and nd series are coincident with the theoretical ones within the experimental errors. Also, the theoretical values for the energy levels of the s and d series are very consistent with the experimental ones.

3.2 The determination of spectral terms

For the single electron transitions of an open shell atom

$$l_1^{w_1}(\alpha_1 L_1 S_1) l_2 (LSJ) \rightarrow \begin{cases} l_1^{w_1}(\alpha_1 L_1 S_1) l_2' (L' S J') & (4) \\ l_1^{w_1}(\alpha_1 L_1 S_1) l_2' (L'' S J'') & (5) \end{cases}$$

the ratio of the line strength (or the relative line strength) is (Cowan 1981):

$$R_5 = \frac{S_4}{S_5} = \frac{[L', J']}{[L'', J'']} \frac{\begin{bmatrix} L & J & S \\ J' & L' & 1 \end{bmatrix}^2 \begin{bmatrix} l_2 & L & L_1 \\ L' & l_2' & 1 \end{bmatrix}^2}{\begin{bmatrix} L & J & S \\ J'' & L'' & 1 \end{bmatrix}^2 \begin{bmatrix} l_2 & L & L_1 \\ L'' & l_2' & 1 \end{bmatrix}^2} \tag{6}$$

R only depends on the quantum numbers of angular momenta.

From the selection rules we know that the Rydberg states of ns series have only two possible spectral terms: $^5F_4$ and $^5F_5$. However the calculated line strength of transition corresponding to $^5F_5$ is one order of magnitude smaller than that of $^5F_4$. Therefore only one line can be measured in the experiment for each Rydberg state of the ns series. The corresponding spectral term is $4d^7(^4F)$ns ($^5F_4$).

For the nd series, each Rydberg state has eight possible spectral terms: $^5H_6$, $^5H_5$, $^5H_4$, $^5G_6$, $^5G_5$, $^5G_4$, $^5F_5$ *and* $^5F_4$. The calculated results show that the line strength of transition corresponding to $^5H_4$ is the strongest one and the ratio of the line strength of this transition to that of $^5F_5$ is 1.91. Other transitions are one order of magnitude or more smaller than that of $^5H_4$. This theoretical result agrees with the experiment very well. Then the spectral terms for the measured lines of the nd series can be identified. They are $4d^7(^4F)nd(^5H_4)$ and $4d^7(^4F)nd(^5F_5)$ respectively.

REFERENCES

Callender C L, Hackett P A and Rayner D M 1988 J. Opt. Soc. Am. **B5** 614–8

Cowan R D 1981 The Theory of Atomic Structure and Spectra (Berkeley: Univ. of California Press) pp402–3

Grant R D 1970 Adv. Phys. **19** 747–69

Lee C M 1986 Electronic and Atomic Collisions (Amsterdam: Elsevier) pp621–29

Moore C E 1971 Atomic Energy Levels NSRDS–NBS 35, Vol. 3 (Washington, D.C.: US Government Printting Office) pp20–23

Paper presented at RIS 90, Varese, Italy, 16–21 September 1990

Resonance ionization mass spectroscopy of uranium

G Herrmann[+], J Riegel[+], H Rimke[+], P Sattelberger[+], N Trautmann[+]
F-J Urban[*], F Ames[*], E-W Otten[*], W Ruster[*], F Scheerer[*]

[+]Institut für Kernchemie, Universität Mainz, D - 6500 Mainz
[*]Institut für Physik, Universität Mainz, D - 6500 Mainz

ABSTRACT: Resonance ionization mass spectroscopy (RIMS) has been used for the sensitive detection of uranium. The apparatus consists of a laser system with three dye lasers and two pulsed copper vapour lasers and a time-of-flight (TOF) mass spectrometer. The uranium atoms are ionized in a three step excitation with the third step leading to an autoionizing state. Several excitation schemes were investigated and for two schemes all three transitions could be saturated with the available laser power. The hyperfine structure splitting (HFS) of ^{235}U, the isotopic shift (IS) between ^{235}U and ^{238}U as well as isotopic ratios in uranium samples were determined.

1. EXPERIMENTAL

For environmental studies a sensitive element- and isotope-selective detection method of actinides is required. With resonance ionization mass spectroscopy (RIMS) this request can be fullfilled. Using RIMS one obtains a selective detection of an element combined with the determination of isotopic abundances via TOF mass spectrometry. The experimental set-up is shown in figure 1.

Two pulsed copper vapour lasers with an average power of 30 W and a pulse repetition rate of 6.5 kHz pump three tunable dye lasers. Depending on the dyes, conversion efficiencies between 10% and 25% have been obtained. The light of the dye lasers is deflected by means of quartz fibres (70% transmission) and focused into the ionization zone, where the laser beams interact with an atomic beam of uranium. The use of fibres provides a homogeneous spatial beam profile and a good spatial overlap of the three laser beams.

The atomic beam of uranium is produced by evaporation either from an oven or from a rhenium filament, on which the uranium is deposited electrolytically and covered with a thin rhenium layer. The ions produced in the interaction zone are accelerated by two electric fields and detected by channel plates after passing a drift tube of 2 m in the TOF spectrometer.

The experimental set-up is decribed in detail by Ruster et al (1989).

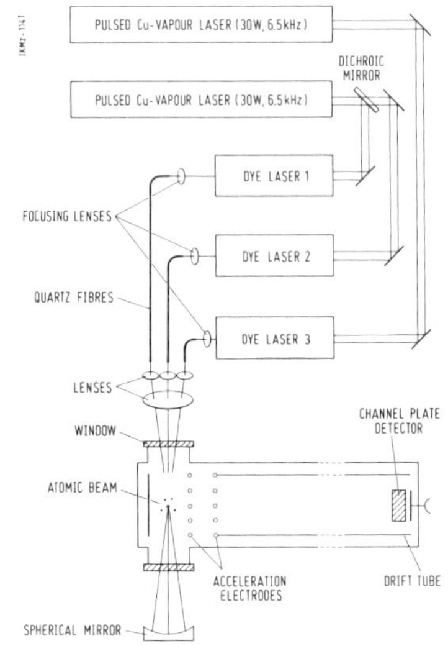

Fig. 1. Experimental set-up for resonance ionization mass spectroscopy

2. RESULTS

The atoms are ionized via three photon excitation with the third step leading to an autoionizing state above the ionization limit. Thereby the ionization rate increases for up to two orders of magnitude compared to ionization into the continuum. The excitation schemes investigated in detail are shown in figure 2. By keeping the first two wavelengths fixed and tuning the third wavelength, autoionizing states for all three excitation schemes could be observed.

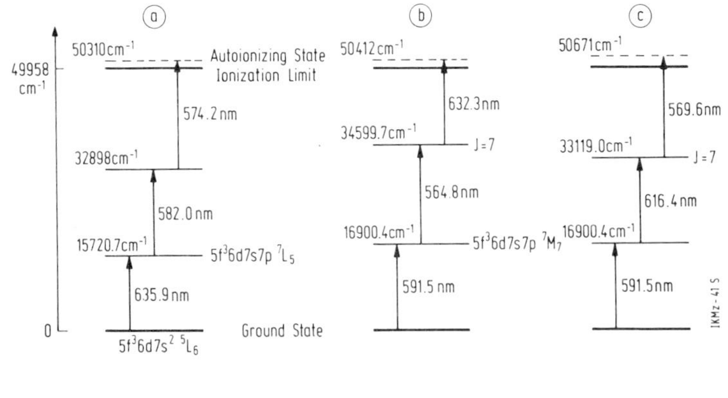

Fig. 2. Investigated excitation schemes of uranium

As an example, figure 3 shows autoionizing states for excitation scheme (b) in figure 2. As can be seen, a large number of such states have been found. The ionization rate in dependence of the laser power for each excitation step of the three schemes was measured. By fitting the data with exponential functions, rate equations and density matrix formalism, saturation powers, cross sections and oscillator strengths were obtained. The results are quite similar for all three fit methods and agree with data reported in the literature (Klose 1975, Corliss 1976 and Palmer et al 1980). For the excitation schemes (b) and (c) in figure 2 all three transitions could be saturated.
The isotopic shift (IS) between ^{238}U and ^{235}U and the hyperfine structure splitting (HFS) of ^{235}U with I_{nucl} = 7/2 were measured. The results are summarized in table 1. The use of an intracavity etalon in the dye lasers reduces the laser bandwidths from 3-6 GHz to 1.2 GHz. Even with this small bandwidth, the HFS of ^{235}U could not be completely resolved. On the other hand, even with the large bandwidth of the laser beams, an uniform excitation of all transitions between the sublevels was not possible. The IS between ^{238}U and ^{235}U in all three steps enables a selective excitation of each isotope.

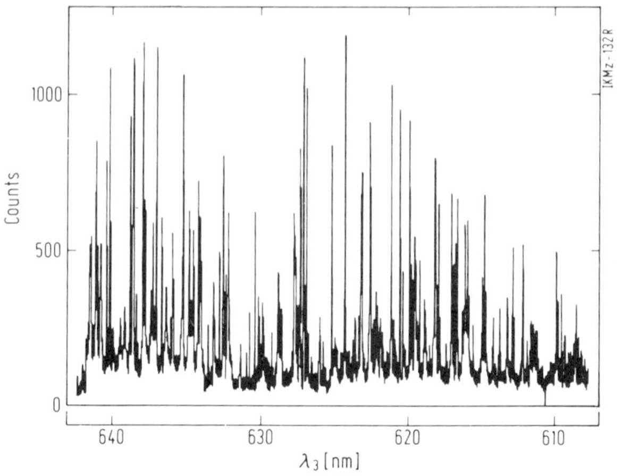

Fig. 3. Autoionizing states found by tuning λ_3 with λ_1 and λ_2 in resonance at 591.5 nm and 564.8 nm

Wavelength [nm]	Bandwidth [GHz]	Linewidth ^{238}U [GHz]	HFS ^{235}U [GHz]	IS ^{238}U - ^{235}U [GHz]
$\lambda_1 = 591.54$	3.7	7.5	8.5	8.5
	1.2*	1.7	5.3	8.5
$\lambda_2 = 564.84$	3.7	8.5	8.5	8.3
	1.2*	5.9	-	-
$\lambda_3 = 632.25$	5.7	7.3	5.7	≤ 3
	1.2*	1.4	1.5	1.6

*with intracavity etalon

Tab. 1. Hyperfine structure splitting and isotopic shift in uranium

Isotopic ratios could be determined by tuning the laser wavelengths to the resonances of each isotope. Non-resonant ionization of uranium oxides provided values of the isotopic ratios ($^{235}U/^{238}U$), which were in good agreement with the natural isotopic abundances.

REFERENCES:

Corliss C H 1976 J.Res. Nat. Bur. Std. **80A**

Klose J Z 1975 Phys. Rev. A **11** 1840

Palmer B A, Keller R A and Engleman R 1980 Los Alamos Scientific Laboratory Report LA-8251-MS

Ruster W, Ames F, Kluge H-J, Otten E-W, Rehklau D, Scheerer F, Herrmann G, Mühleck C, Riegel J, Rimke H, Sattelberger P and Trautmann N 1989 Nucl. Instr. Meth. **A281** 547

Inst. Phys. Conf. Ser. No 114: Section 5
Paper presented at RIS 90, Varese, Italy, 16–21 September 1990

On the improvement of the method for the investigation of giant absorption resonances in rare earths

Antra-Vince Karosiene
The Institute of Theoretical Physics and Astronomy, Lithuanian Academy of Sciences, Vilnius

ABSTRACT: An improved single-electron model is suggested for theoretical investigation of giant absorption resonances of light lanthanides and light actinides in ultraviolet regions. It leads to the reasonable agreement between theoretical resonances and observed ones and to the interpretation of these resonances as the transitions of the type nd→εf. As a result of a more accurate calculation of lanthanide 4f orbitals, there arises a statement that among the final absorption states there are ones in which equivalent 4f electrons may be characterized by considerably different orbitals.

1. INTRODUCTION

Single-electron models in their simplest and most ordinary forms fail to describe cross sections of certain subshells. Therefore, lanthanide 4d resonances as well as actinide 5d resonances were investigated theoretically only in the many-body approaches. We suggest the usage of an improved single-electron model for rare earth giant resonances investigation. As follows from the analysis of the photoelectron spectra and photoabsorption partial cross sections spectra measured by Richter et al. (1989), as well as from photoion-yield spectra of Dzionk et al. (1989) and the discussion therein, the large interchannel interaction in heavy lanthanides occurs. Therefore, we suggest to use the single-electron model, accounting only for intrachannel effects and only for light elements of both groups.

No one can separate the giant resonances from the centrifugal effects (see Connerade (1987)), and, consequently, 4d and 5d resonances of rare earths from the existence of the double-well effective potential for radial f-orbitals. The negligible changes of an inner well of such a potential sometimes cause a considerable alteration of localization of f-orbitals. Therefore, in the next section we are trying to make a more accurate calculation of these sensitive f-orbitals in the final absorption states with nd-vacancy. In the last section, we present our cross section calculation results and carry out discussions.

2. TERM-DEPENDENT ORBITALS FOR CONFIGURATIONS WITH nd VACANCY AND AN OPEN nf SHELL

For the investigations of 3d and 4d giant resonances of Xe, Cs, Ba, and La, the well-known Hartree-Fock term-dependent (HF^{td}) f-orbitals, localized in the double-well effective potential, were successfully used. In the case of rare earths, the substantial improvement of such calculations becomes necessary because of the existence of an open 4f (5f) shell in the initial state. It was proven that for the light lanthanides the term-dependence of radial 4f orbitals in the final state configuration of the type

$$4d^9 4f^N \ (N > 1) \qquad (2.1)$$

becomes distinctly expressed only when the two conditions for 4f orbital calculations are satisfied. The first condition requires the equivalence restriction for equivalent 4f electrons, usually imposed on Hartree-Fock (HF) approximations, to be removed by letting two different orbitals. This condition depends on the two effective potential wells of the f electrons. The second condition requires the usage of such a wave function basis which takes into account the main part of the exchange interaction hidden not only in diagonal but also in off-diagonal matrix elements of the energy matrix.

The effect of such modifications on f-orbitals becomes evident after solutions of the minimally extended term-dependent HF ($MEHF^{td}$) found by Kancerevicius and Karosiene (1989). Such calculations were performed for the cases La I in configuration $4d^9 4f^2$ and Ce I in configuration $4d^9 4f^3$. As a result of the highly lying terms of these atoms, there arises the two different 4f orbitals for the same 4f shell: one orbital characterizes one 4f electron localized in the outer well region, and the second one characterizes all the remaining N-1 4f electrons of the configuration (2.1), which are localized in the inner well region.

However, the efficiency of the usage of $MEHF^{td}$ equations (i.e. significant f-orbital dependence on terms) arises when not only diagonal but also off-diagonal matrix elements of Coulomb interaction are taken into account when solving these equations. Besides, there occurs a negligible difference in term-dependent f-orbitals when either all of the off-diagonal matrix elements or only the ones corresponding to the main exchange interaction $g_1(df)$ are taken into account. Therefore it is proper to use the coupling

$$4d^9 4f' \, (^1P) 4f^{N-1} \ (N > 1) \qquad (2.2)$$

while simplifying the calculations when the orbitals P_{4f} and $P_{4f'}$ for highly lying (2.1) states are searched for. Because of significant distance in effective rare earths' potential between both wells and a positive barrier, if large enough it is even more possible to simplify calculations and to simultaneously ensure the orthogonality principle between all the wave functions for excited states

$$4d^9 4f^{N-1} n(\varepsilon)f \qquad (2.3)$$

of the same symmetry as (2.2):

$$4d^9 n(\varepsilon)f' \, (^1P) 4f^{N-1} \ . \qquad (2.4)$$

When in search for all closed orbitals, including $4d^9 4f^{N-1}$, for atoms with the nuclear number Z, one must solve ordinary HF equations for the average term (HF^{av}) for the ion Z^{+1} with one 4f electron eliminated, and in the frozen field of this ion, using HF^{td} equations to solve $4f'$, nf', and the $\varepsilon f'$ orbitals for the term ^1P of configuration

$4d^9n(\varepsilon)f'$. It is necessary to note, that the orbitals either with prime or without it must not be orthogonalized with each other (see, e.g., Fisher (1977)) if the EHF leads to the large effect. It is valuable to emphasize that the cases of configuration (2.1) and (2.3) investigated by Kancerevicius and Karosiene (1989) are the cases with the most efficiency of EHF and non-orthogonal HF (NHF) (MEHF is a partial case of a more general -- EHF). As an argument, it suffices to say that the integral of overlap of both 4f orbitals, P_{4f} and $P_{4f'}$, is close to zero (compare, e.g., with Fisher (1977)).

3. CROSS SECTION IN SINGLE-ELECTRON MODEL AND DISCUSSION

If such approximate term-dependent 4f′ , nf′ , and εf′ orbitals are used for lanthanide 4d and actinide 5d resonances investigation, a reasonable agreement between theoretical and experimental cross sections is achieved. This agreement manifests itself at first by similar experimental and theoretical cross section maxima positions above the statistical average of the 4d ionization limits I_{4d} and, secondly, by a near assymmetry shape of both maxima. In Figure 1, for Ce both experimental and theoretical cross sections (σ_{4d} theoretical - geometrical average of σ^r and σ^v) are given. Taking into account the experimental results of Richter et al. (1989) and Dzionk et al. (1989), anyone must expect interchannel effects to be negligible in the following lanthanides - Ce, Pr, Nd. Consequently, the single-electron model, in principle, must fit for these electrons, and our calculations testify to that for Pr and Nd, as for Ce in Figure 1.

We suppose such calculations may also be applied for light actinides, but unfortunately, corresponding experimental results testify that the fitting of single-electron models are absent for actinides. Nevertheless, we performed the single-electron calculation (also in a simplified manner, as for lanthanides) for U because of an extremely narrow absorption 5d resonance observed by Pantelouris and Connerade (1983) for U vapour. As a result, our maximum is characterized by the same half-width's order of magnitude as the observed one and by the near position of maximum with respect to 5d absorption edges.

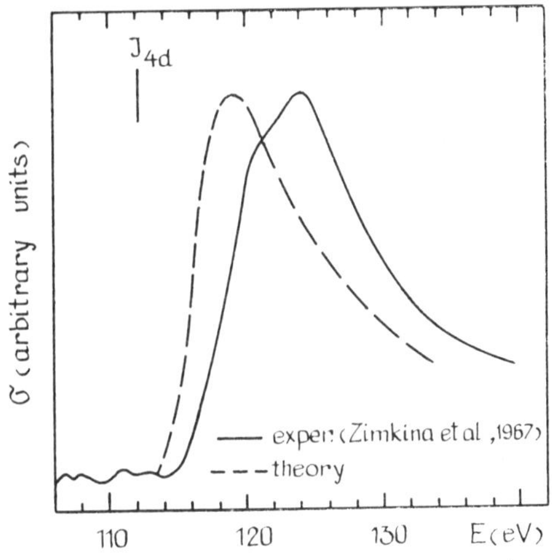

Figure 1. Experimental and Theoretical Cross Sections for Ce

Valence fluctuations and intermediate valence problems are related to the elements, for which giant resonances are expressed most distinctly, as noticed by Connerade (1987) when summarizing many interesting properties inherent in these elements. These elements, namely, light lanthanides and light actinides, are the object of this work. And these elements, according to a quasi-periodic table of d- and f- transition elements, composed by Smith and Kmetko (1983), find themselves on the dividing zone between magnetics and superconductors, according to the behavior of the valence d and f electrons. According to our results of more accurate f-orbital calculations, the definite excited states of these elements exhibit simultaneously both bonding and non-bonding f-orbitals, and anyone must notice a correlation between all of the above-mentioned phenomena.

The author expresses thanks to A. Kancerevicius for discussions and serious help with calculations.

REFERENCES

Connerade, J. P., 1987 *Giant Resonances in Atoms, Molecules, and Solids*, ed. J. P. Connerade, J. M. Esteva and R. C. Karnatak (New York: Plenum), pp 3-23.
Dzionk, Ch., Fiedler, W., Lucke, M., and Zimmermann, P., 1989 Phys. Rev. Let. **62**, 878.
Froese, Fischer, Ch., 1977 *The Hartree-Fock Method for Atoms* (New York: J. Willey and Sons), pp 111-4.
Kancerevicius, A., and Karosiene, A., 1989 Liet. Fiz. Rink. **29**, 159.
Pantelouris, M. and Connerade, J. P., 1983 J. Phys. B: At. Mol. Phys. **16**, L23.
Richter, M., Meyer, M., Pahler, M., Prescher, T., Raven, E., Sonntag, B., and Wetzel, H-E., 1989 Phys. Rev. A **40**, 7007.
Smith, J. L. and Kmetko, E. A., 1983 J. Less-Com. Met. **90**, 83.
Zimkina, T. M., Fomichev, V. A., Gribovskij, S. A., and Zhukova, I. I., 1967 Sov. Phys. - Solid State **9**, 1128.

Paper presented at RIS 90, Varese, Italy, 16–21 September 1990

The measurement of radiative lifetime of Rydberg states of SmI

Yi-qun Lu
Anhui Institute of Optics and Fine Mechanics Academia Sinica, P. O. Box 25, Hefei, Anhui, 230031, P.R. China

ABSTRACT: The natural radiative lifetimes of the 4f 6S nP (n = 44-51) Rydberg states in SmI have been measured using stepwise excitation and a fluorecence monitoring method. The lifetimes (only a few states) are not perturbed by other configurations.

1. Introduction

During the last few years, highly excited states of alkali-metal and alkaline-earth metal atoms have been the subject of extensive studies (Bhatti *et al*; Karlov *et al*; Solarz *et al*). Recently this investigation has been extended to the third group and rare-earth elements (Dobryshin *et al*), but only a few results have been obtained. Rare-earth elements usually have complex spectra; nevertheless for the ground configuration of SmI, the 4f electronic shell has not been filled and the high excited spectra of SmI is complex. In this paper, we report lifetime measurements and the Rydberg states using three-step excitation and fluorecence decay. Highly excited even Rydberg states of SmI have been measured first by Parr and Dobryshin (Parr *et al*). Usually the Rydberg states of SmI are so complex that we cannot distinquish these. Ab initio (MCDF) method calculations indicate that the unperturbed series belongs to highly excited $4f6s[^8F_{1/2}]np$ and $4f5d[^8H_{3/2}]np$ configurations.

2. Experiment

Three pulsed dye lasers (Lambda Physik FL3002) pumped by a XeCl excimer laser (Lambda Physik EMG100) were used to excite the atoms. The pulse had a few mJ/pulse with a linewidth of approximately 0.2-0.3 cm^{-1}. The laser beams were spatially overlapped with 5-10 ns delay for the second and third pulse, and the atom beam was generated by an electronic oven. Figure 1 shows the scheme of stepwise excitation and fluorecence process. The fluorecence-decay signals, released from the investigated states down to the 4f6s7sF state, were detected by a photomultiplier, and it was then fed to a time-scanning boxcar. Most of the resonant fluorecence was blocked by a suitable filter. The ionization signal was detected by a quadrupole mass filter.

3. Measurements and Results

The correct wavelength tuning for the resonance was insured by observing the current variation of a hollow cathode lamp of SmI and the variation of fluorecence intensity from the atomic beam. Our measured lifetime of unperturbed Rydberg states are shown in Table 1. The photoionization spectrum as a function of wavenumber is shown in Fig. 2.

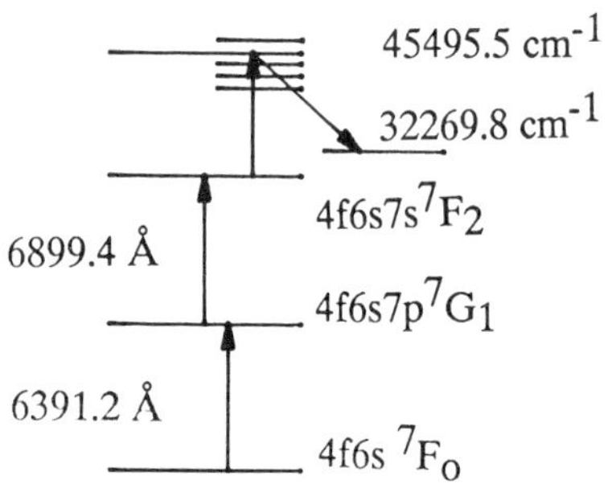

Figure 1. The Scheme of Stepwise Excitation Process

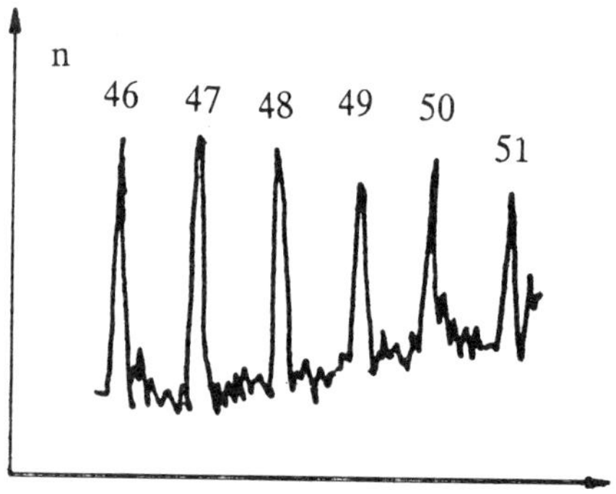

Figure 2. The Photoionization Spectrum

4. Discussion

In the present paper, the lifetimes of Rydberg states of SmI were studied. The optical spectrum of this element is very complex, i.e., contains transitions from both the outer 6s and inner 4f shell. There are lots of channels in it. The study of the Rydberg series in unperturbed regions is so narrow that there are few states in it. The study of Rydberg series in the unperturbed regions also allows one to proceed to the determination of the ionization constant in these regions. The first ionization limit $I_1 = 45495.5\ cm^{-1}$.

Table 1.
The Experimental Values of
Natural Radiative Lifetime

$E(cm^{-1})$	n	T(ns)
45444.3	46	1101
45446.1	47	1270
45448.0	48	1430
45450.1	49	1710
45451.8	50	2080
45453.4	51	3035

5. References

Bhatti, Sa. A., Cooke, W. E., 1983 Phys. Rev. **A28**, 756.
Dobryshin, V. E., Karpov, N. A., *et al.*, 1983 Opt. Spectroscopy, (USSR) 54, 244.
Karlov, N. V., Krynetskii, B. B., *et al.*, 1978 Applied Optics Vol. 17, No. 6, 856.
Parr, A. C., *et al.*, 1975 J.O.S.A. 65, 613.
Solarz, R. W., May, C. A., *et al.*, 1976 Phys. Rev. **A14**, 1129.

The study of resonant enhanced KrII 469.4nm

Zhang Yaoning, Zhang Kai, and Qju Junlin
Institute of Laser, HUST, Wuhan, Hubei, CHINA

ABSTRACT: The dynamics model of plasma decay processes in the afterglow of a He-Kr pulsed HCD (Hollow Cathode Discharge) is set up. Using this model we have calculated the difference numbers populated in the upper and lower states of the spectral line KrII 469.4nm. It was shown that the KrII 469.4nm will be much enhanced with an optimum partial pressure of He and Kr.

It is well known that the Penning ionization and the resonant energy transfer can enhance some spectral lines of ions in a gas discharge tube. This effect is useful to get the laser oscillation of an ion's spectral line. We've determined that in a He-Kr discharge tube, the KrII spectral line, 469.4nm, was stronger than in a pure krypton discharge tube when the parameters of the discharge were the same as the former.

1. The Main Reaction Processes in the Afterglow of Pulsed HCD

The main reaction processes in an afterglow can be divided into three kinds. They are recombination processes of plasma, processes creating new charged particles, and other processes with metastable state atoms or molecules.

A. Recombination processes

(1) $He^+ + 2e \xrightarrow{\alpha_1} He^m + e$

(2) $He^+ + e \xrightarrow{\alpha_2} (He_2')^* \rightarrow He^* + He \rightarrow 2\,He^m + h\nu_1$

$\longrightarrow (He_2'')^* \rightarrow He_2^m(2\,^3\Sigma) + h\nu_2$

(3) $Kr^+ + 2e \xrightarrow{\alpha_3} Kr^m + e$

Where $\alpha_1 \sim \alpha_3$ is the rate of the above three processes, respectively; He^m are metastable He atoms; He_2^* are excited state He_2 molecules; He_2^m are metastable He_2 molecules.

B. Processes creating new charged particles

(1) $2He_2^m + He \xrightarrow{\gamma} He_2^+ + He + e$

$\xrightarrow{\gamma'} He^+ + 2He + e$

(2) $He^+ + 2He \xrightarrow{\Gamma_2} He^+ + 2He + e$

(3) $He^m + e \xrightarrow{\beta_1} He_2^+ + 2e$

(4) $He_2^m + e \xrightarrow{\beta_2} He_2^+ + 2e$

(5) $Kr + e \xrightarrow{\beta_3} Kr^+ + 2e$

(6) $He^m + Kr \xrightarrow{\alpha_P} Kr^+ + He + e$

(7) $He_2^m + He_2^m \xrightarrow{\Gamma_3} He^+ + 3He + e$

(8) $He^m + He_2^m \xrightarrow{\Gamma_4} He^+ + 2He + e$

C. Other processes with metastable atoms or molecules

(1) $He^m + e \xrightarrow{\beta_4} He + e$

(2) $He^m + e \xrightarrow{\beta_5} He^* + e$

(3) $He + e \xrightarrow{\beta_6} He^m + e$

(4) $He^m + Kr \xrightarrow{\alpha_R} Kr^{+*}(6s\,^4P) + He + \Delta E$

(5) $He^m(2\,^1S_0) + e \xrightarrow{\beta_7} He^{m*}(2\,^3S_1) + e$

(6) $He^m + 2He \xrightarrow{\Gamma_5} He_2^m + He$

Where $\beta_1 \sim \beta_7$ is the rate of excited or deexcited processes by electron rates [2-3], [2-4], [2-5], [3-1], [3-2], [3-3], [3-5], respectively; α_P is the rate constant of the Penning transfer process; α_R is the rate constant of the resonant energy ionization process: $\Gamma_1 (\gamma + \gamma')$, Γ_2 - Γ_5 is the rate constant of three-body impact processes [2-1], [2-2], [2-7], [2-8], [3-6], respectively.

Besides these processes the deexcitation or deionization processes of ions impacted to the wall of the tube by diffusion also influences the decay rates of the plasma. For a cylinder of radius R and length L, the characteristic diffusion length Λ is given by

$$\Lambda^2 = [(\pi/L)^2 + (2.4/R)^2]^{-1} \tag{1}$$

Then the rate of the deexcitation processes by diffusion can be expressed as $D/\Lambda^2 P$, where P is the pressure of gas mixture in tube. It is this main difference between the discharge in a gaseous He-Kr mixture and pure Kr gas which explains the Penning ionization process between He^m, Kr and the resonant energy transfer process between He^m, Kr^+. Analyzing those processes listed above, the Penning ionization process would be one of the high reaction rate processes because there are many more Kr atoms in the tube and the rate of the resonant energy transfer process will be close to the rate of the other processes ($\alpha_R = 2^* \ 10^9 \ cm^3 \cdot S^{-1}$) while this process occurs between He^m, Kr^+.

2. The Rate Equations of Dynamics Processes of Decay

Since there is much less Kr than He in the filled tube (generally in order to form laser oscillations, the ratio of partial pressures, He:Kr, is about 200:1), the population of Kr^m and Kr_2^m is at least two orders of magnitude less than the population of He^m and He_2^m. Therefore, the influence of the processes with Kr^m, Kr_2^m can be ignored, and we can set up the following rate equations:

$$\begin{aligned} d[N(He^+)]/dt &= \gamma' N^2(He^m) + \beta_2 N(He^m)n_e + \Gamma_3 N^2(He_2^m) \\ &+ \Gamma_4 N(He_2^m)N(He^m) - \alpha_1 N(He^+)n_e - \Gamma_2 N(He^+)N^2(He) \\ &- [D(He^+)/\Lambda^2 P]N(He^+) \end{aligned} \tag{2}$$

$$\begin{aligned} d[N(He_2^+)]/dt &= \gamma N^2(He^m) + \Gamma^2 N(He^+)N^2(He) + \beta_1 N(He_2^m)n_e \\ &- \alpha_2 N(Kr_2^+)n_e - [D(He_2^+)/\Lambda_2 P]N(He_2^+) \end{aligned} \tag{3}$$

$$\begin{aligned} d[N(Kr^+)]/dt &= \beta_3 N(Kr)n_e + \alpha_P N(Kr)N(He^m) - \alpha_R N(Kr^+)N(He^m) \\ &- \alpha_3 N(Kr^+)n_e - [D(Kr^+)/\Lambda^2 P]N(Kr^+) \end{aligned} \tag{4}$$

$$\begin{aligned} dn_e/dt &= [\beta_1 N(He^m) + \beta_2 N(He^m) + \beta_3 N(Kr)]\, n_e + \Gamma_1 N^2(He^m) \\ &+ \alpha_R N(Kr^+)N(He^m) - [\alpha_1 N(He^+) + \alpha_2 N(He_2^+) + \alpha_3 N(Kr^+)]n_e \end{aligned} \tag{5}$$

$$\begin{aligned} d[N(He^{m1})]/dt &= K_1[\alpha_1 N(He^+) + \beta_6 N(He) + \alpha_2 N(He_2^+)]n_e \\ &- [\beta_4 N(He^m) + \beta_2 N(He^m)]n_e - \alpha_P N(He^m)N(Kr) \\ &- \Gamma_1 N^2(He^m) - [D(He^m)/\Lambda^2 P]N(He^{m1}) \end{aligned} \tag{6}$$

$$\begin{aligned} d[N(He^{m2})]/dt &= k_2[\alpha_1 N(He^+) + \beta_6 N(He) + \alpha_2 N(He_2^+)]n_e \\ &+ \beta_7 N(He^{m1})n_e - [\beta_4 N(He^m) + \beta_2 N(He^m)]n_e - \alpha_R N(He^{m2})N(Kr^+) \\ &- \alpha_P N(He^m)N(Kr) - \Gamma_1 N^2(He^m) - [D(He^m)/\Lambda^2 P]N(He^{m2}) \end{aligned} \tag{7}$$

$$\begin{aligned} d[N(He_2^m)]/dt &= \alpha_2 N(He_2^+)n_e + \Gamma_5 N^2(He)N(He^m) - \beta_1 N(He_2^m)n_e \\ &- \Gamma_3 N^2(He_2^m) - \Gamma_4 N(He_2^m)N(He^m) \\ &- [D(He_2^m)/\Lambda^2 P]N(He_2^m) \end{aligned} \tag{8}$$

where n_e is concentration of electrons; $N(He^{m1})$ and $N(He^{m2})$ are the populations of metastable $He(2\,^1S_0)$ and $He(2\,^3S_1)$, respectively; K_1 and K_2 are the ratios of two branches of transitions from excited states of He atoms to the $He(2\,^1S_0)$ and $He(2\,^3S_1)$. The recombination rates were calculated, according to formulas (9), (10), and (11).

$$\alpha_1 = 5*10^{-10} + 6*10^{-10}\, n_e\, (T_e/293)^{-4} \quad (9)$$

$$\alpha_2 = 5*10^{-10} + 5*10^{-27}\, n_0\, (T_e/293)^{-1} + 4*10^{-20}\, n_e\, (T_e/293)^{-4} \quad (10)$$

$$\alpha_3 = \alpha_1 \quad (11)$$

where n_e is the concentration of electrons, n_0 is the neutral atoms, and T_e is the temperature of electrons. Because the particles' excited or deexcited rates by electron impact are dependent on the distribution of electronic energy and the cross section of various processes by electron impact, the rate should be calculated by formula (12)

$$\beta = (2m_e)^{1/2} \int \sigma(E) f(E) E^{1/2}\, dE \quad (12)$$

where $\sigma(E)$ is a cross section for producing a particular state of the atoms or molecules by electron impact. It is a function of electronic energy, and f(E) is the distribution function of electronic energy. So β is a changeable quantity dependent on f(E).

3. The Solution of This Group of Rate Equations

R. Deloche, M. D. Scheer, F. Dothan have provided the rate constants. The ratio of the two branches $K_1{:}K_2$ was taken from H. Ninomina; then the diffusion coefficient D was obtained from E. W. McDaniel and R. Doloche. The recombination rate, α_1 - α_3, at each time in the afterglow was calculated by formulas (9)-(11) with the numerical numbers of n_e, n_o, and T_e determined by experiment. The rates, β_1 - β_7, were calculated by formula (12) as well, where the quantity $\sigma(E)$ was obtained from M. Janossy. For example, the β_6 calculated at 5 μS and 15 μS after discharge is $2.08*10^{-12}$ cm^3 S^{-1} and $0.9*10^{-13}$ cm^3 S^{-1}, respectively. It can compare with the result that has been calculated by F. Dothen in pulsed normal glow discharge. Under the same temperature of electrons as the temperature of our experiment, their result is $1.05*10^{-12}$ cm^3 S^{-1} and $1.04*10^{-13}$ cm^3 S^{-1}, respectively. So surely the rate changes after the afterglow begins. There is a difference between HCD and normal glow discharge.

The initial population of various particles can be determined from experimental data where the concentration of ions is estimated according to the electrical-neutral ratios. The theoretical calculated result of n_e and $N(He^{m2})$ is shown in Figs. (1-a) and (1-b) by dotted line. The experimental decays are also shown in this figure, simultaneously. Both of them agree fairly well. So it is shown that the simplified dynamics model for afterglow of HCD is a faithful model.

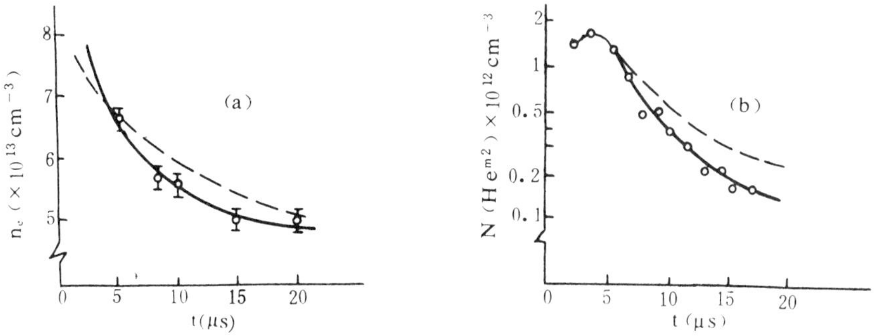

Figure 1: (a) The decay of concentration of electrons (real line) and theoretical calculated curve (dotted line); (b) The decay of population of $He(2\,^3S_1)$ (real line) and theoretical calculated curve (dotted line)

4. Experimental Investigation

We've determined the time-dependent quantity of n_e, T_e, and the time-dependent function of f(E) by the electrical probe method. Then we've also determined the time-dependent quantity of $N(He^{m1})$, $N(He^{m2})$ by the self-absorption method. The same discharge tube was used in both experiments. The hollow cathode was made of a metal cylinder, on which a slot 2mm wide was cut. The anode rod was made of the same metal and was parallel to the cathode. The width of the discharge pulse was 2 μS. In the electrical probe experiment, a probe was made of a ϕ0.4mm tungsten rod enveloped by glass and inserted into the cathode. In the self-absorption experiment, we put two small apertures in the optical path so that the side axis optical line was kept out of sight of the receiver. By means of this method, the signal-noise ratio was improved very much. Lastly, it should be pointed out that we used the discharge pulse as a signal trigger pulse in all of the experimental determinations so that there was the same time base in all signal determinations.

It is clear that the distribution of electronic energy in pulsed HCD afterglow was different from the pulsed normal glow discharge and that the relaxation process was similar to the damping oscillation. While time-dependent population of these particles was used for calculating the difference population of the upper and lower states of KrII 469.4nm, the optimum operating parameters can be chosen and the mechanism for laser oscillations can be explained. For various ratios of He-Kr partial pressures and various total pressures, we can calculate the $(N_0\text{-}N_1)$ in every situation. Fig. 2 shows one of the calculated results, where the operation parameters are P(He) = 12 Torr, P(Kr) = $4*10^{-12}$ Torr, and V_c = 2000V (V_c is voltage charged on reserve capacitor, 0.1 μf). We find at 5 μS of the afterglow that the $(N_u\text{-}N_1)$ reaches a maximum. This agrees with the laser forming time determined by Y. Zhang.

A sealed He-Kr laser tube has now been made by us. The 0.4 mw average output power was measured while the laser was operated at 1000 Hz repetition frequency.

REFERENCES

Deloche, R., et al., 1980 Int. Conf. on Gas Discharge and Their Appl. 6th, Edinberg, 1.2.b.Afterglow, p 82.
Dothan, F. and Kagan, Yu. M., 1981 J. Phys. D: Appl. Phys., 14, p 183.
Janossy, M. and Tuorinen, P., 1979 Acta Physica Acadamics Scie. Hungaricas, 46, p 167.
McDaniel, E. W., 1964 Collisional Phenomena in Ionized Gases (Wiley, New York), p 516.
Ninomina, H., 1981 J. Phys. D: Appl. Phys. 14, p 35.
Scheer, M. D., 1980 Int. Conf. on Gas 1.2.b.Afterglow, p 82.
Zhang, Y., Zhang, K., and Qju, J., 1987 Laser Tech. 11, p 16.

Paper presented at RIS 90, Varese, Italy, 16–21 September 1990

Highly sensitive elemental analysis by laser photoionization spectrometry in vacuum

G.I.Bekov, I.V.Kolpakov, V.N.Radaev, V.A.Veselov

Institute of Spectroscopy USSR Academy of Sciences
Troitsk, Moscow Region 142092, USSR

ABSTRACT: The processes of thermal atomization of element compounds in a vacuum are investigated with different types of atomizers for the purpose of developing a standardless analytical approaches. The conditions of near unity atomization efficiency are found. The laser photoionization spectrometer specially developed for wide use in analytical practice is described.

1. INTRODUCTION

One of the atom detection approaches effectively developed in recent years and most promising for analytical applications is laser analytical photoionization spectrometry in vacuum. The method is based on the thermal atomization of a substance in vacuum followed by resonance stepwise photoionization of element atoms through Rydberg or autoionizing states. This technique makes it possible to detect traces of any element (over 80% of periodic table) with a sensitivity being several orders higher than one actually attainable for most of the analytical methods used now. Besides its high selectivity and the outstanding sensitivity down to the level of single atoms in the laser beam, the LAPIS method is attractive due to the fact that the thermal atomization of samples in the vacuum substantially suppress the influence of the sample form and chemical content on the analytical signal of the element under study (so called matrix effects). The weak manifestation of matrix effects in this technique enables us to set the task of developing a standardless analytical approach based on LAPIS. Its realization will have a revolutionary influence on the introduction of this method in analytical practice. To develop such an approach we should carry out more detailed investigations of kinetic and chemical processes at the thermal atomization of a substance in a vacuum.

2. RESULTS AND DISCUSSIONS

The thermal atomization in a vacuum of Al, Ga and Rh compounds was investigated. The laser spectrometer used in these experiments was arranged by the standard scheme (Bekov and Letokhov 1986). The Al and Ga atoms were excited in two steps to the Rydberg state with $n^* = 17$ and then ionized by electric field pulse. Three step excitation scheme was used for Rh atoms.

The processes of thermal atomization of Al and Ga compounds in the buffer gas are substantially different in spite of belonging Al and Ga to the same group of the periodic table. For instance, the atomization of gallium compounds goes mainly through Ga evaporation from the metallic melt while in case of Al compounds the atomization greatly depends on the presence of oxigen. For the Al and Ga compounds efficiencies of thermal atomization were found to be 0.09 and 0.04 respectively with the use atomic absorption spectrometry (Sturgeon and Berman 1983). This considerable difference in the processes of Al and Ga compounds atomization stimulated our interest to these elements.

We used as the test samples aqueous solutions of various Al and Ga salts - $AlCl_3$, $Al_2(SO_4)_3$, $Al(NO_3)_3$ and $GaCl_3$, $Ga_2(SO_4)_3$, $Ga(NO_3)_3$ with the concentration of the elements equal to 1 mg/l. Solutions of 20 microlitres in volume were introduced with a micropipette on the crucible bottom and dried in the air at 90°C. Another type of samples was prepared through deposition of the element on tantalum foil strips in the vacuum. Deposition was performed in an additional vacuum chamber. The flux of the atomic beam was calibrated by means of depositing the element during 1 hour on a thin plate and its subsequent weighing. The deposition of the definite amount of the element on the foil strips was carried out during 5-10 s. The crucible with the air dried solution or with the piece of element deposited tantalum foil was placed in the atomizer and then, after the vacuum chamber was evacuated, it was heated in some steps to atomize the sample. The crucibles made of porous graphite, pyrolytic graphite and tantalum were used.

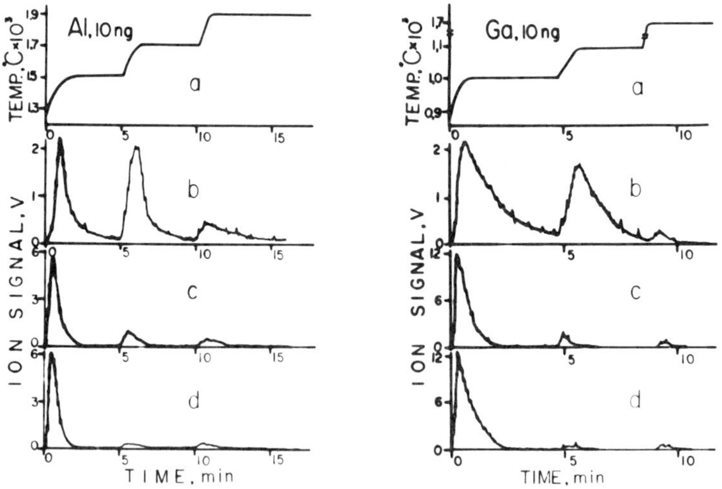

Fig.1. Aluminium and gallium photoion signals as functions of temperature and time: a - time dependance of crucible temperature; b - aqueous solutions of Al and Ga salts in porous crucible; c - aqueous solutions in pyrolytic crucible; d - samples of tantalum foil with deposited Al and Ga in pyrolytic crucible

The photoion signals produced by different samples of Al and Ga in crucibles made of different materials as a functions of temperature and time are presented in Fig.1. The total analytical signal is proportional to the area under the signal curve. The crucible temperature as a function of time is shown in Fig.1a. Figures 1b show signals produced by atomization of air dried solutions in porous graphite crucibles, Fig.1c - for pyrolytic graphite and tantalum crucibles, Fig.1d - atomization of metal samples deposited in vacuum on tantalum foil.

All samples contained the same amount (10 ng) of the element. The analytical signals appeared to be independent on the chemical form in which the element was introduced into the crucible. The signals from porous graphite crucibles (Fig.1b) were 1.5 to 2 times less than from pyrolytic crucibles (Fig.1c). The difference in the signal shape and atomization yield can be explained by penetration of the substance in pores and subsequent diffusion from there at higher temperatures. For pyrolytic graphite this effect is negligible and the signals of element compounds are equal to the signals of metal samples. This means that a near unity efficiency of thermal atomization of Al and Ga compounds is reached in the vacuum conditions. The similar results are obtained for rhodium - the element with different chemical properties. All this enables us to conclude that this phenomenon is universal for thermal atomization in vacuum. The situation changes in the presence of a complex matrix.

We compared the atomization yields of rhodium chloride at radically different conditions: without and with typical rock matrix, as shown in Fig.2,1a and 1b respectively. Near 3 mg of powdered rock with low Rh content was placed in the crucible over dry rhodium chloride residue (Fig.2,1b) and a slower heating procedure was used. The atomization yield appeared to be 1.5 to 3 times less than in the case of pure Rh chlorides (Fig.2,1a) and depended on the type of the rock. Two basic mechanisms can affect the analytical signal of the element under study. Firstly, this is the creation of element volatile compounds during matrix decomposition and vaporization of these compounds from the crucible at low temperatures without atomization. And, secondly, the molecular outflow of particles at a fast matrix vaporization can prevent from complete atomization of primary element compounds.

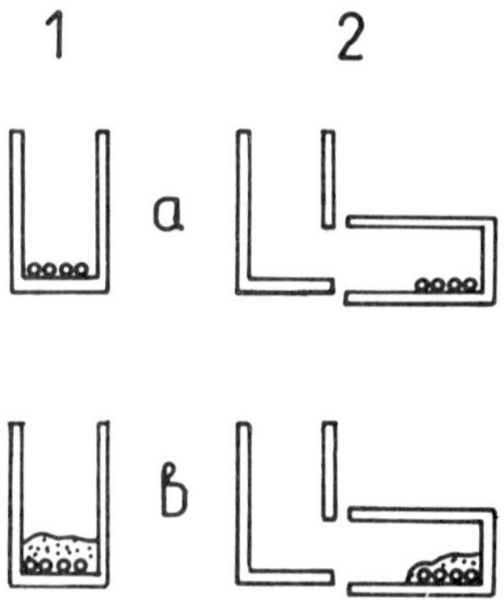

Fig.2. Different atomizers for thermal atomization in vacuum: 1 - standard type, 2 - atomizer with independent vaporization and atomization; a - dry residue of aqueous solution; b - dry residue with rock sample addition

We have developed a new type of atomizer where the vaporization and atomization functions are separated (Fig,2a,2b). The temperature of atomizing crucible was in the experiment near 2000°C, and evaporating crucible was heated in a stepped manner up to 2400°C to completely vaporize rock samples. The atomization yield for complex sample (Fig.2,2b) was equal to the yield for pure rhodium chlorides within 15% experimental accuracy. The results obtained show that this atomizer design allows to eliminate matrix effects that opens new possibilities for standardless direct analysis of a substance.

3. LASER ANALYTICAL SPECTROMETER

A great number of element content determinations has been done during last decade using resonance ionization spectroscopy technique (Bekov and Letokhov 1986, Thonnard et al 1989, Walker et al 1989, Rimke et al 1989). The laser spectrometers used in these experiments were actually laboratory setups which could be exploited only by their creators. However, for the successful introduction of RIS approach in analytical practice one needs the spectrometers with a high degree of automation and reliability much like the commercial conventional analytical equipment. We describe here the laser analytical photoionization spectrometer which satisfies in our opinion all these requirements and can be used in scientific and industrial analytical laboratories.

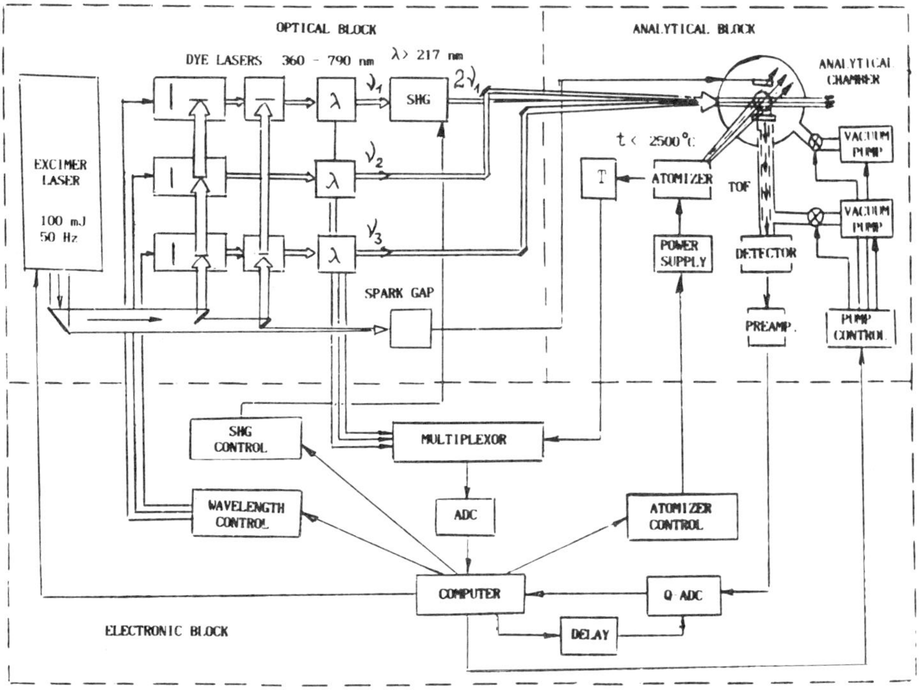

Fig.3. Schematic of automatized laser photoinization analytical spectrometer.

Laser analytical photoionization spectrometer LAPIS-1 consists of three basic parts (Fig.3): optical unit (lasers), analytical unit (a vacuum system and atomizers) and control unit (computer and electronics).

The excimer XeCl laser (100 mJ energy per pulse, 50 Hz pulse repetition rate, 15 ns pulse duration) is used as the pumping source for three dye lasers, which are made as connected oscillator-amplifier units. The generation region with available dyes covers the range 360-790 nm, with linewidth less than 30 pm, pulse energy about 3 mJ and pulse duration around 10 ns. The wavelength is tuned by turning the diffraction grating via the electro-dynamic drive with a position feedback. This drive makes it possible to tune the wavelength within the generation range of most dyes (25-35 nm) providing a precision no worse than 5 pm and a tune time from 0.04 to 0.7 s depending on the displacement value. For calibration all lasers are provided with calibration units consisting of a Fabry-Perot interferometer and an optogalvanic detector, a noble gas glow-discharge tube. Laser calibration consists in computer controlled wavelength scanning with simultaneous recording of the interferometer fringes and optogalvanic spectrum. Then, program identifies optogalvanic repers using tabulated noble-gas spectrum stored in the computer memory and calibrates the laser. Laser radiation frequency doubling is used to widen the wavelength range to UV region. The nonlinear crystal is turned by a computer controlled step motor through a precision sine mechanism. The program tunes the angle of the crystal for the maximum energy in the second harmonic. The total wavelength range is 217-790 nm, and this allows realization of excitation schemes for more than 80 elements.

Analytical unit consist of a vacuum chamber, time-of-flight mass filter and atomizers. The sample is placed in a graphite crucible gripped by massive water-cooled copper leads which form the housing of the atomizer. The main atomizer is installed on the lower flange of the vacuum chamber the lateral walls of which have quartz windows to input laser radiation, a flange to connect time-of-flight mass filter and a flange for high-vacuum pump. There are two metallic plates in the chamber which serve to set up an electric field pulse (5-10 kV/cm field strength, 25 ns pulse duration) ionizing the Rydberg atoms and pushing the ions into the time-of-flight mass filter. The field pulse is formed by a cable charge line with a spark gap triggered by excimer laser radiation. The chamber is evacuated to the vacuum 10^{-5}-10^{-7} Torr by a turbo-molecular pump. The analitycal chamber also has a reference atomizer similar to the main one. It is used for fine tuning of lasers to the transitions of an element, and for testing of the spectrometer. The crucible in the reference atomizer is filled with some amount of pure element the spectrometer is tuned to. This allows to form an atomic beam of element for all necessary testing and tuning. The main atomizer is designed so that the atomizer volume can be isolated from the volume of the chamber. It permits to unseal the atomizer on retention of a vacuum in the chamber which speeds up the change of samples.

Power supply for atomizers comprise a thyristor regulator and a step-down transformer. The power supply control unit stabilizes the power delivered to the atomizer. The power unit supplies a current up to 400 A, a power up to 4 kWt, the temperature of crucible coming to 2700°C. The power supply control provides a temperature reproducibility of no worse than 10% for different crucibles from one set and a temperature maintenance of better than 5%.

An electron multiplier with a 4 kV power supply is used as an ion detector. The signal from the detector enters the preamplifier and then charge-sensitive gated analog-to-digital converter. The system ensures detection of single ions.

The control unit of the spectrometer is based on the microcomputer and CAMAC equipment. The use of PDP-11 compatible computer has made it possible to develop a software that simplifies the work on the spectrometer. Almost all functions of the spectrometer, including the vacuum system control, are automatized. Automatic control over emergency situations increases considerably the reliability of the spectrometer.

The procedure of element content determination consist in the following. First, the crucible with the sample is placed in the atomizer isolated from the chamber volume. After being evacuated (it takes less then 1 min) the atomizer is connected with the chamber, and atomization program is activated. Atomization program consists of several stages with preset temperature and duration for each stage. The number of steps and their parameters are chosen so that firstly the high-volatile part of matrix is evaporated, then the element of interest is atomized and, finally, the most refractive residue is evaporated and the crucible is cleaned. The lasers and detection system are switched on when the atomization temperature of the element is reached. The photoion signal is stored in the computer memory and is used later for calculating the element content in the sample.

The spectrometer described is an instrument for the detection of ultrarare traces of most elements with a ppt-level sensitivity, and is characterized by a high productivity of measurements, a large linear range of calibration, a modern level of automation. All optomechanical and electronic units can be easily reproduced by up-to-date industry. All this allows setting the task of commercial production of the spectrometer.

REFERENCES

Bekov G I and Letokhov V S 1986 Laser Analytical Spectrochemistry ed V S Letokhov (Bristol: Adam Hilger) pp 98-151

Rimke H, Herrmann G, Mang M, Muhlec C, Riegel J, Sattelberger P, Trautmann N, Ames F, Becker A, Kluge H-J, Monz L, Otten E-W, Rehklau D, Ruster W 1989 Resonance Ionization Spectroscopy 1988 eds T B Lucatorto and J E Parks (Bristol: Inst. of Physics) pp 351-356

Sturgeon R E and Berman S S 1983 Anal. Chem. 55 190

Thonnard N, Parks J E, Willis R D, Moore L J and Arlinghaus H F 1989 Surf. Interf. Analysis 14 1

Walker R J, Fasset J D and Travis J C 1989 Resonance Ionization Spectroscopy 1988 eds T B Lucatorto and J E Parks (Bristol: Inst. of Physics) pp 337-342

Inst. Phys. Conf. Ser. No 114. Section 6
Paper presented at RIS 90, Varese, Italy, 16–21 September 1990

The status of real-world analysis problem-solving using RIS

N. Thonnard[1], H.F. Arlinghaus[1], R.D. Willis[1], E.H. Taylor[1], M.C. Wright[1], W.A. Davis[1], M.T. Spaar[1], and L.J. Moore[2]

[1]Atom Sciences, Inc., Oak Ridge, Tennessee, USA; [2]Eastern Analytical, Inc., College Park, Maryland, USA

Abstract: Resonance Ionization Spectroscopy (RIS), with its high element specificity and sensitivity, has been looked upon for more than a decade to providing dramatic advances in analysis capabilities over a wide range of applications. Even though conceptually very simple, the application of RIS to real-world analysis problems requires careful consideration of a number of parameters, not all RIS related. We discuss below practical consideration in implementation of RIS to direct analysis of solid surfaces, analysis of chemically processed samples, and analysis of noble gases.

1. INTRODUCTION

To be a practical, general-purpose analytical tool, instrumentation based on Resonance Ionization Spectroscopy (RIS) requires three basic components: (1) an atomization mechanism that can efficiently provide neutral atoms for analysis, (2) an RIS laser system allowing access to most of the elements, and (3) a detection system capable of separating RIS ions from other ions and providing isotopic information. Only by integrating and optimizing all three of the above items, is it possible to achieve the full analytical promise of RIS.

2. ATOMIZATION

Many approaches to atomization have been used in implementing RIS, including thermal vaporization of samples deposited on a filament, ion sputtering, and evaporation out of a graphite tube. When using pulsed lasers, to achieve efficient sample utilization, the atomization process must also be pulsed. For direct analysis of solids, both Sputter Initiated RIS (SIRIS) and Laser Atomization RIS (LARIS), Figure 1a, are becoming very versatile techniques. As ion beams can be easily focused and rastered across a surface, SIRIS enables spatially resolved trace analyses into the sub-micrometer regime laterally, and to a few nanometers in depth. Using short sputtering pulses, close proximity of the RIS laser beams to the sample, and the appropriate delay between sputtering and RIS, we have achieved useful yields (ions detected/atoms sputtered) as large as 0.3, and detection limits in the low parts per trillion range (Arlinghaus et al., 1990). Substituting a laser beam for the sputtering beam, much higher atomization rates are possible, yielding higher sensitivity, at the expense of less control in the atomization process.

For noble gases, the same principle applies (Figure 1b). By freezing the sample onto a small cold finger, and releasing it by heating the cold spot with a pulsed laser just before firing the

RIS lasers, it is possible to concentrate the noble gas atoms in the RIS laser beams and achieve good throughput and efficiency. Willis (these proceedings) has shown a detection limit of 350 ^{81}Kr atoms in a 6 liter static mass spectrometer volume - a concentration of 0.05 atoms cm^{-3}.

In both examples we indicate a time-of-flight (TOF) detector. With pulsed RIS lasers, this is natural detector choice, but other mass spectrometric methods are also used, with their attendant advantages and disadvantages. Suppression of secondary ions is especially important in SIRIS analyses; this can be achieved by appropriate pulsing of the extraction electrode (V_2 in Figure 1a) or by adding an electrostatic analyzer between V_2 and V_3, or both. Due to the high element selectivity of the RIS process, the mass spectrometer performance can be significantly relaxed, and need only have sufficient resolution to yield good isotope ratio measurements.

In the SIRIS mode, quantitation is much better than with secondary ion mass spectrometry (Parks et al., 1988; Downey, these proceedings) due to the decoupling of atomization from ionization. The practical sensitivity that can be achieved is limited by the maximum ion current that can be focused into a given area. With LARIS, the sensitivity can be much higher, but there are some concerns on how representative of the substrate the atomized plume evolved is. At low power densities, atomization may be related to the vapor pressure of the analyte, and large biases could be introduced. At high power densities, ablation takes place, yielding plasmas and droplets, neither of which are useful for RIS. Beekman and Thonnard (1989) reported mixed quantitative results in LARIS measurements of Ga in Si (OK) and Al in steel (poor). Using 266 nm radiation for atomization in biological substrates, we are now able to achieve identical quantitative results with either LARIS or SIRIS, but with a factor of 100 more signal in the LARIS mode, Figure 2.

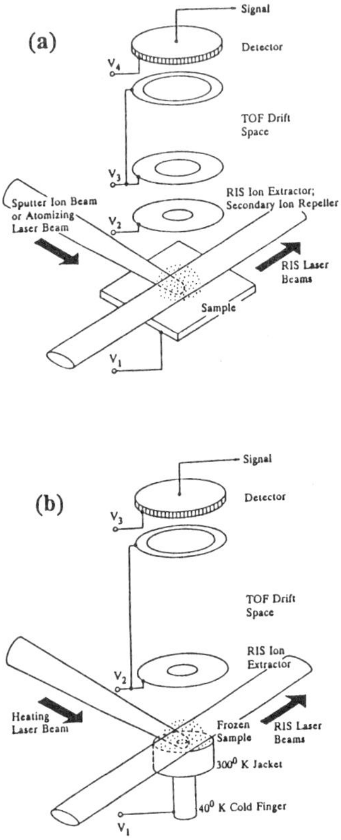

Figure 1.

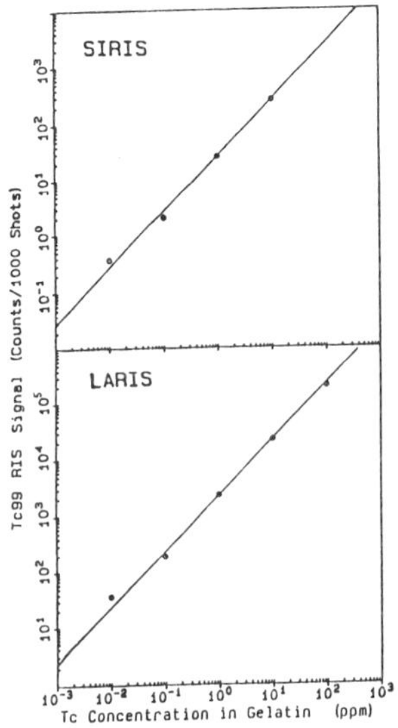

Figure 2.

3. RIS PERFORMANCE CRITERIA

In Figure 3 we illustrate a number of possible RIS schemes. Scheme (a) is very popular due to its simplicity. By doubling the output of a dye laser, more than 40 elements can be reached with a single dye. Unfortunately both sensitivity and selectivity suffer in this scheme, as the same photon is used both for the bound-bound transition and the ionization step, leading to severe over-saturation of the first step while still being unsaturated in the ionization step. With an excimer pumped dye laser, it is possible to use the excimer laser for the second step, thereby reducing the laser intensity in the first step, but frequently the increased sensitivity is marred

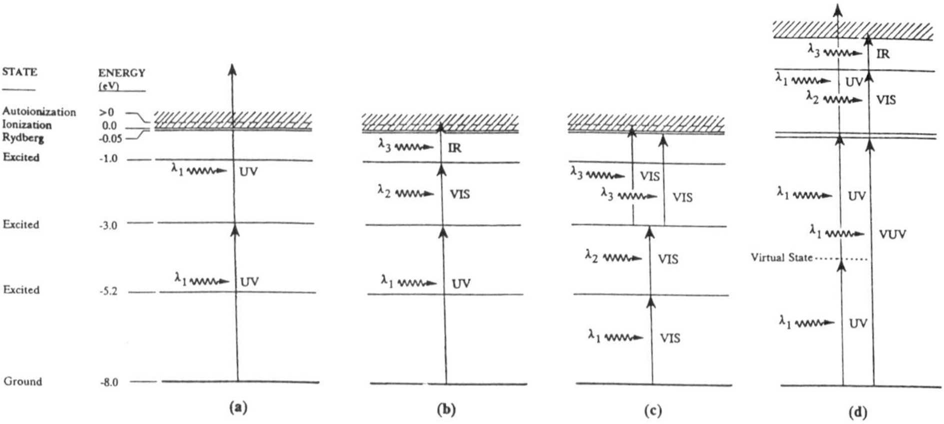

Figure 3. RIS Schemes

by even more non-resonant interferences. Using two dye lasers pumped by a Nd:YAG laser (Figure 4), except for He and Ne, it is possible to achieve RIS for the entire periodic table (Thonnard, et al., 1989). Most elements can be ionized with a scheme similar to what is shown in Figure 3b, in which the intensity of each laser beam can be optimized for each transition. Saturation curves for RIS of vanadium are shown in Figure 5, illustrating that a factor of more than 100 in laser energy is required to saturate the unbound ionizing transition compared to the bound-bound transition. By utilizing the 1.06 μm IR from the Nd:YAG, non-resonant interference from the ionization step is rare. An optimum operating point is at 80% of saturation, thereby maintaining almost full signal at moderate laser intensity.

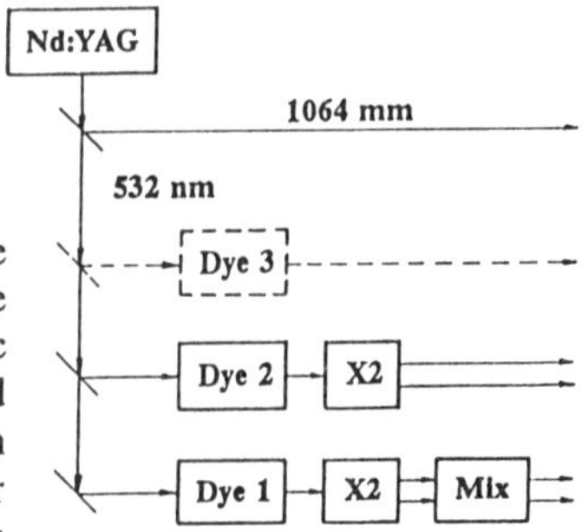

Figure 4.

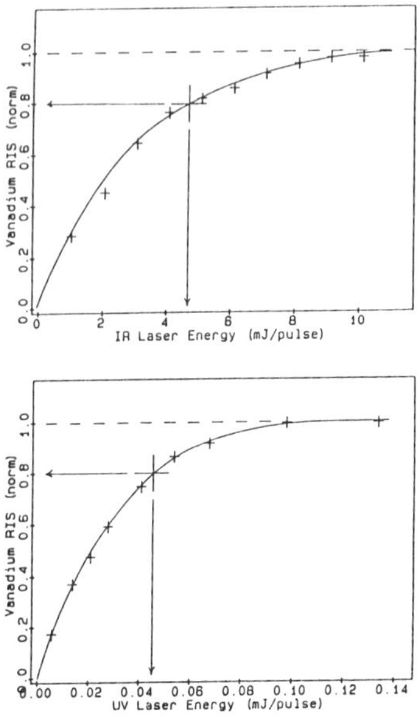

Figure 5.

For 16 elements, the first excited state can be reached with a two-photon transition, Figure 3d, left, at some loss in sensitivity and selectivity. Better still is to generate vacuum ultraviolet radiation (Kramer, 1984), thereby utilizing the RIS scheme of Figure 3d, right, which to date has been implemented for Kr and Xe. By adding a third dye laser, RIS of many elements is possible with no UV photons, and by terminating in either autoionization or Rydberg states followed by field ionization, Figure 3c, unsurpassed selectivity is achieved.

4. DISCUSSION

Even after picking the very best RIS scheme, and optimizing the ionization process, analysis of real samples will still not reach the expected detection limit. At times, this limit is due to effects unrelated to RIS.

In the direct analysis of solid samples, the quality of the sputtering ion beam is of utmost importance. An obvious difficulty is if the ion beam has impurity elements that overlap the analyte. More subtle though, is if the beam profile has low intensity "wings" at a large distance from the main peak. Even if 99% of the current in a sputtering ion beam falls within a 100 μm diameter circle, but only 0.1% falls outside of a 1.0 mm circle, then in a depth profile of a semiconductor in which a 1.0 mm diameter crater is rastered the signal will not decrease below 10^{-3} of the surface concentration, as the stray parts of the beam will contribute sputtered atoms from the surface to the analysis volume. In SIMS, this problem is less severe, as the ions come directly from the surface at a fixed potential and position, thereby enabling rejection of ions that originated from a different position.

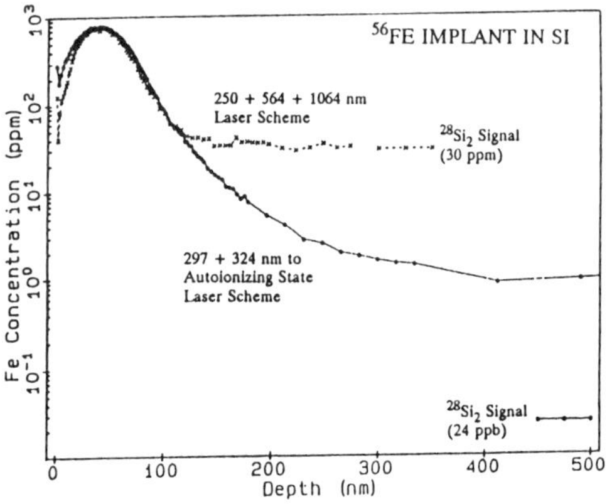

Figure 6.

In Figure 6, we see in a depth profile of Fe in Si, some of the difficulties and solutions that can be applied. Using our "preferred" laser scheme, Figure 2b, the depth profile bottoms out after only 1 1/2 orders of magnitude. By detuning the RIS laser, we realized that the mass 56 signal of 30 ppm is due to non-resonant ionization of $^{28}Si_2$ with the 250 nm photon used in the first excitation step. By using two longer wavelength photons going to an autoionizing level, the mass 56 signal drops to 1 ppm. Detuning the laser gives a measurable signal at 24 ppb, indicating that the limit is now due to the ion beam. Using a three color scheme with still longer wavelength photons (Figure 2c) reduces the non-resonant signal to below 1 ppb. Therefore, by choosing the right laser scheme, facilitated significantly by the compilations of Saloman (1990), it is possible to eliminate interferences from the ionization process. We note that Figure 6 was chosen to illustrate a point; proper adjustment of the ion beam gives us almost 5 orders of magnitude dynamic range in depth profiles.

The extreme sensitivity made possible with RIS makes contamination of the sample a major limitation. In biomedical analyses (Quinn, et al., these proceedings) in which the small-sample capabilities of SIRIS are exploited, exposing a sample during processing to ordinary room air for only one minute will accumulate many ppb's of Al, Si, Fe, etc. from the dust. In noble gas analyses, where we have shown sensitivity to few hundred atoms, the outgassing from the walls of the mass spectrometer will overwhelm the sample signal in a few minutes. Therefore, to fully utilize the capabilities of RIS requires extreme care in sample handling and instrumental background reduction.

5. REFERENCES

Arlinghaus HF, Spaar MT and Thonnard N 1990 JVST A 8 3

Parks JE, Spaar MT, and Cressman PJ 1989 J. Crystal Growth 89 4

Beekman DW and Thonnard N 1989 in Resonance Ionization Spectroscopy 1988 T B Lucatorto and J E Parks ed Inst Phys Conf Ser 94 163

Thonnard N, Parks JE, Willis RD, Moore LJ, and Arlinghaus HF 1989 Surf Interface Anal 14, 751.

Kramer S 1984 in Resonance Ionization Spectroscopy 1984 G S Hurst and M G Payne ed Inst Phys Conf Ser No 71 205

Saloman EB 1990 Spectrochimica Acta 45B Nos 1/2 37

The cosmic ray exposure age of the millbillillie meteorite determined by conventional mass spectrometry and by RIS-TOF

R.D. Willis[1], N. Thonnard[1], O. Eugster[2], Th. Michel[2], and B.E. Lehmann[2]

[1]Atom Sciences, Inc., Oak Ridge, Tennessee, USA; [2]Physics Institute, University of Bern, Switzerland

Abstract: Resonance Ionization Time-of-Flight Spectrometry (RIS-TOF) was used to determine the ^{81}Kr cosmic-ray exposure age of the Millbillillie meteorite. The RIS-TOF exposure age is in excellent agreement with the age determined by conventional mass spectrometry and required a factor of 50 less sample. The enhanced sensitivity of RIS-TOF extends the power of the ^{81}Kr dating technique especially to smaller samples and to material with low ^{81}Kr content.

I. INTRODUCTION

Cosmic ray interactions with extraterrestrial materials produce both radioactive and stable nuclides by spallation processes. In meteorites and lunar rocks typically containing ppm concentrations of Rb, Sr, Y, and Zr, radioactive ^{81}Kr (half-life 213,000 years) and stable Kr isotopes are generated. These cosmogenic Kr isotopes provide the basis for a powerful method for determining cosmic-ray exposure ages of solar system material (Marti 1967, Eugster et al. 1967) and terrestrial ages of meteorites (Eugster et al. 1986). The method consists in the simultaneous measurement of ^{81}Kr and one or more stable Kr isotopes. A detailed discussion of the ^{81}Kr dating technique and a review of results obtained with the method were presented by Eugster (1989) at the previous RIS symposium.

Cosmogenic ^{81}Kr is generated in solar system material at low concentrations, typically about 3×10^5 atoms/g of material. This number of atoms approximates the Kr detection limit for static noble gas mass spectrometers. In instances where the sample quantity is extremely limited, or where the ^{81}Kr concentration is unusually low, e.g., meteorites containing very low levels of Rb, Sr, Y or Zr, or meteorites in which the initial ^{81}Kr levels have decayed during their terrestrial age, the sensitivity of conventional mass spectrometers may be inadequate. Also, as one approaches the detection limit of these instruments, isobaric interferences in the Kr mass region become increasingly troublesome, requiring careful background corrections (Eugster, 1989). The objective of the present study was to demonstrate that the RIS-TOF technique, with its enhanced sensitivity and selectivity, can extend the ^{81}Kr dating method to samples for which conventional sensitivity may be inadequate.

2. EXPERIMENTAL

The noble gases in the Millbillillie meteorite (an Eucrite which fell in Western Australia in 1960) were extracted in an ultrahigh vacuum furnace at the University of Bern. The Kr fraction was separated cryogenically on charcoal. Splits of this gas were analyzed at Atom Sciences using

RIS-TOF and at Bern using a 60-degree sector mass spectrometer with electron bombardment ion source. The Bern group also independently determined the exposure age of Millbillillie by measuring cosmogenic ^{21}Ne and ^{38}Ar.

The RIS-TOF system at Atom Sciences was developed as part of a dedicated facility for the direct-counting of ^{81}Kr and ^{85}Kr (Willis et al. 1989, Thonnard et al. 1987). The RIS-TOF counting system features a pulsed cold finger (Hurst et al. 1984) to concentrate the Kr atoms in front of the RIS laser beams and a second-order time-focusing TOF spectrometer (Thonnard et al. 1984). Data are processed with a 10 ns resolution transient digitizer allowing simultaneous collection of all Kr isotopes. The RIS scheme for Kr (Kramer et al. 1983) utilizes four-wave mixing in an Ar-Xe gas mixture to generate 116.5 nm photons required for the first excitation step.

The sensitivity of the RIS-TOF system is illustrated in Figure 1 showing the RIS-TOF spectrum of Kr which had been extracted from natural gas and processed through two steps of isotope enrichment. In this six-minute integration, the 2σ detection limit in the ^{81}Kr mass region was approximately 350 atoms. These were immersed in a background of about 5x10^{11} atoms illustrating the tremendous selectivity of the RIS-TOF technique.

Three separate splits of the meteorite gas sample were analyzed by RIS-TOF. Each split was about 1.7x10^{-12} cm^3 STP (4.6x10^7 Kr atoms) and was typically analyzed for 20 minutes in successive 80 s (800 shot) runs. Analyses of standard bottled Kr bracketed the data for each split in order to determine correction factors for instrumental mass discrimination.

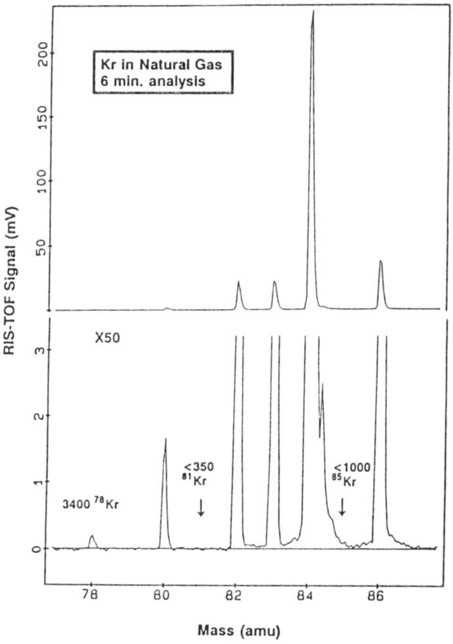

Figure 1

3. RESULTS AND DISCUSSION

Figure 2 shows a representative RIS-TOF spectrum of meteorite gas, integrated for 800 laser shots. Hand-drawn baselines were fitted to each ^{81}Kr peak in order to correct the computed peak area for tails from the neighboring ^{80}Kr and ^{82}Kr peaks.

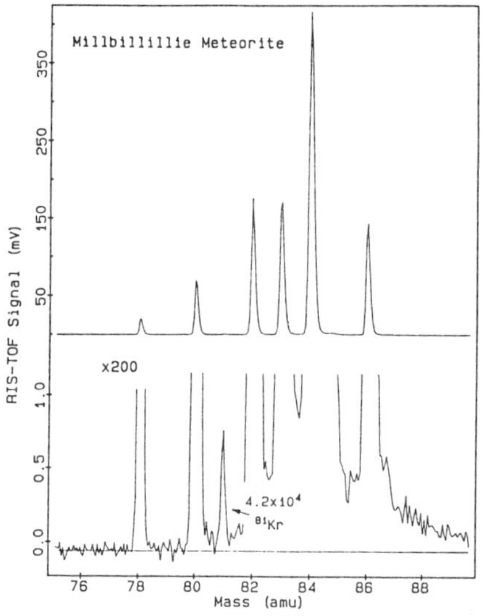

Figure 2

For exposure ages much greater than the half-life of ^{81}Kr, the exposure age is given by:

^{81}Kr Age (years) = 1.4615×10^5 (^{80}Kr/^{83}Kr + ^{82}Kr/^{83}Kr)/(^{81}Kr/^{83}Kr) (Eugster, 1989), where the cosmogenic Kr values are used. Non-cosmogenic Kr was first stripped from the raw meteorite data using the measured ^{86}Kr which has a negligible spallation yield and is presumed here to be purely non-cosmogenic. Mass discrimination correction factors are used to properly subtract non-cosmogenic Kr and to correct the resulting cosmogenic Kr ratios.

Results of the RIS-TOF analyses for the three meteorite splits are presented below.

Split	No. of ^{81}Kr Analyzed	% Uncertainties: Isotope Ratios	Mass Disc. Correction	Est. Total	Exposure Age (Ma)
A	27,000	7.0	3.5	7.8	26.2 ± 2.0
B	42,000	3.9	1.2	4.1	26.6 ± 1.1
C	35,000	2.8	3.4	4.4	23.6 ± 1.0

Comparison of Exposure Ages:

Average RIS-TOF Age =	25.4 ± 1.4 Ma	
Conventional MS ^{81}Kr Age =	22.5 ± 3 Ma	[Univ. of Bern, 1.7×10^6 ^{81}Kr]
^{21}Ne Age =	23 ± 4 Ma	
^{38}Ar Age =	28 ± 4 Ma	

The average exposure age determined by RIS-TOF is in excellent agreement with the ^{81}Kr age determined conventionally and with ages based on cosmogenic ^{21}Ne and ^{38}Ar. Approximately 35,000 ^{81}Kr atoms were analyzed per split using RIS-TOF, a factor of 50 less than the measurement at Bern.

The middle columns in the table above summarize for each split the uncertainties associated with: (1) the measured ^{81}Kr/^{83}Kr ratio (due to the relatively large scatter in the small ^{81}Kr signal, this uncertainty dominates over uncertainties in the ^{80}Kr/^{83}Kr and ^{82}Kr/^{83}Kr ratios), and (2) the measured mass discrimination correction factors for ^{80}Kr/^{83}Kr and ^{82}Kr/^{83}Kr. These corrections ranged from 0.88 to 0.95 and pointed to a slightly lower ionization efficiency for ^{83}Kr relative to ^{80}Kr and ^{82}Kr. We attribute this discrimination to polarization effects in the ionization process (Fairbank et al. 1989) since we have often observed significant changes in the RIS-TOF mass discrimination after retuning the laser. Because of the very low natural abundance of ^{81}Kr (5×10^{-13}), it was not possible to measure a mass discrimination correction factor for ^{81}Kr/^{83}Kr. In future work we will use an isotopic standard for Kr which includes a measurable concentration of ^{81}Kr. For the present results we have simply assumed a correction factor of 1, i.e., ^{81}Kr and ^{83}Kr are ionized and detected with equal efficiency. This assumption may be in error by as much as 10% which would alter the calculated exposure ages accordingly. The error bars assigned to the exposure ages do not include this potential undetermined error.

4. SUMMARY AND FUTURE PLANS

RIS-TOF analyses of the Millbillillie meteorite demonstrate that the technique is capable of quantitative age determinations on samples with very low ^{81}Kr content. Improvements of a factor of 50 or more in sensitivity should be possible while isobaric interferences are eliminated.

The enhanced sensitivity provided by RIS-TOF should extend the ^{81}Kr-Kr dating method to samples of very limited quantity and/or samples containing low ^{81}Kr content. The latter includes classes of meteorites which contain very low amounts of the target elements which generate cosmogenic Kr, meteorites which have lost ^{81}Kr by decay during their terrestrial ages, and meteorites or lunar rocks with non-equilibrium ^{81}Kr concentrations due to brief exposure ages.

Finally, the RIS-TOF data demonstrates the importance of frequent analyses of isotopic standards in order to correct for mass discrimination arising from the laser ionization process.

Future plans include the construction of a new TOF system at Atom Sciences which will provide greater sensitivity and reduced Kr outgassing. The abundance sensitivity of the TOF ($\sim$1000 in the Kr mass range) presently limits the technique to meteorites having ^{82}Kr/^{81}Kr ratios < 10^3. A number of meteorites of interest, however, have ratios exceeding 10^4. Possible ways of improving the abundance sensitivity such as a reflectron design will be investigated.

5. ACKNOWLEDGEMENTS

This work was supported in part by the Swiss National Science Foundation. Partial support for the RIS-TOF analyses was provided by the University of Bern.

6. REFERENCES

Eugster O, Eberhardt P and Geiss J 1967 Earth Planet. Sci. Lett. 2 77
Eugster O, Geiss J, Krahenbuhl U and Niedermann S 1986 Earth Planet. Sci. Lett. 78 139
Eugster O 1989 Resonance Ionization Spectroscopy 1988 ed T B Lucartorto and J E Parks (Bristol: Institute of Physics) pp 343-350
Fairbank W M Jr, Spaar M T, Parks J E and Hutchinson J M R 1989 Phys. Rev. A 40 2195
Hurst G S, Payne M G, Phillips R C, Dabbs J W T and Lehmann B E 1984 J. Appl. Phys. 55 1278
Kramer S D, Chen C H, Payne M G, Hurst G S and Lehmann B E 1983 Appl. Opt. 22 3271
Marti K 1967 Phys. Rev. Lett. 18 264
Thonnard N, Payne M G, Wright M C and Schmitt H W 1984 Resonance Ionization Spectroscopy 1984 ed G S Hurst and M G Payne (Boston: Institute of Physics) pp 213-218
Thonnard N, Willis R D, Wright M C, Davis W A and Lehmann B E 1987 Nucl. Instr. Meth. B29 398
Willis R D, Thonnard N, Wright M C, Lehmann B E and Rauber D F 1989 Resonance Ionization Spectroscopy 1988 ed T B Lucatorto and J E Parks (Bristol: Institute of Physics) pp 213-216

Precision and accuracy of RIS-analysis

R Zilliacus, J Likonen and I Auterinen

Technical Research Centre of Finland, Reactor Laboratory
Otakaari 3A, SF-02150 Espoo, Finland

ABSTRACT: Resonance ionisation methods for element analysis of group IIIa of the periodic table were developed based on the laser excitation in two steps to Rydberg states followed by field ionisation. The precision and accuracy of the results were evaluated by comparison with those made by other analytical techniques and by analysing standard reference samples.

1. INTRODUCTION

Evaluation of the reliability of the results is of uppermost importance when developing new analytical methods. The parameters involved are the precision and accuracy of the results. Good precision is an essential requirement for a method used for analytical work. In trace element analysis, however, dispersion of the results can also be caused by unhomogeneous distribution of the element analyzed in the matrix.

The accuracy of the method is totally dependent on whether the standards are reliable and whether the element in question is behaving similarly in both the samples and the standards. One way of checking this is to analyze standard reference materials of known element concentrations. Such materials, however, are not always available. Another alternative is to compare the results with other independent methods.

In the present work, the analytical results recieved by an RIS analysis of metals in group IIIa of the periodic table, aluminium, gallium, indium and thallium, were studied. Following the atomisation of the samples in electrically heated graphite, the analyses are made using two-step excitation and a field ionization (Bekov et al. 1986). This atomisation method makes it possible to analyse either liquid or solid samples, assuming that the solid sample is e.g. a metal or pure crystalline material. Regarding more complex matrices, such as geological or tissue samples, dissolution of the samples becomes obligatory.

When analysing samples using graphite furnace atomisation, it is essential to be aware of the influence of the chemical environment on the behaviour of the element in the graphite furnace. Solvents and the main components of the sample readily change the atomisation as compared with that of the standards. In graphite furnace atomic absoption spectrometry (GFAAS) a matrix modifier is often used to reduce the volatility of the analyte as compounds at low temperatures. This might be useful in this case too, because the elements analysed all form volatile halides. We looked at the competence of $Mg(NO_3)_2$, a commonly used matrix modifier in GFAAS, in stabilising the atomisation of the elements under study.

Testing of the reliability of the results of the RIS-analysis with other analytical techniques is restricted by the lack of sufficiently sensitive methods. As for indium, neutron activation analysis (NAA) may be used down to a 0.1ng/g level and, also, as a reference method for indium analysis. For other elements GFAAS was only used at the higher concentration level.

2. EXPERIMENTAL

Elements Al, Ga, In and Tl were analysed using two-step excitation to Rydberg states and field ionisation of the atoms (Table 1). The preparation of the samples was performed in a dust poor room. All the samples were dissolved and the sample amount used was 10μl. The acids and water used were distilled in a sub-boiling still of quartz. The samples were dried with an IR lamp. The experiments using $Mg(NO_3)_2$ as a matrix modifier were made by adding 20 μl of a 2% solution of $Mg(NO_3)_2$ on the dried samples and repeating the drying treatment. In the thallium analysis sulfuric acid was added to the sample solutions.

The sample solutions of the reference materials were prepared by dissolving 200mg of material in 2ml of nitric acid.

Table 1. Excitation data

	ground state	λ1 nm	excited state	λ2 nm	excited state
Al	3p	396.15	4s	442.7	19p
Ga	4p	403.3	5s	430.8	19p
In	5p	410.2	6s	456.4	20p
Tl	6p	377.6	7s	446.4	21p

3. RESULTS AND CONCLUSIONS

The precision of the results in water and dilute acids was about 10 % for aluminium, gallium and thallium (Figure 1). By adding $Mg(NO_3)_2$ to the samples we got an increased signal and the precision was somewhat better. The signal was increased and the atomisation temperature was higher, too (Figure 2).The indium results showed differences between the parallel samples in water solutions of more than 50% in some cases and the precision of the results was 10% when matrix modifier was used. The influence of hydrochloric acid could also be compensated for by using the matrix modifier. Sulfuric acid was noticed to stabilize and enhance the thallium signal

Aluminium was analysed parallel by RIS and GFAAS in ground water samples. The results were in good agreement in the case of concentrations over 10 ng/g. The individual results differ less than 15% from each other. In Al analysis at lower concentrations the external contamination is a restricting factor. Although sample preparation was made in a dust poor room the blank contained 1-10pg aluminium. This corresponds to 0.1-1 ng/g when the sample size is 10 ul.

Indium was analysed with NAA and RIS of CdS thin film samples (Samples provided by Microchemistry Ltd). The films were etched from the glass substrate with nitric acid. Indium results of these solutions show good agreement considering the low concentration level (Table 2).

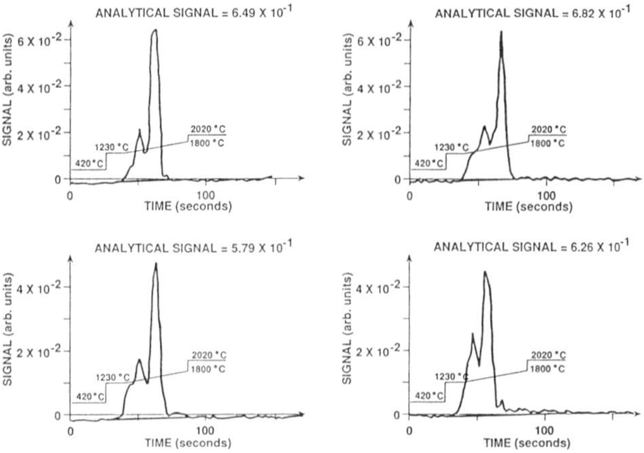

Fig.1. Ga signals of four parallel samples

Table 2. Indium analysis of dissolved CdS thin film samples

Sample	RIS In ng/g	NAA In ng/g
B230	3.47	3.13
B231	1.17	0.95
B232	2.44	1.88
B235	1.18	0.90
B238	0.17	0.28
B241	1.90	1.79
B242	1.44	1.16
B243	5.13	6.99
B247	0.68	1.02
B248	0.88	0.82
B249	0.99	0.64
B252	3.00	3.39

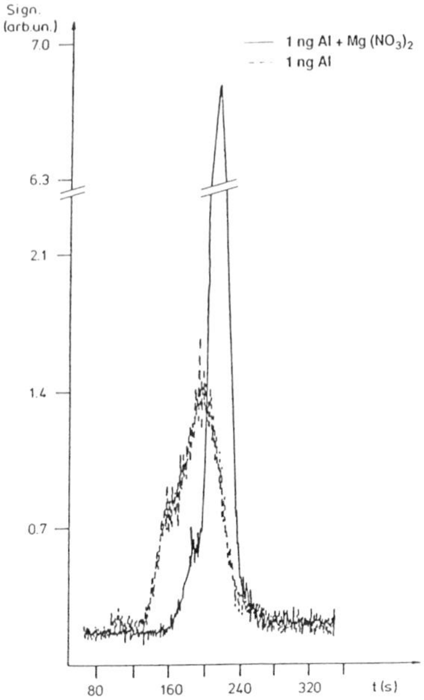

Fig. 2. The effect of $Mg(NO_3)_2$ on the Al signal

Aluminium concentrations of two reference materials were analysed. Kodak s TEG-50-A contains 53 μg/g aluminium while our result showed 50+/-5μg/g. Our result 55+/-10 μg/g Al in NIST's bovine liver was higher than the values previously presented by other analysts, 30+/-20 μg/g (Gladney 1980), 16+/-4μ/g (Gladney et al.

1987), 35.5 μg/g (Nadkarni and Morrison 1978) and 37+/-6 μg/g (Zikovsky and Schwiekert 1977). This was probably caused by contamination of the sample from air dust. The indium result of the Kodak sample was 51 μg/g, the reference value being 53 μg/g. Thallium was analysed from three reference samples (Table 3). The agreement for these results is good as compared to reference values.

Table 3. Thallium analysis by RIS for reference materials

Sample	RIS value μg/g	NIST inf.value μg/g	Kodak ref.value μg/g
Kodak TEG-50A	51+/-9		57
NIST 1570 Spinach	0.037+/-0.006	0.03	
NIST 1577 bovine liver	0.06+/-0.01	0.05	

The detection limit of the aluminium analysis was as high as 10pg because of the high blank. This was mainly caused by the aluminium content of the graphite. Long-term heating of the crucible released a continuous aluminium signal. There was no noticeable contamination in the analyses of In, Ga and Tl. Those analyses had an absolute detection limit below 10^{-15}g.

REFERENCES

Bekov G I,Kudryavtsev Y A, Auterinen I and Likonen J 1986 Resonance Ionization Spectroscopy 1986 ed G S Hurst and C G Morgan (Bristol: The Institute of Physics) pp 97-102

Gladney E S 1980 Anal. Chim. Acta 118

Gladney E S, O Mailey B T, Roelants and Gills T E NBS Special Publication 206- 111

Zikovsky L and Schwiekert E 1977 J. Radioanal. Chem. 37

Inst. Phys. Conf. Ser. No 114: Section 6
Paper presented at RIS 90, Varese, Italy, 16–21 September 1990

A 'rod flame' system in direct atomic-ionization analysis of high-purity substances

N.V. CHEKALIN, V.I. PAVLUTSKAYA and I.I. VLASOV

V.I.Vernadsky Institute of Geochemistry and Analytical Chemistry, USSR Academy of Sciences, 19 Kosygin Str, 117975 Moscow USSR

Abstract Laser enhanced ionization in conjunction with microsample evaporation from electrically heated graphite rod into the flame has been used for the direct determination of copper in Ge and In in high-purity CdHgTe alloys. Samples in the form of both solutions and powders are analysed by the standard addition and the calibration curve methods. A good correlation of the results obtained for the solutions and powders through calibration of the aqueous standard solutions indicates a high accuracy of this approach.The detection limit for In in CdHgTe solutions is 10^{-4} ppm and with solid samples it can be reduced to 10^{-6} ppm. For copper in Ge the DL is higher since it is limited by the rod material purity.

1. INTRODUCTION

At present high-purity materials are finding an ever growing use in science and technology with the requirements for the purity of the materials becoming more and more stringent. In particular, for semiconductor materials and reagents used in microelectronics the admissible level of imrpurities is 10^{-3} - 10^{-7} ppm.A direct checking for the purity of such materials using standard spectral methods such as atomic absorption spectrometry, inductively coupled plasma and so on is impracticable because of an insufficiently high sensitivity and selectivity of these methods. The application of sophisticated procedures relying on analyte preconcentration, matrix modification or removal is time consuming, expensive and also imposes stringent requirements on the purity of the reagents and the equipment used. Moreover, the probable sample contamination or loss of the element under study deteriorate analytical precision and accuracy.

This situation generates a need for direct, high sensitive, indifferent to matrix influences and rapid methods for analysis of high purity materials with a minimum sample preparation. In this context favourable potentials are offered by laser ionization methods, particularly laser enhanced ionization (LEI) in flames. It is applied to high-sensitivity element determinations in liquids and solutions to yield detection limits as low as 10^{-6}ppm [1-5]. At same time, the flame atomizer features some serious drawbacks that limit its application to the analysis of high-purity materials. They are: (1) a small (0.1-0.15) sample utilization factor; (2) impracticability of separating the processes of sample evaporation and atomization; (3) problems associated with handling micro-samples and solid samples; (4) the detection limits when using solutions are no better than 10^{-4} ppm.

This work proposes a hybrid "ROD-FLAME" arrangement featuring the major advantages of the flame atomizer and devoid of the above drawbacks. In this system the microsample is pulse-evaporated from an electrically heated surface of a graphite rod into the flame where the analyte is atomized, laser-excited ans ionized. The temptrature of sample evaporation is normally lower than that of atomization therefore the rod operates in "mild" conditions and with practically no ionization background from the rod-generated thermoemission. Studies of matrix interferences when dealing with solid and liquid samples have made it possible to establish the conditions in which the matrix interferences are negligible and to obtain very low absolute and relative detection limits (DL) in both aqueous solutions and real samples (liquid and solid). The "ROD-FLAME" system is exploited for the direct determination of copper in Ge and indium in CdHgTe alloys with DL of 10^{-3} ppm and 10^{-6} ppm respectively.

2. EXPERIMENT

The block diagram of the LEI spectrometer is shown in Fig. 1. It is composed of two tunable dye lasers pumped by an excimer laser, an atomizer-ionizer and a recording system. The atomizer-ionizer consists of a 20 mm slot burner, a 4 mm dia pyrographite rod inserted into the flame at a height of 10 mm above the burner and heated on the programme selected, and a charged particle collector. The collector comprizes two parallel water-cooled titanium plates which are placed in the flame at a 10 mm spacing above the graphite rod. One of the plates is grounded and the other is at a negative voltage of 1.2 kV.

The spectrometer operates as follows. The analyte in the form of solution, powder or solid 1-20 mg in weight is placed into a groove on the rod surface. Measurements are started with heating the rod in a stepwise manner for sample drying and, if necessary, sample preheating. On switching the evaporation mode for the analyte compounds the rod is inserted into the flame. The analyte atoms appearing in the flame are laser-excited and ionized by collisions.

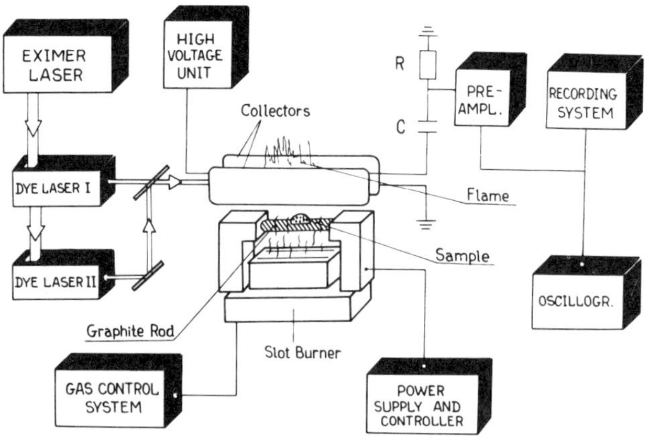

Fig. 1. Block-diagram of the LEI spectrometer

A pulse electrical signal formed across the collector is supplied to a boxcar integrator input. The output analytical signal integrated over the evaporation time and proportional to the element content in the sample is measured with a digital voltmeter.

3. RESULTS

First, schemes for stepwise excitation of indium and copper have been studied (Fig. 2) and methods of these elements determinations in aqueous solutions developed.

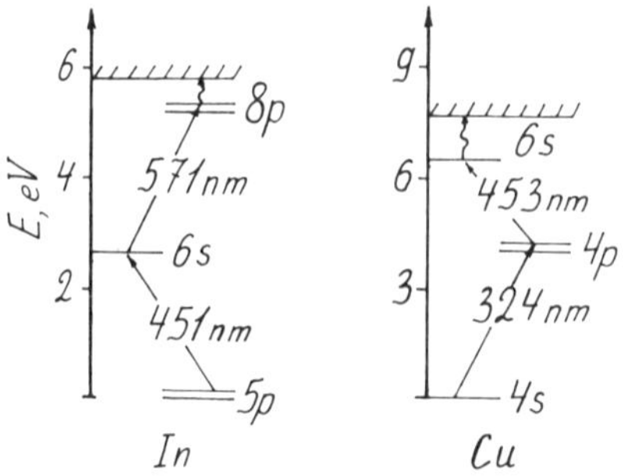

Fig. 2. Schemes for stepwise excitation of In and Cu

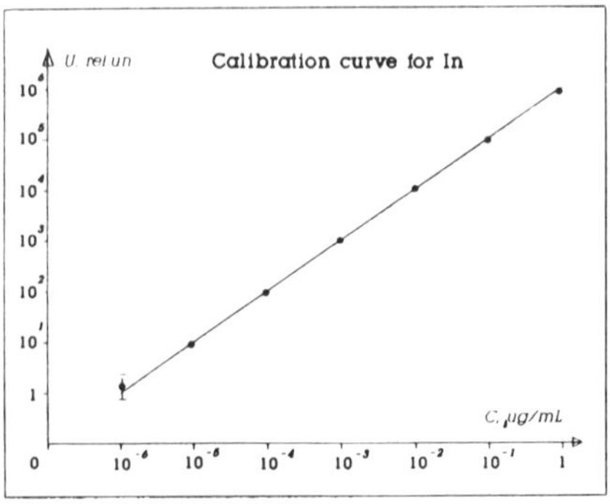

Fig. 3. Calibration curve for In solutions.

Calibration curves, linear in a wide concentration range (Fig. 3) were obtained to give absolute and relative DL (Table 1).

For copper the DL is shown to be governed by its content in the rod, for indium the DL is dependent on the recording system noises. The next step consisted in experimenting with the real samples.

Table 1. RESULTS OF In AND Cu DETERMINATIONS IN THE AQUEOUS SOLUTIONS (10 mg sample)

El.	(nm)	DL abs.(C 10^{-15}g)		DL rel.(pg/ml)
		Instrum.	True	True
In	451 571	4	4	0.4
Cu	324 453	200	2000	200

COPPER DETERMINATION IN GERMANIUM

At the first stage use was made of germanium solutions (10 g/l) prepared as follows: 1) without removal of the matrix and 2) with its partial removal.

The effect of the sample matrix on the analytical signal was studied with the standard addition technique. An increase in the analytical copper signal, observed as the additives were introduced, was compared with the corresponding value for the aqueous standard solutions. In all cases the signal increments were equal within the experimental errors (Table 2). Hence two important conclusions can be made:

1) in the solutions at germanium concentrations as high as 10 g/l the matrix has no effect on the analytical copper signal and 2) calibration can be performed with the aqueous standard solutions. Nevertheless, the matrix manifests itself through the following two factors that are responsible for a detection limit rise: 1) stronger flame noises associated witn the thermal matrix ionization and 2) the appearance of an additive background signal due to multiphoton ionization of matrix components.

Table 2. STUDES OF MATRIX INTERFERENCES BY THE STANDARD ADDITION METHOD

El.	Matrix	Introduced (ng/ml)	Found
In	CdHgTe (10 g/l)	0.050	0.051±0.004
Cu	Ge (10 g/l)	10.0	9.7±0.8

For the solutions without matrix removal the stronger flame noises predominated (Fig. 4a). Attempts to markedly diminish these noises by selecting the heating conditions were not success. The resultant DL was 1 ppm.

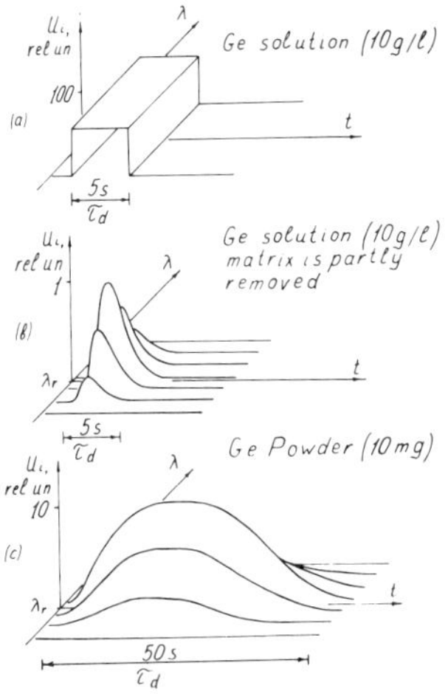

Fig. 4. Analytical (b, c) and background (a) signals versus the exciting radiation wavelength and time at the recording stage (τ_d), c - Integral analytical signal increases in proportion to copper content in the sample.

On passing to the solutions with partially removed matrices in the form of GeF the flame noises were noticeably weaker and the background signal caused by multiphoton ionization was reduced to the recording system noise level (Fig. 4b). In this case the DL is restricted by a copper content in the rod. To minimize the influence of this factor, kinetic studies were made to understand the appearance of signals from copper in the sample and in the rod with respect to the heating temperature and rate (Fig. 5). It is evident that two dependences obtained differ from each other. The signal from the copper in the sample depends on the heating rate and appears before that generated by the rodcontained copper. This fact was used to accomplish the temporal signal selection. As a result of optimization the copper detection limit in the solutions was lowered to 10^{-2} ppm for a 30 μl sample.

To further improve the detection limits, we tried to apply solid samples themselves onto the rod. In this case there was no sample dilution and the concentration of the copper and matrix in the solid sample were two orders of magnitude as high as that in the solutions. In so doing, the following two problems emerged:

(1) whether the copper undergoes a complete evaporation and atomization when solid samples are taken and (2) how do the background matrix-generated signals behave?

Studies with the use of multistep heating indicate that it is possible to find and accomplish conditions in which (1) the analytical signal increases in proportion to copper content in the sample (in other words, there is no loss of the analyte during the preheating, evaporation and atomization) and (2) the background signal from the matrix is far from increasing. In fact it even decreases below the recording system noises. (Fig. 4c)

As a result of this, in the case of solid samples, one can observe a dramatic growth of the signal-to-noise ratio and improvement of the DL to 5.10^{-4}ppm for a 10 mg sample.

Table 3 presents the analytical results for the germanium samples.

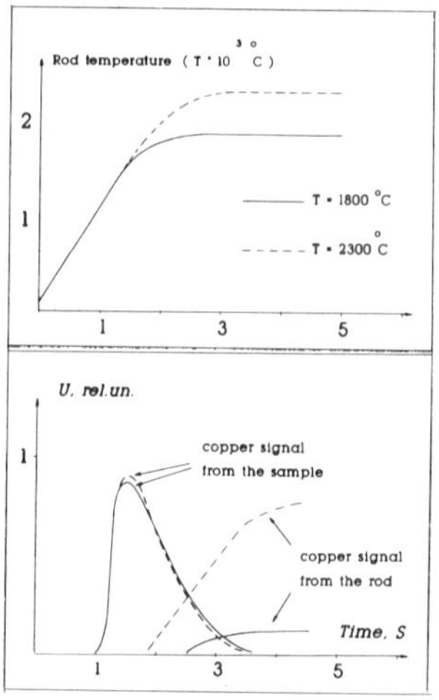

Fig. 5. Kinetics describing the appearance of copper ionization signals.

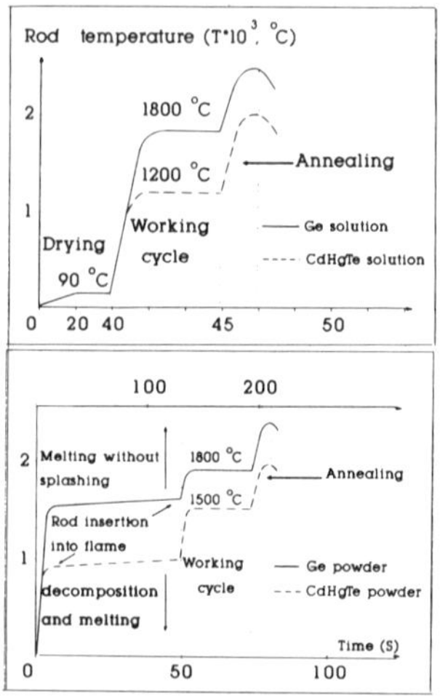

Fig. 6. Rod heating conditions of Cu and In determinations in Ge and CdHgTe.

INDIUM DETERMINATION IN CdHgTe ALLOYS

The detection limit obtained is not the only one minimally possible for this spectrometer design since it is restricted to a large blank signal from rod-contained copper. A more favourable situation exists with indium determination in the CdHgTe alloys. As in the previous case, the experiments were first made with the solutions and then with the solid samples.

As in the case of germanium, no signal suppression by the matrix occurs in the solutions (Table 2), and it is possible to perform calibration using aqueous standard solutions. The laser-stimulated nonselective ionization background from the matrix was small and its fluctuations were of the same order of magnitude as the recording contained indium is negligible. As a result, the DL for the system noises. The signal from the rod- solutions is 10^{-4} ppm. When passing to the solid samples, one can observe that the optimal rod heating conditions change greatly. Fig. 6.As in the case of germanium, the indium signal in the solid CgHgTe sample is proportional to indium content therein as there is no loss of the analyte. The background signal due to thermal matrix ionization was not observed.

The laser-stimulated ionization background from the matrix does not increase when passing to the solid sample. More than that it diminishes below the recording system noise level, that is likely due to a variety of molecular forms in which the matrix evaporates in both the solution and solis sample. As a result, the lowest, for a 10 mg sample, detection limit of 10^{-6} ppm was attained. The data for the liquid and solid samples correlate well (Table 3) to suggest the accuracy of the analysis.

Table 3. RESULTS OF GERMANIUM AND CdHgTe ANALYSIS FOR COPPER AND INDIUM CONTENT, RESPECTIVELY ($C \cdot 10^{2}$ ppm)

Sample		Solution, 10 g/l		Solid, 10 mg	
		Content	DL	Content	DL
			Copper		
Ge	1	7.9±0.6	1	7.1±0.7	0.05
	2	3.7±0.6		3.0±0.4	
			Indium		
CdHgTe	1	3.2±0.2	0.3	3.4±0.3	0.0001
	2	2.9±0.2		2.6±0.3	

CONCLUSIONS

1. Using the LEI technique in conjunction with the "ROD-FLAME" system makes it possible to directly analyze liquid and solid microsamples with high sensitivity, speed, precision and practically without sample preparation.
2. The matrix interferences can be completely eliminated and the spectrometer calibrated by the aqueous standard solutions.
3. The method proposed can be used to advantage in the construction of simple, reliable and cheap laser spectrometers that are applicable to the solution of a wide range of problems.

REFERENCES

1. Yu.Ya.Kysyakov, N.B.Zorov, CRC Crit. Rev. Anal. Chem. 4/20, 223 (1988).
2. J.C.Travis, G.C.Turk, Q.R.De Voe, P.K.Schenk, C.A.Van Dijk Progr. Anal.At. Spectr. 7, 199 (1984)
3. A.G.Marunkov, N.V.Chekalin Zh. Anal. Khim. 42, 638, (1987)
4. A.G.Marunkov, N.V.Chekalin, E.I.Tikhomirova Zh. Prikl. Spectr. 48, 542 (1988)
5. S.U.Bachin, A.G.Marunkov, V.I.Pavlutskaya, I.I.Vlasov, N.V.Chekalin Vysokochist. Veshch. 4, 186 (1989)

Paper presented at RIS 90, Varese, Italy, 16–21 September 1990

RIS of technetium in a laser ion source for a solar-neutrino experiment

F. Ames, H.-J. Kluge, B.M. Suri*, A. Venugopalan*
Institut für Physik, Universität Mainz, D-6500 Mainz, FRG
H. Rimke, N. Trautmann
Institut für Kernchemie, Universität Mainz, D-6500 Mainz, FRG
R. Kirchner
GSI, D 6100 Darmstadt, FRG

*Permanent address: Bhabha Atomic Research Centre, Bombay-400085, India

ABSTRACT: A very sensitive technique is presented for trace analysis of technetium. The method involves resonance ionization in a hot cavity, where atoms are confined to increase the probabillity for interacting with laser light, followed by mass spectrometry of the photo ions. With this setup, we obtained an efficiency of up to 13% for the photo ionization and extraction of technetium ions using minimum sample sizes of $5 \cdot 10^8$ atoms. The aim of these experiments is the determination of the amount of ^{98}Tc and ^{97}Tc produced by solar neutrinos in molybdenum ore and hence an integral measurement of the solar - neutrino flux over the past several million years.

Isotopic analysis of small amounts of technetium chemically separated from molybdenite can throw light on the integral ^{8}B solar-neutrino flux over the past several million years (Cowan 1982). Such an experiment involves the detection of about 10^8 atoms of ^{97,98}Tc produced from molybdenum by an inverse β-decay induced by solar neutrinos, accompanied by about 10^{11} atoms of ^{99}Tc from the spontaneous fission of uranium. Furthermore such a sample will contain about 10^{15} atoms of molybdenum even after chemical separation. So besides high sensitivity high elemental and isotopical selectivity is required. In principle this can be done by RIMS.

In conventional RIMS experiments a thermal atomic beam is produced and crossed by perpendicular laser beams (Ruster 1989). In this case the sensitivity is limited by the spatial and temporal overlap of the atomic beam and the pulsed laser beams. In case of a laser ion source, a hot cavity is used with a small hole to inject the laser beams and to extract the photo ions (Kluge 1985, Andreev 1986). Thus, the atoms are confined inside the cavity and have the chance to cross the interaction region several times. An atom released from the wall in such a cavity has three possibilities to escape:

1. It can be photoionized by the laser light and be extracted by an electric field.

$$P_I = \frac{V_{int}}{V_{tot}} \nu_{rep} \tau_{flight} \epsilon_{photo} (1 - \epsilon_{sur}) \epsilon_{ex}^{photo}$$

with V_{int} beeing the interaction volume, V_{tot} the total volume, ν_{rep} the pulse repetition frequency of the lasers, τ_{flight} the mean time of flight the atoms have inside the cavity, ϵ_{photo} the probability for photo ionization, ϵ_{sur} the probability for surface ionization and ϵ_{ex}^{photo} the extraction probability for photoions.

2. It can diffuse as an atom out of the hole during pulse pauses.

$$P_D = \frac{A_{hole}\bar{v}\tau_{flight}}{4V_{tot}}(1-\epsilon_{sur})$$

A_{hole} is the area of the extraction hole and $\bar{v}$ the mean thermal velocity.

3. It can leave the surface as an ion (ϵ_{sur} = surface ionization probability) and be extracted with a probability ϵ_{ex}^{sur}.

$$P_S = \epsilon_{sur}\epsilon_{ex}^{sur}$$

The later process is undesired in a laser ion source as it is non - selective. Thus the total efficiency is given by

$$\epsilon_{total} = \frac{P_I}{P_I + P_D + P_S}$$

In case of Tc with an ionization potential of 7.2 eV one can neclect surface ionization in a tungsten cavity ($\epsilon_{sur} \simeq 10^{-6}$ at 2400 K). Considering a cylindrical cavity with 2R = l the total efficiency comes to

$$\epsilon_{total} = \frac{P_I}{P_I + P_D} = \frac{\epsilon_{photo}\nu_{rep}\epsilon_{ex}^{photo}}{\epsilon_{photo}\nu_{rep}\epsilon_{ex}^{photo} + \frac{\bar{v}}{4l}}$$

Assuming saturation for the photoionization ϵ_{photo} will be equal to the ground - state population. Hence in case of technetium at 2500 K the total efficiency is calculated to be 17% for $\epsilon_{photo} = 0.6$, $\nu_{rep} = 6.5$ kHz and l =1 cm.

We have build up such an experiment for the detection of trace ammounts of technetium. It consists of a tungsten or graphite cavity with 2R = l = 1 cm, heated by electron bombardement up to a maximum temperature of 2900 K. This design was developed at GSI for the ion sources at the GSI on - line mass seperator. The cavity has a hole with a diameter of 2r = 2.7 mm to inject the laser beams into the cavity and to extract the photoions, which are detected at the exit of a Mattauch - Herzog mass spectrometer. The laser system consists of two copper vapor lasers (Oxford Lasers CU40) with an average output power of 30 W at a pulse repetition rate of 6.5 kHz pumping three dye lasers (Lambda Physik FL2001). RIS of technetium is done by a three - step, three - colour resonant excitation to an autoionizing state. The scheme we have choosen is λ_1 = 313.12 nm ($4d^5 5s^2\ {}^6S_{5/2} \rightarrow 4d^6 5p\ {}^4P_{5/2}$), λ_2 =821.13 nm ($\rightarrow 4d^6 6s\ {}^4D_{7/2}$) and λ_3 = 670.74 nm. UV light required for the first step is obtained in a BBO crystall by second - harmonic generation from the dye laser output. The laser light is coupled into one common quartz fiber ($\o$= 200μm) for transport of the laser beams to the experiment. This procedure results in an excellent, very homogeneous spatial overlap for the three beams at the interaction region much better than obtained by use of prisms or mirrors.

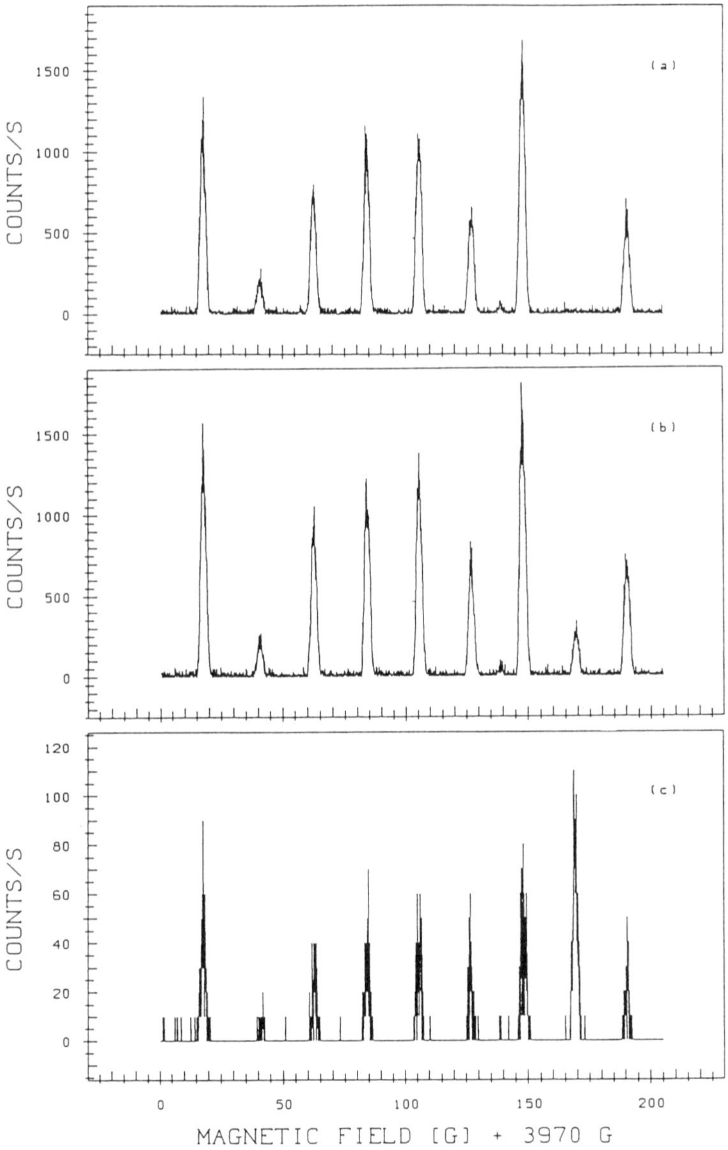

Figure 1: Part (A = 92 to A = 100) of a mass spectrum obtained with a graphite cavity containing $1.5 \cdot 10^9$ atoms ^{99}Tc at 2270 K. a) Lasers blocked, b) Lasers set on resonance for technetium, c) Gated detection

Samples containing 10^9 to 10^{14} atoms ^{99}Tc were deposited electrolytically on small tungsten or rhenium filaments (Sattelberger 1989) which were introduced in the cavity and heated up. Figure 1 shows part of a mass spectrum obtained from a graphite cavity at 2270 K with a ^{99}Tc sample containing $1.5 \cdot 10^9$ atoms. The first part shows the signal without lasers. Only thermally ionized molybdenum can be seen arising from impurities in the cavity material. The spectrum in the middle shows the signal with lasers set on resonance for ^{99}Tc. The count rate of thermal molybdenum ions stay at the same level as before but there is a tremendous increase in the signal at mass 99. To reduce the signal of surface ionized molybdenum and thus to increase the signal - to - background ratio even more, one can make use of the pulse structure of the photo ions. This is demonstrated in the bottom part of the figure. Here, a 4.5 μs coincidence gate at the detection system is used. A suppression of the background rate of thermal ions by a factor of 30 is obtained in this way.

The efficiency of the laser ion source was measured by introducing a known amount of ^{99}Tc (e.g. 10^9 atoms) into the cavity and taking the integral photo ion signal while heating up in several steps to 2750 K for a tungsten cavity and to 2340 K for graphite. In both cases we obtained a total efficiency for the whole set up of $\epsilon_{tot} = 4 \cdot 10^{-4}$ including the efficiency of the mass spectrometer and the ion detector of $\epsilon_{spec} = 3 \cdot 10^{-3}$. This results in an efficiency for the laser ion source to be ϵ_{exp} =13%, very close to the theoretical value given above. This satisfies the requirements of the solar neutrino experiment in respect to sensitivity. However, the detection of ^{97}Tc and ^{98}Tc is hindered by a strong interference of surface - ionized isobaric molybdenum. This problem might be solved by choosing other construction materials for the cavity with (i) lower workfunction than tungsten and/or, (ii), with less contamination by molybdenum. For instance, ultra - pure graphite and tantalum carbide will be tested. Especially the release of technetium from these materials has to be investigated.

This work was funded by the Deutsche Forschungsgemeinschaft under contract number Kl409-2.

References:

Andreev S V, Mishin V I, Letokhov V S 1986 Opt. Com. **57** 317
Cowan G A, Haxton W C 1982 Science **216** 51
Kluge H J et al 1985 Proc. on Accelerated Radioactive Beams Workshop ed Buchmann L and D'Auria J M TRIUMF Proc. TRI-85-1 119
Ruster W et al 1989 Nucl Instr Meth **A281** 547
Sattelberger P et al 1989 Radiochimica Acta **48** 165

Collinear resonance ionization spectroscopy for the quantitative detection of strontium-90 and strontium-89 in environmental samples

K. Wendt, K. Christian, H.-J. Kluge, L. Monz, E.-W. Otten, G. Passler, J. Stenner, K. Stratmann,, K. Zimmer, G. Herrmann, N. Trautmann, K. Walter

Institut für Physik und Institut für Kernchemie, Johannes-Gutenberg Universität, D-6500 Mainz

ABSTRACT: A set-up is presently under construction for the fast and quantitative trace detection of the radioisotopes ^{89}Sr and ^{90}Sr in environmental samples. For this purpose the strontium atoms are chemically extracted from the sample, surface ionized and accelerated to 50 keV for mass separation. The resulting fast ion beam is neutralized and resonance ionization in collinear geometry is realized via excitation into Rydberg states and subsequent field ionization. The extremely high selectivity and efficiency of this technique forms the basis for trace detection of $^{89,90}Sr$ contaminations as low as 10^8 atoms in a surplus of more than 10^{18} stable strontium atoms.

1. INTRODUCTION

The outstanding features of resonance ionization spectroscopy, i.e. its sensitivity and selectivity in respect to element or isotope under study, makes it a valuable tool for many applications in trace detection and analysis of toxic and radioactive material (Letokhov 1987). One case of high interest and importance is the fast and sensitive detection of the radioactive isotopes ^{89}Sr and ^{90}Sr. Both isotopes are produced in the maximum yield of fission from uranium-235 and appear through the atmosphere as fall-out during nuclear weapon tests and in case of a reactor accident. As chemical equivalent to calcium strontium is accumulated in the human body and therefore already a small contamination of radioactive isotopes causes serious danger to health.

The radiochemical detection of the most dangereous isotope, the long-living pure β-emitter ^{90}Sr ($T_{1/2} = 28.5a$) is complicated as its activity is not easily distinguished from the one of the shorter-lived isotope ^{89}Sr. Therefore the conventional detection must be carried out via a time-consuming separation of its daugther product yttrium-90 and requires altogether about two weeks for a reasonable sensitivity limit. As a consequence a fast, sensitive and thus radioactivity-independent detection method for both isotopes is desirable. Unfortunately its experimental realization must struggle against the high abundance of stable strontium in the air sample, mainly ^{88}Sr. Following the numbers which have been measured in Germany shortly after the Tchernobyl reactor accident in 1986, we can formulate the requirements for a competitive trace detection system for radioactive strontium isotopes as follows:

1. The overall efficiency must surpass a value of 10^{-5} for the realization of a reasonable detection limit of about 10^8 atoms per sample and for quantitative results.
2. The selectivity in the suppression of stable isotopes must exceed a value of $\approx 10^{11}$.
As further requirement the overall measuring time should not exceed one day.

In this paper we present the application of resonance ionization spectroscopy to this task. Its experimental realization is found in the combination of collinear fast-beam laser spectroscopy (Kaufman 1976) with resonance ionization detection as proposed by Kudriatsev and Letokhov (1982) already in 1982.

2. THEORETICAL CONSIDERATIONS

In general the selectivity of any laser spectroscopic technique is based on the sharpness of the optical transitions involved and the low level density. For the case of a Lorentzian line shape it can easily be shown that the probability for optical excitation off-resonance is given by $(2\Delta\nu/\Gamma)^2$, where $\Delta\nu$ reflects the frequency shift from resonance and Γ the line width. While the elemental selectivity is generally guaranteed by the low chance of random coincidence of transition wave lengths, the isotopic selectivity is limited by the size of the isotope shift or hyperfine structure of the transitions under study to values around $S = 10^4 - 10^6$ per step. Usually only the ground state transitions show significant splittings or shifts, which in the case of the strontium isotopes only slightly exceed the size of the natural line width. Thus nearly no selectivity can be achieved in an experiment on a thermal atomic beam of strontium even by applying multi-step resonant excitation, in which case the individual selectivities can be multiplied to give the overall selectivity.

One solution to this quite general problem to reach extremely high isotopic selectivities is found in the acceleration of all isotopes under study to a high energy of fixed value. As the Doppler shift is given by the velocity and depends therefore on the mass of the different isotopes, this results in an artificial mass shift of the order of a few GHz for neighbouring isotopes in the A=90 region. In that way single-step selectivities in the order of $S = 10^9$ are obtained. In addition to this the collinear excitation of a fast atomic beam also results in a strong narrowing of the Doppler width of the optical transition. This fact can easily be understood by considering the conservation of phase space during electrostatic acceleration. Due to this the energy spread remains constant and the introduction of the strong Doppler shift $\Delta\nu_D$ results in a corresponding reduction of the Doppler width $\delta\nu_D$. This allows the observation of experimental line widths close to the natural line width and increases the selectivity if laser excitation of sufficiently narrow band width is used.

While the achievement of the selectivity necessary for trace detection of ^{90}Sr and ^{89}Sr can be guaranteed in collinear laser spectroscopy, special attention must be paid to the efficiency and the background of the method. Optical detection is out of question, due to its limited solid angle, high background from laser stray light and low quantum efficiency as well as dark current of the multiplier. A better choice is particle detection. For this the strontium atoms must be ionized after selection via the optical excitation. This can either be done by direct optical ionization, by field ionization of highly excited Rydberg states or by state selective reionization in collisions. The latter process is applied very succesfully and background-free for the study of pure, weak beams of extremely short-lived isotopes at ISOLDE, CERN (Borchers et al, 1989). However, it fails in the presence of strong contaminations, due to the small difference of cross sections for collisional ionization from excited and ground state, respectively. Similarly the direct ionization by excitation into the continuum or via auto-ionizing states is hampered by the strong background produced by residual gas collisions of the high number of stable strontium atoms in the beam. For maximum suppression of this back ground the $^{90,89}Sr$ atoms are excited to high-lying Rydberg states. These Rydberg atoms can be ionized by field ionization well localized and energetically marked in a small region. In that way the number of collisionally ionized background ions is strongly reduced by installation of an energy selector and suitable ion beam optics in front of the particle detector.

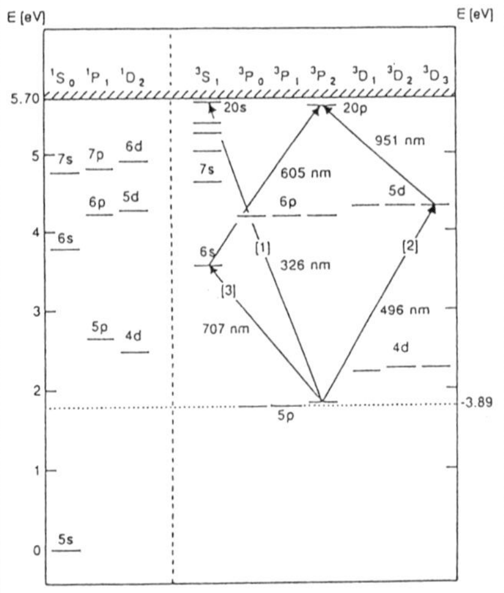

Fig. 1: Part of the level scheme of Strontium

The interesting part of the strontium level scheme is given in Fig. 1. Three different optical excitation schemes are indicated for excitation into a n = 20 Rydberg state, starting from the metastable $5s5p\,^3P_2$ state. The efficient population of this state is achieved in the neutralization process of the initial strontium ion beam to the fast atomic beam by choosing cesium as charge exchange medium with its low ionization energy of only 3.89 eV. To meet the requirements given above the direct single-step excitation scheme (1) needs an additional isotope selective step, e.g. via a conventional mass separator. Nevertheless it is the most simple approach concerning laser stabilization and frequency adjustment and will be used in the beginning of the experiment.

For realization of higher efficiencies a two-step excitation with visible or infra-red lasers is desirable. Scheme (3) is hampered by strong optical pumping into the $5s5p\,^3P_{0,1}$ states. Thus it can only be realized using pulsed lasers, resulting in high duty cycle losses. Most promising is scheme (2), which avoids optical pumping by choosing the 5s6d $^3D_3$ state as intermediate level.

3. EXPERIMENTAL

A schematic drawing of the experimental set-up is given in Fig. 2. The measuring sequence starts with the chemical extraction of strontium from the environmental sample. This step is carried out via the use of different crown ethers as carriers for strontium atoms and to remove the other alkaline earth elements. The strontium content of the sample can then be electrolysed onto a tantalum foil, which is directly introduced into the surface ionization source. Quantitative ionization of the macroscopic sample is realized in a two-step process of evaporation and subsequent ionization. The resulting strontium ions are accelerated to form a 50 keV ion beam by suitable ion optics. After mass separation in a 60° separator magnet the beam is overlapped collinearly with the laser beam and charge exchanged in the cesium vapour cell. At vapour pressures around 10^{-3} *mbar* complete neutralization is achieved without significant change of the beam quality. The resulting fast atomic beam is cleaned from still remaining ions in a set of three different ion filters, also removing Rydberg atoms populated in the charge exchange process. Excitation into the Rydberg states for the isotope under study is performed by resonant laser radiation in a 1 m long field-free excitation region. In this way sufficiently long interaction time is obtained. Subsequently the selected atoms are downstream ionized in a field ionization region by an overcritical electric field of about 10 kV/cm. The design of this region is crucial, as it determines the number of background counts from collisionally produced ions. For this purpose the vacuum in this part of the set-up is required to be better than 10^{-10} mbar and ionization is carried out in a field region as short as 5 mm and at an elevated potential of about 6 kV for energy selection. The latter is performed in a 40° electrostatic deflector in front of the particle detector.

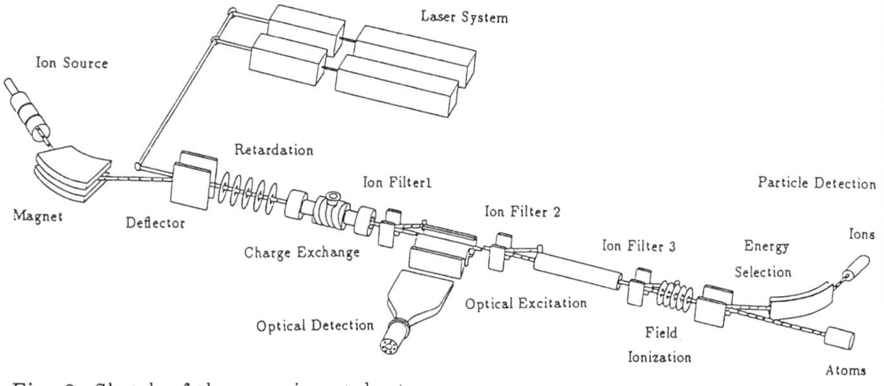

Fig. 2: Sketch of the experimental set-up

With the system shown we expect to perform quantitative measurements on concentrations as low as 10^8 radioactive atoms in a sample of about 10^{18} stable strontium atoms. For higher selectivity or larger samples pre-enrichment in the existing medium-current mass separator facility SIDONIE/GSI at Darmstadt is foreseen. To avoid the complicated determination of overall efficiencies, relative measurements to a known number of tracer atoms are planned by cyclic switching of masses and laser frequencies. Including sample preparation the full measuring sequence will need less than one working day for reasonable accuracy. The experimental apparatus presently undergoes the testing phase. First results are expected for 1991.

This work has been funded by the Deutsche Bundesministerium für Umwelt, Naturschutz und Reaktorsicherheit.

REFERENCES

Borchers W et al 1989 Phys. Lett. **B 216** 7
Kaufman S L 1976 Opt. Comm. **17** 309
Kudriatsev Y A and Letokhov V S 1982 Appl. Phys. **B 29** 219
Letokhov V S 1987 *LaserPhotoionizationSpectroscopy* (Orlando: Academic)

Inst. Phys. Conf. Ser. No 114: Section 6
Paper presented at RIS 90, Varese, Italy, 16–21 September 1990

Ultrasensitive analysis of Au in mineral samples using resonance excitation and autoionization mass spectrometer

Q. Hui, D. Y. Chen, J. G. Niu, Y. Cheng, X. Y. Xu, W. Z. Zhao

Department of Modern Applied Physics, Tsinghua University, Beijing, PRC

ABSTRACT: The first application of autoionization state of ^{197}Au for trace analysis of gold in mineral samples is described. The detection limit is 0.1 ppb (1×10^{-10}) which is two orders better than that in two–step resonance excitation and photoionization scheme used in our lab before. The expriment results are proved to be linear and reproducible.

1. Introduction

Resonance Ionization Mass Spectrometer (RIMS) is of the advantages as (1) high selectivity due to the multistep resonance excitation, (2) high sensitivity, (3) high efficiency of ion detection, (4) low background level due to the two–dimension selective detections of element and mass. The hypersensitive analytical technique based on RIMS possesses very important applications for the geological prospecting and marine resources exploiting. The mean concentration of gold in earth surface is 5ppb (5×10^{-9}). The concentration in mineral samples is often of the order of 10^{-9} or even lower. For traditional trace analysis methods such as atomic absorption spectrometry and atomic fluorescence spectrometry, such a low concentration has approached to their detection limits. Recently, some new sampling methods have been applied to geological exploration. For example, sampling from air or water, the concentration of gold in such samples is of the order of 10^{-12}, and the volume of samples is very small. The traditional methods cannot satisfy the needs of analysis. RIMS would be a new selection.

In this paper, we report trace analysis of gold in mineral samples by means of RIMS and autoionization states. In the following sections, the experimental setup of RIMS is described firstly. Then the detection efficiency between photoionization and autoionization is compared. The background and detection limit are discussed. Finally a short conclusion is given.

2. Experimental Setup

The schematic diagram of the RIMS is illustrated in Fig.1. The experimental setup includes several parts: electrothermal atomizer, sample autochanger, laser system, time–of–flight mass spectrometer and signal collector.

The laser system consists of a excimer laser (model EMG202 from Lambda Physik Inc.) and three dye lasers. Two of dye lasers are model 3002 E and 3002EC which are from Lambda Physik Inc. too. The third one is made in our lab.

The residual pressure in the vacuum chamber was about 10^{-6}Torr. By means of the sample autochanger, 20 samples can be measured continuously in each experiment without damaging the vacuum. As a result, the measurement was carried out efficiently and the experimental conditions were uniform for all samples. The crucible was heated to 2500℃ by an alternative current generated by a program–controlled power supply which can provide same heating program for every sample.

To suppress a strong background from thermal ions and electrons, there were two D shape boxes at electric potentials of ± 24V above the atomizer.

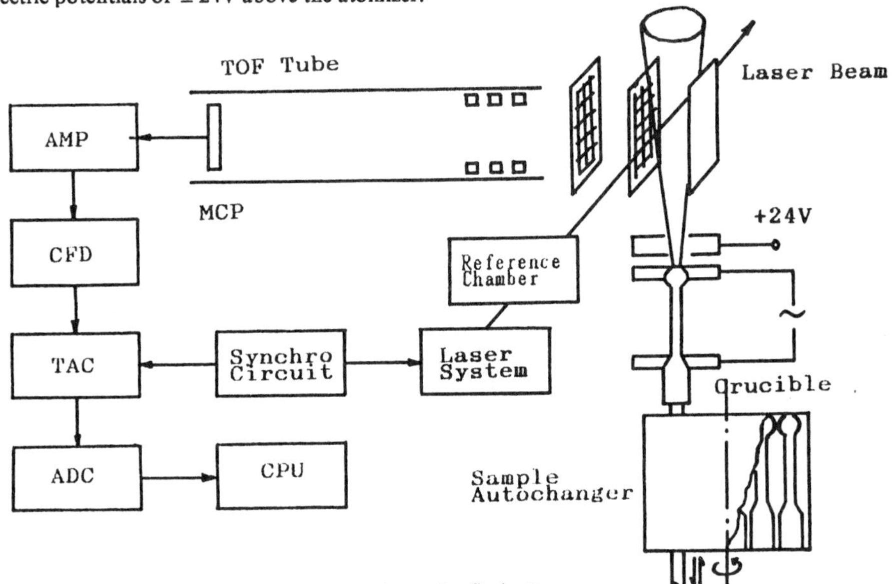

Fig.1 Expriment Setup

The atom beam intersected the laser beams between the two electrodes propendicularly. The atoms under analysis were excited and ionized by lasers. The RIS ions were subsequently collected and accelerated by electrodes with grid. After focused by a set of lens, the RIS ions were detected by a microchannel plate (MCP) detector at the end of about 1.5m field–free drift tube. The time–of–flight spectrum was obtained by TAC and multichannel analyzer (MCA) synchronized with the laser pulses by time–controlled circuits. The mass resolution was above 300.

The reference chamber similar to the main one was used to maintain the lasers at resonance wavelengths. Instead of the mineral sample, pure gold was put in graphite crucible of the reference chamber and lower temperature was used to heat it.

3. Two–step Resonance Excitation and Photoionization

The scheme was:

$$5d^{10}6s^2S_{1/2} \xrightarrow[10\mu J]{267nm} 5d^{10}6p^2P_{1/2} \xrightarrow[2\sim8mJ]{406nm} 5d^{10}6d^2D_{3/2} \xrightarrow[2\sim8mJ]{406nm} Au^+ + e \qquad (1)$$

The main advantage of this scheme is that only two lasers are required. The typical diameter of the two lasers was about 5mm. The second step excitation was saturated but the ionization was not. Several hundreds millijoule of pluse energy was needed for the saturation ionization. This is unavailable for our dye laser. Focusing the laser light can make the ionization saturated but the detection volume was decreased. The first step excitation was also unsaturated due to the non–resonance ionization background discussed in section 5. The overall efficiency of excitation and ionization was estimated as less than 1% with laser pluse repetition rate 20Hz.

We used this scheme for the analysis of solid samples(1.32ppm). The weight of the sample was

12.5mg. Integrated ion counts in the Au peak were 979 with 87 background counts in it. The variation of 5 samples was determined less than 30%.
The concentration of gold in mineral samples is very non–uniform because of its existence as simple substance and its extensibility. Then 1g~20g samples is required for the analysis of very low concentration of gold, for exemple, 100ppb~1ppb. In our experiments, 5g solid sample of gold concentration 1ppb~100ppb was mixed with aqua–regia and heated to dissolve gold. Then the deionized water was put in to 5ml and the acidity was adjusted to about 2.0. Only 10 or 20μl of the 5ml liquid sample was injected into a graphite crucible for analyzing.

4. Three–step Excitation and Autoionization

Some of the autoionization states of Au atom have been searched with three–step resonance excitation by Krönert et al. [1]. In our experiments, it was found that the excitation cross–sections of these autoionization states of Au were one order higher than that of Rydberg states ($n^* \sim 40$). So we were interested to use these states to realize the trace detection of Au in mineral samples.
The scheme was:

$$5d^{10}6s^2S_{1/2}\xrightarrow[10\mu J]{267nm}5d^{10}6p^2P_{1/2}\xrightarrow[100\mu J]{406nm}5d^{10}6d^2D_{3/2}\xrightarrow[3\sim 10mJ]{490\sim 650nm}Autoionization\ states \qquad (2)$$

Scanning the third laser in the range between 490nm–650nm, several resonance peaks were found. The most intensive one, $\lambda_3 = 592$nm, was chosen for our experiments. With the pulse energy presented in (2) and typical beam diameter of 5mm, both the second and third step excitations were satuated, The efficience of the first step excitation was about 30%, the same in scheme (1).The integrated ion counts in the Au (10μl of 13ppb sample) peak was 941 with 20 background counts in it. It showed that one ion can be detected per 3×10^8 Au atoms and the overall efficiency was two orders higher than that in scheme (1).
The efficiency of about 3×10^{-9} can be estimated to have the following contributions: the duty factor of 10^{-4} with laser pulse repetition rate 20Hz, the spatial factor of 10^{-2}, the excitation and ionization efficiency of about 15%, the rest coming from the efficiency of atomization, ion transmission and detection etc.
In order to increase the efficiency of the third step excitation, another scheme was searched in our lab [2]. The scheme was:

$$5d^{10}6s^2S_{1/2}\xrightarrow[10\mu J]{243nm}5d^{10}6p^2P_{3/2}\xrightarrow[100\mu J]{479nm}5d^{10}6d^2D_{5/2}\xrightarrow[3\sim 10mJ]{522-644nm}Autoionization\ states \qquad (3)$$

Several new autoionization states were found in the wavelength range 522–644nm. The curves of saturation pulse energy for three states were showed in Fig.2. One of the three states, in $\lambda_3 = 595$nm, can also be searched from $D_{3/2}$ states by means of the third laser in $\lambda = 592$nm that has heen used in scheme (2). It seems that the excitation efficiencies corresponding to the resonances in 588nm and 622.5nm were higher than that in 595nm. The further investigation is in progress.
The linearity of the spectrometer was investigated using two series of liquid sampls supplied by the Institute of Rock and Mineral Analysis, Chinese Academy of Geological Sciences.
The first series called standard series was made from $AuCl_3$. The second, called mineral sample series, was made from solid mineral standard sample GAu–x series according to the method described in section 3. The results were showed in Fig.3. The curve was found to be linear in the range lppb–25ppb.

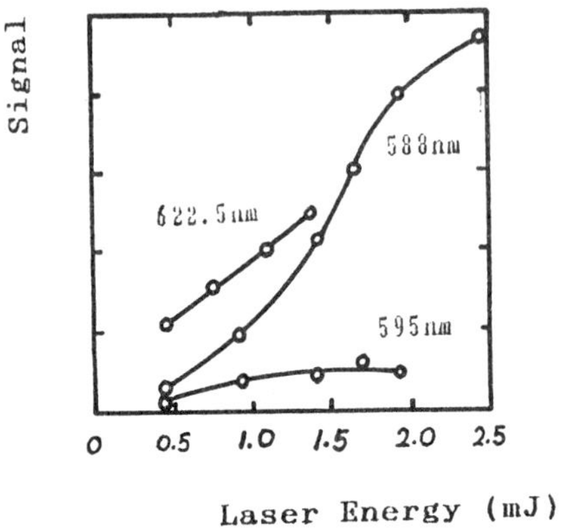

Fig.2 Saturation Curves

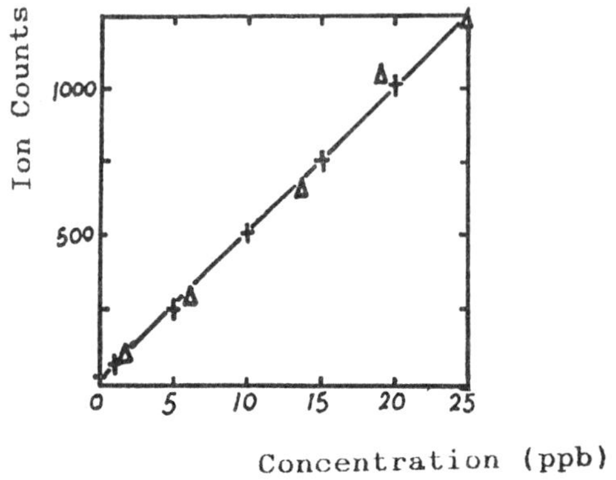

Fig.3 Linearity

+ Standard Series
Δ Mineral Sample Series

5. Background and Detection Limit

In the experiment showed in Fig.3, the counts of blank sample were about 20. It was estimated to have the following contributions:

(1) Non–resonance multiphoton ionization: Mainly came from the multiphoton ionization of the pump oil molecules which have mass number about 200. For the UV laser pulse energy of about 10μJ and the beam diameter of about 5mm, there were 6~ 10 multiphoton ions in the overall background of each sample.

(2) Memory effect: Came from the resonance ionization of the residual gold atoms left in the used crucibles previously although these crucibles were overheated up to about 2500℃ for 2min. This is the main background source.

According to the general definition and Fig. 3, the detection limit was estimated as 0.1ppb (10^{-10}). Increasing the sample volume to 50μl and concentrating it by extraction, the detection limit can be improved to 10ppt (10^{-11}) easily. In other way, new crucible can be used to avoid the memory effect and better vacuum condition would help us to decrease the non–resonance ionization of pump oil molecules. So the detection limit would be improved furtherly without higher efficience.

6. Conclusions

RIMS is the most hopeful method for the detection of gold in mineral samples. Gold can be determined in concentration down to 0.1ppb now and no complicated pre–processing is required. The analytical results are reproduceable and linear. Both the detection limit and linear rang will be improved furtherly in next experiments.

References

1. **U. Krönert, St. Becker, Th. Hilberath, H. J. Kluge, and C. Schulz:** Appl. Phys. A44, 339–345(1987)

2. **W. Z. Zhao et al. :** Autoionization States of Au Atom, to be published.

Paper presented at RIS 90, Varese, Italy, 16–21 September 1990

Collisional and electric-field ionization of laser-prepared Rydberg states in an ion trap mass spectrometer

J. M. Ramsey, W. B. Whitten, D. E. Goeringer, B. T. Buckley

Oak Ridge National Laboratory, P.O. Box 2008, Oak Ridge, TN 37831-6142

ABSTRACT: Rydberg states of rubidium are selectively generated by one- and two-photon laser excitation in a quadrupole ion trap mass spectrometer. Collisional and electric-field ionization of these states is investigated in a trapping device. CCl_4 is studied as a target for ionization of Rydberg states through electron attachment.

1. INTRODUCTION

Resonance Ionization Mass Spectroscopy (RIMS) combines the selectivity of resonance ionization spectroscopy (RIS) to produce ions of a specific kind (species, state, translational energy, etc.) and of mass spectroscopy to isolate ions of specific mass (energy, momentum)(Smith 1989, Young 1989). The combined selectivity has allowed the measurement of isotope ratios where severe isobaric interferences exist and the extension of the abundance sensitivity provided by either technique alone. The resonance ionization process involves the absorption of two or more photons through resonant transitions with sufficient energy to remove a valence electron from the atom or molecule. The laser produced ions are then mass analyzed using any of a number of mass spectrometers such as magnetic sector, time-of-flight, and linear quadrupole devices (Lubman 1990). The ability of time-of-flight instruments to produce a complete mass analysis at each firing of a pulsed laser is particularly attractive.

We have recently investigated the use of the quadrupole ion trap mass spectrometer (QITMS) for RIMS experiments (Goeringer 1990). This mass spectrometer has the ability to store all ions above a given cut-off m/z and then sequentially eject them, from low to high mass, to yield a mass spectrum (March 1989). The QITMS has several attractive features for RIMS experiments including efficient collection of all ions generated independent of excitation duration (provided space charge effects are not encountered) and efficient detection of the trapped ions. Both of these features lead to high sensitivity as has been demonstrated for both electron impact ionization (EI) (Stafford 1983) and RIS (Goeringer 1990). In addition, the stored ions can be further manipulated in the QITMS allowing for collision induced dissociation or photodissociation experiments. Optical spectroscopy can also be used to further study the trapped ions.

A disadvantage of ion trapping devices, both the QITMS and Fourier transform ion cyclotron resonance mass spectrometer, is dynamic range which is typically limited to $\approx 10^4$ (March 1989, McCrery 1986). The dynamic range of the QITMS has been improved by roughly 10^2 through the use of what has been termed "automatic gain control" (Stafford 1987, Yost 1987). In this mode of operation the electron beam used for ion formation is turned on for a duration sufficient to fill the trap with ions. The optimum duration will obviously depend on the density of species present in the trap thus, detection of species at low density is enhanced by integrating over a longer period of time. The same approach to dynamic range improvement might be employed when RIS is used to generate the ions. To effectively implement a variable time integration scheme, either a CW or quasi-CW laser should be used for excitation because of the finite storage times possible in the QITMS. The use of CW laser excitation will also allow more efficient use of samples that are generated in a continuous manner such as atoms from hot filaments. Although the ionization efficiency using CW lasers can approach 10% using multiple wavelength excitation (Bushaw 1989) it is generally much lower. The resonant bound-bound transitions of atoms can be saturated with CW lasers. The optical absorption cross section for the bound-free transition in the RIS process is typically 10^3 smaller than the bound-bound transitions making it difficult to saturate with commercially available CW lasers.

Alternative approaches for generation of ion pairs from highly excited states include electric field ionization and collisional ionization (Hurst 1979, Ivanov 1975). Collisional ionization mechanisms, as related to RIS experiments, have been studied by others in cells and using linear quadrupole mass filters. Whitaker (1981) examined the collisional ionization of Rb Rydberg states in $\approx$ 10 torr of Kr buffer gas and observed an order of magnitude increase in ion signal following excitation to the 20 D level versus threshold photoionization. Ionization of optically populated excited states with electron collisions has also been studied (Bushaw 1986). There have been numerous applications of field ionization to RIS and RIMS experiments (Bekov 1983).

Both of these effects are interesting candidates for ionization in the QITMS. The QITMS operates at relatively high pressures, typically with 1 mtorr of He buffer gas, for damping of the ion trajectories (Stafford 1984). In addition, electric fields are present in the trapping volume as a natural consequence of its operation. An ion can experience field strengths from a fraction to several kV/cm depending on m/z. We are currently investigating these two alternative ionization mechanisms in a QITMS and present here some of our preliminary results. Rubidium atoms from an oven are introduced into the trap where Rydberg states are selectively produced by laser excitation of allowed one- and two-photon transitions. The effects of He and He doped with 3% CCl_4 on ion yield are studied versus principal quantum number. Evidence of electric field ionization from such states is also presented.

2. EXPERIMENTAL

A commercial hyperbolic quadrupole ion trap mass spectrometer (Finnigan MAT ITD model 800), hereafter referred to as ITD, was used in our experiments. Electrodynamic trapping is accomplished by the application of a radio frequency voltage (1.1 MHz) to the ring electrode with the end caps grounded. Mass analysis is provided by the mass selective instability mode of operation (Stafford 1984). In this mode all ions above a given cutoff m/z are trapped by setting the rf voltage to the appropriate amplitude. The applied voltage is then ramped with ions being ejected through the end caps, from low mass to high, as they are excited into unstable trajectories. The ejected ions are detected by an electron multiplier. The rf voltage is ramped at a rate corresponding to 5.55 amu/ms. The commercial software supplied with the instrument was used to control the ion trap's operating conditions. This software was run on an IBM PC/XT through the interface supplied with the instrument. Modifications to the vacuum chamber and ring electrode for laser excitation have been described in a previous report (Goeringer 1990). Briefly, collinear holes were drilled through the center of the ring electrode and vacuum can; 3 mm and 6 mm respectively to provide clear passage of the laser light through the center of the trap. The holes in the vacuum chamber were closed by pyrex windows, 25 mm in diameter, held in place by low vapor pressure epoxy.

A Quanta Ray PDL-2 dye laser, with Kiton Red dye dissolved in methanol, was used to produce radiation for excitation of rubidium ground state atoms to s and d Rydberg levels. This light was frequency doubled to produce p state Rydberg atoms. The dye laser was pumped with frequency doubled (532 nm) light from a Quanta-Ray DCR 2A Nd:YAG pulsed laser. The pulse duration of the dye laser was nominally 6 ns with a bandwidth of ≈ 8 GHz. The laser pulse energies were ≈ 1 mJ for the dye fundamental and < 0.1 mJ at the second harmonic when generated. The laser light was turned with quartz prisms and focussed to the center of the trap with a 25-cm focal length quartz lens to produce a beam waist of ≈ 100 μm diameter with the polarization axis aligned with the rotational symmetry axis of the trap. The laser firing was synchronized with the ITD by using the EI-gun gate pulse to trigger the laser through a buffer amplifier.

The rubidium was introduced into the ion trap via a Knudsen cell. The electron filament and gate assembly, normally used to ionize the analyte in the trap, was replaced by a quick disconnect vacuum fitting which allowed a glass u-tube, containing rubidium metal, to be attached to the can. The rubidium vapor was passed from one arm of the u-tube into the trap by warming the entire u-tube assembly with a pair of clamshell heaters. The Knudsen cell was typically operated at ≈ 100 °C.

He buffer gas was used for all experiments and maintained at a pressure of 1 mtorr in the trap chamber. The gas was passed through a oxygen scrubber (model OT3-2 by R+D Separations) for all experiments which did not involve the introduction of CCl_4 into the trap. When CCl_4 was used the oxygen scrubber was replaced by a 38 mm O.D. x 150 mm sealed tube containing a few mL of liquid CCl_4. The He buffer gas flowing through the sealed tube carried vapor phase CCl_4 into the trap. The vapor pressure of CCl_4 at ambient temperature fixed its concentration in the carrier at ≈ 3%.

A mass spectrum from m/z 10 to 100 is recorded for each laser pulse at the rate of ≈ 6 Hz. The laser is typically scanned from 600 nm to 594 nm at the rate 0.01 nm/s. An excitation spectrum contains nominally 3400 mass spectra taken in ≈ 10 minutes. Spectra are further manipulated after data collection by integrating over five laser pulses and selected masses.

3. RESULTS AND DISCUSSION

Fig. 1 shows a typical two-photon excitation spectrum of Rb in the ITD with He buffer gas. The spectrum includes ions detected at m/z 85 and 87, the two isotopes of Rb and covers the energy range from the 19d state to n ≈ 73 at 594 nm. The two-photon ionization threshold is at 593.64 nm. The ns resonances are also observable to the high energy side of the (n-2)d states. Rydberg states can be resolved to n ≈ 40, the ion signal decaying with n due to reduction of both the Rydberg excitation and photoionization (PI) probability. The rise in the baseline between 596 nm and 594 nm is due to incomplete resolution of the Rydberg states. The ion signal becomes constant in the unresolved region. The probability of generating Rydberg states should be a constant in this region as the excitation cross section is decreasing as n^{-3} and the density of states in increasing as n^3. Ions generated in this region presumably result from a number of phenomena including PI, collisions, electric fields, and blackbody radiation.

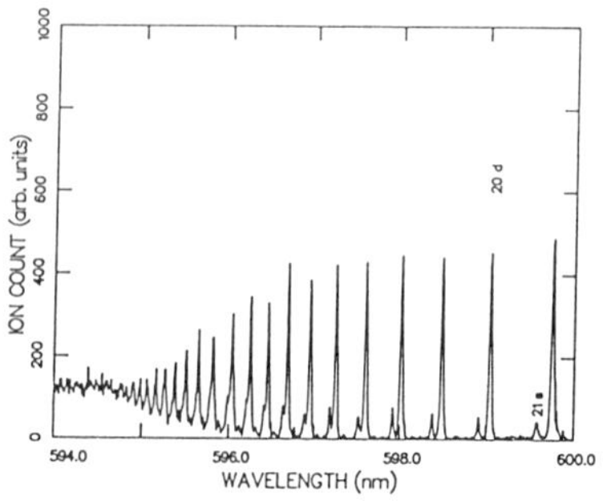

Fig. 1. Two-photon excitation spectrum of Rb; m/z 85 and 87

The s states are not normally observed in 2+1 PI (Noordam 1990). Their appearance in this spectrum is suspected to be due to electric-field induced state mixing (Klots 1985) and partial saturation of the nd PI (see below). Additional ionization mechanisms are likely to be operative to varying extents. It is unlikely that field ionization can explain the occurrence of the ns ions.

The electric-field distribution, E(r,z), in the ion trap is given by

$$\mathbf{E} = -\phi/2\ (\mathbf{r} - 2\mathbf{z})/r_0^2, \qquad 1)$$

where ϕ is the electric potential, r the radial dimension, z the direction along the rotational symmetry axis and r_0 is the minimum radius of the ring electrode (1 cm). The applied potential oscillates in a sinusoidal fashion at 1.1 MHz with variable amplitude. From simulations (SIMION) we have found that thermal velocity ions generated within $r_0/2$ of the trap origin are efficiently trapped. The rf potential amplitude required to eject a given m/z in this trap is ≈ 9.3 V/amu. The laser fires at a time when m/z 21 is ejected from the trap. Thus the field strength that a Rydberg atom can experience within the trapping volume is < 195 V/cm. Such a field is sufficient to ionize only those states with $n^* > 36$ ($n^* = n - \delta$, where δ is the quantum defect). Collisional induced ionization from the He buffer gas present in the trap is possible but probably not dominate at 1 mTorr (kT energy will ionize states with $n^* > 23$).

Experiments were also performed under the same trap conditions but with the dye laser radiation directed through a frequency doubling crystal before entering the ITD. With the frequency doubled light we can expect to populate the np states through one-photon transitions from the 5s ground state. Fig. 2 shows an excitation spectrum for the m/z 85 and 87 ions where both the dye fundamental and second harmonic are allowed to enter the trap. The abscissa is labeled with the fundamental wavelength. Ion signals are easily observed for both the np and nd states and to a lesser degree from the ns states. Ionization from the low np states is suspected to be largely from 1+1 PI including a blue photon to the p state and a red photon to ionize. The second harmonic photons will be relatively ineffective for PI due to their lower flux and greatly reduced PI cross section (Aymar 1976). The reduced relative signal from the ns versus nd states as compared to Fig. 1 is thought to be due to the reduced fundamental intensity. The change in relative amplitude and reduced linewidth in Fig. 2 suggests that the spectrum of Fig. 1 is partially saturated. The most apparent change in the spectrum with the addition of the second harmonic is the growth in ion signal at high energy. This increased signal is believed to be due to the presence of electrons produced by photoemission from the electrode and window surfaces. The generation of photoelectrons in the ITD has been observed in previous experiments involving near-UV light (Goeringer 1990). At high n where the n manifold is not resolved, the Rydberg excitation probability should be constant as discussed above. The EI cross section, based upon a binary-encounter model, increases with ≈ n^2 and scales roughly inversely with electron energy (Flannery 1983). The broken line in Fig. 2 between 596 and 594 nm shows the increase in impact ionization cross section predicted by this model from n^*=30-70 for thermal electrons. This model accurately predicts the general trend of ion yield versus wavelength in this region. The energy distribution of the electrons is not known. The photoelectrons generated at the surface of the trap presumably have energies < 1 eV but will in general be quickly accelerated by the trapping fields. The collision cross section for hot electrons has the same basic dependence with wavelength (n) as that shown in Fig. 2 but with reduced magnitude.

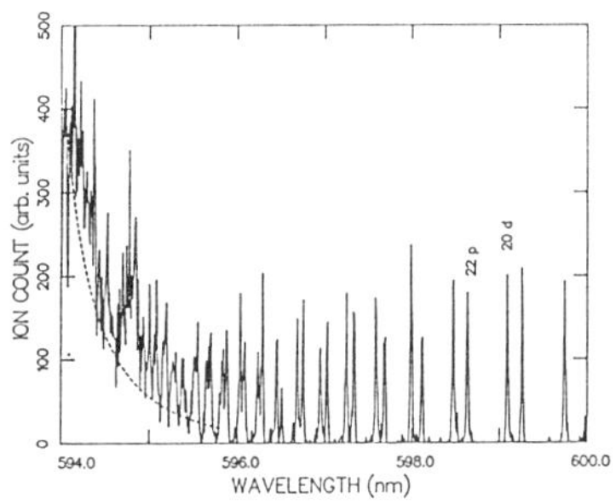

Fig. 2. One- and two-photon excitation spectrum of Rb; m/z 85 and 87

Fig. 3A shows an excitation spectrum taken over the same energy range with only the second harmonic radiation entering the trap. The abscissa is again labeled with the dye laser fundamental wavelength. The amplitudes of the np resonances are greatly reduced and the ns and nd resonances are absent. The small ion yield at the np resonances is attributed to collisional ionization with both He and electrons in addition to blackbody radiation PI. The unresolved manifold in the region of $n^* \approx$ 40-70 again shows the increasing ion yield with n. The same experiment was repeated using He saturated with CCl_4 vapor ($\approx$ 3%) as the buffer gas in the ITD with the results presented in Fig 3B. The CCl_4 has two immediately noticeable effects on the ion

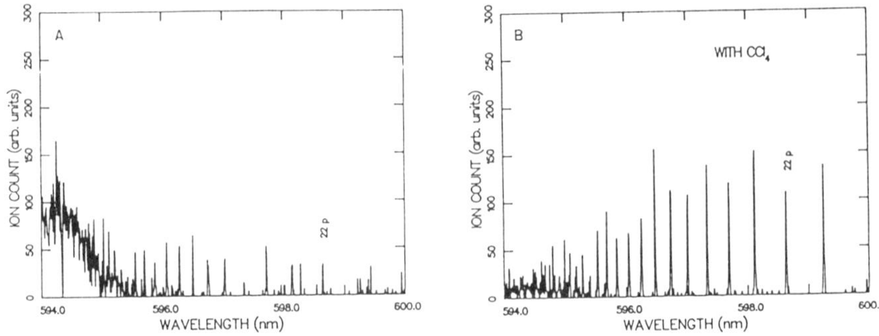

Fig. 3. (A) One-photon excitation spectrum or Rb; (B) One-photon excitation spectrum with CCl_4 added to He buffer gas; m/z 85 and 87. The fundamental wavelength is shown

yield; the np resonances produce more ions and the ion yield at high n is decreased. The increased ionization of the np states was expected due to electron attachment by the CCl_4. The electron attachment rate constant for thermal CCl_4 and Xe(25f) has been measured to be 2.6 x 10^{-7} cm^3s^{-1} (Foltz 1977). The CCl_4 density in the trap is $\approx 10^{12}$ cm^{-3} giving a collision rate of 2.5 x 10^5 s^{-1}. A Rb Rydberg atom will thus have, on average 4 electron attaching collisions with CCl_4 as it transits from the center of the trap to $r_0/2$. The electron attachment rate constant for Rydberg atoms and electron attaching targets is essentially constant for n = 25-100 (Zollars 1985) and thus should be an efficient ionization mechanism for Rydberg states in general. The actual ionization efficiency is not known in the present experiment because of the uncertainty in both the Rb density and the number of Rydberg atoms generated. The drop in ion yield at high n implies that the CCl_4 is effecting the laser generated free electron density. The electron attachment cross section for CCl_4 is $\approx 10^{-14}$ cm^2 for thermal electrons and drops markedly above 0.1 eV (Orient 1989). Even for thermal electrons the mean free path is an order of magnitude larger than the trap dimensions at our CCl_4 density. Therefore it appears that the CCl_4 is affecting the production of electrons rather than scavenging them.

Some preliminary evidence of electric-field ionization is presented in Figs. 4 and 5. This experiment involved only the dye laser fundamental which was scanned from 600 to 594 nm. The laser trigger was delayed an additional 11 ms so as to fire when m/z 83 is ejected from the trap. The data in Figs. 4 and 5 are from the same scan but include ions at m/z 85 and 83 respectively. (The m/z 83 ions actually include all ions ejected promptly from the trap but will be referred to as mass 83 ions.) The m/z 85 ions are trapped while the 83 ions are ejected immediately upon formation. Rydberg atoms that ionize within the trapping volume (within $r_0/2$ of the center) can experience a field as large as 770 V/cm or sufficient to ionize $n^* > 26$. The corresponding two-photon wavelength is $\approx$ 596.5 nm. A noticeable rise in the baseline ion signal is present at wavelengths shorter than this threshold in Fig. 4. The nd resonances at longer wavelength are attributed to 2+1 PI. The prompt ions are not trapped, thus excited Rb atoms can travel to the end cap before ionizing and still be detected. The field strength at the end cap during the laser pulse is 1090 V/cm and can ionize states of $n^* > 24$; two-photon wavelength of $\approx$ 597 nm. The data in Fig. 5 again show a rise in the baseline at $\approx$ 597 nm. More interesting is the appearance of the ns resonances that were not present in the trapped ion spectrum (Fig. 4). The nd transitions are also present in the prompt ion spectrum. Although ions generated by 2+1 PI at

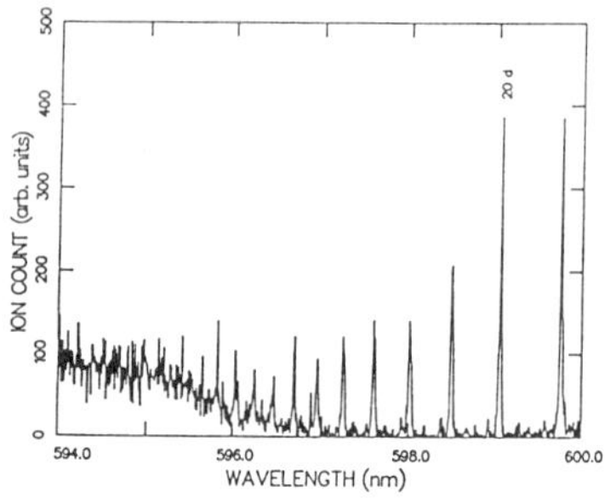

Fig. 4. Two-photon excitation spectrum with laser delayed 11 ms; m/z 85

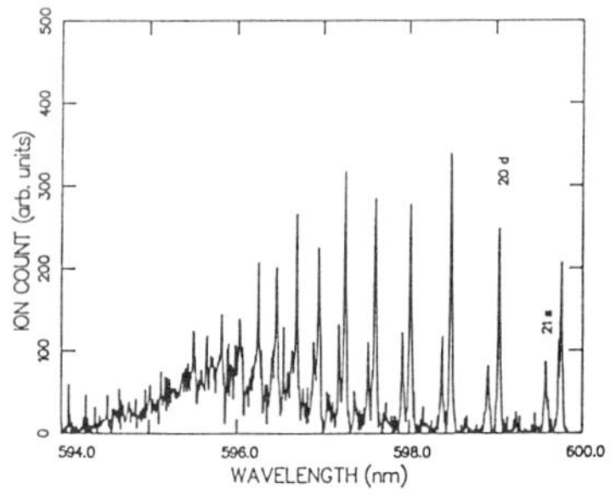

Fig. 5. Two-photon excitation spectrum with laser delayed 11 ms; m/z 83

large r along the laser beam could be rapidly ejected through the detector end cap, this is probably an inefficient process and does not account for the appearance of the ns resonances. A critical field strength of ≈ 3 kV is required to ionize the 19d state. Small imperfections in the electrode surface or the effects of the edges of the ion exit holes might account for the necessary fields to ionize the Rydberg atoms with $n^* > 24$.

REFERENCES

Aymer M, Luc-Koenig E and Combet Farnoux F 1976 *J. Phys. B* **9** 1279

Bekov G I 1983 *Appl. Phys. B* **30** 161

Bushaw B A, Cannon B D, Gerke G K and Whitaker T J 1986 *Opt. Lett.* **11** 422

Bushaw B A 1989 *Prog. Anal. Spectrosc.* **12** 247

Flannery M R 1983 *Rydberg States of Atoms and Molecules* ed R F Stebbings and F B Dunning (London: Cambridge Univ Press) pp 393-472

Foltz G W, Latimer C J, Hildebrandt G F, Kellert F G, Smith K A, West W P, Dunning F B and Stebbings R F 1977 *J. Chem. Phys.* **67** 1352

Goeringer D E, Whitten W B and Ramsey J M 1990 *Int. J. Mass Spectrom. Ion Process.* in press

Hurst G S, Kramer S D, Payne M G and Young J P 1979 *IEEE Trans. Nucl. Sci.* **NS-26** 133

Ivanov L N and Letokhov V S 1975 *Soviet J. Quantum Electron.* **5** 329

Klots C E and Compton R N 1985 *Phys. Rev. A* 525

Lubman D M 1990 *Lasers and Mass Spectrometry* (New York: Oxford University Press)

March R E and Hughes R J 1989 *Quadrupole Storage Mass Spectrometry* Vol. 102, "Chemical Analysis", ed J D Winefordner (New York: Wiley-Interscience)

McCrery D A, Peake D A and Gross M L 1986 *Anal. Chem.* **57** 1181

Noordam L D, de Boer M P and van Linden van den Heuvell H B 1990 *Phys. Rev. A* **41** 6267

Orient O J, Chutjian A, Crompton R W and Cheung B 1989 *Phys. Rev. A* **39** 4494

Smith D H, Young J P and Shaw R W 1989 *Mass Spectrom. Rev.* **8** 345

Stafford G C, Kelley P E and Bradford D C 1983 *Am. Lab.* **June** 51

Stafford G C, Kelley P E, Syka J P, Reynolds W E and Todd J F J 1984 *Int. J. Mass Spectrom. Ion Phys.* **60** 85

Stafford G C, Talyor D M, Bradshaw S C and Syka J E P 1987 *Proc. 35th Ann. Conf. Amer. Soc. Mass Spectrom.* 775

Whitaker T J and Bushaw B A 1981 *Chem. Phys. Lett.* **79** 506

Yost R A, McClennen W and Snyder A P 1987 *Proc. 35th Ann. Conf. Amer. Soc. Mass Spectrom.* pp 789

Young J P, Shaw R W, and Smith D H 1989 *Anal. Chem.* **61** 1271A

Zollars B G, Higgs C, Lu F, Walter C W, Gray L G, Smith K A, Dunning F B and Stebbings R F 1985 *Phys. Rev. A* **32** 3330

Excitation schemes for isomer selective RIS in niobium based dosimetry

H Lauranto*, T Kajava*, T Lönnroth**, R Salomaa* and J Tulkki*

*Helsinki University of Technology, Department of Technical Physics,
Rakentajanaukio 2 C, SF-002150 Espoo, Finland
** Åbo Akademi, Department of Physics,
Porthansgatan 3, SF-20500 Åbo, Finland

ABSTRACT: Several three-step excitation schemes for resonance ionization of ^{93}Nb were investigated to apply RIS to determination of ^{93m}Nb/^{93}Nb isomer ratio for fast neutron fluence measurements. Hyperfine structures of the states $4d^35s5p\ ^6F^\circ_{1/2}$, $4d^35s6s\ ^6F_{1/2}$ $4d^35s5p\ ^4D^\circ_{1/2}$, $4d^35s5p\ ^6D^\circ_{1/2}$ and $4d^46s\ ^6D_{1/2}$ as well as Rydberg spectra and corresponding convergence limits of the schemes were measured. The differences in hyperfine splittings of electronic states of ^{93}Nb and ^{93m}Nb and the isomer shifts were estimated theoretically using self-consistently calculated nuclear and electronic wave functions. The feasibility of the excitation schemes is discussed.

1. INTRODUCTION

The fast neutron radiation in nuclear reactors induces material damage, especially embrittlement of the pressure vessel. For safety and economic reasons there are needs to accurately estimate and predict the accumulated neutron dose and the embrittlement of the pressure vessel.

Neutron dosimetry is based on neutron induced threshold reactions in the active material. Due to restrictions caused by, e.g., half-lives and threshold energies there is only very limited number of suitable dosimeters available, especially in the low energy region (1–4 MeV). The neutron scattering reaction ^{93}Nb(n,n')^{93m}Nb is of special importance in neutron dosimetry due to its low threshold energy ($\approx$ 0.9 MeV), large reaction cross section, and the long half life of the isomer (Bärs and Karnani 1986). Ordinarily the isomer ratio ^{93m}Nb/^{93}Nb, which is typically of the order of 10^{-6}, is determined by measuring the activity of the daughter nucleus, but this approach includes some technical problems, e.g., difficulties in determination of the total mass of the irradiated Nb specimen. Complementary methods for the measurement of the isomer ratio — possibly with higher accuracy — are thus highly desirable.

Because the nuclear spin of the stable isotope (I = 9/2) and that of the isomer (I = 1/2) are very different, the hyperfine splittings (hfs) of the corresponding electronic energy levels of the two atom species considerably differ from each other. The deviations in the charge distributions of the nuclei cause additional shifts of the isomer levels with respect to those of the stable isotope. Due to these spectral changes we have proposed RIS, as a very sensitive and spectrally selective method, for the determination of the niobium isomer ratio (Auterinen *et al* 1989, Lauranto *et al* 1990).

Niobium is a highly refractory element which melts at 2470°C and boils at 4742°C. Its ground state is $4d^45s\ ^6D_{1/2}$. Because there are several electrons outside the closed shells in Nb atom, the spectrum of Nb is very rich: the element has circa one hundred tabulated spectral terms with over 350 fine structure levels (Moore 1971). Its first-ionization potential is 54513.8 cm^{-1} (Rayner *et al* 1987). Thus, niobium atoms can be excited stepwise close to the ionization limit either with one UV and one blue photon or with three visible photons. In this paper we present a study of several excitation schemes of niobium with three resonant steps and subsequent field ionization. Observed hyperfine splittings of the intermediate states of the schemes are reported, and the feasibility of the schemes to determination of the isomer ratio is discussed.

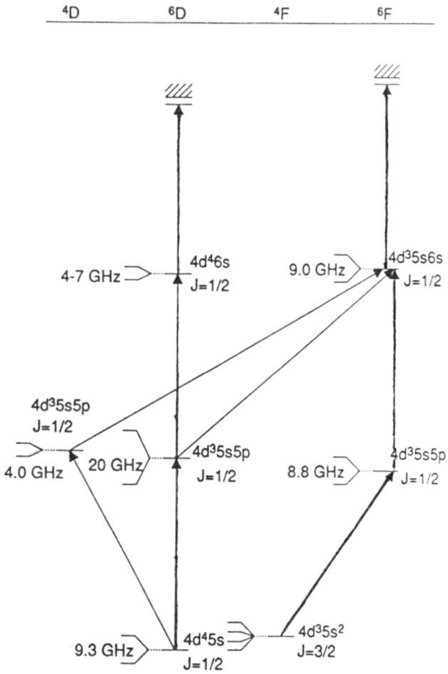

Figure 1. The investigated excitation schemes with measured hyperfine splittings.

2. EXPERIMENTS

The experiments were performed with our RIS apparatus (Bekov *et al* 1987) consisting of three dye lasers pumped by an excimer laser, a vacuum chamber, data collection and control electronics. The laser line widths were 0.15 cm^{-1} and could be narrowed with an intracavity etalon to 0.04 cm^{-1}. The beam diameters were adjusted to ca. 5 mm. The wavelengths were calibrated to neon lines using the optogalvanic signal of a hollow cathode lamp. During the hyperfine measurements the pulse energies (of the order of 10–100 μJ) were adjusted below the measured saturation values.

The atomic beam was generated by electrically heating a graphite crucible containing pieces of niobium wire. Typical evaporation temperatures were 2200°C. Thermal ions were suppressed electrostatically. The atomic beam and the laser beams were crossed in the vacuum chamber between two electrode plates. Following the simultaneous laser pulses, a high voltage pulse with an amplitude of ca. 9 kV and duration of ca. 20 ns was applied to field-ionize the highly excited atoms and to sweep them subsequently towards the ion detector.

The investigated excitation schemes are presented in Figure 1. As the starting level for the first excitation scheme we used the state $4d^35s^2\,{}^4F_{3/2}$ (1142.79 cm^{-1}). Intermediate levels in this scheme were $4d^35s5p\,{}^6F^{\circ}_{1/2}$ (18791.09 cm^{-1}) and $4d^35s6s\,{}^6F_{1/2}$ (37871.30 cm^{-1}). With the two lower steps tuned to exact resonance, the scanning of the wavelength of the third step yielded a clean Rydberg spectrum with a convergence limit at 56871 cm^{-1} corresponding to the state $4d^35s\,{}^5F_1$, which lies 2356.76 cm^{-1} above the ionic ground state. In hyperfine measurements we used the Rydberg state at 56382 cm^{-1} as the uppermost level. The measured hfs of the lower and upper intermediate states were 8.8 and 9.0 GHz, respectively. The Rydberg state showed a splitting of 26 GHz, which was presumably the fine structure of the state. Without intracavity etalon in the first step laser the hfs of the starting level remained unresolved.

In the next two schemes we used the same upper intermediate level as previously, $4d^35s6s\,{}^6F_{1/2}$, and the ground state $4d^45s\,{}^6D_{1/2}$ as the starting level. As the lower intermediate level we used both $4d^35s5p\,{}^4D^{\circ}_{1/2}$ (20107.36 cm^{-1}) and $4d^35s5p\,{}^6D^{\circ}_{1/2}$ (19623.96 cm^{-1}), which showed hfs of 4.0 GHz (Figure 2) and 20 GHz, respectively. As the uppermost level the same Rydberg level as previously was used. The hfs of the ground state was verified to be 9.3 GHz, in agreement with the value reported by

Büttgenbach *et al* (1975).

We also investigated the hfs of the level $4d^46s\ ^6D_{1/2}$ (37410.17 cm^{-1}). This belongs to the other of two tabulated even parity spectral terms above 30000 cm^{-1} and is thus the only alternative to $4d^35s6s\ ^6F_{1/2}$ to be used as the upper intermediate state (when only states with $J = 1/2$ are considered). The ground state was used as the starting level and $4d^35s5p\ ^6D^\circ_{1/2}$ as the lower intermediate level. The Rydberg spectrum arising from the level $4d^46s\ ^6D_{1/2}$ was very tangled and had the convergence limit approximately at 55310 cm^{-1} corresponding to the ionic state $4d^4\ ^5D_3$ (801.38 cm^{-1}). In the hfs measurement the Rydberg level at 54854 cm^{-1} was used. Due to the complicated structure of the Rydberg states the hfs of the upper intermediate level, 4–7 GHz, could be determined only approximately.

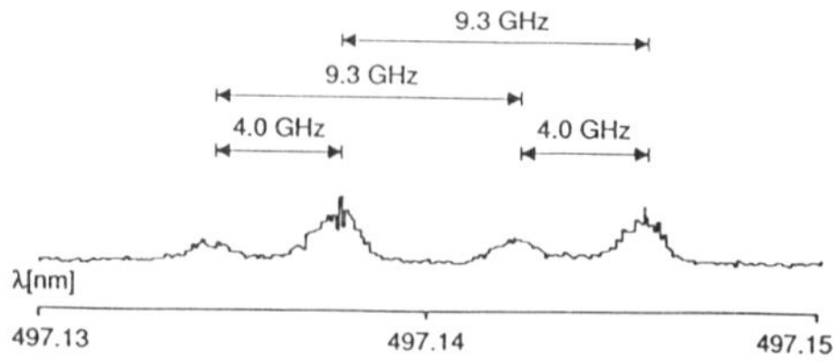

Figure 2. Hyperfine splitting of the transition $4d^45s\ ^6D_{1/2}$—$4d^35s5p\ ^4D^\circ_{1/2}$.

3. THEORETICAL ESTIMATION OF THE ENERGY LEVELS OF THE ISOMER

The differences in the optical spectra of ^{93}Nb and ^{93m}Nb are accounted for by two effects: deviations in the hyperfine splittings due to differing nuclear spins and magnetic moments and the isomer shifts caused by the different charge distribution of the nuclei. The hfs of the fine structure states reads (Armstrong 1971)

$$E_{IJF} = A\bar{I}\cdot\bar{J} + B\frac{3(\bar{I}\cdot\bar{J})^2 + 3\bar{I}\cdot\bar{J}/2 - I(I+1)J(J+1)}{2I(2I-1)J(2J-1)}, \tag{1}$$

where $\bar{I}$ is the nuclear spin, $\bar{J}$ the electronic angular momentum, $\bar{F}$ the total angular momentum of the atom and $\bar{I}\cdot\bar{J} = [F(F+1) - I(I+1) - J(J+1)]/2$ for coupled states. The coefficients A and B depend on the electronic wave function and the nuclear state, A is proportional to the nuclear gyromagnetic ratio μ_I/I and B to the nuclear electric quadrupole moment Q. The isomer shifts between the electronic levels of the two nuclei can be approximated by the formula (King 1984)

$$\Delta E_{\text{iso}} = C\rho_e(0)\delta\langle r^2\rangle, \tag{2}$$

where the coefficient C is a function of the atomic number, $\rho_e(0)$ the electron density in the nucleus and $\delta\langle r^2\rangle$ the difference in mean square radii of the nuclei.

To estimate the differences in the spectra of the two atoms, the electronic wave functions of the states considered, nuclear moments and the mean square radii of the nuclei have to be known. The nuclear parameters were calculated with a code which solves the nuclear wave functions self-consistently by minimizing the total energy functional based on Skyrme interaction (Skyrme 1956) and Bardeen–Cooper–Schrieffer -type pairing correlations. The procedure has been presented in detail by Vautherin (1973). The relativistic electronic wave functions were also solved self-consistently using an atomic multiconfigurational Dirac–Fock package (Grant *et al* 1980). According to the calculations, the ground state of the nucleus is oblately deformed from spherical shape and the isomeric state prolately. Both deformations are very modest so that the calculated isomer shifts in the transitions under consideration are less than 300 MHz. The calculated value for the magnetic moment of the isomer is −0.29 nuclear magnetons. This means that the magnitude of the hfs in the isomeric atom is less than one tenth of the corresponding hfs in the stable atom.

4. SELECTIVITIES OF THE EXCITATION SCHEMES

The selectivity of an excitation scheme can be defined as the ratio of the ionization probability of the isomer to that of the stable isotope when the laser wavelengths are tuned to the isomer transitions. Because the isomer concentration in a typical sample is of the order of 1 ppm, an excitation scheme applicable to isomer selective RIS of niobium is required to have a selectivity clearly greater than 10^6. We have estimated the selectivities of the investigated excitation schemes using rough, lowest order rate equation calculation. Each rate coefficient for the stable isotope is supposed to be the corresponding coefficient of the isomer multiplied by a detuning factor calculated as the convolution of the Gaussian laser line with the Lorentzian absorption line. In the calculations the Rydberg states were assumed to have no hyperfine or any other structure.

The selectivity of the excitation scheme having the starting level $4d^3 5s^2\ {}^4F_{3/2}$ is 10^4, which is insufficient for our purposes. The selectivity is relatively poor because the hfs of intermediate states are almost of equal magnitude and also because one of the six components of the first step multiplet overlaps the isomer transition. The excitation schemes with the ground state as the starting level and $4d^3 5s6s\ {}^6F_{1/2}$ as the upper intermediate level have the selectivities of 10^8 and 10^{11} corresponding to the lower intermediate levels $4d^3 5s5p\ {}^4D^\circ_{1/2}$ and $4d^3 5s5p\ {}^6D^\circ_{1/2}$, respectively. The selectivity of the fourth scheme through the upper intermediate level $4d^4 6s\ {}^6D_{1/2}$ is 10^9.

5. CONCLUSIONS

When using three-step excitation in isomer selective RIS of niobium, sufficient selectivity can be easily achieved, provided that all the levels involved have a simple two component hfs and the hfs of successive levels in the scheme differ from each other. When using pulsed lasers the main difficulty is to achieve a high enough sensitivity to detect the isomer signal from the backgound fluctuations. These are caused by electrical noise and by signal from the stable isotope due to amplified spontaneous emission (ASE) background of the lasers, the influence of which has not been taken into account in the selectivity calculations. A rough order-of-magnitude calculation shows that, if we assume the selectivity to be of the order of 10^9 and shot-to-shot variation in number of ions produced 50 per cent, the isomer signal can be detected from the background fluctuations by using averaging over 10^3 pulses. This implies that, if the errors in the theoretical estimates of the energy levels of the isomer are small compared to the line widths of the lasers, the measurement of a spectrum with a clear isomer peak takes less than one hour.

REFERENCES:

Armstrong L Jr 1971 *Theory of Hyperfine Structure of Free Atoms* (New York: Wiley-Interscience) pp 70–1

Auterinen I, Bärs B, Lauranto H, Nyholm K and Salomaa R 1989 *Inst. Phys. Conf. Ser. No. 94* ed G Hurst and M Payne (Bristol: IOP Publishing) pp 281–4

Bekov G, Radaev V, Likonen J, Zilliacus R, Auterinen I and Lakomaa E-L 1987 *Anal. Chem.* **59** 2472

Büttgenbach S, Dicke R, Gebauer H, Herschel M and Meisel G 1975 Z. Phys. A **275** 193

Bärs B and Karnani H 1986 *Kernenergie* **29** 369

King W H 1984 *Isotope Shifts in Atomic Spectra* (New York: Plenum Press) pp 38–41

Lauranto H M, Auterinen I H, Kajava T T, Nyholm K M and Salomaa R R E 1990 *Appl. Phys. B* **50** 323

Moore C E 1971 *Atomic Energy Levels, NSRDS-NBS* **35**/*Vol. II* (Washington, D.C.: National Bureau of Standards) pp 216–20

Rayner D M, Mitchell S A, Bourne O L and Hackett P A 1987 *J. Opt. Soc. Am.* **4** 900

Skyrme T H R 1956 *Phil. Mag.* **1** 1043

Vautherin D 1973 *Phys. Rev. C* **7** 296

Paper presented at RIS 90, Varese, Italy, 16–21 September 1990

Thorium RIMS for geochronological and geochemical applications

B.L. Fearey, S.G. Johnson, N.S. Nogar, M.T. Murrell, and C.M. Miller
Los Alamos National Laboratory, Los Alamos, New Mexico 87545

ABSTRACT: Resonance Ionization Mass Spectrometry (RIMS) is being developed to measure thorium isotopics from geological samples. These measurements, in conjunction with uranium measurements by thermal ionization, permit geochronological dating in the timescale of 10,000 to 350,000 years.

1. INTRODUCTION

Detailed geochronology is critical to understanding both the mechanisms and frequency of dynamic geologic processes. The geochronology of the last million years takes on a special significance with the need to address questions regarding geologic hazards and the influence of man on his environment. One specific concern is the survivability of long-term storage sites for nuclear waste over geologic time. Unfortunately, there are few techniques for dating events on the million-year timescale.

The measurement of uranium-series disequilibrium is a well-established and valuable approach for the study of geochronology and geochemistry in relatively young geologic materials (Ivanovich and Harmon 1982). The longer-lived members of the series can be used to date events over the last 100 to 1,000,000 years. For example, disequilibrium between ^{238}U and ^{230}Th produced during magma melting and crystallization or from redistribution by geothermal fluids can be used to date samples younger than 350,000 years (Allegre and Condomines 1976, Condomines *et al.* 1982, Ivanovich and Harmon 1982). In addition, the $^{230}Th/^{232}Th$ ratio can be used as a geochemical tracer, in order to follow the evolution of magmatic or aqueous source regions.

Unfortunately, such applications have been limited due to the relatively large sample size requirements and the poor precision obtainable by alpha spectrometry, the most common analytical technique. Emphasis is currently being placed on mass spectrometric techniques (Edwards *et al.* 1987, Edwards 1988, Goldstein *et al.* 1989). The reasons are clear: 1) smaller sample size and purer mineral separates (critical for internal isochron determinations); 2) improved analytical precision (thus more precise ages); and 3) significantly shorter measurement times.

Using thermal ionization mass spectrometry, Goldstein *et al.* (1989) have been able to reduce sample size and improve precision both by about factors of ten compared to alpha counting. These techniques have been applied to the $^{238}U/^{230}Th$ disequilibrium in bulk samples of ocean basalts. However, to apply the internal isochron technique for crystallization ages, the use of mineral separates is required. The uranium and thorium contents of many separated phases are quite low; the current thermal efficiency for thorium is not sufficient. For this reason, we are presently investigating the use of RIMS as a measurement technique for thorium, with the goal of increased ionization efficiency.

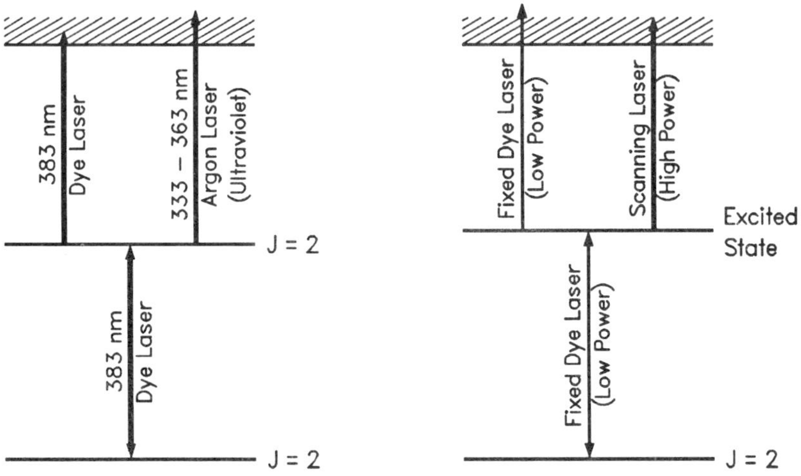

Fig. 1. Schematic diagrams of the resonance ionization schemes for thorium. The diagram to the left is applicable to the cw laser ionization studies; that to the right applies to the pulsed laser spectroscopy studies.

2. EXPERIMENTAL

The applicable energy level diagram for thorium is shown in Figure 1. As it indicates, both pulsed and continuous-wave (cw) laser experiments have been done. For pulsed laser work, a single excimer laser-pumped dye laser was used in combination with a time-of-flight mass spectrometer, producing ions through a single-color (1+1) process. Subsequent pulsed laser experiments have used two pulsed dye lasers, pumped by the same excimer laser. The first, operating at a low pulse energy, was fixed at wavelengths corresponding to known thorium atomic transitions (Palmer and Engleman 1983). The second, operating at high pulse energy, was scanned in wavelength as the ionization signal was recorded. In this fashion, both the onset of ionization (as the combined photon energies exceeded the ionization potential) and autoionization structure could be observed.

Continuous-wave experiments were done to demonstrate a practical resonance ionization method for thorium. The advantages of cw lasers for analysis have been discussed previously (Miller and Nogar 1983a, 1983b). A single broadband dye laser, pumped by an Ar^+ laser, was used to produce efficient ionization. A second, non-resonant Ar^+ laser, operating in the ultraviolet at high power levels, could be added to create additional ion signal (Fearey *et al.* 1989). A magnetic sector mass spectrometer was used for these studies, in order to be able to distinguish between the abundant ^{232}Th and the rare ^{230}Th. Samples consisted of both natural thorium and a thorium standard enriched in ^{230}Th to the $\sim$ 400 ppm level.

3. RESULTS

The pulsed experiments were done initially to find the optimum sample preparation method to produce thorium atoms for resonance ionization. It was found that co-depositing $\sim$ 1 μg Th with colloidal graphite produced a stable, reproducible atom source. All experiments described here were done with samples of this type.

The majority of the pulsed laser experiments were done using the two-color configuration. Numerous autoionization resonances were observed; most could be seen using different resonant levels for the first excitation step. Of particular interest was a level at 52,227 cm^{-1} above the thorium atomic ground state. This corresponds to almost exactly twice the energy of the 383 nm resonant transition, and is in effect doubly resonant. This permits highly efficient ionization to occur in the cw experiments described below.

Experiments were also done at lower total photon energy, scanning the second dye laser across the possible energies corresponding to the ionization potential (Sugar 1974, Hildenbrand and Murad 1974). Using two different resonant intermediates, a value for this potential of 6.31 ± 0.01 eV was obtained. This value is somewhat higher than those reported in the references (6.08 ± 0.12 eV and 6.2 ± 0.2 eV), but does explain why no ionization could be observed in one-color cw experiments using the relatively strong 395 nm and 404 nm thorium resonance lines: two photons at these wavelengths contain insufficient energy to reach the ionization continuum from the ground state.

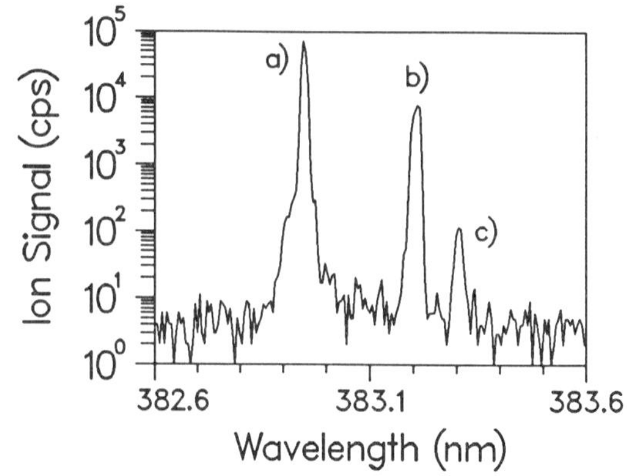

Fig. 2. Thorium ionization spectrum near 383 nm. Transitions are: a) 26113 cm^{-1} J=2 → J=2; b) 26097 cm^{-1} J=2 → J=3; c) 26091 cm^{-1} J=0 → J=1.

Experiments with cw lasers demonstrated that efficient ionization was readily obtainable with only a single dye laser. Using the resonance at 383 nm, ionization signals greater than 10^9 counts per second were seen at a filament temperature of 1850 °C with a 1 μg sample and ~ 50 mW dye laser output. A relatively strong signal was also observed at 393 nm, originating from a thermally-populated excited electronic level. By scanning the dye laser, a large number of transitions could be observed; Figure 2 displays a small section of the spectrum, including the 383 nm transition. Generally, the resonant transitions could be found in the thorium atlas (Palmer and Engleman 1983); transitions from low-lying, thermally-populated excited electronic levels were common.

By varying the power of the dye laser at 383 nm, we observed that the overall ionization process was not saturated at the 50 mW level, in spite of the large ionization signal. When the Ar^+ laser was added, an increase in efficiency was seen. However, the increase was much less than would be expected for the Ar^+ laser power level, but ionization was not yet saturated. With consideration of the pulsed laser autoionization data, we infer that the significant signal with 383 nm alone reflects both the strong bound-bound transition and a large cross section ionization process.

The thorium standard sample, with $^{230}Th/^{232}Th$ at 400 parts per million, was used to measure the fidelity of the RIMS isotope ratio results at 383 nm (Miller *et al.* 1989). As expected, the isotope shift between the two species, 0.393 cm^{-1} (Engleman and Palmer 1983), leads to a bias of about a factor of two at the peak of the ^{232}Th signal, with the exact value depending upon the precise wavelength and bandwidth of the laser. The presence of the autoionizing level at twice the photon energy may also contribute to this bias. No fundamental difficulties with practical isotope ratio measurements are expected because of the bias, however.

4. FUTURE WORK

Practical application of resonance ionization to thorium analysis is near. The demonstrated ionization efficiency of the RIMS process exceeds that presently available with thermal ionization; isotopic bias in the laser-based process occurs at a manageable level. Demonstration of RIMS analyses with a high-abundance sensitivity two-stage magnetic sector mass spectrometer, using the sub-nanogram-sized samples required by the mineral separates studies, should occur in a short time. Spectroscopic investigations concerning the ionization potential of thorium, autoionizing levels, and the wavelength dependence of isotope ratio data will continue as well.

Our preliminary experiments have suggested several improvements that would benefit the thorium analytical problem. First, the use of an external cavity, described elsewhere in this volume (Fearey *et al.* 1990), is expected to significantly increase the ionization efficiency beyond the present level. This is accomplished by constructing an optical cavity around the source region of the mass spectrometer, thus increasing the circulating power. Given our experience with the 383 nm resonance, a cavity operating at this wavelength would be extremely efficient at producing thorium ions. The addition of a cavity using an argon ion laser for the non-resonant step would also increase the efficiency, although at perhaps a lower rate.

Separated samples of ^{230}Th and ^{232}Th are available in macroscopic quantities. This would permit the construction of hollow-cathode resonance lamps containing the separated isotopes, which could be used as frequency standards and laser wavelength locking reference sources. A system is envisioned where each isotope is measured with the laser locked to a specific wavelength. This could be used to provide either fixed measurement biases, or expressed another way, selectivity for minor isotopes. Such a system would be very useful for precise isotope ratio determinations.

References

Allegre C J and Condomines M 1976 *Earth Planet. Sci. Lett.* **28** 395

Condomines M, Tanquy J C, Kieffer G and Allegre C J 1982 *Geochim. Cosmochim. Acta* **46** 1397

Edwards R L, Chen J H and Wasserburg G J 1987 *Earth Planet. Sci. Lett.* **41** 175

Edwards R L 1988 *Earth Planet. Sci. Lett.* **90** 371

Engleman Jr. R and Palmer B A 1983 *J. Opt. Soc. Am.* **73** 694

Fearey B L, Parent D C, Keller R A and Miller C M 1989 *Resonance Ionization Ionization Spectroscopy 1988* eds T B Lucatorto and J E Parks (Bristol: Inst. of Physics) 263

Fearey B L, Johnson S G and Miller C M 1990 *Resonance Ionization Spectroscopy 1990* (submitted, this volume)

Goldstein S J, Murrell M T and Janecky D R 1989 *Earth Planet. Sci. Lett.* **96** 134

Hildenbrand D L and Murad E 1974 *J. Chem. Phys.* **61** 5466

Ivanovich M and Harmon R S, eds 1982 *Uranium-Series Disequilibrium: Applications to Environmental Problems* (Oxford: Oxford Univ. Press) 571

Miller C M and Nogar N S 1983a *Anal. Chem.* **55** 481

Miller C M and Nogar N S 1983b *Anal. Chem.* **55** 1606

Miller C M, Fearey B L, Palmer B A and Nogar N S 1989 *Resonance Ionization Ionization Spectroscopy 1988* eds T B Lucatorto and J E Parks (Bristol: Inst. of Physics) 297

Palmer B A and Engleman Jr. R 1983 *Atlas of the Thorium Spectrum*, Los Alamos National Laboratory Rep. LA-9615

Sugar J 1974 *J. Chem. Phys.* **60** 4103

Inst. Phys. Conf. Ser. No 114: Section 6
Paper presented at RIS 90, Varese, Italy, 16–21 September 1990

LEI spectroscopy in real samples—problems and remedies

O. Axner, M. Lejon, M Norberg, M. Persson

Department of Physics, Chalmers University of Technology, S-412 96 Göteborg, Sweden

Abstract: Laser Enhanced Ionization is a very sensitive technique for trace element analysis, but its performance is severely decreased when measuring samples with high content of easily ionized elements (EIE:s). In this paper we have tried to determine the limits of its usability, and also investigated some methods to overcome the problems with EIE:s.

1. INTRODUCTION. THE PROBLEMS.

Laser Enhanced Ionization in flame has in many papers been shown to be an extremely sensitive and selective method for trace metal analysis in pure samples. Since the method depends on detection of laser-created charges, it is naturally rather sensitive to matrixes with a high content of easily ionized elements (EIE). These elements will ionize thermally in the collision-rich environment in the flame, and a charge-cloud will build up, which will severely affect the charge collection efficiency. This problem can in most cases be overcome by the use of an immersed electrode of the type suggested by Turk (1981). The increased amount of charges in the flame will also increase the DC-current through the flame, and this will give rise to an increase in the noise. This noise will in many cases affect the detection limit. The only way to get rid of this problem is to decrease the amount of EIE:s in the sample.

Another problem is that many elements, and in particular some of the EIE, have low-lying energy levels which will be thermally populated, and if UV light is used, the atom will photoionize from those levels, and thus give rise to a non-resonant signal. Even if the photoionization cross section is rather low, this will give rise to a broad-band background signal if the content of these matrixes is high enough. One example of such an element which exists in high concentrations in realistic samples of both biological and medical nature is Na. In most cases this is a larger problem than the increased noise or the reduced collection efficiency. The Na problem has been treated extensively by Axner (1990a) et al.

Calcium is another element with high abundance in especially biological samples. We have found that Ca gives significant LEI signals in the entire UV region (220 nm – 300 nm). The signal increases monotonically when scanning toward shorter wavelengths (Figure 1). We believe that this broad-band Ca background originates from calcium oxides. The Ca background signals are significantly larger than those from Na in the far UV region (220 nm – 240 nm).

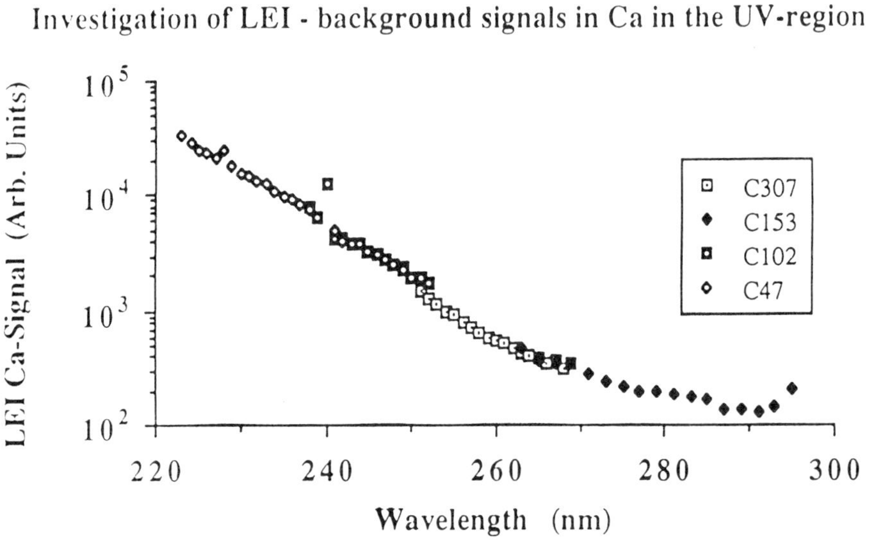

Figure 1: Background signals from Ca in the UV- region

2. POSSIBLE SOLUTIONS.

Some possible solutions to the problems mentioned above will be presented

2.1 Intelligent choice of wavelengths.

For a few elements it is possible to choose a visible excitation wavelength which is not hampered by the abovementioned problems. This is unfortunately not possible for most elements, because they have too high lying ionization limits to be reached with one or two photons from the visible spectrum. The only elements which are possible to analyze with visible light is in principle the easily ionized elements, i.e. the elements from group 1 and 2 in the periodic system.

2.2 Removal of EIE:s by preionization.

In Axner (1990b) it is suggested that spectral interferences from Na can be reduced by the use of a pre-ionization technique, where a significant amount of the Na atoms are ionized by laser light just before the measurement of the analyte is done. The Na atom are ionized by first exciting them by one or two visible steps, and then photoionized with 308 nm light from the excimer laser pumping the dye laser. It was demonstrated that up to 80% of the Na could be ionized by a two-step excitation prior to the measurement of other elements.

2.3 Removal of EIE:s by chemical methods.

In Turk (1990) and Epler (1989) it is suggested to use ion exchange in a chelating resin to remove EIE:s. This has been shown to work rather well, but it requires great care in the choice of materials in the setup, and the used reagents must be of the highest purity, since any contamination in the reagents can accumulate in the ion exchanger, and thus give rise to large errors in the measurements. This method is not so very effective in removing large amounts of Ca, and some other methods which are more effective on this are possible, for example extraction methods. These have however the same problems as the ion exchange methods with contamination and so on.

2.4 Measurement in ultrapure samples

The LEI method is probably most useful in samples with a low content of EIE:s for example: analytical grade solvents, analytical grade acids and other reagents, process liquids from processes requiring extreme purity, for example nuclear power plants or process industries with special demands. Another possibility is measurements on ultrapure natural samples, for example polar ice or rainwater from non coastal areas. The content of Ni in analytical grade ethanol has been measured (Figure 2) with good results. The detection limit in this solution is a few nanograms per liter, and the content in the ethanol is determined to be below 45 nanograms per liter.

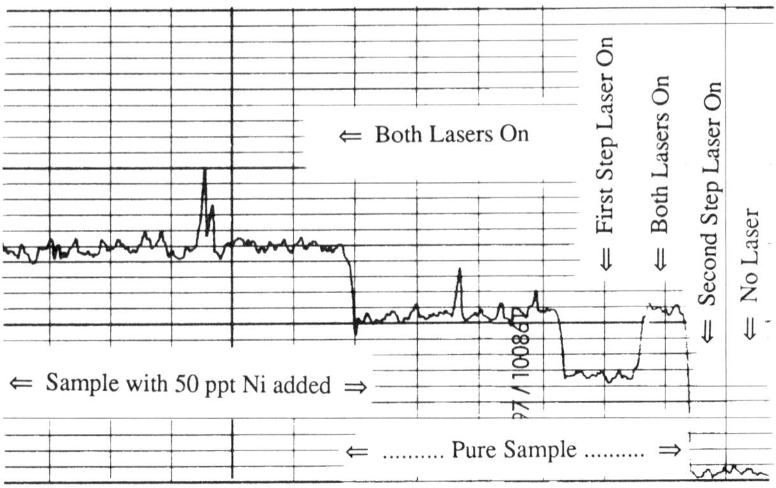

Figure 2: Ni measurement in ethanol. First step wavelength is 229 nm and second step is 547 nm.

REFERENCES

Axner O , Norberg M and Rubinsztein-Dunlop H 1990a to appear in Appl. Spectr. aug 1990

Axner O , Norberg M, Persson M and Rubinsztein-Dunlop H 1990b to appear in Appl. Spectr. aug 1990

Epler K S 1989 Thesis: Liquid Chromatography – Laser Enhanced Ionization Spectrometry: A Synergistic Technique

Turk G C 1981 Anal. Chem. **53**, 1187

Turk G C and Kingston H M 1990 To be published in J. Anal. Atom. Spectr.

Inst. Phys. Conf. Ser. No 114: Section 6
Paper presented at RIS 90, Varese, Italy, 16–21 September 1990

Simultaneous detection of O atoms by resonant three-photon ionization and two-photon laser induced fluorescence

G. SULTAN, G. BARAVIAN and J. JOLLY

Laboratoire de Physique des Gaz et des Plasmas (Associé CNRS), Université de Paris Sud, 91405 Orsay, France

ABSTRACT: A method is developed which allows in situ measurements of absolute number densities of oxygen atoms in discharges and in post-discharges.

1. INTRODUCTION

Since U.V. tunable lasers have been available, several experiments are nowadays devoted to the determination of O atom number density in post-discharges, in plasmas and essentially in flames. The laser-induced fluorescence (LIF) is the main non-intrusive diagnostic technique used principally in flames (Dyer and Crosley, 1989) and more scarcely the MPI (multiphotoionization) technique so-called optogalvanic detection (Goldsmith, 1983)

The LIF technique used alone leads only to relative information about the concentration of atoms. Also a complementary experiment is necessary to reach by calibration the absolute concentration. For example the calibration for two-photon excited fluorescence signals in flames was made by using measurements of known concentrations of atoms generated in a discharge system (Meier et al, 1986) by photolysis of O_2 via the Herzberg continuum absorption (Walkup et al, 1986) or by electron paramagnetic resonance measurements of O and O_2 concentrations downstream from discharge (DiMauro et al,1984). In all cases the complication in the calibration procedure limits the accuracy. In this paper we present a different calibration technique for the determination of O atom density by LIF in a discharge. This calibration uses the detection of the ionization signal in post-discharge near saturation conditions.

2. EXPERIMENT

The arrangement is given in fig.1. and is almost the same that used for the MPI of atomic hydrogen (Baravian et al, 1989). Briefly, the laser used was a Quantel Datachrom-5000 model. The emitted UV photons around 226 nm were generated by mixing the residual 1064 nm radiation from the Nd:YAG laser and the doubling of the dye laser output around 574 nm generated with a mixture of rhodamines R590 and R610. The laser provided radiation pulses of 2.5 mJ of maximum energy and 5 ns duration (FWHM). Typical conditions of discharge were a flow of 0.1 l/mn (NTP) and a pressure of about 1 Torr. The current intensity in the discharge was in the range 5-50 mA. In the first part of the experiment the O atoms were created in a dc discharge between two electrodes and were carried along by the gas-flow and detected downstream in a steel vessel where the laser beam was focused. The O^+ ions resulting from the interaction with the laser were collected by a dc electric field. Moreover the fluorescence at 844 nm was simultaneously observed. The resulting signals (photoions and fluorescence) were then analyzed using a 9400 Lecroy digital storage oscilloscope and a 4420 EG§G boxcar averager. In the second part of the experiment the discharge was created in the interaction volume and the laser beam was focused directly into the plasma.

3. RESULTS AND DISCUSSION

The laser interaction with O atoms is represented in fig.2. The first step is the two-photon excitation of the ground level which can be followed by REMPI with 1 photon more or LIF at 844 nm or quenching and LIF at 777 nm.

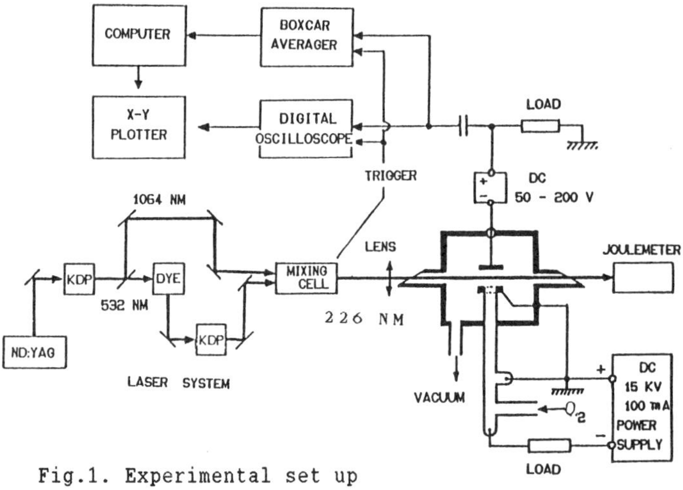

Fig.1. Experimental set up

(i) Post-discharge

When the laser beam is tuned in the range 224-227 nm, the $3p\ ^3P$ level is populated by a two-photon process and the ionization takes place with a third photon. A typical resonance enhanced ionization profile is shown in fig.3a. There are three peaks due to the fine structure splitting of the $2p\ ^3P$ state. The curve in fig. 3b shows the LIF signal in the post-discharge.

(ii) Discharge

The REMPI technique is not directly transposable in discharge because there are much more ions in the initial plasma than produced by MPI. Therefore the LIF technique was chosed exclusively in this case. The fig.3c shows the results.

The study of the ion yield versus the energy of the laser beam shows that the slope in log-log coordinates (fig.4) is equal to 2.6 for low laser energies and to 1 for higher energies. These values are similar to that given by Bischel et al (1982) equal to 2.6 and and to that obtained by Walkup et al (1986) for highest energies (slope equal to 1); Aldén et al (1982) find a slope equal to 2. The change of the slope at about 1 mJ corresponds to the depletion effect, roughly speaking we can say that the whole focal volume is depleted for this energy value and all the O atoms contained before interaction in this volume are ionized or tranferred by LIF or by two-photon excitation followed by collisional desexcitation (quenching) to an other excited level.

(iii) Number of ions

The circuit detection for ions is the same that used one year ago (Baravian et al,1989). At the depletion value of 1 mJ of the laser energy the ionization of the level $^3P(J=2)$ leads to $2.5\ 10^8$ ions. The total number of ions is obtained by measuring the ions issued from the three levels of the ground state. We find $3.6\ 10^8$.

(iv) Number of photons

The population of the resonant level n_2 can be written as follows

$$dn_2/dt = \sigma_1\ \Phi^2\ n_1\ - \sigma_2\Phi n_2 - (1/\tau + k\ [O2])\ n_2 \qquad (1)$$

where σ_1 is the cross-section for the two-photon excitation of the ground state whose population is n_1, σ_2 is the cross-section for the ionization of the excited state ,Φ is the laser flux, τ is the radiative lifetime of the ^3P level, k is the quenching term of this last level with the molecules of oxygen whose the population density is represented by [O2].

(v) Absolute density of O atoms in the focus point in the post-discharge

For a pressure equal to one Torr we obtain $1/\tau + k\ [O2] = 5.4\ 10^7\ s^{-1}$, and the two terms (radiative and quenching) are almost identical. The ionization term can be calculated for a laser energy equal to 1mJ which corresponds to the saturation conditions with a focal length = 36 cm for 226 nm. For the divergence of the laser beam ϑ equal to 10^{-3} rd, Φ is equal to $2\ 10^8\ W/cm^2$ and $\sigma_2\Phi = 10.2\ \ 10^7\ s^{-1}$; the number of ions is about two times larger than the desexcitation terms. With the experimental values the relative importances for the three processes above mentioned are respectively in the proportion 10.2 : 2.8 :2.6. As the interaction volume is fully depopulated of O atoms in the ground state, these atoms are transformed either in ions (65%), or give LIF at 844 nm. (18%), or transfer their excitation by collision (17%), then the total number of O atoms in the ground state which were present in the interaction volume before the interaction is equal to $15.6/10.2\ N = 5.5\ 10^8$. In a first approximation we can consider that these atoms were present in a spherical volume v whose radius is equal to $f\vartheta/2$ and then $v = 2.44\ 10^{-5}\ cm^3$. The O atom density at the focus point located 40 cm downstream from the discharge is then equal to $2.25\ 10^{13}\ cm^{-3}$.

(vi) Absolute density of O atoms in the focus point into the discharge

The previous determination allows the calibration of the setup for the LIF. The respective numbers of photoions and photons at 844 nm in the post-discharge are in the proportion 65% and 18%, i.e when 100 photoions are produced 28 photons are emitted by LIF at 844 nm.The obtained LIF signal for the strongest transition is 20 mV which corresponds to 800 mV for the ion signal and a neutral atom density equal to $2.25\ 10^{13}$. When the laser beam is focused into the discharge it is not yet possible to measure the number of photoions but only the fluorescence signal which is found equal to about 80 mV at the same laser energy. The equation governing the population of the excited level is the same that equation (1) plus a stationary term for n_2 due to the plasma which leads to a continuous signal for the 844 nm transition. We can then conclude that the O atom concentration into the discharge at a distance from the cathode equal to about 1 cm is $9\ 10^{13}\ cm^{-3}$. The ratio O/O2 is equal to $6.4\ 10^{-4}$ in the post-discharge and to $2.5\ 10^{-3}$ in the discharge.The values here given are obviously subject to incertainties due to the facts that firstly the interaction volume is not accurately defined and secondly it is necessary to take into account the spatio-temporal distribution of the laser flux.

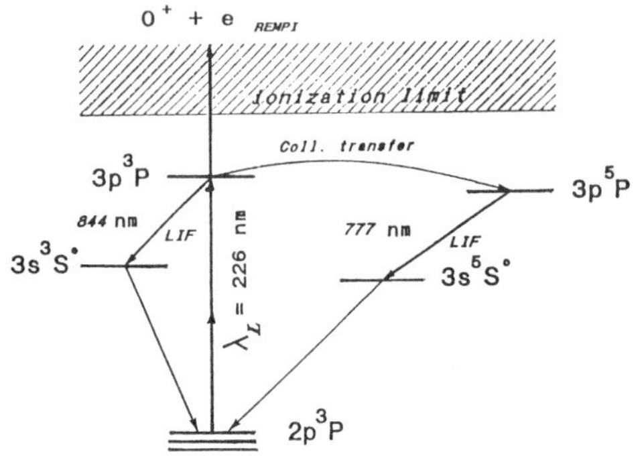

Fig. 2 - Interaction diagram

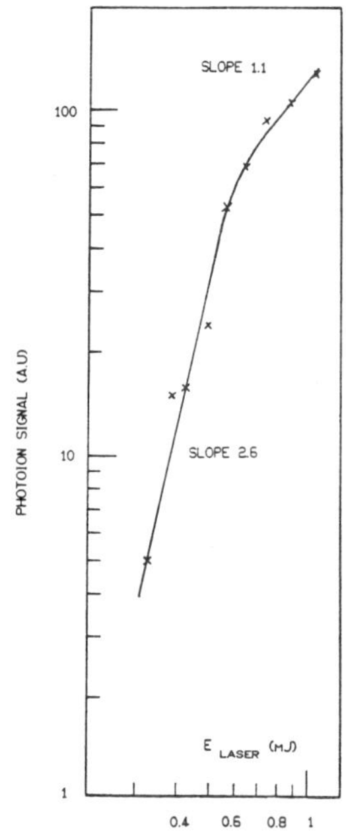

Fig.4- Ion yield

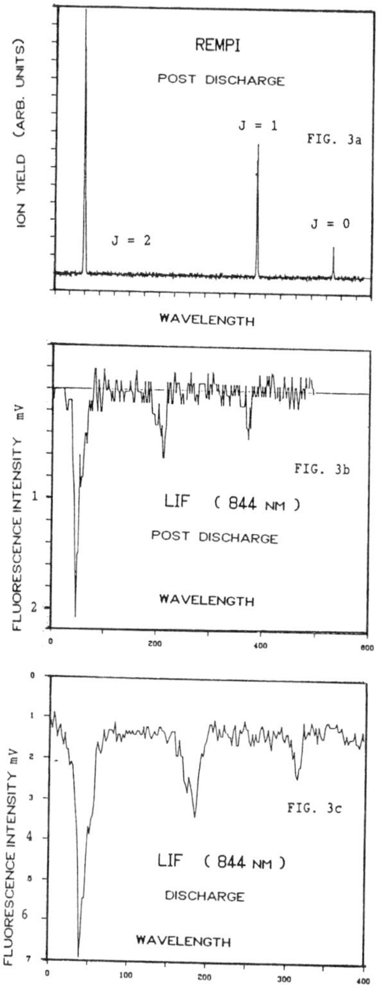

Fig.3- REMPI and LIF

REFERENCES

Aldén M, Edner H., Grafström P. and Svanberg S. 1982 Opt. Commun. 42 244
Baravian G.,Jolly J.,Persuy P.and Sultan G. 1989 Chem. Phys. Let. 159 361
Bischel W. K., Perry B. E.and Crosley D. R. 1982 Appl. Opt. 21 1419
DiMauro L. F., Gottscho R. A. and Miller T. A. 1984 J.A.P. 56 2007
Dyer M. J. and Crosley D. R. 1989 Opt. Letters 14 12
Goldsmith J.E.M. 1983 J. Chem. Phys. 78 1610
Meier U., Kohse-Höinghaus K. and Just T. 1986 Chem. Phys. Let. 126 567
Walkup R.E., Saenger K.L. and Selwyn G.S.1986 J.Chem.Phys. 84 2668

The progress of the applied study of ultra-trace detection by RIS–TOFMS in noble metal ores prospecting

D. Y. Chen, K. L. Wen, X. Y. Xu, W. Y. Ma, Q. Hui, W. Z. Zao
Laboratory of Applied Research of Laser Single Atom Detection, Department of Physics, Tsinghua University, Beijing, China

ABSTRACT: The research work we are pursuing in geological prospecting is an applied ultra-trace analysis technique -- a combination of Resonance Ionization Spectroscopy and Time of Flight Mass Spectroscopy (RIS-TOFMS). It has an important significance to science and a practical economical value when applied to the direct prospecting of concealed orebeds of gold, platinum and other rare and noble metals. Research work has already made important progress.

Here is an important contribution of RIS technology to the national economy. One of the important trends and demands in the science and technology of geological prospecting is the acceleration of the prospecting of noble metal ores and petroleum gas resources in the direction of macro-prospecting by micro-geology. The noble metals ores are gradually becoming impoverished, depleted, and "blindfolded," making the degree of difficulty in prospecting greater and greater day by day. In addition it is expensive. There is therefore an urgent need to develop new and effective methods of detection which have extremely high sensitivity and selectivity, in combination with corresponding chemical detection methods and theory of the formation of ores.

Methods of geological primary halos, single-ore body construction, inclusion explosion, isotope ratio, water chemistry, etc., all necessarily demand the development of the technology of trace quantity and micro-regional measurement methods so as to combine the theory of formation of ores with the technique of prospecting, to combine field work with laboratory work, and to combine the macro with the micro. The characteristics of RIS-TOFMS satisfy to a considerable degree the above mentioned requirements of geological prospecting, thus it facilitates the combination of the different aspects and makes RIS mature rapidly. It becomes practical to the process of directly serving the national economical construction. The reasons why RIS-TOFMS can be applied to geological science and prospecting work are:

(1) This technology can afford the above mentioned ultrasensitive detection method and can furnish striking improvement to the efficiency and quality of geological prospecting, especially that of a general survey;

(2) RIS-TOFMS technology, with its Z and A two-dimensional, super-high selectivity and high resistance to interference, makes possible the precise determination of elemental microquantities and surmounts the peculiar

serious matrix interference effect; thus improving the working efficiency of the determination of isotope ratios by hundredfolds and leading to an enormous reduction in cost;

(3) RIS does not require complicated chemical pretreatment of the specimen before it is subjected to direct on-the-scene determination, and the specimens may be large or small (order of mg); their state may be solid, liquid, gas, or in suspension, thus greatly relieving us from the difficult problems of treating specimens and the different states encountered in geological prospecting;

(4) It is not limited, in principle, by the kind of element to be determined, therefore it can be used to solve the difficulty of the determination of the micro-content of "problem elements," such as Au, Os, Ru, Rh, Ir, Pt, Sm, As, etc.;

(5) It can give out the statistical average value of bulk specimens, and also the distributed quantities in microstructures of ores.

The above mentioned characteristics possessed by RIS in geological prospecting may bring great changes to geological prospecting and an enormous economic benefit. For instance, of all the existing microanalysis methods, the problem of determining gold content is very difficult, but it can be solved by this method because:

* the abundance of gold in the upper lithosphere and soil is about 5ppb (5×10^{-9}), i.e. the average gold content contained in 10 mg of rock sample is about 1.526×10^{11} atoms;

* the average content of gold in natural freshwater is about 0.03 ppb (3×10^{-11}), i.e. there are on the average 4.58×10^{8} atoms in 50 μl of freshwater;

* the average content of gold in seawater is 0.012 ppb (1.2×10^{-11}), i.e. there are on the average 1.83×10^{8} gold atoms in 50 μl of seawater.

It is seen from the average gold content data of the above listed specimens that the gold content in samples of 10 mg of rock, soil, freshwater and seawater can be directly, quickly, and precisely determined. This is, no doubt, of important scientific value and economical significance because:

(1) To a single atom, an ore crystal constitutes a "big universe." This technology takes a view from the angle of atomic structure, so it can change the appearance of mineralogy and even that of geology greatly. Of course, the crux of the matter lies in linking it closely with the geological sciences. Its significance cannot be over-estimated.

(2) It is generally considered that gold itself forms the best indication element in prospecting, i.e. "To prospect for gold by gold." The difficulty lies in low sensitivity. Now RIS-TOFMS possesses extrahigh sensitivity, so the prospecting of gold ore can be done by directly determining the gold content and distribution in single ore grain. This is by itself a breakthrough. To look for gold ores which are deeply concealed or terrain covered with transported overburdens with chemical prospecting has serious limitations as the efficiency is low. There is an extremely small amount of gas indicated in such layers, which can be precisely measured with RIS-TOFMS that sieves and delimits abnormal zones and then discovers gold ores.

(3) Places that are very sparsely populated or that are not easily accessible may be zones for good prospecting. In such regions there is a need for remote

prospecting to collect specimens. Determination of elements in water suspension with RIS-TOFMS is a very effective technical method. This method can greatly increase the area of coverage. It is extremely difficult to make this determination by previously existing methods. The microcontent of elements like Au, Sn, etc. in water (especially in suspension particles) are simply impossible. But RIS-TOFMS can measure their content directly, needing only a few scores of microliters of a water sample. This is very significant because it enables us to precisely determine the gold content from flowing water collected from streams that may be 10 km away from the target area. This is in fact remote sensing of a certain region.

(4) Prospecting with water chemistry is now considered an effective means of prospecting, for instance, in the region of an alluvial skirt which may extend to a belt 50-60 km wide. Chemical prospecting methods are used to collect samples of water seeped into ground from both the upper and lower ends of the zone to see if there is any abnormality. But during the past it could not pass the test due to insufficent sensitivity (especially on Au, Sn, W, and rare-earth elements). The problem can be solved with RIS-TOFMS sensitivity. It has changed the appearance of water chemistry.

(5) The determination of isotope ratio has become more and more important in modern geological sciences. It can be quickly and conveniently done with this two-dimensional spectroscopy. In addition, it affords the isotope ratio measurement in microregions. This is another breakthrough, having great significance to prospecting and the theory of formation of ores.

(6) RIS-TOFMS can be used to measure microquantities of samples without requiring inconvenient chemical or physical pretreatments. This does not only quicken and simplify the process, but also has practical and academic significance. More important is the fact that under and after long and tedious pretreatment processes, the original structure of the ore must be unrestorably changed. They can give only average values, so it does not reflect the real structure. However, RIS-TOFMS can overcome these deficiencies.

STATE OF PROGRESS OF OUR LABORATORY

We have installed four sets of experimental research equipment, i.e. one set consisting of a sputtering atomization resonance ionization time-of-flight mass spectrometer (SARI-TOFMS) and three sets consisting of electro-thermal atomization resonance ionization time-of-flight mass spectrometers (ETRI-TOFMS). One of these is a large one with quick and continuous automatic sample-changing for research and production, and includes an advanced computerized data acquisition and processing system. It also includes a high power, high precision pulsed laser system composed of a Lambda Physik high power excimer laser (West Germany model EMG202MSC) which is used to pump two sets of dye lasers (Lambda Physik model FL3002E and FL3002EC). Other experimental details and the spectroscopy employed are described elsewhere in these proceedings by Hui *et al.*

After having accomplished the tasks of designing and fabricating the instruments and equipment and having completed the theoretical preparatory work, our laboratory has now entered the stage of methodical experimental and applied developmental research work and has already made important progress, such as:

(1) For the first time we successfully made direct microquantity geological sample determinations of ultra-trace gold content by autoionization resonant spectroscopy. A direct detection limit of <0.1 ppb (10^{-10}) was demonstrated,

which means a total detection limit of <0.1 ppt (10^{-13}) was made after chemical pretreatment;

(2) We have found some new autoionizing resonant states of Au, and Ru with high oscillator strengths and with very sharp spectrum lines which have not been reported yet;

(3) We have successfully demonstrated field ionization from highly excited Rydberg states of Au, Ru, Pt atoms, with principal quantum numbers, $n > 70$;

(4) Measurements have been made of the isotopic ratios of the platinum family elements Pt and Ru.

CONCLUDING REMARKS

As for the outlook of its development, we are optimistic that as the results of our research mature, its practical development has a bright future. It will:

(1) Develop new instrumentation and equipment for prospecting and surveying underground concealed ores of rare noble and rare-earth metals;

(2) Develop effective shipborne instruments and equipment for prospecting undersea mineral deposits;

(3) Develop hightech research methods in materials science (including superconducting science), thin film science, biology, medicine, etc.;

(4) Develop effective research methods of autoionizing Rydberg resonance spectroscopy, high-Rydberg atoms molecules and high-excited chemical reaction dynamics.

We shall endeavor to the best of our efforts to realize the above goals and are looking forward to the day when the above objectives are fulfilled!

Paper presented at RIS 90, Varese, Italy, 16–21 September 1990

Laser spectroscopy applied to energy, environmental and medical research

Sune Svanberg

Department of Physics, Lund Institute of Technology, S-221 00 Lund, Sweden

ABSTRACT: The rapid development of coherent optical sources and laser spectroscopic techniques is having a major impact on basic atomic and molecular spectroscopy but also on applied fields related to energy, environment and medicine. A survey of applied laser spectroscopy in the fields of combustion diagnostics, remote sensing of the environment, and cancer and atherosclerotic plaque diagnostics is given with illustrations from work performed at the Lund Institute of Technology.

1. INTRODUCTION

Laser spectroscopic techniques are finding important applications in many applied sciences. Energy, environment and medicine are all fields of major relevance, and analytical and diagnostic techniques based on lasers are increasingly being used for improved characterization of many processes. The success of applied laser spectroscopy is related to strong signals, high sensitivity, near real-time data production and the non-intrusive nature of most of the techniques. The field is treated in monographs such as those by Radziemski *et al* (1987) and Svanberg (1990a). Shorter reviews have previously been made by Svanberg (1988a,b). In the present paper the topic is revisited with illustrations chosen from recent research work at the Lund Institute of Technology.

2. COMBUSTION DIAGNOSTICS

A large number of techniques has been developed for non-intrusive measurements on aggressive media such as burning or exploding gases in combustion processes. The goal is to use spectroscopic techniques together with high-temperature kinetics and turbulence theory to provide greater insight in the processes, thus enabling better energy conversion and pollution management. Imaging of the natural emission, or laser-induced fluorescence (LIF), can be used for mapping the spatial distribution of flame radicals. We have developed a new technique of multi-colour imaging as illustrated in Fig. 1. Four individually filtered and electronically intensified images of the object are placed in the four quadrants of a ccd detector, or are placed adjacent to each other as shown for the case of a Bunsen burner flame in Fig. 2. Selecting the spectral bandpasses properly and using image processing the technique can be used for simultaneous multi-species imaging or for detecting the radical emission only, also in the presence of a heavy continous background due to soot particles (Andersson-Engels *et al* 1990a). By using spectral coincidences several species can be actively imaged using LIF. Absorption measurements provide line-of-sight data that can be inverted to 3D images by tomographic techniques. An absorption recording of water vapour (1 % abs.) using two-tone frequency modulation (FM) spectroscopy with a diode laser at 789 nm is shown in Fig. 3 (Kauranen *et al* 1989).

Degenerate four-wave mixing (DFWM) is a new technique in combustion diagnostics, supplementing Coherent Anti-Stokes Raman Scattering (CARS). Recordings of DFWM signals from a flame seeded with a 2000 or 20 ppm sodium solution are shown in Fig. 4 (Bengtsson *et al* 1989).

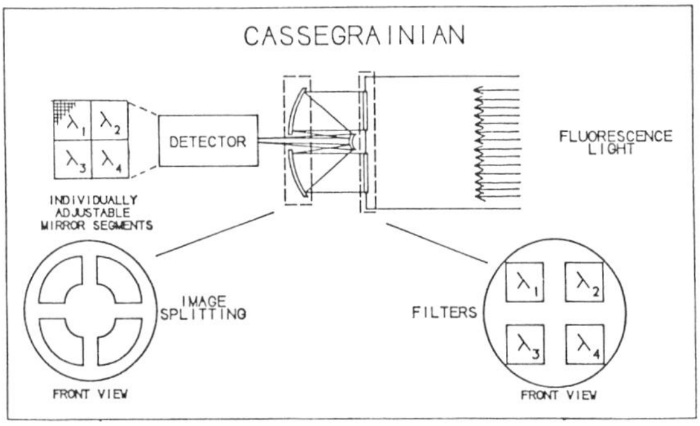

Fig. 1. Set-up for multi-colour imaging

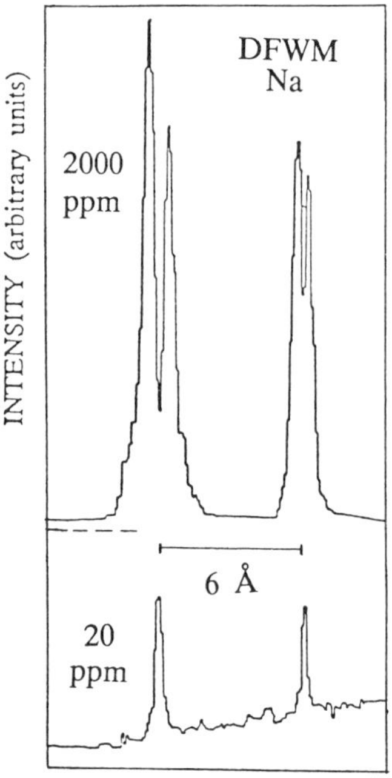

Fig. 4. DFWM spectroscopy of sodium-seeded flame

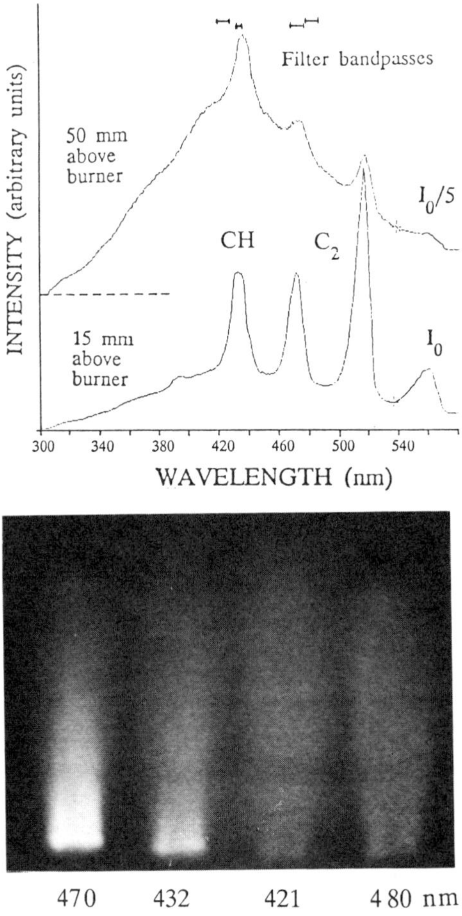

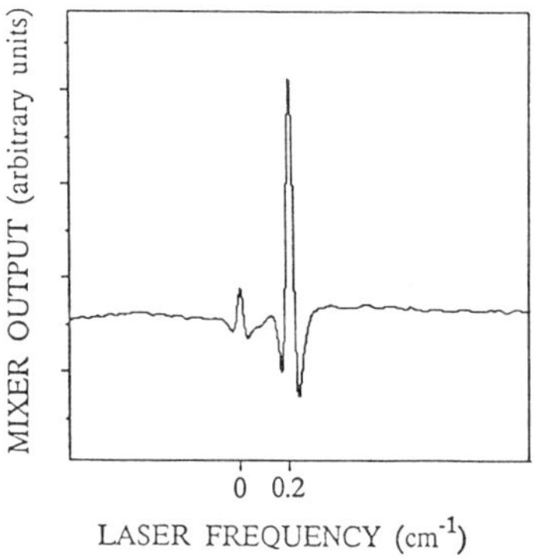

Fig. 3. Two-tone FM diode laser spectroscopy of H_2O

Fig. 2. (left) Multi-colour imaging of flame radical emission

Different techniques for imaging measurements of gas flow are also being developed. One technique relies on the observation of LIF using a narrow-band laser, tuned to the steep slope of a molecular absorption line (Westblom and Svanberg 1985). We are now considering a simpler, passive technique using gas filter correlation as illustrated in Fig. 5. We have verified the technique (that should be particularly useful for fast, hot flows) in measurements on a thermal atomic sodium beam, observed through a sodium absorption cell (Larsson *et al* 1988).

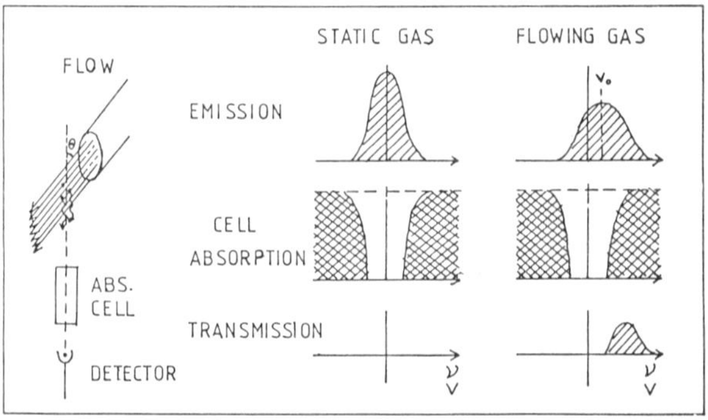

Fig. 5. Principle of gas correlation flow imaging using thermal emission and an optically dense absorption cell

3. ENVIRONMENTAL REMOTE SENSING

Environmental pollution is of increasing concern and laser techniques provide remote sensing capabilities for atmosphere, land and sea. Gases are being probed by long-path absorption and differential absorption lidar techniques. Recently, special attention has been given to stratospheric as well as tropospheric ozone. Geophysical studies using atomic mercury as a tracer gas have also become possible. Different measurement scenarios are shown in Fig. 6. LIF methods provide means for studying marine oil slicks, algal blooms and the state of stressed land vegetation. An illustration of the latter aspect is given in Figs. 7 and 8. Our fluorescence lidar system is shown and a remote recording (50 m) of maple leaves, excited with about 25 mJ of frequency tripled Nd:YAG radiation (355 nm) is given, featuring chlorophyll-*a* fluorescence peaks at 690 and 735 nm (Edner *et al* 1990). A review of environmental remote sensing with active optical techniques has recently been completed (Svanberg 1990b).

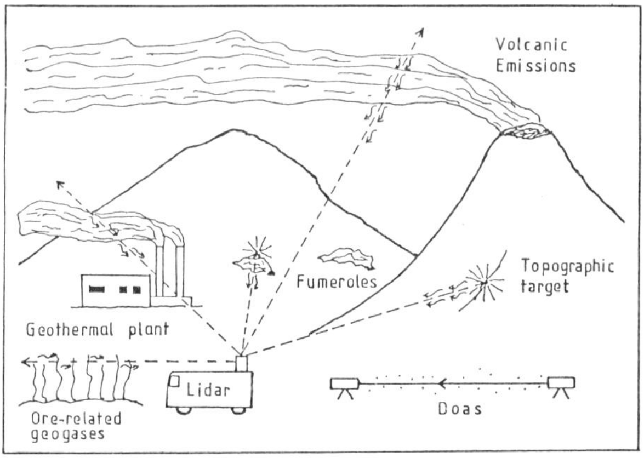

Fig. 6. Scenarios for detection of atomic Hg of geophysical origin using lidar and differential optical absorption spectroscopy (doas) techniques

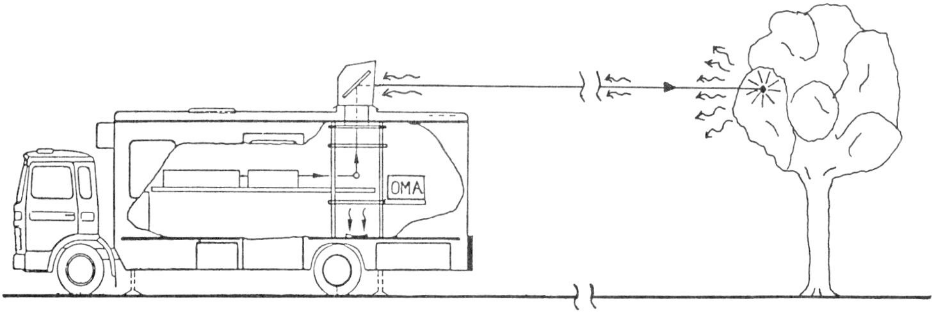

Fig. 7. Fluorescence lidar monitoring of vegetation

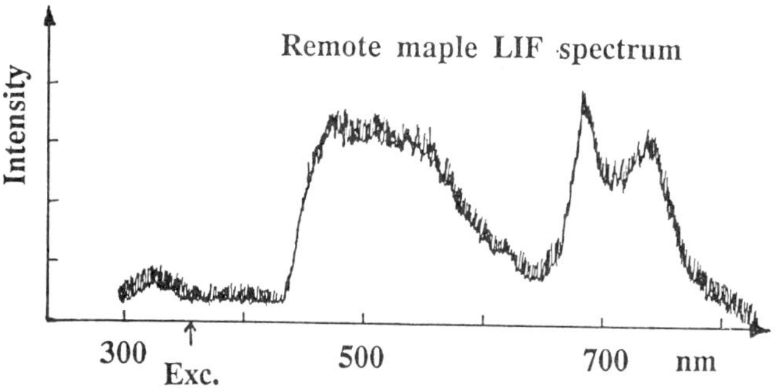

Fig. 8. Recording of vegetation fluorescence at 50 m distance

4. MEDICAL DIAGNOSTICS AND TREATMENT

In the medical area laser methods are being used for diagnostics as well as photochemical treatment of cancer and circulatory system diseases. For recent reviews of our work in medical laser spectroscopy we refer to Svanberg (1989) and Andersson-Engels *et al* (1989). LIF from injected tumour-seeking agents or from natural chromophores can be employed to detect and demarcate the extent of the diseased area. Point-monitoring as well as imaging instrumentation are being developed for clinical use. Using a set-up of the kind shown in Fig. 1 together with object illumination with a pulsed nitrogen laser, multi-colour fluorescence imaging of cancer tumours can be performed. A tumour is characterized by an increase of red fluorescence at 630 nm, due to injected hematoporphyrin derivative (HPD), and a decrease in the blue (470 nm), endogenous tissue fluorescence. By also imaging the fluorescence at 600 nm and subtracting it the HPD-signal can be recorded background-free. In Fig. 9 a processed image of a rat tumour is shown employing the dimensionless contrast function [I(630nm)-I(600 nm)]/I(470 nm). Fully processed images in false-colour coding can now be obtained at an update rate of 5 Hz. Thus, "the TV that only sees cancer" has largely been achieved (Andersson-Engels *et al* 1990b).

Time resolved spectroscopy provides further possibilities for discriminating tumours, or vessel atherosclerotic plaque, from surrounding healthy tissue. Atherosclerotic plaque exhibits a longer fluorescence decay time

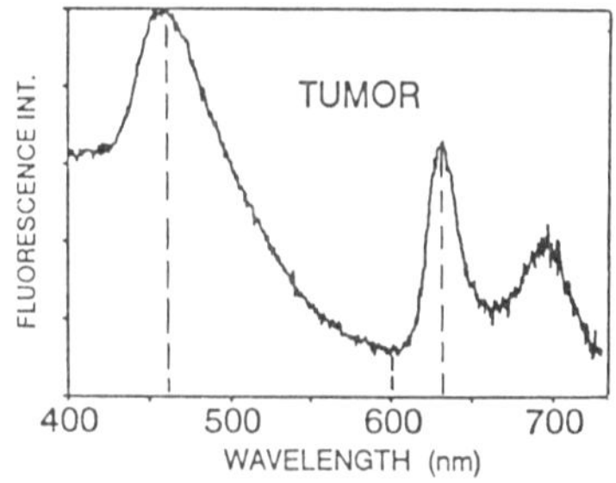

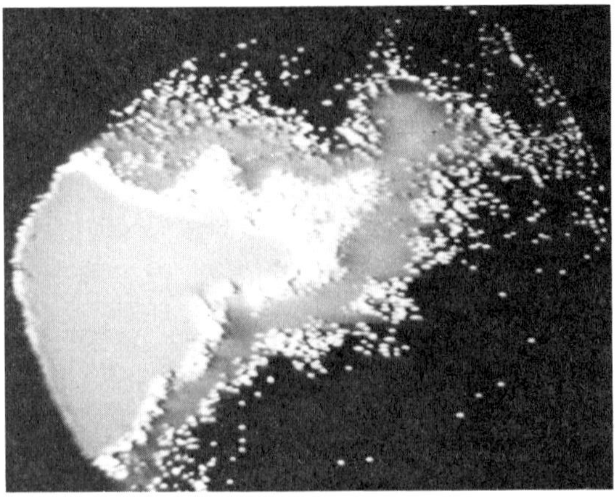

Fig. 9. Tumour spectral signature and contrast-enhanced image of tumour

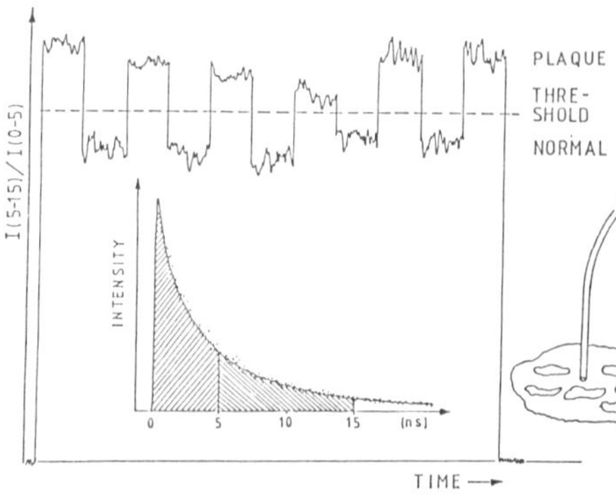

Fig. 10. Plaque detection using time-resolved laser-induced fluorescence

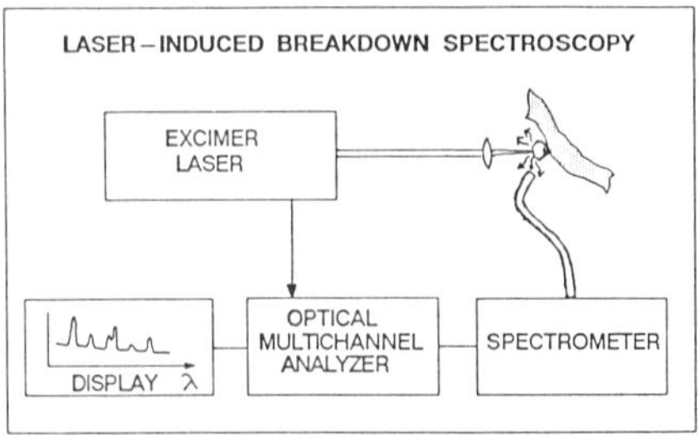

Fig. 11. Set-up for monitoring of plaque plasma emission

after pulsed UV excitation than does normal artery wall, as illustrated in Fig. 10 (Andersson-Engels *et al* 1989). By dividing "late" fluorescence by "early" fluorescence a signal can be obtained to spectroscopically steer laser ablation of plaque by high-power excimer radiation delivered through the same transluminal optical fibre. Spectroscopic guidance of laser angioplasty can also be provided by observing the plasma emission from the ablation plume as illustrated in Fig. 11. When a calcified plaque has been penetrated the otherwise strong Ca and Ca^+ plasma lines disappear.

The topic of light penetration, absorption and scattering in tissue is a very important one, both in view of developing optical dosimetry for photodynamical therapy (PDT) using HPD and laser light, and for developing improved optical transillumination (diaphanography). By using time gated detection of picosecond laser pulses and only observing the photons first emerging from the tissue, image blurring due to strong multiple scattering can be reduced. This is illustrated in Figs. 12 and 13 for the case of a human hand. By restricting the observation to photons arriving "early" the bones give rise to much more pronounced shadows (Anderson-Engels *et al* 1990c). Presently, different groups are pursuing also other techniques including heterodyne detection, holography and phase-shift recording, giving new prospects for, e.g., mammography using non-ionizing radiation.

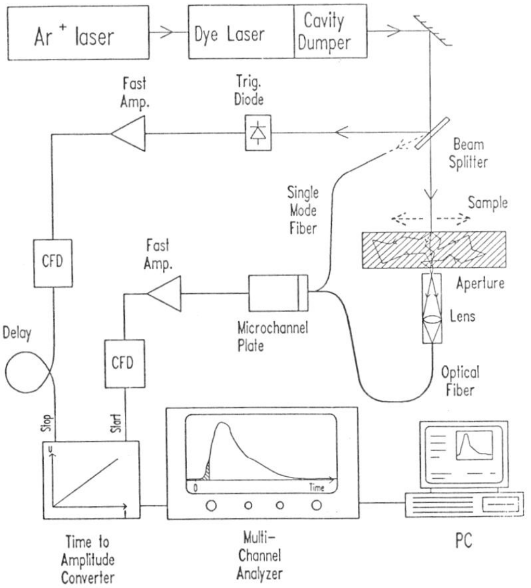

Fig. 12. Set-up for time-resolved tissue transillumination

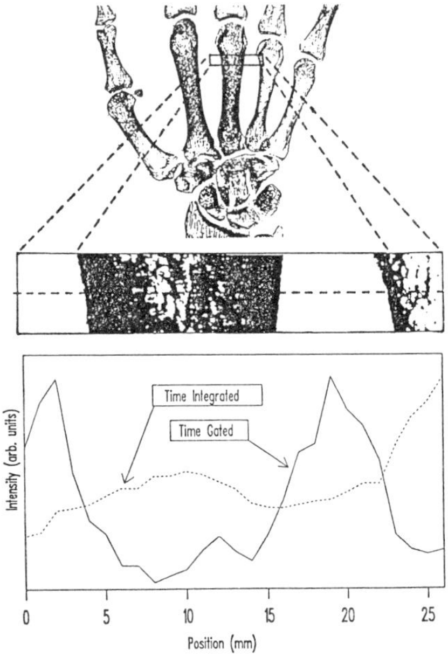

Fig. 13. Enhanced contrast for bones obtained by gated viewing

ACKNOWLEDGEMENTS

The author gratefully acknowledges a most stimulating cooperation with a large number of collegues and graduate students. This work was supported by the Swedish funding agencies STU, DFR, NFR, FRN, RMC, MFR, and the Wallenberg foundation

REFERENCES

Andersson-Engels S, Johansson J, Svanberg K and Svanberg S 1989 *Anal. Chem.* **61** 1367A (Part I); 1990 *Anal. Chem.* **62** 19A (Part II)
Andersson-Engels S, Kauranen P and Svanberg S 1990a (to appear)
Andersson-Engels S, Johansson J, Montán S, Svanberg K and Svanberg S 1990b (to appear)
Andersson-Engels S, Berg R, Jarlman O and Svanberg S 1990c *Opt. Lett.* (to appear)
Bengtsson J, Bengtsson P, Larsson J and Svanberg S 1989 (unpublished)
Edner H, Johansson J, Svanberg S and Wallinder E 1990 (unpublished)
Kauranen P and Svanberg S 1989 (unpublished)
Larsson J, Sturesson L and Svanberg S 1988 (unpublished)
Radziemski L J, Solarz R W and Paisner J A eds 1987 *Laser Spectroscopy and its Applications* (New York: Dekker)
Svanberg S 1988a *Appl. Phys.* **B46** 271; 1988b *Phys. Scr.* **T23** 281
Svanberg S 1989 *Phys. Scr.* **T26** 90
Svanberg S 1990a *Atomic and Molecular Spectroscopy, Springer Ser. in Atoms and Plasmas Vol 6* (Heidelberg: Springer)
Svanberg S 1990b *Applied Laser Spectroscopy* ed M Inguscio and W Demtröder (New York: Plenum)
Westblom U and Svanberg S 1985 *Phys. Scr.* **31** 402

Biomedical analyses using resonance ionization mass spectrometry

T.L. Quinn, T.J. Murphy and L.J. Moore Eastern Analytical, Inc., College Park, Maryland 20742

H.A. Arlinghaus, M.T. Spaar, E.H. Taylor and N. Thonnard, Atom Sciences, Inc., Oak Ridge, Tennessee 37830

Unique and potentially powerful analytical attributes are embodied in the concepts of resonance ionization mass spectrometry (RIMS) technology. Prominent among these attributes are elemental ionization sensitivity and selectivity. During the last few years, a program has been established to demonstrate the technology potential through its application to the solution of significant biomedical problems. The purpose of this paper is to summarize these developments over the preceding two years, with particular emphasis on the chemical and atomization characteristics associated with the analysis of 'real-world' samples.

Instrumentation used in most of the analytical and biological studies discussed here has been previously reported (1). Basically the RIMS system consists of two sub-systems: a) a Nd:YAG-pumped tunable dye laser system; b) a mass spectrometer with electrostatic and magnetic focusing. Atomization was accomplished by using a primary argon ion beam that was pulsed synchronously with the laser. A second RIMS system was used for a portion of the most recent measurements of nickel and chromium isotopes. This system is similar to the original, but based on a time-of-flight mass spectrometer capable of detecting all ions formed during a given laser pulse.

The sensitivity of RIMS derives primarily from the ability to ionize the analyte element with essentially unit efficiency (2). Ionization selectivity derives from the tunability of the laser system and the consequent ability to access element-specific energy levels to the exclusion of other elements and potential interferences. State-specific ionization selectivities have been demonstrated in the range 10^9 - 10^{12} (e.g.,3). New levels of sensitivity permit the crossing of new measurement threshholds and the attendant titillation of biomedical investigators' imagination to pose new problems. Analytical selectivity portends the ability to simplify sample preparation procedures compared to other techniques, while retaining much of the sensitivity. In fact, these two attributes are mutually reinforcing. Selectivity permits the simplification of sample processing procedures that would otherwise limit the ability to utilize the extreme sensitivity available from the technology.

For the analysis of discrete biological samples, the sample preparation and atomization steps may be said to collectively form the Achilles' heel of the RIMS technology. Although the resonance ionization technology itself has not been a particularly routine development, the resonance ionization process has now been demonstrated for most elements of biomedical interest. In contrast, the sample preparation is dependent in part upon the type and complexity of the matrix. In the history of analytical technique development, there still is no panacea for the direct analysis of trace elements and/or isotopes in biological samples.

A case in point is the determination of chromium in skin tissue. This tissue has been proposed as an indicator of chromium status in the study of insulin resistance in diabetes (4). The wet-ashing of 10 mg skin samples in nitric acid yielded a residue on gold foil whose atomization was attempted directly in the RIMS system. The major chromium-containing ion species produced as a result of this process was a metallo-organic species, CrM^+. Further treatment of this residue by dry-ashing at 450-500 ^oC produced a signal that was essentially total atomic chromium, Cr^+. This simple but critical modification in the sample preparation procedure was sufficient to permit the successful completion of a pilot study of chromium in insulin resistance. This study utilized 10 mg skin biopsy samples and has now progressed to a full-scale study in which the RIMS technology will be the primary analytical vehicle.

A second example of simplified sample preparation is that of the determination of copper in serum (5). Sample processing for this element in this matrix required a simple wet-ashing of the serum, followed by direct cathodic electrodeposition of the copper onto gold foil. This sample preparation procedure yielded an essentially pure Cu^+ signal from the resonance ionization process. Similar results have been observed for the direct cathodic electrodeposition of molybdenum as the oxide from neutral to ammoniacal solution (6).

Clearly these examples represent the opportunity to simplify the sample preparation compared to, say, thermal ionization mass spectrometry, for which a relatively pure sample is required to prevent poisoning of the surface ionization process. Although we anticipate that the selectivity of the technology will find an appropriate place in analytical chemistry as a means to simplify sample preparation, it is also clear that the evolution of the most efficient methodologies will be complex and will occur as they have for other new analytical technologies: slowly.

In addition to the capacity of RIMS for direct trace element determinations in minimally prepared samples, a second major functional measurement thrust has been initiated that employs RIMS as a tracer of isotope tags in biomedicine. Radioactive species have been employed predominantly in nuclear medicine as a means of tracing various metabolic processes. Increasing concern over the long-term effect of even occasional or short-term exposure to radiation, and the inability to employ such techniques in infants, children and pregnant women, have increased the desire to substitute safe stable isotopes as tracers. Concomitantly, there is a trend in analytical chemistry and biomedicine toward the use of small samples, particularly for biopsies, infants and other biomedical samples for which there may be very limited amounts of material available. The determination of minute amounts of stable isotope tracers in small samples is ideally suited to RIMS technology.

An additional feature of stable isotope tagging measurements is that whereas the analytical blank (contamination) directly limits the trace element level determinable, the blank plays a less significant role in the measurement of stable isotope tags. In the latter, the stable isotope tag typically is resolved above natural background by measuring an isotope ratio (two isotope ratios, if available, are useful for internal calibration) and computing the excess (tag) isotope. This general analytical approach has been used successfully with RIMS for the demonstration of iron uptake by red blood cells in ten microliter whole blood samples (7). This project has been sponsored by NASA, in which there is an interest in understanding the mechanics behind the loss of red blood cell mass under microgravity conditions. There is a strong desire to eliminate the use of ^{59}Fe (and other radioactive) tracers in astronauts and in the general spaceflight environment. Iron-58 is of sufficiently low abundance in nature that it provides significant leverage in being able to resolve the contribution of a few nanograms of excess ^{58}Fe in the otherwise isotopically natural blood sample.

In addition to the measurement of iron uptake, the determination of red blood cell survival using stable chromium isotope tags is being studied. This is a more demanding measurement, since it requires the resolution of picogram quantities of the enriched isotope in ten microliter whole blood samples. Preliminary calibration experiments with synthetic ^{53}Cr enriched isotopic standards (a limited system calibration) has demonstrated that the isotope ratios of two nanograms of chromium may be determined at the projected biological tagging levels with nominal precisions and accuracies of 1-3 %. At one end of this calibration range, the $^{52}Cr/^{53}Cr$ ratio approaches 10; therefore this measurement represents the measurement of approximately 200 pg of ^{53}Cr.

Other ongoing applications of RIMS technology to biomedical problems includes those of silicon in nutrition and nickel in toxicology. The determination of silicon in biological materials has been reported previously (8). This work has now been extended to successful determinations of silicon in milk powder and mixed diet materials. The accurate determination of silicon has traditionally been difficult, and the present methods permit its nearly direct determination in relatively complex matrices with only wet and dry-ashing.

Preliminary calibrations of a ^{62}Ni enriched spike solution have yielded isotope ratios measurement precisions and accuracies in the range of 1-3%, relative standard deviation. These measurements are currently being extended to the determination of nickel at extremely low levels in natural materials. Preliminary indications of measurement sensitivity suggest that measurements at the few hundred picogram level should be easily attainable.

Several conclusions may be drawn from the RIMS biomedical applications work pursued thus far in our and other laboratories:

- RIMS can be used for picogram and sub-picogram isotopic determinations

- RIMS selectivity can be used to determine elements and isotopes in minimally prepared samples

- Useful isotope ratio precisions in the 1-5% (relative standard deviation) range are readily achievable

- RIMS can assist in providing unique solutions to important biomedical problems

Based on these and related studies, we are confident that the RIMS technology is reaching a level of technical maturity that has the potential to effect a substantial impact on biomedical element and isotope determinations.

<u>References</u>

1) Parks JE, Schmitt HW, Hurst GS and Fairbank WM "Sputter-initiated resonance ionization spectroscopy, "Thin Solid Films <u>108</u>, 69 (1983)

2) Hurst GS, Payne MG, Kramer SD and Young JP "Resonance ionization spectroscopy and one-atom detection," Rev Mod Phys <u>51</u>, 767-819 (1979)

3) Fassett JD, Moore LJ, Travis JC and Lytle FE "The characterization of thermally-produced metastable excited-state atomic species using resonance ionization mass spectrometry," Int J Mass Spectrom & Ion Phys <u>54</u>, 201-216 (1983)

4) G. Brewer, University of Michigan, personal communication

5) Moore LJ, Parks JE, Spaar MT, Beekman DW, Taylor EH and Lorch V "Ultra-trace elemental and isotopic quantification for neonatal nutrition studies," J Res NBS <u>93</u>, 328-321 (1988)

6) Moore LJ, Parks JE, Taylor EH, Beekman DW and Spaar MT "Medical and biological applications of resonance ionization spectroscopy," Inst Phys Conf Ser No 84, pp. 239-44. Paper presented at RIS 86, Swansea, Wales, Sept. 7-12, 1986

7) Atom Sciences' Phase I SBIR report NASA, 1988

8) Moore LJ, Spaar MT, Taylor EH, Parks JE and Beekman DW "The microdetermination of silicon using resonance ionization spectroscopy," Inst Phys Conf Ser No 94, pp. 273-276. Paper presented at RIS 88, Gaithersburg, Maryland, March 10-15, 1988.

Inst. Phys. Conf. Ser. No 114: Section 7
Paper presented at RIS 90, Varese, Italy, 16 21 September 1990

UV laser induced desorption/ionization of peptides, polypeptides and proteins from fibrillary polymers

I K Perera *, P Håkansson, G Brinkmalm and B U R Sundqvist,

Department of Radiation Sciences, Box 535, Uppsala University, Uppsala, Sweden.

J M Candy and A E Oakley,

MRC Neurochemical Pathology Unit, Newcastle General Hospital, Newcastle upon Tyne, U.K.

ABSTRACT: We report the disaggregation and mass analysis of several biologically important peptides, poly-peptides and proteins, present as ß-pleated sheet fibrillary polymers, by UV laser induced desorption/ionization mass spectrometry (LIDI-MS). Intense molecular ion signals have been obtained with mass resolutions ~ 650, using a focussed 266 nm laser beam on the dried fibrillary aggregates in a ferulic acid matrix. Furthermore, the desorption of mixed fibrillary polymers has also enabled the molecular weights of the individual constituents to be determined with a high accuracy (~ 0.05%).

1. Introduction

Matrix assisted UV laser induced desorption/ionization mass spectrometry (Karas et al 1988, Tanaka et al 1988), has been demonstrated as a highly sensitive, precise and an accurate method (Beavis et al 1990), for determining molecular weights of large biomolecules. The technique, combined with biochemical methods, have also provided insights into the structural composition of highly glycosylated enzymes (Perera et al 1990a). Proteins upto a mass of about 300,000 u (Karas et al 1989), have already been studied employing several UV light absorbing matrices (Beavis et al 1989) and by switching to an infra-red absorbent matrix the wavelength range of useful lasers has recently been extended to the infra-red (Overberg et al 1990). In this approach, a dilute solution of the sample molecules are mixed with a large molar excess of a suitable matrix which absorbs resonantly at the laser wavelength, and the resulting mix is deposited on a metallic substrate and air dried to form a thin surface layer. The deposit is then desorbed by means of pulsed laser radiation and the ions produced are detected by time-of-flight (TOF) mass spectrometry. It has been suggested that the matrix molecules which assist in isolating the analyte molecules from each other and perform as a sink for the incident laser energy also play a vital role in the ionization process of the sample molecules via possible energy transfer from electronically excited states (Vertes et al 1990a).

Although this method has successfully been applied for the analysis of a variety of different bio-organic molecules that could not be studied by other mass spectrometric techniques (Hillenkamp et al 1990a, Salehpour et al 1990, Spengler et al 1990, Vertes et al 1990b), all investigations to date appear to have been mainly concerned with non aggregated peptides, polypeptides and proteins. We now report, for the first time, that the technique of LIDI-MS can be used to disaggregate and determine the molecular weights of biologically important peptides, polypeptides and proteins, present as ß- pleated sheet fibrillary polymers which are resistant to disaggregation. High quality mass spectra with resolutions ~ 650 have been obtained from samples containing femto-mole amounts of the fibrillary proteins and their fragments and the respective molecular weights (< 6000 u) have been determined with an accuracy ~ 0.05%.

2. Experimental

We have investigated the ability of LIDI-MS to disrupt the H-bonding involved in the β-pleated sheet confirmation and disaggregate fibrillary polymers of the neuropeptide substance P, the C-terminal fragments (2-11, 4-11 and 7-11) of substance P, a fibrillary form of bovine insulin (Oakley et al 1981), and the N-terminal fragments (1-28 and 12-28) of the fibrillary protein A4 amyloid 1-42, a major pathological feature in the Alzeheimer's disease and Down's syndrome (Glenner 1981). Substance P and its C-terminal fragments and bovine insulin were obtained from Sigma while the N-terminal fragments of the A4 amyloid protein were from Peninsula Laboratories. The synthesis of the fibrillary aggregates were then carried out as described by Glenner (1981) and Perera et al (1990b).

The experimental set up used has been described elsewhere (Perera et al 1990a, Salehpour et al 1990b). Briefly, it is comprised of a linear TOF mass spectrometer with a 1 m flight path coupled to an excimer laser (XeCl) pumped dye laser (Coumarin 307) system. Mass spectra were recorded by means of a Lecroy transient digitizer and an IBM PC was used for data acquisition and storage. The pulsed (10 ns) beam of the dye laser was frequency doubled into the UV (266 nm) and was focussed onto the sample (beam diameter ~ 150 μm); the maximum irradiance on the sample being ~10^8 W/cm^2. The sample was maintained at a high potential (16-20 kV). The ions produced are accelerated towards a grid at ground potential, then drift through a field free region and are detected by a tandem assembly of channel electron multiplier plates.

The samples for mass spectrometric studies were prepared by dissolving the fibrillary aggregates in 0.1% TFA and ultra-pure water (deionized) at concentrations in the range 0.01-1.0 g/l. The matrix, ferulic acid, was dissolved in 50 % ethanol and water at the optimum concentration of 15 g/l. The sample was then prepared by spreading 0.5 - 6.0 μl of the mixture of the fibrillary protein solution and ferulic acid on etched stainless steel backings (~ 3 mm^2) and allowing the drop to dry. Electron microscopic studies have demonstrated the fibrillary nature of the aggregates investigated and have also revealed that the treatment of the aggregates with 0.1% TFA and ferulic acid at 15 g/l have not resulted in any noticeable disaggregation of the fibrillary polymers studied.

3. Results and Discussions

Figure 1 compares the time-of-flight mass spectra of (A) bovine insulin and (B) insulin fibrils in a ferulic acid matrix. Almost similar signals are observed in both spectra obtained from about 4 picomoles of the analyte, with the molecular ion signal M^+, forming the base peak. In addition, the doubly charged monomer M^{2+} and the cluster ion $2M^+$ are also observed. No fragment ion signals are observed and the signals in the low mass region (< 500 u) can unambiguously be assigned to the matrix molecules. The results indicate that in contrast to PDMS, where the ion yield of insulin fibrils was found to be atleast an order of magnitude lower than that due to bovine insulin (Perera et al 1990b), the fibrillary nature of the protein does not inhibit the ion formation process in LID-MS to any noticeable extent. In the molecular weight determination, mass calibrations of several spectra employing the photon signal and the singly charged molecular ion signal of a bovine insulin spectrum obtained under identical experimental conditions were employed. Accordingly, the measured molecular weight of insulin fibrils, based on the peak centroids, was found to be 5734.6 u, the uncertainity of measurement being < 0.1%.

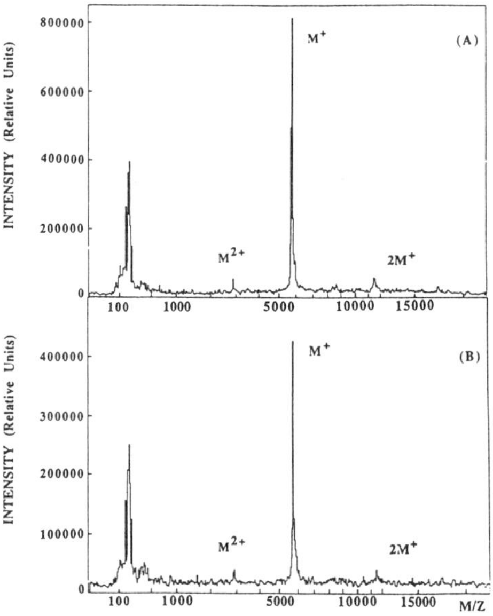

Fig. 1 TOF mass spectra of (A) bovine insulin and (B) insulin fibrils. Samples were prepared by mixing 0.5 μl of the respective analyte at 0.05 g/l with 2.5 μl of the ferulic acid solution.

The next sample to be investigated was a mixture containing the substance P fibrils and three of its fibrillary C-terminal fragments (2-11, 4-11 and 7-11), and the TOF spectrum obtained from about 10 pico-moles of this mixture with ferulic acid as the matrix is shown in fig. 2.

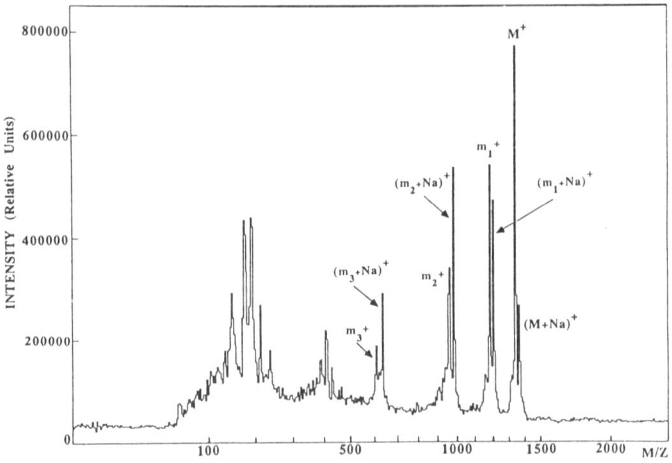

Fig. 2. TOF spectrum from a mixture of substance P fibrils and three of its fibrillary fragments, averaged over 20 laser shots.. Samples were prepared by mixing 0.5 µl of 0.02 g/l fibrillary protein mixture with 2.5 µl of the ferulic acid solution.

Intense molecular ion signals are observed from the substance P fibrils (M^+) and its three fragments 2-11 (m_1^+), 4-11 (m_2^+) and 7-11 (m_3^+). No multiply charged ions or cluster ions or even satallite signals of matrix attached analyte ions are observed. Instead in all cases, in addition to the parent singly charged molecular ions, prominent quasimolecular ion signals corresponding to the attachments of sodium ions, are observed. The results also indicate that the presence of Na salts in the sample has not affected the signal to noise ratio nor has it created any observable fragmentation of the constituents of the mixture into their subunits, as has been observed in the desorption of large proteins (Hillenkamp et al 1990b). Similar results have also been obtained from a sample containing only the fibrillary peptide, substance P, with its TOF spectrum illustrating an intense molecular ion (M^+) peak and a signal corresponding to the metal cation ion adduct $(M + Na)^+$. The molecular weights obtained, using the photon signal and a known mass peak from the ferrulic acid matrix as the calibration standard, are given in table 1A. The results indicate that the measured molecular masses are within 1 mass unit (u) of their calculated values.

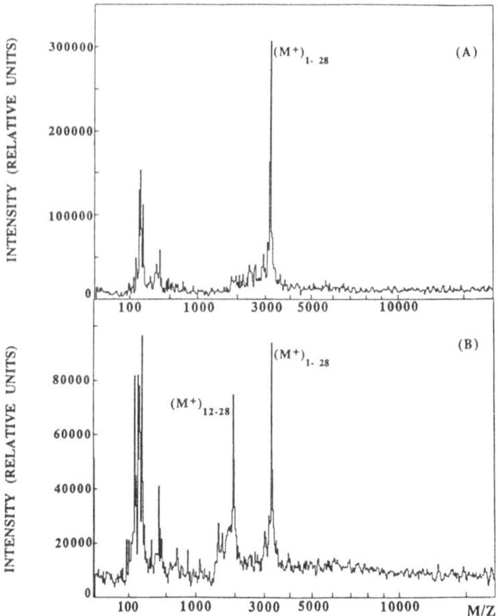

Fig 3 TOF spectra of (A) A4 amyloid fragment 1-28 and (B) mixture of the fragments 12-28 and 1-28. Samples prepared by adding 1 µl of 0.01 g/l protein solution to 2.5 µl of the matrix.

Fig. 3 shows the TOF spectra of (A) A4 amyloid fragment 1-28 and (B) a mixture of two A4 amyloid fragments 1-28 and 12-28 in a ferulic acid matrix. Intense singly charged molecular ion signals are observed from both fragments, $(M_{1\text{-}28})^+$ and $(M_{12\text{-}28})^+$, while no other signals including those corresponding to metal cation ion adducts are observed. In the desorption of the mixed sample both ferulic acid and sinapinic acid were found to produce uniform responses from the two fragments in the mixture, while the use of nicotinic acid as the matrix realized only a single peak that was relatively broad and was of low intensity, corresponding to the fragment A4 Amyloid 12-28. The measured molecular weights of the two fragments are given in Table 1B; the uncertainty of measurements again being ~ 0.1 %.

Table 1 Comparison of measured and calculated molecular weights.

	Fibrillary Aggregate	Molecular Weights (u)	
		Measured	Calculated
A	Substance P	1348.6	1348
	Substance P fragment 2-11	1192.1	1191
	Substance P fragment 4-11	966.8	966
	Substance P fragment 7-11	614.1	613
B	Amyloid A4 fragment 12-28	1955.8	1954
	Amyloid A4 fragment 1-28	3262.3	3259

Excellent mass spectra have been obtained from samples containing 300 fmol - 10 picomole of the respective analyte, the signal to noise ratio being somewhat lower at very low protein concentrations. In the determination of the peak centroids of all these signals a precision of < 0.1 % has been attained. In addition, a fibrillary form of the peptide eledoisin, the A4 amyloid fragment 8-21 and a sample containing a mixture of the two A4 amyloid fragments 8-21, and 1-28 have also been studied recently, by this method (Perera et al 1990b).

4. Conclusions

We have demonstrated that laser induced desorption/ionization mass spectrometry, can be used to disaggregate and determine the molecular weights of β-pleated sheet fibrillary polymers of peptides , polypeptides and proteins. Furthermore, it was possible to determine the molecular weights of the monomeric constituents of both individual and mixed fibrillary polymers accurately and at subpicomolar concentrations making the laser technique attractive for characterising polypeptides and proteins associated with both isolated and in situ amyloid deposits in a number of diverse disease states.

* Correspondence Address: Department of Geology, The University, Manchester M13 9PL, U.K.

5. References

Beavis R and Chait B, 1989, Rap. Comm. Mass Spectrom., **3,** 432; and **3,** 436.
Beavis R and Chait B, Anal Chem, 1990 (in press).
Glenner G G, 1981, Progr. on Histochem. Cytochem., **13**, 1.
Hillenkamp F, Karas M, Ingendoh A and Stahl B, 1990a, Proc. Second Intl. Sym. on Mass Spectrom. in Health & Life Sciences, Eds. A. Burlingame & J. McCloskey, Elsevier, (in press).
Hillenkamp F and Karas M, 1990b, Methods in Enzymology (in press).
Karas M and Hillenkamp F, 1988, Anal. Chem., **60,** 2299.
Karas M, Bahr U, Ingendoh A and Hillenkamp F, 1989, Angew Chem. Int. Ed. Engl., **28,** 760.
Oakley A E, Parry R H, Candy J M and Perry E K, 1981, Neuropeptides, **2,** 1.
Overberg A, Karas M, Bahr U, Kaufmann R and Hillenkamp F, 1990, Rapid Commun. Mass Spectrom., 4, 293.
Perera I K, Uzcategui E, Hakansson P, Brinkmalm G, Pettersson G, Johansson G and Sundqvist B U R, 1990a, Rapid Commun. Mass Spectrom, **4,** 135.
Perera I K, Candy J M, Hakansson P, Oakley A E, Brinkmalm G and Sundqvist B U R , 1990b, to be submitted to Rapid Communications in Mass Spectrom.
Salehpour M, Perera I K, Kjellberg J, Hedin A, Islamian A, Håkansson P and Sundqvist B U R, 1989, Rap. Comm. Mass Spectrom., **3,** 259;
Salehpour M, Perera I K, Kjellberg J, Hedin A, Islamian A, Håkansson P and Sundqvist B U R, 1990, Proc. IFOS - V, Eds. A. Hedin et al (Wiley , Chichester), pg. 119.
Spengler B, Ping Y, Cotter R J and Kan L S, 1990, Rapid Commun. Mass Spectrom., **4**, 99.
Tanaka K, Waki H, Ido Y, S. Akita, Yoshida Y and Yoshida T, 1988, Rap. Comm. Mass. Spectrom., **2,** 151.
Vertes A, Gijbels R and Levine R D, 1990a, Rapid Commun. Mass Spectrom., **4,** 228.
Vertes A, Balazs L and Gijbels R, 1990b, Rapid Commun. Mass Spectrom., **4,** 263.

Use of RIS to significantly increase the speed of sequencing the human genome

H.F. Arlinghaus[1], N. Thonnard[1], M.T. Spaar[1], and K.B. Jacobson[2]

[1]Atom Sciences, Inc., Oak Ridge, Tennessee, USA; [2]Biology Division, Oak Ridge National Laboratory, Oak Ridge, Tennessee, USA

Abstract: Since a fast, sensitive and selective detection method is available, large numbers of stable isotopes can be used to label DNA, thereby multiplexing the separation process, and providing a new, much faster procedure for localizing DNA after electrophoresis. Sputter-Initiated Resonance Ionization Spectroscopy (SIRIS) and Laser Atomization Resonance Ionization Spectroscopy (LARIS) were used to demonstrate detection of labeled DNA with excellent lateral and mass resolution. Both SIRIS and LARIS have the potential of making a strong contribution to DNA sequencing.

I. INTRODUCTION

The genome, which determines the genetic make-up of an organism, is characterized by the sequence of the four nucleotide bases, A, G, C and T in its DNA. Most present techniques [Maxam and Gilbert (1977), Sanger et al. (1977), Church and Kieffer-Higgins (1988)] use gel electrophoresis to separate ^{32}P or ^{35}S radionuclide labeled DNA that has been randomly cut (or synthesis terminated) into different length fragments, all terminating with a specified base and detecting the labeled DNA fragments with autoradiography. Since the human genome consists of approximately 3×10^9 nucleotide pairs, and current sequencing methods typically can sequence at most 10^4 bases (10^5 with some new developments) in 24 hours, the challenge of sequencing the human genome is significant indeed.

To reduce the steps in electrophoresis, multiplexing has been introduced. Church & Kieffer-Higgins (1988) have demonstrated a method of combining the A, G, C and T-terminated fragments of 40 or more DNAs on the same four adjacent gel lanes for electrophoresis. Another multiplex method is employed in DNA sequencing in which the labels are fluorescent moieties attached to dideoxynucleotide of the DNA fragments that terminate in A, G, C and T and allows all four types to be combined in a single lane [Smith et al. (1986)].

We describe here the development of a method that combines aspects of both of these multiplexing methods by utilizing stable isotopes instead of radioisotopes or fluorescent labels [for more details see Jacobson et al.(1990), Arlinghaus et al.]. Attachment of four isotopes from one element to the A, G, C and T terminated fragments of the DNA allows these fragments to be combined in one gel lane, since the isotopes would have no differential effect during electrophoresis. Using sets of four isotopes from several elements allows several DNAs to be combined together in that same gel lane. Since the isotopes could be analyzed in the electrophoresis gel directly, there would be no need for subsequent hybridization, as in the Church & Kieffer-Higgins method.

Detection of isotopes of tin on DNA was achieved by sputter-initiated resonance ionization spectroscopy (SIRIS) and laser atomization RIS (LARIS). The SIRIS/LARIS technique, and differences in response as a function of atomization parameter, substrate, and analyte will be discussed.

2. EXPERIMENT

The present SIRIS/LARIS system consists of a Perkin Elmer duoplasmatron microbeam ion gun, a Quanta Ray Nd:YAG laser for laser atomization, a Kimball Physics pulsed flood electron gun, an RIS laser system consisting of a Quanta Ray DCR-2A pulsed Nd:YAG laser (repetition rate 30 Hz) pumping two Quanta Ray PDL-1 pulsed dye lasers having a typical bandwidth of 0.3 cm^{-1}, a computer-controlled (x,y,z,ϕ) sample holder, two HeNe lasers for exact target positioning, a video imaging system, a sample interlock system, and a mass spectrometer detection system (for more details, see Arlinghaus et al.). Secondary ions produced by the sputtering process are suppressed by a combination of the relative timing between the ion sputtering pulse and the firing of the RIS lasers, timed extraction voltage switching, and electrostatic energy analysis. For the SIRIS experiments, charge compensation was achieved by using low energy pulsed electrons (20-70 eV) in combination with pulsed target and extraction voltages. Imaging was achieved for the DNA experiments by changing the x or y target position (x-scan, y-scan) while the atomization laser beam or ion beam position remains fixed.

3. RESULTS AND DISCUSSION

A three color RIS scheme was found to be best for tin [$5p^2\ {}^3P_0$ - $6s\ {}^3P^o$ (286.332 nm) + $6s^3\ P_1{}^o$ - $7p\ {}^3D_2$ (614.965 nm) + 1064.1 nm]. Both resonant transitions could be saturated with very low laser power (10-50 μJ/pulse), while several millijoules of IR photons were required to saturate the ionization step.

To demonstrate that SIRIS/LARIS can localize and quantify labeled DNA bands, a ^{116}Sn-labeled 17-mer DNA (50 pmoles) was placed in a 20 mm wide well on a 0.4 mm thick polyacrylamide gel and subjected to electrophoresis. Figures 1 a-d show ^{116}Sn SIRIS (a,b) and LARIS (c,d) y-scans obtained from the same sample. For each scan, a prominent band of ^{116}Sn was observed at the ultraviolet absorption site of the DNA band. The y-scan shown in Figure 1b, a repeat of the first experimental run, indicates the very good reproducibility of the band structure. In the first laser y-scan (Figure 1c) a much higher background signal was observed as compared to the second laser y-scan (Figure 1d). This background is presumably caused by creeping of Sn compounds along the surface during or after electrophoresis. The broad width of the DNA band at lower signal voltages is probably due to saturation effects caused by overloading the acrylamide with ^{116}Sn-labeled-17-mer oligonucleotides. After removing the surface background with the first laser scan, subsequent laser scans were able to reproduce the bands with only minor signal and structure changes (not shown). The second band observed at the surface (Figure 1c, y = $\sim$8.5 mm) is presumably caused by a shorter fragment. In the SIRIS y-scans (Figures 1 a,b) a small bump is seen at the same position, but with much less signal. This observation suggest that quite different desorption and/or fragmentation mechanism are involved in the ion sputtering and laser atomization processes.

A new polyacrylamide gel sample with a ^{116}Sn-labeled 17-mer DNA band ($\sim$0.5 nmol) was used to repeat the above described LARIS experiments, but now using a 532 nm atomization laser beam (instead of 266 nm). With SIRIS, a prominent band of ^{116}Sn was observed at the site of the DNA band, while with LARIS, no band could be detected. The results indicated that wavelengths shorter than 532 nm are required to produce neutral Sn atoms with laser atomization.

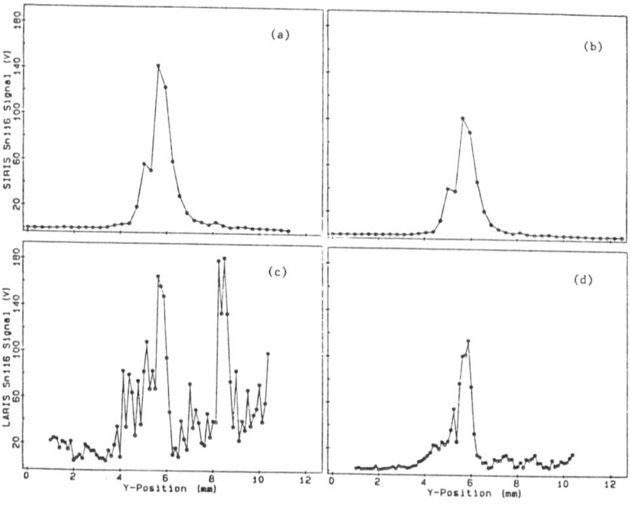

Figure 1. SIRIS (a,b) and LARIS (c,d) analysis of a ^{116}Sn-labeled 17-mer DNA band on polyacrylamide gel (total amount = ~0.5 nmol). The beam direction of the ion and laser beam was parallel to the electrophoresis direction. (b) and (d) are a repeat of the first experimental runs using the same track. The SIRIS data results from 100 shots/data point at an ion current of 2.15 μA (spot size = ~ 100 μm). For LARIS, 15 laser shots/data point of 266 nm light at 140 μJ/pulse (spot size = ~ 150 μm) were used.

Figure 2 shows a SIRIS/LARIS scan of a ^{116}Sn-labeled 2000-mer DNA band which was transferred onto a nylon membrane. This experiment was chosen because the multiplexing procedure proposed by Church and Kieffer-Higgins, and many other DNA analyses, occur on this type of material. With both methods, a prominent band of ^{116}Sn was observed at the site of the DNA band. The LARIS signal intensity is much higher (~ factor of 200) than the SIRIS signal. This demonstrates that LARIS is more suitable to detect Sn-labeled DNA on nylon membranes. The structure and intensity of the band was well reproduced with a second LARIS y-scan (not shown). The ^{116}Sn peak signal to ^{116}Sn background signal ratio is only ~5, which may be caused by Sn contamination, or by the electrophoresis and/or transfer process. For all the Sn experiments, no attempt has yet been made to reduce the background Sn contamination, e.g. by using cleaner chemicals. The irregularity of the data that define the peak may be done to the irregularity of the nylon surface, which consists of a mat of fibers. The limited signal obtained by SIRIS from the nylon membrane is presumably due to penetration of the oligonucleotides below the surface. Therefore LARIS analyzes results in higher signal than SIRIS which analysis only the first few monolayers.

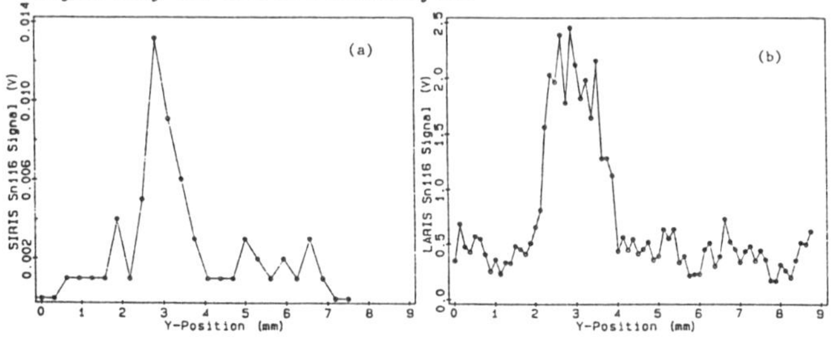

Figure 2. SIRIS (a) and LARIS (b) analysis for of a ^{116}Sn-labeled-2000-mer DNA band on nylon membrane (total amount = ~8 pmol). 500 shots/data point at an ion beam current of 2.15 μA were used in the SIRIS mode, while 15 shots/data point at 266 nm and 105 μJ/pulse were used with LARIS.

Figure 3 demonstrates that we can resolve neighboring oligonucleotides very well (Figure 3a) and that the DNA band can be tightened down to 300 μm by reducing the gel thickness from 400 μm to 50 μm (Figure 3b). Employing smaller amounts of DNA material for electrophoresis, and longer oligonucleotides (50-500), it should be possible to reach a band width of 100 μm or less. Since the ion beam and the laser beam diameter used in the two atomization processes can be focused to less than 10 μm, the resolving power of the system may permit a reduction in the gel length necessary for sequencing. Reducing the gel from the 50-100 cm length currently used to ~10 cm, or smaller, might be possible, leading to shorter electrophoresis times, smaller DNA samples, less background, and faster analysis. Sensitivity tests have demonstrated that subattomole amounts of Sn-labeled DNA are readily detected with SIRIS. We conclude that RIS has the potential to make a strong contribution to DNA sequencing, and for any other method that requires detection of the location and amount of DNA or an oligonucleotide that hybridizes to DNA. Comparison of the SIRIS and LARIS techniques indicated that both methods represent viable options for significantly improving the rate of DNA sequencing. Using higher repetition rate lasers, (10 kHz using Cu vapor) and considering that only a few atomizing laser pulses on an optimized substrate could produce a signal that is far above background, it should eventually be possible to detect more than 10^7 bases per day.

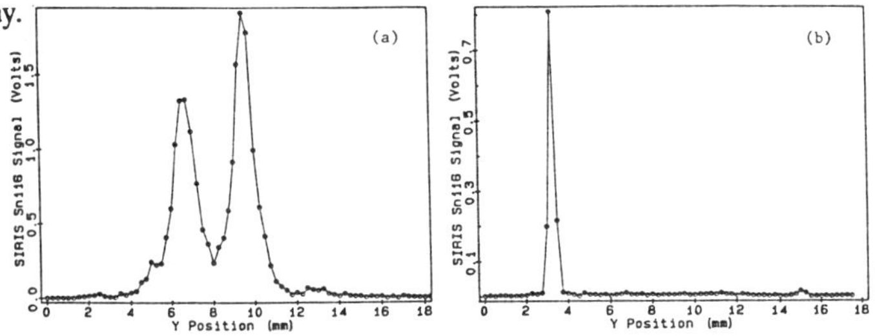

Figure 3. SIRIS analysis (a) of ^{116}Sn-labeled 25- and 26-mer DNA bands on 400 μm thick 12% polyacrylamide gel and (b) of ^{124}Sn-labeled 28-mer DNA band on 50 μm thick 12% polyacrylamide gel. Both analyses consisted of 150 shots/data point at 2.0 μA ion current.

4. ACKNOWLEDGEMENTS

This work was supported in part by the United States Department of Energy under contract No. DE-AC-05-84OR214000 to Martin Marietta Energy Systems, Inc., and under contract No. DE-AC-05-89ER80735 to Atom Sciences, Inc., and the Oak Ridge National Laboratory Exploratory Studies Program.

5. REFERENCES

Arlinghaus, H.F.; Thonnard, N.; Spaar, M.T.; Sachleben, R.A.; Larimer, F.W.; Foote, R.S.; Woychik, R.P.; Brown, G.M.; Sloop, F.V.; and Jacobson, K.B; submitted to Anal. Chemistry.

Church, G.M. and Kieffer-Higgins, S., Science (1988) 240, 185-188.

Jacobson, K.B.; Arlinghaus, H.F.; Schmitt, H.W.; Sachleben, R.A.; Brown, G.M.; Thonnard, N; Sloop, F.V.; Foote, R.S.; Larimer, F.W.; Woychik, R.P.; England, M.W.; Burchett, K.L.; and Jacobson, D.A.; Genomics, (1990) in press.

Maxam, A.M. and Gilbert, W., (1977) Proc. Natl. Acad. Sci. US 74, 560-564.

Sanger, F.; Nickelen, S.; and Soulson, A.R.; Proc. Natl. Acad. Sci. US (1977) 94, 5463-5467.

Smith, L; Sanders, J.Z.; Kaiser, R.J.; Highes, P.; Dodd, C.; Connell, C.R.; Heiner, C.; Kent, S.B.H.; and Hood, L.E.; Nature, (1986) 321, 674-679.

Inst. Phys. Conf. Ser. No 114: Section 8
Paper presented at RIS 90, Varese, Italy, 16–21 September 1990

Effect of breaking cylindrical symmetry on photoelectron angular distributions resulting from resonance-enhanced two-photon ionization

Katharine L Reid, David J Leahy, Sarah W Allendorf,* and Richard N Zare
Department of Chemistry, Stanford University, Stanford, CA 94305, USA

Abstract. A general expression is presented for the photoelectron angular distribution (PAD) following resonance-enhanced two-photon ionization with linearly polarized light beams. When the two polarization vectors make a nonzero angle, cylindrical symmetry is broken and coherence effects manifest themselves in the form of the photoelectron angular distribution, which can show a ϕ dependence as well as a θ dependence. This theoretical treatment is applied to $(1+1')$ REMPI–PAD data of nitric oxide (NO), from which it is possible to deduce the phase shifts and amplitudes of the partial waves that characterize the photoelectron wavefunction.

1. INTRODUCTION AND THEORY

An intimate probe of molecular ionization dynamics is the measurement of energy-resolved photoelectron angular distributions (PAD's) following ionization from a single rovibrational quantum level. Both the branching ratios of the different ionic states formed and the shape of each PAD contain information on the partial wave composition of the outgoing photoelectron. Particularly revealing is the resolution of molecular rotational states in photoelectron energy spectra (Viswanathan *et al* 1986, Allendorf *et al* 1989).

Consider a two-photon resonance-enhanced ionization of a molecular target in which both photon beams are linearly polarized. If the two polarization vectors point in the same direction, the PAD has the simple form

$$\begin{aligned} I(\theta) &\propto a_0 + a_2 P_2(\cos\theta) + a_4 P_4(\cos\theta) \\ &\propto \beta_{00} + \beta_{20} Y_{20}(\theta,\phi) + \beta_{40} Y_{40}(\theta,\phi), \end{aligned} \quad (1)$$

where θ is the angle between the direction of photoelectron ejection and the polarization vector of the ionizing photon beam and ϕ is the azimuthal angle. Note that the PAD has *cylindrical symmetry* about the ionizing polarization vector. However, if the two polarization vectors make a nonzero angle, then this cylindrical symmetry is broken (Leuchs and Walther 1984). In this case, the PAD has the more complicated form

$$I(\theta,\phi) \propto \beta_{00} + \beta_{20} Y_{20}(\theta,\phi) + \beta_{40} Y_{40}(\theta,\phi) + 2[\beta_{22} Y_{22}(\theta,0) + \beta_{42} Y_{42}(\theta,0)] \cos 2\phi. \quad (2)$$

Examination of Eq. (2) shows that the ϕ dependence of the PAD is brought about by the contribution of the $Y_{L2}(\theta,\phi)$ terms with $L = 2$ and 4, since the $Y_{L0}(\theta,\phi)$ terms have no ϕ dependence. This ϕ dependence comes about as a result of M_{J_i} coherence.

* Present address: Lawrence Livermore National Laboratory, M/S L-483, P. O. Box 808, Livermore, CA 94550, USA

Let us specialize these considerations to the (1 + 1′) REMPI of NO, that is, to a situation in which the first photon causes a transition from a ($v = 0, J$) level of the $X\,^2\Pi$ state to a ($v_i = 0, N_i J_i$) level of the intermediate $A\,^2\Sigma^+$ state, and the second photon ionizes this level to produce the NO^+ $X\,^1\Sigma^+$ ion in some distribution of ($v^+ = 0, N^+ = J^+$) levels. We denote the orbital angular momentum quantum number of the photoelectron as ℓ, which has the projection λ on the molecular axis and the projection m_ℓ on the lab-frame quantization axis. The β_{LM} coefficients can be expressed as follows (spin-generalized form of Eq. (29), Dixit and McKoy 1985):

$$\begin{aligned}
\beta_{LM} =& \left(\frac{4\pi}{3}\right)(2N_i+1)(2N^+ +1) \\
&\times \sum_{J_i} \sum_{M_N^+, M_{Si}} \sum_{\ell,\lambda,\ell',\lambda'} \sum_{m_\ell, m_\ell'} \sum_{M_{Ji}, M_{Ji}'} \sum_{N_t, N_t'} (-1)^{1+\lambda+\lambda'+m_\ell-M_{Ji}-M_{Ji}'} \\
&\times \left[\frac{(2\ell+1)(2\ell'+1)(2L+1)}{4\pi}\right]^{1/2} (2N_t+1)(2N_t'+1) \\
&\times \begin{pmatrix} \ell & \ell' & L \\ m_\ell & -m_\ell' & M \end{pmatrix} \begin{pmatrix} \ell & \ell' & L \\ 0 & 0 & 0 \end{pmatrix} \begin{pmatrix} N_i & \frac{1}{2} & J_i \\ M_{Ni} & M_{Si} & -M_{Ji} \end{pmatrix} \begin{pmatrix} N_i & \frac{1}{2} & J_i \\ M_{Ni}' & M_{Si} & -M_{Ji}' \end{pmatrix} \\
&\times \begin{pmatrix} N^+ & N_i & N_t \\ -M_N^+ & M_{Ni} & -m_\ell \end{pmatrix} \begin{pmatrix} N^+ & N_i & N_t' \\ -M_N^+ & M_{Ni}' & -m_\ell' \end{pmatrix} \begin{pmatrix} N_t & 1 & \ell \\ m_\ell & 0 & -m_\ell \end{pmatrix} \begin{pmatrix} N_t' & 1 & \ell' \\ m_\ell' & 0 & -m_\ell' \end{pmatrix} \\
&\times \begin{pmatrix} N^+ & N_i & N_t \\ 0 & 0 & 0 \end{pmatrix} \begin{pmatrix} N^+ & N_i & N_t' \\ 0 & 0 & 0 \end{pmatrix} \begin{pmatrix} N_t & 1 & \ell \\ 0 & \lambda & -\lambda \end{pmatrix} \begin{pmatrix} N_t' & 1 & \ell' \\ 0 & \lambda' & -\lambda' \end{pmatrix} \\
&\times r_{\ell\lambda}\, r_{\ell'\lambda'} \cos(\eta_{\ell\lambda} - \eta_{\ell'\lambda'}) \; {}^{J_i}\rho_{M_{Ji}, M_{Ji}'}(\theta_T),
\end{aligned} \tag{3}$$

where

$$\begin{aligned}
{}^{J_i}\rho_{M_{Ji}, M_{Ji}'}(\theta_T) =& \sum_{\mu,\mu',M_J} (-1)^{M_{Ji}+M_{Ji}'} D^{1*}_{\mu\mu_0}(\phi_T,\theta_T,\chi_T)\, D^{1}_{\mu'\mu_0}(\phi_T,\theta_T,\chi_T) \\
&\times \begin{pmatrix} J & 1 & J_i \\ -M_J & -\mu & M_{Ji} \end{pmatrix} \begin{pmatrix} J & 1 & J_i \\ -M_J & -\mu' & M_{Ji}' \end{pmatrix}.
\end{aligned} \tag{4}$$

In Eq. (3), $r_{\ell\lambda}$ and $\eta_{\ell\lambda}$ are the magnitude and the phase of the radial dipole matrix element connecting the intermediate state to the $\ell\lambda$ outgoing partial wave. In Eq. (4), ϕ_T, θ_T, and χ_T are the Euler angles associated with the transformation from the excitation frame to the ionization frame. The excitation photon makes the projection μ_0 on the z axis of the excitation frame and μ, μ' in the ionization frame. Our application of the theory is for $\mu_0 = 0$ as well as $\phi_T = \chi_T = 0°$, as in the experiment of Allendorf *et al* (1989).

The ${}^{J_i}\rho_{M_{Ji},M_{Ji}'}(\theta_T)$ are density matrix elements that describe the populations of ($M_{Ji} = M_{Ji}'$) and coherence relations ($M_{Ji} \neq M_{Ji}'$) between the M_{Ji} levels of the intermediate state (Blum 1981, Dixit and Lambropoulos 1983, Dixit and McKoy 1985). When $\theta_T = 0°$, all off-diagonal elements of the density matrix vanish, *i.e.*, no coherence effects exist in the intermediate state, causing the PAD to be cylindrically symmetric [see Eq. (1)]. Here, only channels involving different values of ℓ, λ, and N_t interfere in the ionization process. However, when $\theta_T \neq 0$, the $M_{Ji} = M_J \pm 1$ sublevels may originate from a common ground state sublevel M_J. As a result, an M_N^+ sublevel of the ion may be reached from the same M_J sublevel by more than one route. This causes the m_ℓ channels that bring this about to interfere constructively or destructively depending upon their phase relationship. Hence, terms with $M \neq 0$ appear [see Eq. (2)]. Therefore, the appearance of a noncylindrically symmetric PAD is a manifestation of this additional coherence.

The β_{LM} have direct physical significance. In particular, β_{00} gives the angle-integrated cross section for the $J \to J_i = N_i \pm \frac{1}{2} \to J^+ = N^+$ branch in question, and the β_{L2} represent coherence effects arising from breaking the cylindrical symmetry between the excitation and ionization steps. This noncylindrical symmetry appears as a ϕ dependence of the PAD.

More subtly, for $\phi = 0°$ (which is the common way PAD's are recorded), the M_{Ji} coherence also alters the form of the θ dependence. Specifically, coherence effects cause the appearance of a $\cos^4\theta$ contribution to the PAD when $\theta_T = 54.7°$ (the "magic angle"). This result might at first glance be unexpected because the M_{Ji} sublevels are equally populated for this value of θ_T. However, this result is another consequence of coherence effects caused by breaking of the cylindrical symmetry. This $\cos^4\theta$ contribution arises from the $Y_{42}(\theta, 0)$ term in Eq. (2).

2. $(1+1')$ REMPI–PAD OF NITRIC OXIDE

The resonance-enhanced two-photon ionization of NO via its $A\,^2\Sigma^+$ state provides a good illustration of the effects of breaking the cylindrical symmetry between the excitation and ionization steps. Allendorf *et al* (1989) described such PAD experiments on NO in which the polarization vectors of the two photon beams were parallel, perpendicular, and at the magic angle of 54.7°. Figure 1 shows the normalized PAD's for the transition $P_{21} + Q_{11}$ (25.5) at (a) $\theta_T = 0°$ and (b) $\theta_T = 90°$. These data were analyzed by Allendorf *et al* (1989) neglecting coherence effects for $\theta_T \neq 0°$. Using Eq. (3) we have reanalyzed these data using a modified least-squares method (Levenberg–Marquardt) (Press *et al* 1986) different from that used previously. These results, as well as those obtained from *ab initio* calculations (Rudolph and McKoy 1989), are displayed in Table 1.

The p wave, which is the strongest contributor to the photoelectron wavefunction, is very well determined by our fit. In general, there is fairly good agreement between experiment and theory, although for the fraction of parallel character it is not fully satisfactory.

Using the results of our fit, we have predicted PAD's for the transitions Q_{11} (2.5) and R_{21} (1.5). The former are shown in Figure 2 for $\theta_T = 54.7°$ with (a) coherences included and (b) coherences not included. As discussed previously, the photoelectron angular distribution cannot be described by the familiar expression (Cooper and Zare 1968)

$$I(\theta, \phi) = I(\theta) = \alpha[1 + \beta P_2(\cos\theta)].$$

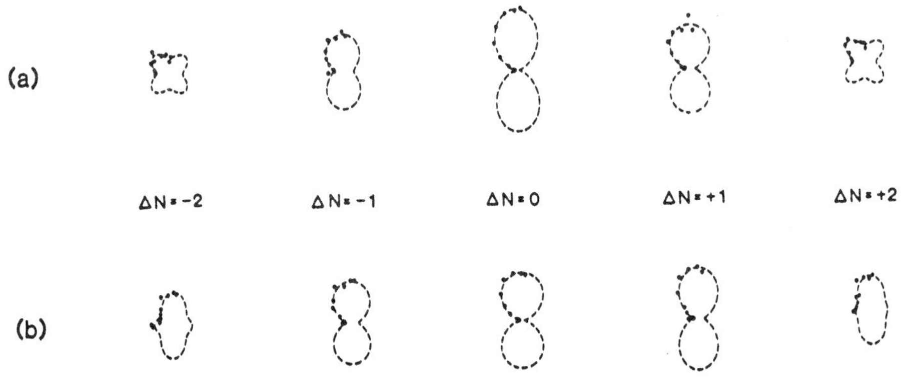

Figure 1. Experimental data of Allendorf *et al* (1989) for $(1+1')$ REMPI of NO via $A\,^2\Sigma^+$ $(v'=0)$ $P_{21}+Q_{11}$ (25.5) at (a) $\theta_T = 0°$ and (b) $\theta_T = 90°$; $\Delta N = N^+ - N_i$. Dashed lines are fits to Eq. (1).

Table 1. Parameters resulting from the fit of $\theta_T = 0°$ and $\theta_T = 90°$ data for $(1+1')$ REMPI of NO via $A\,^2\Sigma^+$ $(v'=0)$ $P_{21}+Q_{11}$ (25.5), using coefficients that include or omit coherence terms. Error bars represent 1σ uncertainties. Also shown are the *ab initio* results of Rudolph and McKoy (1989). The parameters are defined as follows: $F_\ell \equiv \sum_\lambda |r_{\ell\lambda}|^2 / \sum_{\ell'\lambda} |r_{\ell'\lambda}|^2$, $\Gamma_\ell \equiv |r_{\ell\sigma}|^2 / \sum_\lambda |r_{\ell\lambda}|^2$, $\delta_{\ell\lambda-\ell'\lambda'} \equiv \eta_{\ell\lambda} - \eta_{\ell'\lambda'}$.

	Fit *with* coherence terms	Fit *without* coherence terms	*ab initio*
F_s	0.048±0.006	0.047±0.008	0.025
F_p	0.714±0.010	0.728±0.010	0.655
F_d	0.025±0.006	0.026±0.008	0.050
F_f	0.212±0.007	0.199±0.008	0.273
Γ_p	0.368±0.040	0.357±0.053	0.119
Γ_d	0.542±0.110	0.524±0.109	0.984
Γ_f	0.314±0.070	0.258±0.099	0.471
$\delta_{p\sigma-p\pi}$	12.7°±2.3°	10.8°±2.8°	9.8°
$\delta_{d\sigma-d\pi}$	52.7°±8.9°	51.8°±42.4°	93.3°
$\delta_{f\sigma-f\pi}$	14.3°±13.1°	0.0°±3610°	1.6°
$\delta_{s\sigma-d\sigma}$	1.1°±21.7°	0.0°±25.2°	6.3°
$\delta_{p\sigma-f\sigma}$	124.3°±7.4°	112.3°±2409°	103.7°

The coherence terms cause the appearance of a $\cos^4\theta$ dependence of the PAD. This effect is particularly pronounced for the $\Delta N = \pm 2$ transitions, where $\Delta N = N^+ - N_i$. A similar result is seen for the R_{21} (1.5) transition. The coherence effects are magnified at low J_i because of the relative magnitudes of the diagonal and off-diagonal elements of the density matrix. However, the nonzero contribution of $Y_{42}(\theta, \phi = 0)$ for $\theta_T = 54.7°$ persists at high J_i.

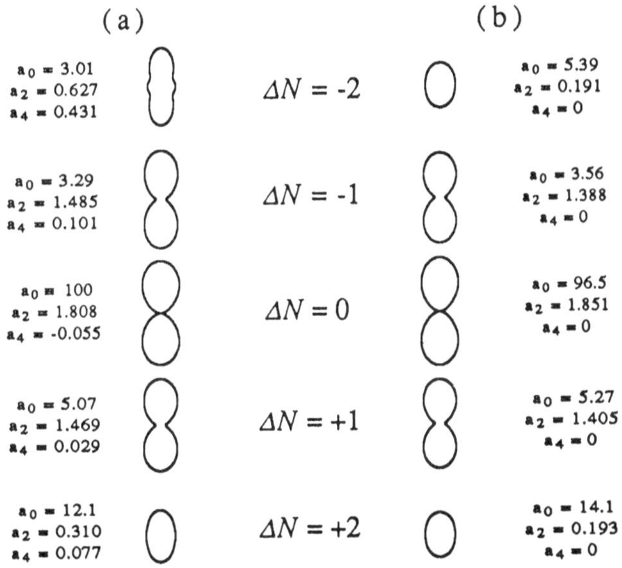

Figure 2. Q_{11} (2.5), $\theta_T = 54.7°$. Calculated using fit parameters with (a) coherence included, (b) coherence omitted.

We have assessed the importance of the coherence terms in two ways. Firstly, in Table 1, we show the results of fitting the data if we use coefficients that do not include the coherence terms. We see that the magnitudes of the radial dipole matrix elements are equally well determined by both fits but that some phases become undetermined if the coherence terms are omitted.

Secondly, we fit data for the R_{21} (20.5) transition at $\theta_T = 0°$ and $\theta_T = 90°$ and use the results to predict PAD's for $\theta_T = 54.7°$ using coefficients that (a) include and (b) omit coherence terms. These can then be compared with the experimental results of Allendorf *et al* (1989). We find that, in this case, the coherences introduce a correction smaller than experimental scatter.

3. CONCLUSION

We have presented a treatment that includes the coherence terms that result when cylindrical symmetry is broken in a $(1 + 1')$ REMPI–PAD experiment. We find that these cause $Y_{2\pm2}(\theta, \phi)$ and $Y_{4\pm2}(\theta, \phi)$ contributions to the PAD's that alter both their θ and their ϕ dependence. When the two polarizations vectors are at the magic angle, we find a $\cos^4 \theta$ contribution to the PAD's. We predict that the coherence terms have the most significant effect on the PAD's for transitions at low J_i.

From the analysis of NO two-photon PAD's (Allendorf *et al* 1989), we have deduced the magnitudes and phases of the radial dipole matrix elements for the partial waves. These are well determined when data for more than one laser geometry are fit simultaneously. When we attempt to fit this experimental data using coefficients that do not include coherence effects, we find that, although we get good results for the magnitude of the radial dipole matrix elements, we are unable to determine all relative phases.

When the angle between the polarization vectors of the lasers is varied, we predict that measurement of the ϕ dependence of the PAD's (a feature as yet unobserved) will provide a more intimate probe of the ionization dynamics.

References

Allendorf S W, Leahy D J, Jacobs D C, and Zare R N 1989 *J. Chem. Phys.* **91** 2216

Blum K 1981 *Density Matrix Theory and Applications* (New York: Plenum Press)

Cooper J and Zare R N 1968 *J. Chem. Phys.* **48** 764

Dixit S N and Lambropoulos P 1983 *Phys. Rev. A* **27** 861

Dixit S N and McKoy V 1985 *J. Chem. Phys.* **82** 3546

Leuchs G and Walther H 1984 *Multiphoton Ionization of Atoms* ed S L Chin and P Lambropoulos (Toronto: Academic Press) p 109

Press W H, Flannery B P, Teukolsky S A, and Vettering W T 1986 *Numerical Recipes* (Cambridge: Cambridge University Press) pp 523–528

Rudolph H and McKoy V 1989 *J. Chem. Phys.* **91** 2235

Viswanathan K S, Sekreta E, Davidson E R, and Reilly J P 1986 *J. Phys. Chem.* **90** 5078

Acknowledgments

We are grateful to P H Vaccaro for providing us with a computer program for the least-squares fit. KLR thanks the SERC for a NATO postdoctoral fellowship. This work was supported by the US Air Force under Contract AFOSR-89-0264.

Inst. Phys. Conf. Ser. No 114: Section 8
Paper presented at RIS 90, Varese, Italy, 16–21 September 1990

Resonant ionization in solution: Initiation of ion-molecule chemistry by REMPI

Joseph J. BelBruno* and Gary Siuzdak

Department of Chemistry, Dartmouth College, Hanover, NH 03755 USA

ABSTRACT: REMPI is employed to create ions in solution. The radical cations are then involved in ion-molecule chemistry with the solvent or added co-solutes. The formation of the ion is confirmed by pulsed conductivity measurements and transient absorption spectroscopy. Long-term photochemical products are also identified. The ion kinetics and a likely reaction mechanism are presented.

1. INTRODUCTION

Considerable research effort has been concentrated on determining the effects of ionizing radiation on organic solutions. Typically in radiation chemistry, an electron beam interacts with the solvent, so that the observed reactions involve a large quantity of reactive solvent molecules interacting with a much smaller concentration of neutral solute molecules. The technique described in this report overcomes this limitation of conventional radiation chemistry. Multiphoton induced ion-molecule chemistry is initiated by the creation of reactive *solute* species, that is, the selectivity of the REMPI technique permits ionization of the species of interest, rather than that present in the greatest amount. In this report, we describe our initial results (BelBruno, et al., 1990a,1990b) applying this technique to phenol and nitrobenzene in polar and non-polar solvents.

2. EXPERIMENTAL

Experiments involving long-term photo-pumping, pulsed conductivity measurements and transient absorption spectroscopy were conducted. In all cases, the light source was a Quanta Ray DCR11 Nd: YAG laser with appropriate third or fourth harmonic crystals. Photo-pumping studies utilized a specially constructed 2 cm^3 stainless steel cube cell. Irradiation time and laser intensity were selected so as to produce a reactant consumption of 10% or less. The solutions of phenol were 0.001M, but increased to 0.1M for nitrobenzene. The pulsed conductivity measurements a modified version of the same stainless steel cell. Routine measurements involved solutions of 10^{-5}M and fields of the order of 2kV/cm to 10kV/cm. Reaction intermediates were observed in real time (ns scale) by means of transient absorption spectroscopy. The spectrometer consists of a 0.25m monochromator and an optical multichannel analyzer. The probe flash is produced by a 0.5 ms flashlamp. Sample solutions were 10^{-4}M in target molecule and the flow rate through the cell was such that the entire volume was replaced between flashes.

3. RESULTS

No ionization or ion-molecule chemistry was observed for irradiation of phenol at 355nm or nitrobenzene at 266nm. For nitrobenzene, the excitation of the $\pi \rightarrow \pi^*$ transition was presumed to efficiently produce the benzyl radical, while for phenol the coherent two-photon absorption at 355nm produces radicals with subsequent recombination.

At 355nm the initial absorption for nitrobenzene in hexane is in the wings ($\varepsilon = \sim 100\ M^{-1}cm^{-1}$) of the $n \rightarrow \pi^*$ transition. Photochemical experiments indicate the production of nitrosobenzene, phenol and hexyloxybenzene with quantum yields (at 8mJ/pulse) of 0.021, 0.0053 and 0.00048, respectively. The nitrosobenzene observed in the current study depends upon the square of the laser intensity and is a primary product. No products are observed for the weakly (50cm focal length lens) focused pump beam nor is the observed product concentration dependent upon the time between production and analysis. Dark reactions and monophotonic chemistry are unimportant under the conditions of the experiments. Two-photon excitation would be expected to populate a high-lying singlet of the neutral molecule, which would decay, after internal conversion to a vibrationally excited ground state molecule, to nitrosobenzene. Since, a one-photon absorbance is known at 355nm, the likely mechanism for the two-photon excitation to the precursor state is sequential. The remaining two products depend upon the cube of the laser intensity and the mechanism for production of these products is also sequential. Pulsed conductivity experiments at this wavelength resulted in the typical trace shown in Figure 1. A plot of the area under the curve vs. laser intensity yields a quadratic dependence, indicating that the intensity regime probed was intermediate between that of a totally saturated process and the unsaturated (power dependence of 3) pathway. These data are consistent with a sequential multiphoton induced process and are further evidence of the ion-molecule nature of the reaction which leads to phenol and hexyloxybenzene. Transient absorption spectra were recorded. There are considerable experimental difficulties, since the ground state absorption is so intense in the 240-290nm range. However, in the region about 300nm, it was possible to observe a weak, but reproducible absorption, which has previously been attributed to the phenoxy cation. This is direct evidence for the production of ionic reactants.

Figure 1. Pulsed conductivity of nitrobenzene in hexane.

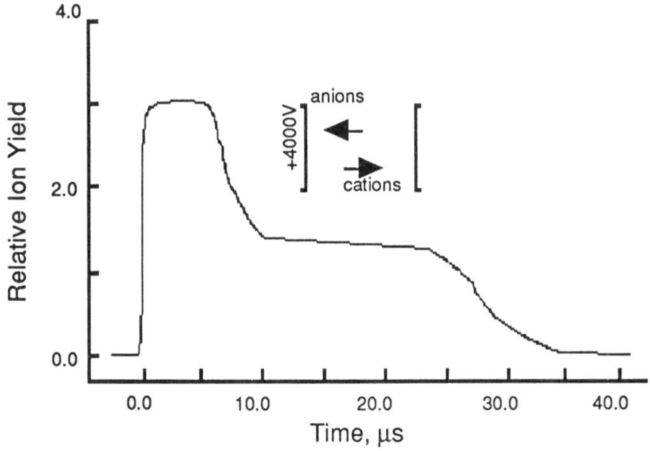

Preliminary photochemical results for the methanolic system were reported in an earlier communication (BelBruno, et al., 1990a). We have observed the production of nitrosobenzene, phenol and benzyl alcohol, all of which are consistent with the products reported above in the hexane solutions. The corresponding quantum yields at 8 mJ/pulse incident radiation were 0.034, 0.0075 and 0.00048, respectively. These yields are consistent with those found in hexane. The nitrosobenzene may again be attributed to a neutral photochemical reaction initiated by production of a highly excited singlet state. The small increase in quantum yield over that found in hexane is attributable to triplet formation from the photostate produced by absorption of the first photon in the two-step process. The remaining products were shown to depend upon the cube of the laser intensity, indicating an ionic precursor for the observed photochemistry. Transient absorption measurements detected the cation absorption in the region about 300nm.

It has been previously shown that 266nm multiphoton induced chemistry of phenol in hexane leads to phenoxyphenol and hexyloxybenzene with quantum yields of 0.022 and 0.026 at 0.5mJ/pulse. The laser intensity dependence of these products was shown to be I^2, confirming the multiphoton nature of the process and indicating that radical chemistry was not the source of the observed products. The qualitative effect of polar solvents on the 266nm photochemistry of phenol was studied by means of methanolic solutions. Although no quantitative measurements were attempted, analysis of the products indicated only the formation of o- and p-methoxyphenol. Concentration dependence studies proved that the phenoxyphenol product was second order in the concentration of phenol, while that of hexyloxybenzene was first order in the same reagent. Ion yields from pulsed photoconductivity measurements and the quantum efficiency indicate that less than 10^{-5} of the incident photons result in free ions. We estimate by extrapolation, that the absolute quantity of free ions produced at 500µJ/pulse and 10^{-5}M solutions to be 10^{13} ions/pulse or 0.1%. Calculations also indicate that of these free ions only approximately 0.1% participate in ion-molecule chemistry. Photolysis of 10^{-4}M solutions of phenol with 2.5ns pulses at 266nm resulted in intermediates which exhibited absorption maxima at 247nm and 300nm in hexane and 250nm and 290nm in methanol. Typical transient data, as a function of the delay time between photolysis and probing, is shown in Figure 2 for hexane. The band at 247nm is readily assigned to the phenoxy radical. The 300nm feature has not been previously reported, but is assigned to a cationic species.

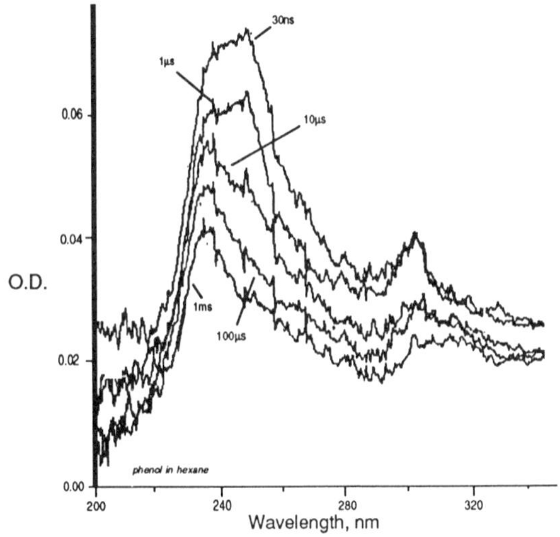

Figure 2. Transient absorption spectra of phenol in hexane

Longer wavelength transient data for methanolic solutions of phenol are assigned to the spectrum of the solvated electron. The absorption at these wavelengths is readily quenched by electron scavengers. The extinction coefficient of the cation was determined using the absorption of the solvated electron as a calibration. In methanol, 30ns after the laser photolysis pulse, the optical density at 630nm was 0.11. The extinction coefficient of an electron in methanol is known to be 1.15 x $10^4 M^{-1}cm^{-1}$ yielding an electron concentration of 9.6 x 10^{-6}M. Given an optical density for the phenoxy cation (at 300nm) of 0.012, an upper limit for the extinction coefficient is found to be 1.3 x $10^3 M^{-1}cm^{-1}$ assuming a 1:1 correspondence between electrons and cations. The order of the photoprocess at both 247nm and 300nm was determined by measuring the absorption intensity vs. laser pulse energy. A linear relationship was found for the 247nm band, but the 300nm band was observed to follow an $I^{3/2}$ dependence indicating a saturated two-photon ionization process.

4. DISCUSSION

In nitrobenzene, all of the evidence points to competitive absorption in the parent molecule leading to branching between neutral and ionic product channels. One pathway leads to nitrosobenzene via neutral bond cleavage, while the other leads to formation of the phenoxy cation after absorption of additional photons by the initially created radical cation. A total of three photons is necessary to ionize nitrobenzene which has a reported ionization potential of 7 eV in solution. Given this data, one concludes that it is the radical cation which is the reactive precursor for production of the new products, phenol and hexyloxybenzene. The radical cation would clearly be produced with a large excess energy, ~3.5eV, and fragmentation would be readily possible. The phenoxy cation is readily produced by MPI in the gas phase. The implication is that, in the liquid as well, fragmentation results in production of the phenoxy cation and that this is the ion involved in the ion-molecule chemistry leading to production of phenol, by hydrogen abstraction from the solvent, and hexyloxybenzene, by reaction with the solvent. Two additional considerations support this conclusion: photolysis of nitrobenzene in CCl_4 results in nitrosobenzene formation, but no other products and studies involving the phenylium cation reacting with methane indicate that hydrogen abstraction leading to formation of benzene is significantly favored, by a factor of ~20, over production of toluene. Rate constants for the cation with the solvent are obtained from data in Figure 2. The values for decay of the cation, by all processes, are $5.4 \times 10^2 s^{-1}$ and $5.0 \times 10^2 s^{-1}$, for methanol and hexane. Given that the uncertainty in these measurements is estimated to be 20%, the two rate constants are identical. The small rate constant may be attributed to the apparent lack of ion reactivity with the parent molecule.

For phenol, results show that the final fate is ionization via a 1+1 REMPI process. Two photons at 266nm supply 9.3eV of excitation energy. The ionization potential of gas phase phenol is 8.5eV; however, in solution previous studies have shown that it is reduced by a minimum of 2eV, and perhaps as much as 4eV. The net result is that the phenol radical cation is formed with ~4eV excess energy. The gas phase appearance potential for $C_6H_5O^+$ and H• is 3.8eV. Given the amount of excess energy in the radical cation, formation of the phenoxy cation is the probable fate of the 1+1 REMPI produced ions. A good case may be made on thermodynamic grounds for three-photon production of the phenoxy cation; however, this is inconsistent with our observations. Similarly, one may make a case for the active ion being the phenol radical cation. Again, this alternative is inconsistent with experimental observations, most especially those experiments involving anisole as the target molecule. In those studies using methanol as the solvent, the only products were methoxy phenol and methoxy anisole. These may only be derived from the phenoxy cation. All evidence points to the phenoxy cation as the final cationic species. The photoproducts are attributed to reactions of this cation with neutral phenol molecules or solvent.

The assignment of the two transient bands to different reactive species may also be made on the basis of the disappearance of the absorption band as a function of time in hexane and methanol. In the case of the phenoxy cation, the loss of the signal is attributed to reaction with the solvent or the parent compound. In the case of the phenoxy radical, the most probable fate is hydrogen abstraction from the solvent leading to regeneration of the parent molecule, phenol. This path is confirmed by the detection of phenol in the photolysis of anisole. The rate constant for the loss of the cation in methanol was determined to be $2.3 \times 10^5 s^{-1}$ and in hexane the same rate constant was measured as $0.8 \times 10^5 s^{-1}$. Similar measurements for the decay of the phenoxy radical (spectral feature at 247nm) were $0.6 \times 10^5 s^{-1}$ and $1.0 \times 10^5 s^{-1}$ in hexane and methanol, respectively. Measurements of rate constants via the transient absorption spectra are necessarily prone to error and these values should be taken as estimates of the relative rates of radical versus cation reactivity, rather than absolute values. However, the approximate factor of 3 difference in the rate constant was reproducible.

Finally, the conductivity measurements may be used in a time-of-flight mode. The measurements yield ionic mobilities and may, if sufficiently different, be used to distinguish reactant ions from product ions. The radius of the solvated cation for phenol in hexane was found to be 2.9Å, which is comparable to that for quinine (2.8Å). Furthermore, the anion was found to have a radius of 2.3Å which agrees with that for O_2 in hexane (2.2Å). Oxygen is the expected anion, even in degassed solutions, since a small, but not negligible, quantity remains.

5. REFERENCES

BelBruno J J, Siuzdak G and North S 1990a, *Laser Chem.* **10**, 177
BelBruno J J, Siuzdak G and North S 1990b, *Chem. Phys. Lett.* **166**, 167

Inst. Phys. Conf. Ser. No 114: Section 8
Paper presented at RIS 90, Varese, Italy, 16–21 September 1990

Photofragment imaging: The photodissociation of bromomethane, bromoethane and bromoethanol

David W. Chandler [1], John W. Thoman Jr. [2], Wayne P. Hess [3]

[1] Combustion Research Facility, Sandia National Laboratories, Livermore, CA 94550
[2] Department of Chemistry, Williams College, Williamstown, MA 01267
[3] Batelle Pacific Northwest Laboratories, Richland, WA 99352

ABSTRACT: Bromomethane, bromoethane and bromoethanol are photolyzed with 205-nm light and the velocity of the bromine atoms is recorded by the technique of photofragment imaging. The velocity distribution of the bromine atoms is a direct reflection of the internal-state distribution of the methyl, ethyl, and hydroxy-ethyl radicals and the orientation of the transition moment in the parent molecule.

1. INTRODUCTION

Photolysis of halogenated hydrocarbons is routinely used to produce radicals for kinetic studies, but the details of this process are not completely understood (Van Veen 1985). We describe the photolysis of three brominated organics and the measurement of the kinetic energy distributions of the bromine fragments. We use a variation of the resonance-enhanced-multiphoton-ionization time-of-flight (REMPI-TOF) technique called photofragment imaging (Chandler 1990 and Thoman 1988) in order to determine the bromine velocity following the 205-nm photolysis of bromomethane, bromoethane and bromoethanol. The partitioning of energy between the translational, vibrational and rotational degrees of freedom of the methyl, ethyl and hydroxy-ethyl radicals is discussed.

2. EXPERIMENTAL

The apparatus and data treatment is described in detail in the literature (Chandler 1990). A vacuum-distilled, brominated sample (5% in He) is expanded in a pulsed, skimmed, supersonic molecular beam into the ionization region of a time-of-flight mass spectrometer. A 205-nm, 5-ns, 1.5-mJ, polarized laser pulse photolyzes the brominated compound and creates neutral fragments in various translational, rotational, vibrational, and electronic states. A second, tunable, probe-laser pulse intersects the photofragments a few nanoseconds after photolysis. Light at 250.98 nm or 277.77 nm is used to resonantly ionize ground ($^2P_{3/2}$), or excited-state ($^2P_{1/2}$), bromine atoms, respectively. The photofragments move off the molecular-beam axis due to the translational energy release of the dissociation. Ionizing the fragments does not substantially change their velocity because of the small mass of the electron. Ions are detected by a position-sensitive ion detector (microchannel plate / phosphor screen / CCD camera combination) located at the end of the time-of-flight tube. Accelerating potentials are set such that the arrival times result in ~1.5-cm-diameter images. By measuring the arrival time and the distance the photofragments have moved off-axis, we obtain the speed of the fragments. The polarization vector of the photolysis laser (which is the symmetry axis of the photodissociation) is parallel to the face of the detector, and perpendicular to the time-of-flight axis (which is coincident with the molecular-beam axis). With this geometry, the three-dimensional spatial distribution of the selected photofragment is uniquely determined by taking the inverse Abel transform of the two-dimensional image recorded by the detector (Chandler 1990). Planar cuts through the reconstructed three-dimensional velocity distributions of the Br $^2P_{3/2}$ and Br $^2P_{1/2}$ obtained in the 205-nm photolysis of bromomethane, bromoethane and bromoethanol are presented in Fig.1.

3. RESULTS AND DISCUSSION

The first absorption continuum of these brominated organics consists of three repulsive electronic states, collectively called the $\tilde{A}$ band. In order of increasing energy they are the $^3Q_1$, $^3Q_0$ and ^1Q states. Excitation to these states leads to subpicosecond dissociation of the C-Br bond. The angular distribution of the bromine fragments therefore reflects the position of the transition moment in the parent molecule. For bromomethane, photodissociation experiments performed at wavelengths both greater and less than the 205 nm, the absorption maximum, indicate that excitation is primarily to the $^3Q_0$ state (Sparks 1981, van Veen 1984, and Barry 1984). This state correlates to the formation of Br $^2P_{1/2}$, and is coupled to the ground state via a parallel transition moment. Absorption via a parallel transition leads to Br $^2P_{1/2}$ with a predominantly $\cos^2\theta$ angular distribution (Fig. 1a), where the angle θ is measured with respect to the polarization vector of the photolysis light. The ^1Q state correlates to formation of Br $^2P_{3/2}$. Absorption to the ^1Q state is via a perpendicular transition and thus the angular distribution of Br $^2P_{3/2}$ should show a dominant $\sin^2\theta$ distribution. For bromomethane this is indeed the case (Fig. 1d), but this is less so for the bromoethane and bromoethanol (Figs. 1e,f). Excited-electronic-state curve crossing as the molecules dissociate mixes the states and adds a $\sin^2\theta$ component to the Br $^2P_{1/2}$ distribution and a $\cos^2\theta$ component to the Br $^2P_{3/2}$ distribution. The angular distributions of each Br $^2P_{3/2}$ and each Br $^2P_{1/2}$ are similar in shape and magnitude, indicating that the nature of the electronic states of the three different parent molecules that are excited at 205 nm are similar.

In Table 1 we list the maximum and minimum kinetic energies (total and Bromine) obtained from the data displayed in Fig. 1. The fact that the velocity distributions of the Br atoms are almost identical for all dissociations is somewhat surprising, because from simple conservation of linear momentum one would expect that the Br atoms from bromoethanol would be moving faster than those coming from bromomethane. The photon energy (140 Kcal/mole) minus the total kinetic energy of the system associated with the fastest Br $^2P_{3/2}$ atom, gives 67 kcal/mol, 81 kcal/mol and 94 kcal/mol for the bromomethane, bromoethane and bromoethanol systems, respectively. Time-of-flight measurements (van Veen 1985) indicate a C-Br bond strength in bromomethane of 66±0.5 kcal/mol. The C-Br bond strength is expected to be slightly stronger in bromoethane (67 kcal/mol) and stronger still for bromoethanol (68 kcal/mol) (Hintsa, 1990). Why do the fastest bromine atoms correlate to the C-Br bond strength in the photodissociation of bromomethane but not in the photodissociation of bromoethane or bromoethanol? We believe the answer to this question is that the ethyl and hydroxy-ethyl fragments are always produced in rotationally excited states. We detect the methyl fragments from the bromomethane photodissociation and find that some methyl radicals are produced in the ground vibrational state for both the $^2P_{3/2}$ and the $^2P_{3/2}$ channels. We have made extensive studies of the rotational excitation of the methyl fragment following the photodissociation of iodomethane through the equivalent electronic transitions (Chandler 1989) and find that there is very little rotational energy imparted to the methyl radical in the dissociation. This is expected since the halogen atom pushes against the center-of-mass of the methyl fragment during dissociation. These observations are consistent with the correlation of the fastest bromine atoms to the bond dissociation energy of the C-Br bond.

In the photodissociation of bromoethane and bromoethanol, the impulse is not directed at the center-of-mass of the radical fragment and substantially more rotational energy is expected to be deposited into the radical fragment. Because ground-vibrational-state methyl fragments are formed in the photodissociation of bromomethane, a Franck-Condon argument would indicate that comparable amounts of vibrational-ground-state ethyl and hydroxy-ethyl radicals are formed in the dissociation of their brominated analogs. If the fastest bromine atom detected correlates with the vibrationless radical, as is the case in bromomethane photodissociation, then the minimum amount of rotational energy imparted to the radical fragment equals the photon energy minus both the bond dissociation energy and the maximum total kinetic energy observed. The minimum E_{rot} for the ground state Br channel is ~15 kcal/mol for bromoethane and ~27 kcal/mol for bromoethanol. A simple impulsive dissociation model, treating the bromoethanol molecule as a Br-(CH_2)-(CH_2OH) pseudo-triatomic with a bend angle of 93° and a (CH_2)-(CH_2OH) bond length of 1.3 Å, corresponding to the *cis* configuration, predicts an average rotational energy imparted to the hydroxy-ethyl radical of 42 kcal/mol. For the *trans* configuration an average rotational energy of 27 kcal/mol is predicted. These values are consistent with the minimum value of ~27 kcal/mol obtained from our time-of-flight data. This same model applied to bromoethane dissociation predicts that on average 27 kcal/mol of rotational energy will be deposited in the ethyl fragment, consistent with the minimum measured value of ~15 kcal/mol we determine from our Br time-of-flight data. We take this agreement as partial confirmation of our interpretation that the difference in energy is due to rotational excitation. Because we are measuring the velocity of the heavy fragment and inferring the energy of the lighter fragment (which takes away most of the kinetic energy) small errors in measurement of either the arrival time or size of the image leads to significant error in the deduced minimum rotational content. Our estimated 5% error in the velocity of the bromine fragment from CH_3Br leads an uncertainty of ~10 Kcal/mole error in the measured CH_3Br dissociation energy. This is less of a problem for the heavier radicals.

The velocity distributions in Fig. 1 for Bromomethane show are distinctly broader on the equator than at the poles. This is a reflection of the different dissociation channels (Br and Br*) leading to different amounts of excitation in the methyl fragment. The more energetic Br channel leading to more internal excitation of the CH_3 is formed by a perpendicular transition and therefor is found primarily at the equator. Because of curve crossing the enhanced width at the equator is observed in both channels.

In Table 2, we list the percentage of energy in the translation, rotation and vibration of the radical fragments. The assumption that the fastest bromine atoms correspond to the minimum amount of rotational energy deposited in the organic radical is implicit in the values given in Table 2. For our data we can determine the amount of translational energy in each fragment but we are unable to determine the partitioning of the remainder of the energy between the vibrational and rotational degrees of freedom in the organic radical.

4. ACKNOWLEDGEMENTS

This work is supported by the U. S. Department of Energy, Office of Basic Energy Sciences, Chemical Sciences Division. JT thanks the Associated Western Universities, Inc. under a contract with the Department of Energy for an AWU-DOE faculty fellowship.

5. REFERENCES

Barry M. D. and Gorry P. A. 1984 *Mol. Phys.* **52** 461

Chandler D. W., Thoman J. W. Jr, Janssen M. H. M., and Parker D. H. 1989 *Chem. Phys. Letts.* **156** 151

Chandler D. W., Janssen M. H. M., Stolte S., Strickland R. N., Thoman J. W. Jr., and Parker D. H. 1990 *J. Phys. Chem.* **94** 4839

Hintsa E. J., Zhao X., and Lee Y. T. 1990 *J. Chem. Phys.* **92** 2280

Sparks R. K., Shobatake K., Carlson L. C., Lee Y. T. 1981 *J. Chem. Phys.* **75** 3838

Thoman J. W. Jr., Chandler D. W., Parker D. H., and Janssen M. H. M. 1988 *Laser Chem.* **9** 27

van Veen G. N. A., Baller T., and DeVries A. E. 1985 *Chem. Phys.* **92** 59

van Veen G. N. A., Baller T., DeVries A. E., and van Veen N. J. A. 1984 *Chem. Phys.* **87** 405

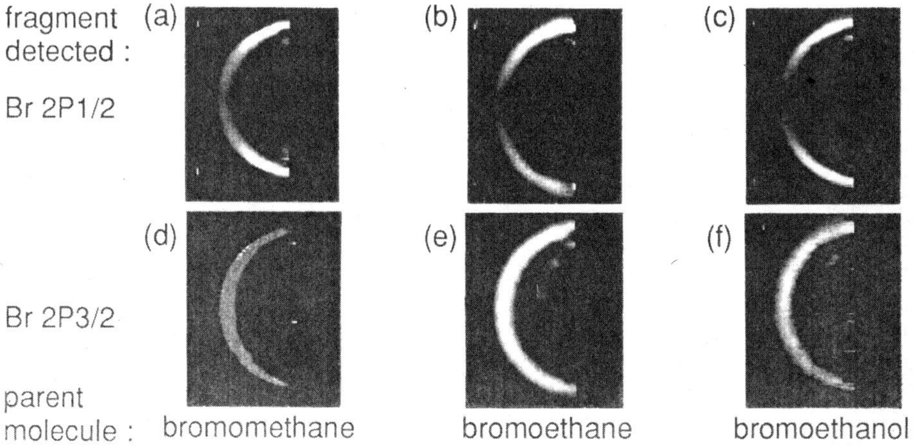

Fig.1: Cross section through the velocity distributions of the electronically-excited state, (Br $^2P_{1/2}$) and ground-state, (Br $^2P_{3/2}$) bromine atoms following the photolysis of bromomethane, bromoethane, and bromoethanol. These distributions were obtained by performing a inverse Abel transform on the two-dimensional images recorded in the lab.

Table 1: Kinetic Energy maximum / minimum (in kcal/mol) for the Br and Br* and calculated total kinetic energy for both fragments.

	bromomethane	bromoethane	bromoethanol
Br $^2P_{3/2}$ K.E.	11.5 / 8.6[(a)]	15.7 / 15.8	16.7 / 9.1
Total K.E.	73 / 45.8	59 / 31	46 / 25
Br $^2P_{1/2}$ K.E.	10.3 / 8.6	14.3 / 9.4	15.1 / 9.9
Total K.E.	65 / 45.7	54 / 35	42 / 27

[(a)]Error estimate for the absolute values of the measured energies for the Bromine fragments are ±5%. This translates into larger errors for the total K. E. See text for discussion.

Table 2: Partitioning of Excess Energy into the Degrees of Freedom of the Radical (%)

Radical	Minimum E_R[(a)]	Minimum E_T	Remaining Energy[(b)] $E_V + E_R + E_T$
CH_3	0	65	35
CH_3CH_2	15	45	40
CH_2CH_2OH	34	35	31

(a) Assuming that some vibrationless radical is formed (see text). This was confirmed for the case of CH_3Br by observation of the $3p_z \leftarrow X\ 0^0_0$ band of the methyl fragment.

(b). This quantity is the percentage of energy that can not be ascribed to a particular degree of freedom for the photodissociation. In our experiment we measure the translational energy of each Br atom directly and can ascribe a translational energy to each bromine atom but are unable to determine the partitioning of the residual energy between vibration and rotation.

Inst. Phys. Conf. Ser. No 114: Section 9
Paper presented at RIS 90, Varese, Italy, 16–21 September 1990

The optical parametric oscillators of bariumborate: A new source for powerful tunable laser radiation in the ultraviolet, visible and near infrared

A. Fix, T. Schröder, and R. Wallenstein
Fachbereich Physik
Universität Kaiserslautern

Since the first demonstration of an optical parametric oscillator (OPO) by Giordmaine and Miller(1) in 1965 the OPO has been subject to detailed theoretical and experimental investigations (2,3). The OPO is considered as a source of powerful, broadly tunable coherent radiation. The development and the scientific application has been hampered, however, by the scarcity of nonlinear optical materials with suitable optical and mechanical properties.

Recently the new nonlinear material Bariumborate (BBO) became available. With the unique properties (i.e. high nonlinearity, wide transparency range and high damage threshold) BBO should be a very useful material for an OPO. This OPO combines the advantages of an all–solid–state tunable source with a wide tuning range in the ultraviolet, visible and near infrared, high peak and average power and high conversion efficiency.

Pumped by the harmonics of the Nd:YAG laser, the BBO–OPO generates tunable laser radiation in the spectral range between 240 and 3000 nm. Fig. 1 shows the calculated tuning regions of the signal and idler waves as function of the phase–matching angle. The OPO's are pumped by 532 nm, 355 nm, 266 nm and 213 nm Nd:YAG laser radiation.

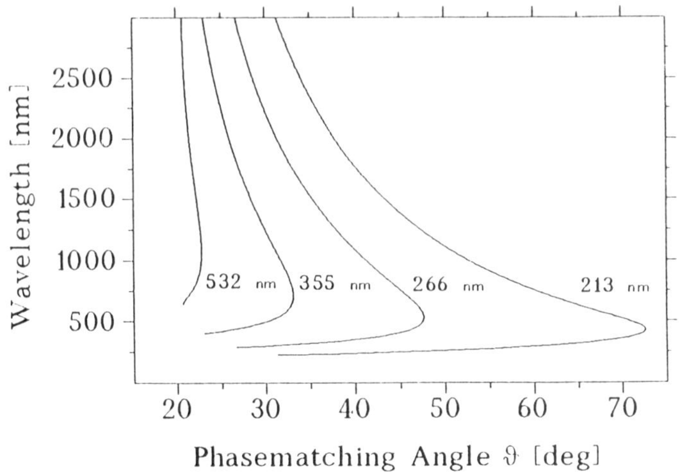

Fig. 1: Wavelength of the signal and idler wave of the BBO–OPO as function of the phasematching angle. The OPO is pumped by the second, third, fourth and fifth harmonic of a Nd:YAG laser.

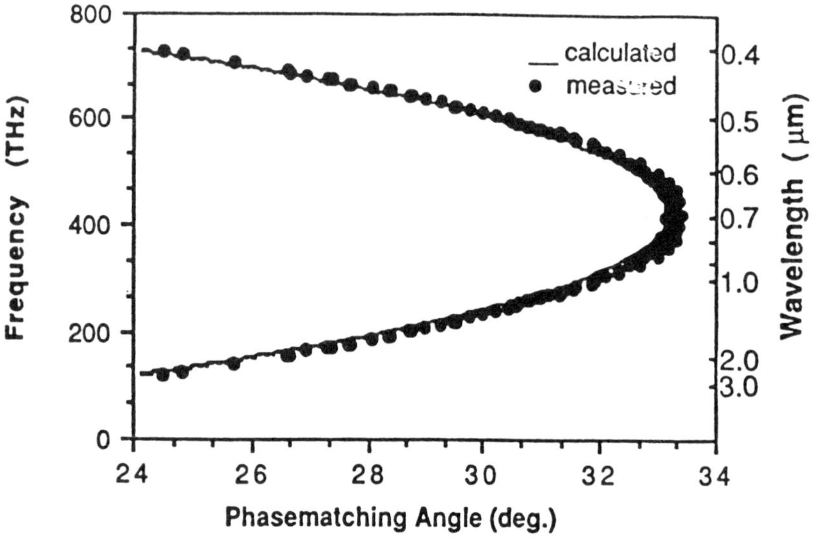

Fig. 2: Measured and calculated wavelength of the signal and idler wave as function of the phasematching angle of the BBO optical parametric oscillator (Ref. 5).

In a first experimental demonstration a BBO–OPO was pumped by the 532 nm second harmonic of a Nd:YAG laser [4]. The wavelength tuning (940 – 1220 nm) was limited by the used mirrors. The tuning range could be extended substantially into the UV and the near infrared (330 nm – 2550 nm) by pumping with the 355 nm third or the 266 nm fourth harmonic of a Nd:YAG laser or a 308 nm XeCl excimer laser. This was demonstrated in several experiments first reported in 1988 at the CLEO conference [5] [6] [7].

In the experiments performed at the Stanford University[5] the BBO–OPO was pumped by single axial mode 355 nm third harmonic Nd:YAG laser radiation. The Nd:YAG pump laser with unstable resonator (Spectra Physics, Model DCR3) was injection seeded to obtain single–mode operation (Spectra Physics Model 6300 Injection Seeder). The doughnut shaped output beam was spatially filtered in the far field by a suitable pinhole and frequency tripled with KD*P. The 355 nm radiation with almost Gaussian intensity distribution (diameter d = 2.8) provided 6 ns long light pulses with an energy of 30 mJ/pulse and a repetition rate of 30 Hz.

The 3.2 cm long resonator consisted of two flat mirrors with a reflectivity of 98 percent (input mirror) and about 80 percent (output mirror). The transmission at 355 nm and at the idler wavelength exceeded 80 percent. The BBO crystals (size 6 x 6 x 12 mm^3) were cut at an angle of 25 and 35 degrees. Fig. 2 displays the measured wavelengths of the signal and idler radiation as function of the type I phasematching angle. The estimated experimental accuracy of each angle and wavelength measurement was about ±1 degree and ±1 nm, respectively. Within these uncertainties the measured values were in good agreement with those calculated from the Sellmeier formula reported by Kato[8]. The observed tuning range extended from 412 nm to 710 nm (signal wave) and 710 nm to 2.6 μm (idler wave). This operating range corresponds almost to the maximum possible tuning range which is limited by the increasing absorption of BBO at wavelengths larger than 2.5 μm [9]. As shown in Fig. 2 the measured phasematching angle varies between 24.5 and 33.2 degrees.

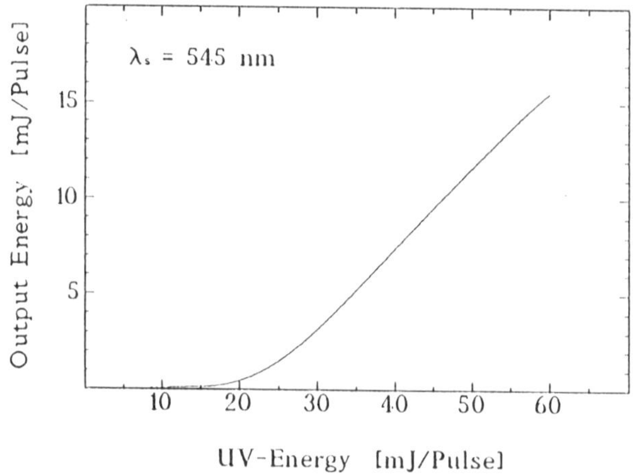

Fig. 3: Output energy as function of the energy of the 355 nm pump pulse. The BBO crystal is 12 mm long; the wavelength of the signal wave is 545 nm.

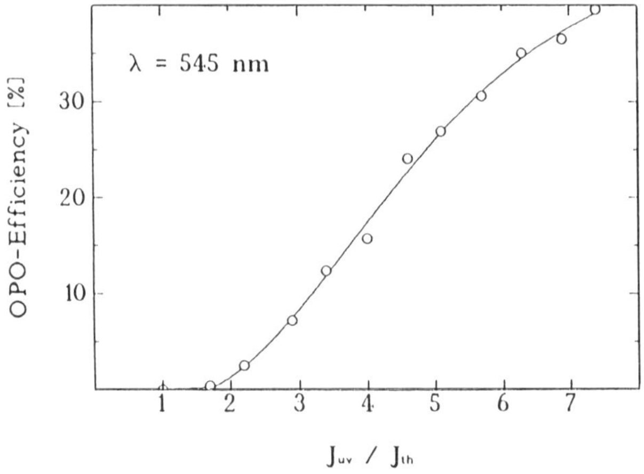

Fig. 4: Crystal–internal efficiency as function of the ratio of the pump energy J_{uv} and the pump energy J_{th} at threshold. The length of the BBO crystal is 12 mm; the wavelength of the signal wave 545 nm.

Besides the wavelength tuning, parameters like threshold power, conversion efficiency and the spectral width of the generated light are of special importance.

Fig. 3 shows the total output energy as function of the energy of the 355 nm pump radiation. At an input pulse of 60 mJ the output exceeds 15 mJ. The corresponding internal conversion efficiency – which takes the losses at the mirrors and the crystal surfaces into account – is close to 40 % (see Fig. 4). With higher input powers the output could be increased considerably. In recent experiments(10) the BBO–OPO was pumped by the third harmonic of a powerful Nd:YAG laser (Spectra Physics, Model DCR–4). At input powers of 220 mJ the OPO output exceeded 100 mJ. The total efficiency was 45 %, the slope efficiency as high as 60 %.

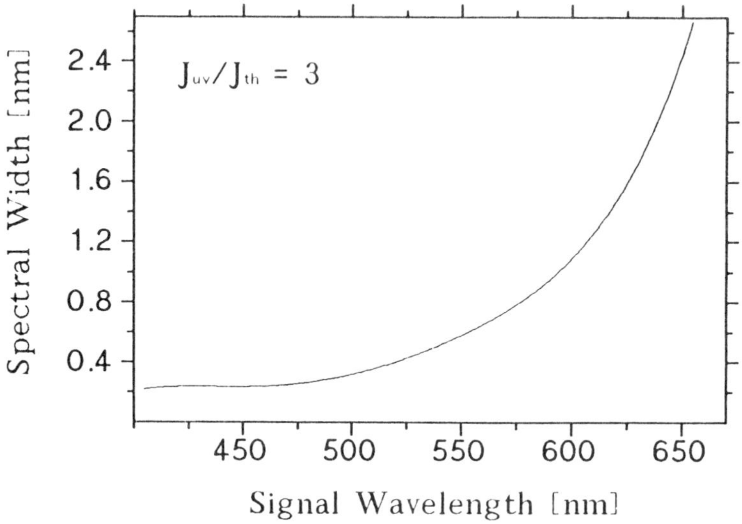

Fig. 5: Spectral linewidth of the unseeded OPO at wavelengths between 410 and 650 nm.

The bandwidth of the OPO output depends on the wavelength, the parameters of the resonator, the linewidth of the pump radiation and on the pump power.

At a ratio of the pump energy J_{uv} and the pulse energy at threshold J_{th} of $J_{uv}/J_{th} = 3$ the linewidth increases with wavelength from about 0.2 nm at 410 nm to almost 4 nm at 650 nm (see Fig. 5).

Many applications require narrowband operation. Pumping with a single–axial mode laser source (5) the output contained typically 6 axial modes of the 3.2 cm long OPO resonator indicating a linewidth of about 23 GHz. Single mode operation was achieved at 532 nm or 1,06 μm (and at the corresponding idler and signal wavelength) by injection seeding with light of the second harmonic or the fundamental of the single mode Nd:YAG pump laser. The injection seeding not only provided single–mode operation but also reduced the oscillator build–up time from 4 to 2.2 nsec. Because of the reduced build–up time the pulse length increased from 2.5 nsec to 4 nsec. Simultaneously the OPO threshold power decreased by a factor of 3.

In a detailed investigation the OPO was seeded with the visible radiation of a single–mode cw dye laser. A seed power as low as 0.3 mW was sufficient for stable single–mode OPO operation. As an example Fig. 6 shows recordings of the linewidth of signal and idler–wave monitored with a pressure–tuned Fabry–Perot. Taking the contribution of the low–finesse Fabry–Perot into account the linewidth of the OPO is less than 500 MHz. As demonstrated recently (10), the seeded OPO can be tuned continuously over wavelength ranges of up to 10 nm.

In addition to the systems mentioned so far, synchronously pumped BBO–OPO devices have been investigated. With the second harmonic of a mode–locked pulsed Nd–YAG laser (11) or the third harmonic of a Nd:YAP system (12) the BBO–OPO produced 75 psec long light pulses tunable in the regions of 680 nm to 2.4 μm (11) or pulses of about 20 psec duration at 406 nm to 3.17 μm (12), respectively. More recently synchronously excited BBO–OPO was pumped with a continuously mode–locked and Q–switched Nd:YAG laser system (13).

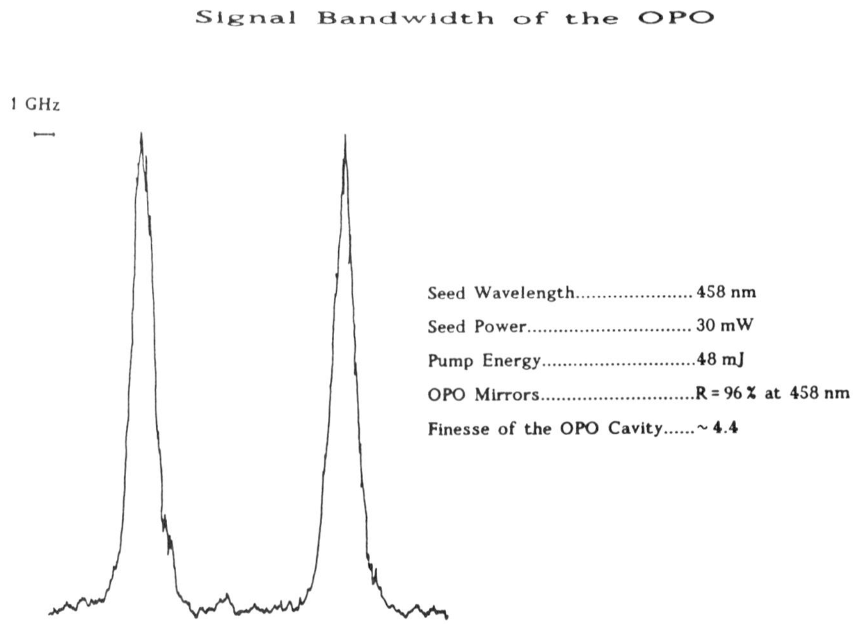

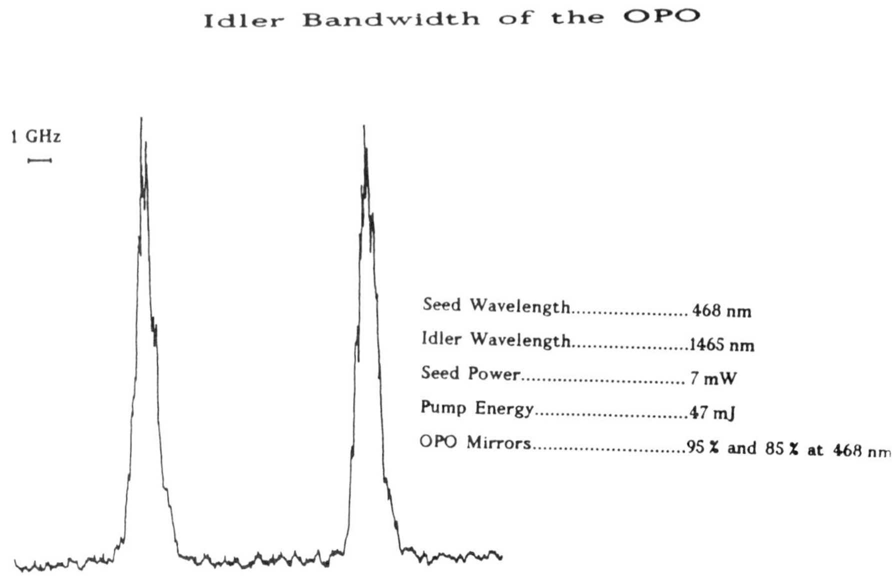

Fig. 6: Spectral width of the signal– and idler wave of a BBO–OPO seeded with a single–mode cw dye laser (for further details: see text).

In summary these first experiments clearly demonstrate the advantages of the BBO–OPO as tunable coherent light source. The most attractive features are the high power capability, high conversion efficiency, and in particular the large tuning range.

The OPO tuning range can be extended into the UV and VUV as well as into the infrared by nonlinear optical methods. The second–order sum–frequency generation of UV radiation and of the OPO–infrared idler wave should generate in LBO UV radiation at wavelengths as short as 160 nm [14]. A further extension into the VUV and XUV is provided by frequency mixing in gases (15–19). In the infrared powerful tunable radiation of narrow spectral width will be produced at 3 – 30 μm by stimulated Raman scattering in H_2.

The results obtained so far are certainly very promising for the further development of this new tunable source which should provide coherent radiation continuously tunable from the XUV to the far infrared.

References:

(1) J.A. Giordmaine and R.C. Miller, Phys. Lett. 14, 973 (1965)
(2) S.E. Harris, Proc. IEEE 57, 2096 (1969)
(3) R.L. Byer, in Treatise in Quantum Electronics, edited by H. Rabin and C.L. Tang (Academic, New York, 1973) pp. 587–702 see also: Y.X. Fan and R.L. Byer in SPIE Proceedings Vol. 461, 27 (1984)
(4) Y.X. Fan, R.C. Eckardt and R.L. Byer, Conference on Lasers and Electro–Optics (CLEO) 1986 (San Francisco), postdeadline paper ThT4
(5) Y.X. Fan, R.C. Eckardt, R.L. Byer, J. Nolting and R. Wallenstein, Appl. Phys. Lett. 53, 2014 (1988)
(6) H. Komine, Opt. Lett. 13, 643 (1988)
(7) L.K. Cheng, W.R. Bosenberg, D.C. Edelstein and C.L. Tang Phys. Lett. 54, 13 (1989)
(8) K. Kato, IEEE J. Quant. Electron., QE–22, 1013 (1986)
(9) D. Eimerl, L. Davis, S. Velsko, E.K. Graham and A. Zalkin, J. Appl. Phys. 62, 1968 (1987)
(10) A. Fix, J. Haub, B.J. Orr, T. Schröder and R. Wallenstein (to be published)
(11) L.J. Bromley, A. Guy and D.C. Hanna, Opt. Commun. 67, 317 (1988)
(12) S. Burdulis, R. Grigonis, A. Piskarskas, G. Sinkevicius, V. Sirutkaitis, A. Fix, J. Nolting and R. Wallenstein, Opt. Commun. 74, 398 (1990)
(13) A. Piskarskas, V. Smilgevicius, A. Umbrasas, A. Fix and R. Wallenstein, Opt. Commun. 77, 335 (1990)
(14) Ch. Huang, A. Borsutzky, R. Brünger und R. Wallenstein Appl. Phys. (in print)
(15) R. Hilbig, G. Hilber, A. Lago, B. Wolff and R. Wallenstein Comments At. Mol. Phys. 18, 157 (1986)
(16) G. Hilber, A. Lago and R. Wallenstein, JOSA B 4, 1753 (1987), and references therein
(17) G. Hilber, A. Lago and R. Wallenstein, Phys. Rev. A 36, 3827 (1987)
(18) G. Hilber, D.J. Brink, A. Lago and R. Wallenstein, Phys. Rev. A 38, 6231 (1988)
(19) J. Nolting, H. Kunze, I. Schütz, and R. Wallenstein Appl. Phys. B 50 331 (1990)

Paper presented at RIS 90, Varese, Italy, 16–21 September 1990

Copper vapour laser pumping of dye lasers in atomic and nuclear research

C E Webb
The Clarendon Laboratory, Parks Rd.,Oxford, OX1 3PU, England.

ABSTRACT: Copper Vapour Lasers offer unique advantages as pump sources for pulsed dye lasers, especially in applications to RIMS and AVLIS where their high repetition rate and efficiency is important. The power and efficiency capabilities of currently available copper laser pumped dye laser systems will be reviewed.

1. INTRODUCTION

Copper Vapour Lasers (CVLs) possess unique advantages as sources of pulsed radiation for pumping dye lasers. They operate simultaneously at two wavelengths - 510.6 nm (green) and 578.2 nm (yellow) - producing pulses 10-50 ns in duration with energies of several mJ per pulse, at repetition rates in the range 2-32kHz. The overall efficiency of commercially available CVL systems is typically 0.5 - 1%, so that an air-cooled CVL with a mean output power of 15W can draw the 3kW it needs from a domestic single phase electricity supply. Water-cooled versions demand only about a tenth of the cooling water supply of typical ion lasers despite the fact that commercial CVLs are typically more than twice as powerful. The most common power output range for CVLs is 40-60W but units giving 100W are commercially available. Table 1 lists the performance characteristics of some of the CVL units manufactured by Oxford Lasers Ltd.

TABLE I Copper Vapour Laser Parameters

Model No.	**ACL25**	**ACL45**	**ACL 100**
Average power (W)	25	45	100
Beam Diameter (mm)	25	42	60
Max Pulse Energy (mJ)	4	9	20
Repetition Frequency			
Standard Range (kHz)	8-20	5-8	4-6
Extended Range (kHz)	2-32	3.5-20	4-10
Pulse Width* (ns)	15-80	20-60	30-65
Power Consumption (kW)	4.5	6.5	14

* Depending on repetition frequency selected

2. THE IMPORTANCE OF HIGH REPETITION RATE

The reason why the CVL is so well matched to the requirements for RIMS applications is evident from the following considerations:

The average transit time T of atoms moving with thermal velocity v through a region of laser irradiation of dimension a is $T=a/v$. The probability P that any given atom will experience irradiation during its flight through the irradiated region is given by $P=fT$,where f is the pulse repetition frequency. In order that all of the atoms should have a good probability of being irradiated (i.e. that P should be of order unity) f must be of order $1/T$.

In a typical RIMS apparatus, the laser beam diameter is of order 1-10mm, and thermal velocities are typically of order $100ms^{-1}$. Thus, to satisfy the requirement that every atom passing through the irradiation zone has a chance of experiencing irradiation requires that the pulse repetition frequency f should be of order 10-100 kHz. Among currently available pulsed laser systems, the CVL uniquely has the ability to satisfy this requirement with an appreciable energy per pulse.

3. SATURATION ENERGY FLUENCE

Although CW lasers can easily satisfy the requirement for a high probability of irradiation of each atom, the intensities available from CW lasers are such that the excitation rates are generally smaller than spontaneous decay rates from the levels involved. The populations which can be achieved in the intermediate levels therefore tend to be much smaller than their saturated values, and thus CW irradiation can only be used when the highest excitation efficiencies are not required.

An estimate of the average power required in order to provide efficient excitation of the intermediate steps in a RIMS scheme employing CVL pumped dye lasers can be made as follows: In order to achieve saturation of population on a transition of optical cross-section Q cm^2 it is necessary to supply a fluence of order $1/Q$ photons cm^{-2} per pulse at the irradiation zone (Letokhov 1987). Assuming a beam diameter of 3mm at this zone, for radiation near 600nm (dye laser fundamental) a laser pulse of 1J produces a fluence of about 3.10^{19} photons cm^{-2}. For a CVL pump repetition rate of 6kHz, the tuned dye laser power W thus required is given by:

$$W = \frac{6.10^3}{3.10^{19}} \cdot \frac{1}{Q} = \frac{2.10^{-16}}{Q} \text{ watts.}$$

Since fully allowed atomic transitions have cross sections in the range 10^{-11} to 10^{-13} cm^2, the required condition can be easily met both at the dye laser fundamental and at the second harmonic which would require about 20 times the dye laser power assuming a doubling efficiency of 10% (which is readily achievable). It is only in the case of non-resonant ionization from excited states, that cross-sections of $10^{-17}cm^2$ or smaller are encountered. The energy per pulse available from CVL pumped dye lasers then fails to satisfy the above condition implying that the efficiency falls below the ideal value. However, good ionization efficiency is still possible in most cases where a autoionizing state in the continuum can accessed from an intermediate excited state with radiation within the the tuning range of the CVL pumped dye laser. Such autoionizing resonances can have cross-sections of $10^{-16}cm^2$ or larger, and a dye laser with output of order 2 watts can therefore ionize them with close to ideal efficiency.

It is this type of scheme, involving two or three excitation steps using fundamental or frequency doubled radiation from a CVL pumped dye laser, in which the final step is excitation to an autoionizing state or to a Rydberg state which can be ionized by application of a weak Stark field, which has been employed by Letokhov and co-workers for much of their elegant work on trace element detection (Letokhov 1987,1986). Similar ionization routes were also employed by Trautmann and co-workers at the University of Mainz who recently reported (Ruster *et al* 1989) a considerable advance in the detection of trace quantities of plutonium and technetium using a RIMS technique in which two 40W CVL units were used to pump three commercial dye lasers (Lambda Physik FL2001).

4. CVL RESONATOR DESIGN FOR DYE LASER PUMPING

In order that the dye solution in the pumped region of the couvette should maintain good optical homogeneity, it is necessary that the flow velocity should be high enough to clear the dye in the pumped stripe and to replace it with fresh dye in the interval between successive pulses. Efficient operation of the dye laser also requires that all of the CVL pump power should be concentrated into a stripe which is less than 1mm wide in the direction of the dye flow. In order to focus the CVL output beam this tightly with a beam waist long enough to cover the thickness of the dye stream, requires the replacement of the CVL's normal plane-plane resonator with an unstable resonator. Although, as indicated in Fig.1, the overall power of the CVL is reduced by the addition of the unstable resonator, its beam divergence and hence its focusability is vastly improved.

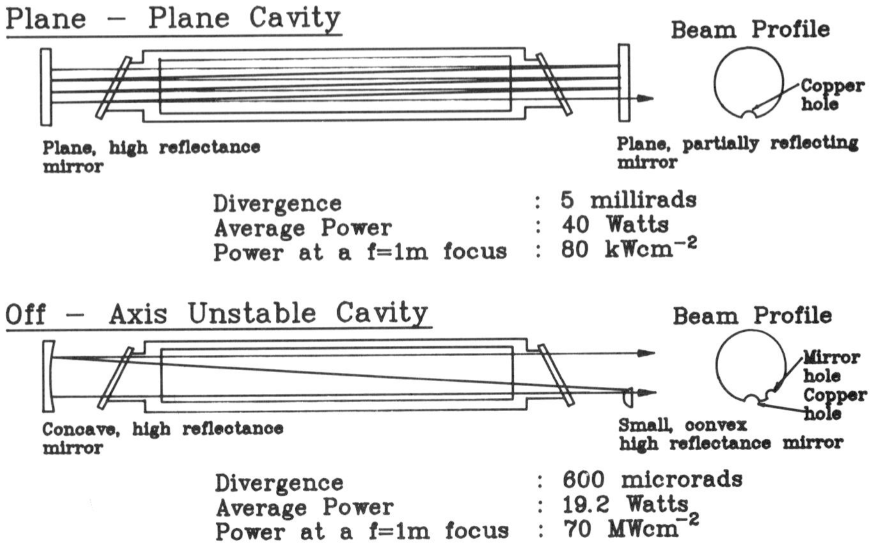

Fig.1. Resonator Designs for Copper Vapour Lasers

5. CVL PUMPING OF COMMERCIAL DYE LASERS

Two commercially available pulsed dye lasers (Lumonics Hyperdye-300, and the Lambda Physik FL 3001) have been successfully modified (from their original configuration for Nd:YAG/excimer pumping) to use CVL pumping.

The fact that the pump wavelengths are in the visible region of the spectrum implies that the fundamental tuning range of a CVL-pumped dye laser is restricted to wavelengths longer than 530nm. Of course, using pump radiation of relatively low photon energy and short duration brings the corresponding advantage that the dyes are subject to much less rapid photolytic degradation than when they pumped by an ultraviolet laser. As shown in Table 2, conversion efficiencies of 25% at the tuning peak of Rh6G have been measured. Matching the polarization of the pump laser to that favoured by the dye laser improves the efficiencies by some 20-30%. Excellent beam quality (important for frequency doubling) and frequency stability have been obtained at dye laser output power levels of several watts.

TABLE 2 CVL PUMPED DYE LASER PERFORMANCE

	Hyperdye-300	**FL3001**
Bandwidth (GHz)	4	4-7*
Conversion Efficiency (%)	25%	25%
Beam Divergence (mrad)	< 5	n/m
Frequency Stability (GHz)	< 1	n/m

* Can be reduced to less than 2GHz with intracavity etalon.
n/m Not measured.

6. ATOMIC VAPOR LASER ISOTOPE SEPARATION PROCESS - AVLIS

The CVL is also the pump laser selected for the AVLIS process pioneered in the USA at the Lawrence Livermore National Laboratory from the early 1970's. An important feature in this selection was the ability of the CVL to be used as an amplifier with even better extraction characteristics than it possesses as an oscillator. A chain of CVLs can be built in which the beam from the primary oscillator CVL passes down a chain of amplifier units, extracting from each a power which would be greater than the power extracted from the individual units used as oscillators. In this way CVL beams with several kilowatts of mean power can be generated. Dye laser systems again involving oscillator amplifier configurations pumped by such high power CVL arrays are under construction in several countries including the US, UK, Israel, France and Japan for the exploration of the commercial potential of the AVLIS process. The UK effort is centred at BNF plc, Capenhurst, and employs 100W CVL modules supplied by Oxford Lasers Ltd.

7. REFERENCES

Letokhov V S 1986 *Laser Analytical Spectrochemistry* (Adam Hilger)
Letokhov V S 1987 *Laser Photoionization Spectroscopy* (A P)
Ruster W,Ames F, Kluge H-J, Otten E-W, Rehklau D, Scheerer F, Herrmann G, Muhleck C,Riegel J, Rimke H, Sattleberger P, and Trautmann N 1989 Nuc. Inst. and Methods in Phys. Res. **A281** 547

Paper presented at RIS 90, Varese, Italy, 16–21 September 1990

Characterization of the laser enhanced ionization and photoionization techniques as photon detectors

N. Omenetto[1)], B.W. Smith[2)] and P.B. Farnsworth[3)]

1) Joint Research Centre, Environment Institute, Ispra (Varese), Italy
2) Department of Chemistry, University of Florida, Gainesville, Fla., USA
3) Brigham Young University, Provo, Utah, USA

ABSTRACT: A photon detector consisting of an air-acetylene flame containing magnesium atomic vapour is theoretically discussed and characterized in terms of its **quantum efficiency** and **ion yield**. It is shown that, under optimized experimental conditions, the quantum yield of the magnesium flame ionization detector approaches unity.

1. Introduction

Because of its versatility, a conventional atmospheric pressure air-acetylene flame has been most widely used as atom reservoir in analytical laser enhanced ionization (LEI) spectroscopy. An attractive use of the flame in combination with the ionization technique was proposed and evaluated several years ago as well as more recently (Matveev *et al.*, 1979; Smith *et al.*, 1990). The basic idea is that the atomic system and the ionization technique combined together form what can be classified as a *photon detector based upon ionization*. The practical implementation of such detector, whose principle is illustrated in Figure 1, can be as follows: a given metal atom, M, emits resonance photons of energy $h\nu_M$ as a consequence of some excitation process. These photons are collected and transferred to the flame containing pure M vapour, mainly in its ground state. The absorbed photons will then create M* excited atoms. If the flame is simultaneously illuminated with one or two lasers which ionize M*, the flame will act as an ionization detector for the primary photons emitted at $h\nu_M$. These photons can be the result of thermal excitation (e.g., a plasma discharge), electrical excitation (e.g., a glow discharge), or radiative excitation (e.g., fluorescence or Raman photons). The primary figure of merit of a photon detector must be its "quantum efficiency", which is the ratio between the number of charges created in the detector and the number of incident photons. Another pertinent parameter is the "ion yield", i.e., the ratio between the number of charges created during the interaction and the number of atoms present in the probe volume. The two concepts are different, since a given atom could experience 10^4 excitation and quenching collisions before a collisional ionization occurs (Havrilla *et al.*, 1982). This explains the different basic requirements that the ionization method has to fulfill: (i) if it is used as an *analytical* method, the ion yield should be unity, irrespective of the number of absorption transitions (i.e., primary photons) necessary to form an ion pair; (ii) if, on the other hand, it is used as a *photon detector*, its quantum efficiency should be unity, so that one ion pair is created for every primary photon absorbed.

In this paper, an attempt is made to characterize theoretically a photon detector based upon the ionization of magnesium in an air-acetylene flame.

2. Simplified theoretical model and definitions

The atomic level diagram pertinent to our ionization detector is shown in Figure 2. As shown by the shaded wavy lines in the figure, two-photon photoionization is disregarded. Non resonant photoionization (308 nm) and photoionization into an autoionizing level at 300.9 nm (Bradley *et al.*, 1976) are also considered, in addition to collisional ionization from the 5 d level.

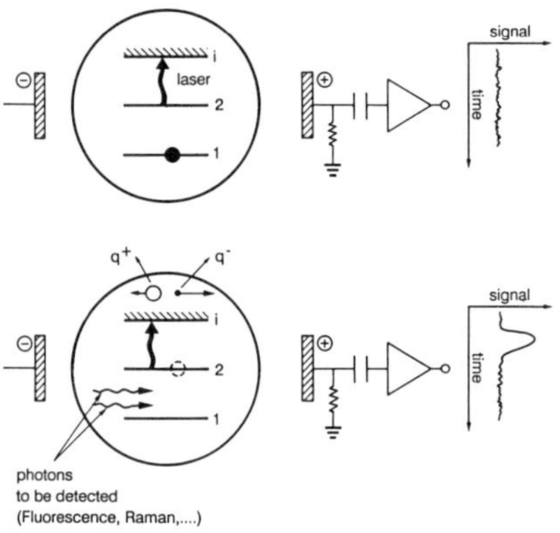

Fig. 1 Oversimplified scheme of the operating principle of the ionization detector.

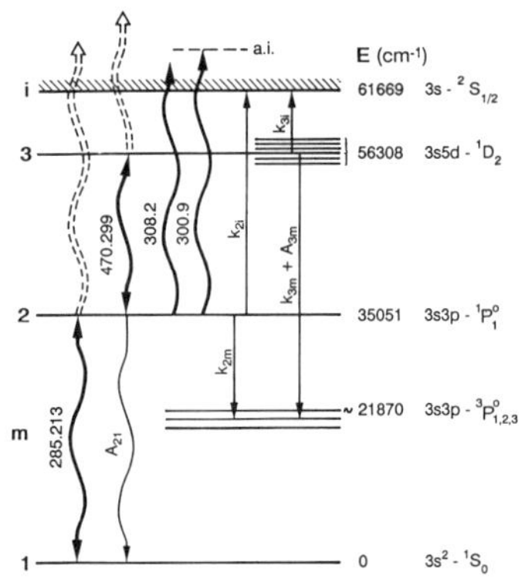

Fig. 2 Energy level diagram for magnesium. Collisional (k's) and radiative (A's) rate coefficients (s^{-1}) are indicated.

In terms of the rate coefficients given in Figure 2, and in the limiting case in which *all incident photons are absorbed*, we can define the quantum efficiency, η_i, of the detector as the ratio between the total ionization rate coefficient pertinent to the scheme adopted, R_i, and the sum of all the rates of depopulation of the level, i.e., $(R_i + R_m + R_{21})$. The quantity $\xi_i = [R_i/(R_i+R_m)]$ can be defined as the "*ion branching ratio*", i.e., the relative proportion of atoms which are lost by ionization with respect to the total number of atoms lost in the two "traps" (ionization continuum + mestastable level). In absence of a metastable trap, ξ_i is unity ($R_m = 0$) and η_i correctly represents the fractional probability of ionization, in complete analogy with the quantum efficiency of resonance fluorescence (see also Axner *et al.*, 1989).

The above definitions can be combined to give the expression:

$$\eta_i^* = \frac{\xi_i^*}{1 + \xi_i^* \dfrac{R_{21}}{R_i}} \tag{1}$$

As seen in Eq. (1), η_i^* cannot exceed ξ_i^*.

3. Ion yield and fluorescence expressions

By considering the magnesium scheme shown in Figure 2 as a 5-level system, a large spectral bandwidth laser of duration T_p, and using the rate equation approach, the temporal behaviour of the various number densities of the excited atoms as well as of the ions can be obtained by solving the system of differential equations with standard mathematical procedures. The *instantaneous ion yield* of the detector is given by the expression:

$$Y_i(t) \equiv \frac{n_i(t)}{n_T} = \xi_i^* \left[1 - \exp \left(-R_{12} \frac{\eta_i^*}{\xi_i^*} t \right) \right] \tag{2}$$

which again emphasizes the difference between the two definitions, since Y(t) can still approach unity even if $\eta_i \ll 1$. For the *integrated* ion yield, one simply needs to replace the variable t with T_p. In a similar way, one can derive the instantaneous as well as the integrated fluorescence signals. The analytical expressions obtained show that, by time resolving the saturated fluorescence waveform *within* the laser pulse, in the presence and in the absence of the second (ionizing) laser, the quantities $(k_{2i} + k_{2m})$ and $(R_i + R_m)$ can be *directly* obtained. In addition, these measurements allow the evaluation of the decrease in the fluorescence signal due to the presence of the ionizing laser, i.e., of the fluorescence dip (Omenetto *et al.*, 1987). It is interesting to relate the *integrated* fluorescence dip, Δ', to the integrated ion yield. The results of this exercise are shown in Figures 3 (which refers to the photoionization case) and 4 (which refers to collisional ionization). Several considerations can be made after inspecting these figures. Firstly, in both cases, the highest value of Y_i cannot exceed ξ_i. For $Y_i = \xi_i$, $\Delta' = 1$. Secondly, in the photoionization case, for $Y_i = 0$, $\Delta' = 0$, while in the collisional ionization case, for $Y_i = 0$ $\Delta' = [g_3/(g_1+g_2+g_3)]$. Thirdly, if the metastable level is absent, i.e., $\xi_i = 1$, an experimental measurement of Δ' would give *directly* Y_i.

It is worth stressing that the equations given in the figures refer to the case of low quenching to the metastable, and are valid for short probing times, i.e., $T_p \sim 10$ ns.

4. Evaluation of the quantum efficiency

From the preceding discussion, an experimental evaluation of η_i will require the following measurements: i) *an absorption measurement*, to find the magnesium concentration which gives 100% absorption; ii) *an absolute measurement of the integrated ion yield*. If ξ_i is less than unity,

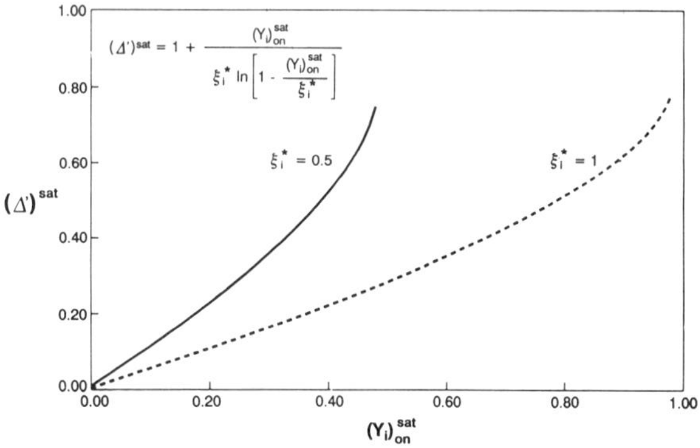

Fig. 3 Theoretical dependence of the integrated fluorescence dip (under saturation) versus the integrated ion yield in the case of *direct photoionization* from level 2. The behaviour of the dip is given for two values of the ion branching ratio.

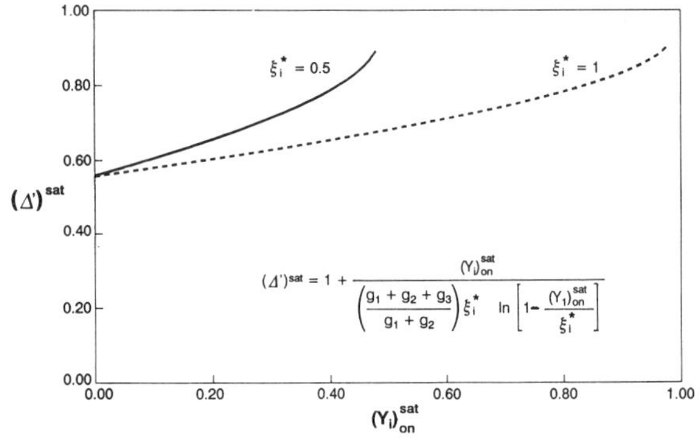

Fig. 4 Theoretical dependence of the integrated fluorescence dip (under saturation) versus the integrated ion yield in the case of *collisional ionization* from level 3. The behaviour of the dip is given for two values of the ion branching ratio.

this parameter cannot be obtained from a measurement of the integrated fluorescence dip. It must therefore be measured directly (see, e.g., Smith *et al.*, 1986); iii) a *time-resolved measurement of the resonance fluorescence* waveform within the laser interaction time.

Preliminary absorption, fluorescence and ionization measurements were performed with two excimer-pumped dye laser (Lambda Physik, Göttingen FRG and Jobin Yvon, Longjumeau, France) tuned at the bound-bound transitions shown in Figure 2. The total loss rate out of level 3, if essentially dictated by A_{3m}, is approximately 10^8 s^{-1}. Since the 5d level reached with the second laser (470.299 nm) lies 0.66 eV from the ionization potential, k_{3i} is likely to be at least 5 times greater (Smith *et al.*, 1986, Axner *et al.* 1989). Indeed, no fluorescence could be observed, at the many possible transitions coupling the 5d level and the 3p level, after excitation with the second laser at 470.299 nm. It is therefore concluded that $\xi_i \geq 0.83$. An integrated fluorescence measurement, obtained in the presence and in the absence of the second laser step, gave a relative fluorescence dip of 83.5%. With $\xi_i = 0.83$, and referring to Figure 5, the ion yield during the probing time is found to be 0.76. If a quenching rate, R_{21}, of ~ 10^8 s^{-1} is appropriate for our flame, then, with (R_i/R_{21}) ~ 5, η_i is calculated to be $\geq$ 0.73.

Clearly, these are theoretical speculations and a more rigourous experimental evaluation is still warranted. However, it can be seen that, in order to approach unity ionization efficiency, the detector must be characterized by a *quenching-free* environment and by the *absence of traps* other than the ionization continuum. Work is now in progress in order to verify these predictions.

5. Literature

Axner O and Berglind T 1989 Appl. Spectrosc. **43** 940

Bradley D J, Dugan C H, Ewart P and Purdie A F 1976 Phys. Rev. A **13** 1416

Havrilla G J, Weeks S J and Travis J C 1982 Anal. Chem. **54** 2566

Matveev O I, Zorov N B and Kuzyakov Y Y 1979 J. Anal. Chem. USSR **34** 654

Omenetto N, Turk G C, Rutledge M and Winefordner J D 1987 Spectrochim. Acta **42B** 807

Smith B W, Hart L P and Omenetto N 1986, Anal. Chem. **58** 2147

Smith B W, Omenetto N and Winefordner J D 1989, Spectrochim. Acta B, Special Supplement, 101

Design and test of a Daly-type detector for RIMS

P. Benetti*, G. Fossati*, M. Rossella*, A. Tomaselli and F. Sigon#

University of Pavia (Italy)
* also INFN, sezione di Pavia
# ENEL-Crtn, Milano

ABSTRACT: A Daly type ion detector has been assembled for a Rims-Tof instrument. The design criteria have been initially tested with Magplot program for the numerical simulation of the magnetic field, then with Simion for the simulation of the trajectories of ions and electrons in the crossed magnetic and electric fields. The magnetic field is adjustable, coil-generated outside the vacuum system and applied through a C-type configuration. The same instrumentation as for the previous MCP's transducers has been utilized. The experimental results obtained in the photoionization of an Al atomic beam are discussed.

1. BASIC CONSIDERATIONS AND DESIGN

A Rims-Tof instrument has been assembled at University of Pavia. In particular, microchannels plates (MCPs) are used as the ion-electron transducer.

In the meanwhile we have been looking for a device that maintaining the large useful area, time resolution and single ion detection sensitivity of the MCPs, would also be reliable, rugged and with quantum efficiency nearly 100%. In principle, a Daly-type detector (Daly 1960) seems to fit all these requests and therefore it has been decided to built one prototype.

In our instrument, see Figure 1, ions coming from the Tof equipotential drift tube (-3 kV), enter the (Daly-type) detection system and are then further accelerated without modifying their trajectory until they strike a dynode (for instance at -10 kV) with an energy sufficient to get suitable secondary electron emission. The emitted electrons are themselves accelerated within the same electric field configuration and deflected by means of a magnetic field normal to the ion trajectories. In this way the electrons are directed toward a β^- scintillator with the addition of a small time jitter; of course the higher masses of parent ions lead to only small perturbation in their trajectories.

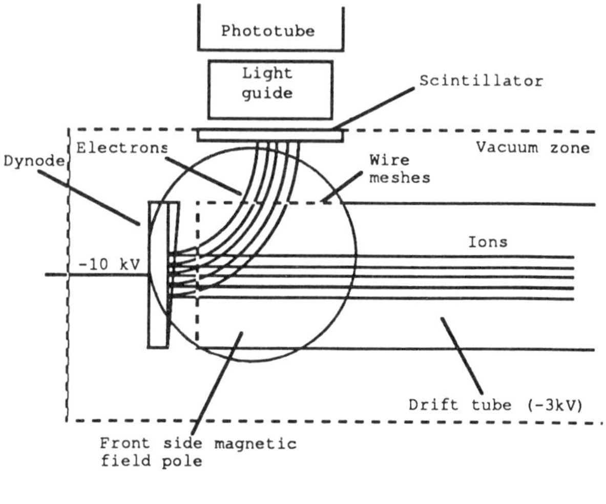

Fig.1. Layout of the Daly detector

The photon yield from the scintillator is detected by a photomultiplier tube placed outside the vacuum system, Figure 2.

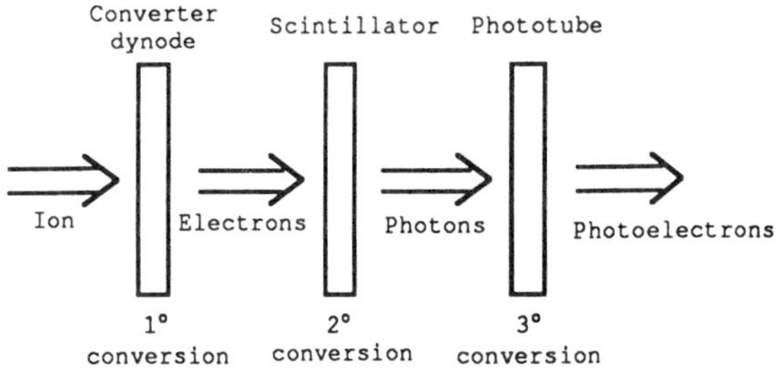

Fig.2. Conversion chain in Daly detector

The design of a Daly detector comprises essentially the following steps: a) evaluation of the conversion processes involved and b) study of the electric and magnetic field configuration needed to optimize the space and time electron collection on the scintillator. In the following these points will be briefly discussed.

As about the conversion processes involved in the detection system, special care was taken in the choice of the ions excited secondary electrons metallic emitters, according to the literature (Lichten 1967, Kaminsky 1965). On a general ground it can be stated that, for a given ionic species, the yield grows as the energy goes up to few hundreds of keV; for different ionic species, at low energies the yield decreases with their masses while this tendency is inverted at high energies.

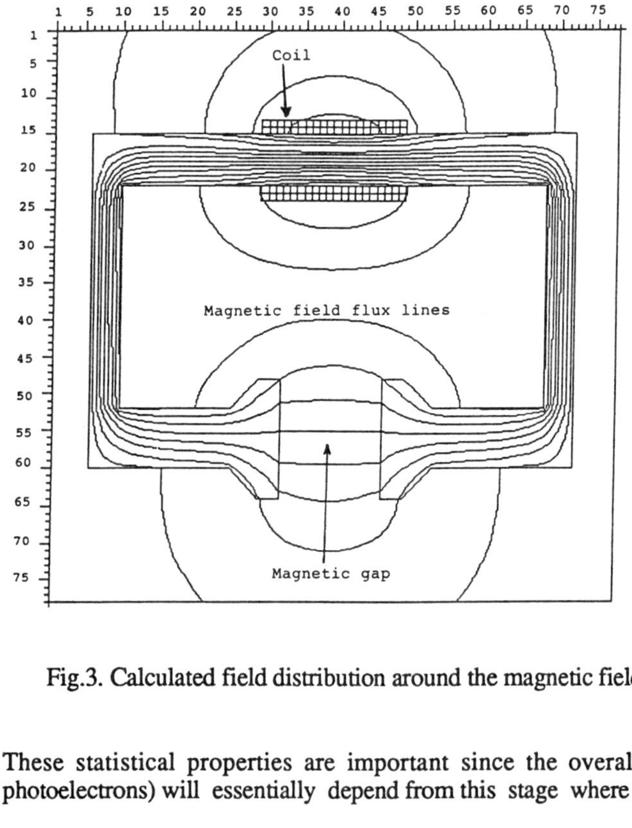

Fig.3. Calculated field distribution around the magnetic field

Furthermore, molecular ions are more conversion effective than atomic ions of the same energy and mass. Finally, the average yield of secondary electrons per ion is a function of the target material and of the incidence angle. Referring to the specific literature for the details, we only observe that if the target composition involves only one chemical element the probability distribution of the emission process is Poissonian (Dietz and Sheffield 1973) while for a "bi-element" target it is a Polya distribution (Dietz and Sheffield 1975).

These statistical properties are important since the overall conversion efficiency (ion-photoelectrons) will essentially depend from this stage where the probability of at least one emitted electron per ion from the Polya distribution is given by:

$$1-P_0 = \begin{cases} 1-(1+b\mu)^{-1/b} & 0<b<1 \\ 1-e^{-\mu} & b=0 \end{cases}$$

P_0 = probability of no-electron emission

μ = mean value of the distribution

b = parameter controlling the shape of the distribution

As a dynode we have used an oxidized aluminum (easier to get and safe to manipulate rather than Be) for which an at-normal-incidence conversion efficiency larger than 99.3 % was measured at 30 keV (Dietz and Sheffield 1973).

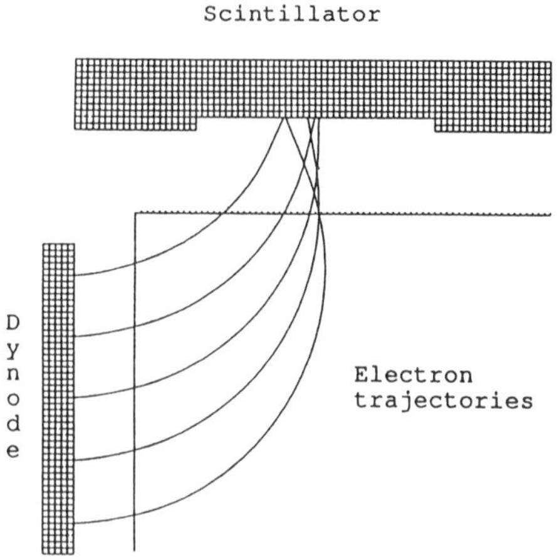

Fig.4. Calculated electrons trajectories

A C type magnetic circuit, external to the vacuum chamber, has been chosen as the source for the magnetic field, while the dynode and scintillator are in vacuum.
This decision has been determined by some worry in placing a coil into the vacuum system because of outgassing and by considerations about field confinement in the useful region of the electrons flight coupled with geometrical and space constraints. The detector has been designed with the help of numerical simulation programs, taking into account the geometrical and electrical features of the preexisting Rims-Tof apparatus.The first step was to evaluate the proper magnetic field for the electron deflection by means of MAGPLOT program (Saltek Services Van Nuys, USA). The output of this program, whose an example is shown in Figure 3, becomes the input for a second program SIMION (Dahl and Delmore 1988) which performs a full electrical and magnetic analysis of the trajectories of the ions and the electrons, Figure 4. The study and evaluation of these trajectories allowed for the optimization of the scintillation positioning, the dimensions of the magnetic pole pieces and some minor modifications noticeably affecting the collection efficiency and the transit time spread of the electrons. Last but not least, for a given mechanical configuration, a bijection between optimal values of electric and magnetic fields can be tabulated.

2. EXPERIMENTAL AND RESULTS

Once completed the simulations, the detector has been built around a six ports cubic cross. The plastic scintillator used, with suitable response to low energy β^-, is Bicron 400 disc (ϕ= 38 mm, thickness 1 mm); the rise and fall times are 0.9 and 2.7 ns, respectively.
The surface striked by electrons is coated with a 54 nm aluminum film to prevent detection of optical photons coming from the Tof.

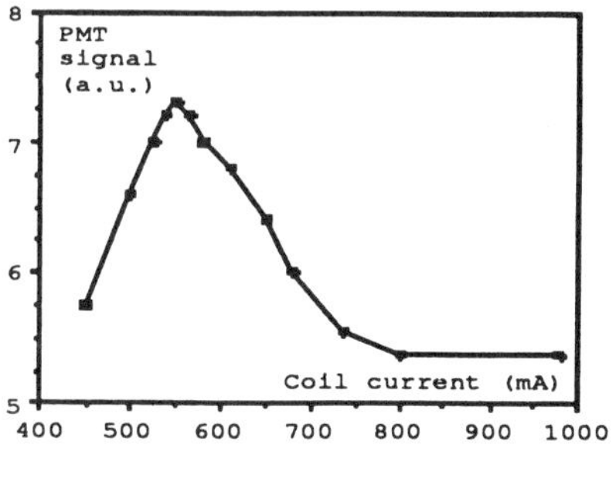

Fig. 5. Ion current vs coil current, at -9 kV dynode voltage

A cylindrical plastic light guide (ϕ =35 mm, h = 40 mm) is placed between the scintillator window and the PMT input (EMI 9964).This light guide, in addition to μ-metal shielding, avoids perturbations inside the PMT from the magnetic field.
First of all, the validity of the simulation has been checked. To this purpose, all but the central area of the scintillator has been masked, then a voltage of -9 kV set for the dynode. According to the simulations, only a small, well defined range of magnetic field (or coil current) values should be operative: a good agreement has been found, as shown in Figure 5. The unit has been operated up to - 12 kV; at higher voltages erratic internal microdischarges took place.
Finally, a simple Rims experiment has been done photoionizing an atomic Al beam through one color (396.1 nm), two steps process.
A record of the obtained Tof spectrum is shown in Figure 6. The resolving power of the system turns out to be R=400, slightly larger than obtained using MCPs in one equivalent experiment.

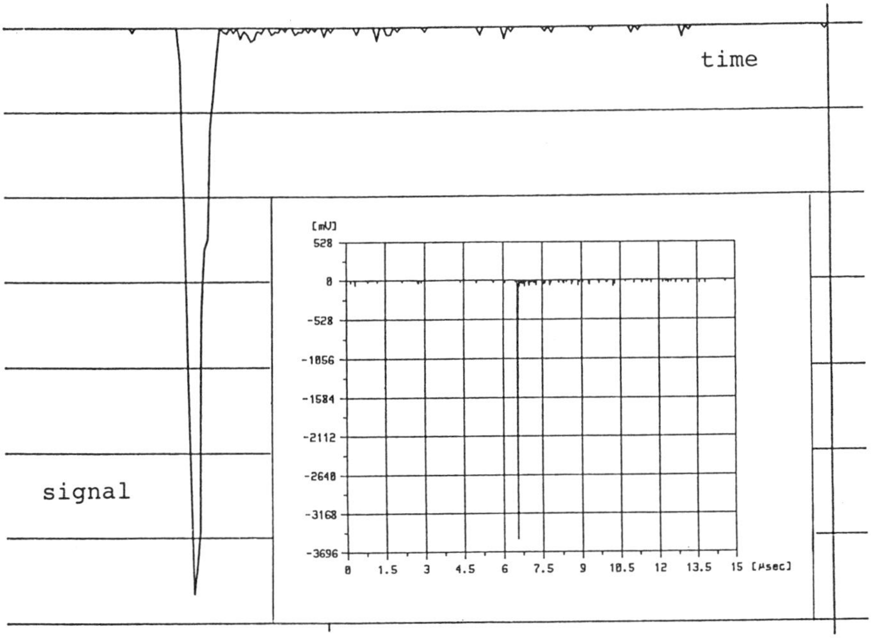

Fig.6. Part of the time of flight spectrum of photoionized Aluminum .
In the inset the full recorded spectrum

This work is still in progress; in particular it is necessary to further test the overall sensitivity of the system since, at first analysis, it appears worse than expected.

References

Daly N R 1960 Rev.Sci.Instrum. 31 264
Lichten W 1967 Phys.Rev. 164 131
Kaminsky M 1965 "Atomic and ionic impact phenomena on solid surfaces" (New York: Academic)
Dietz L A and Sheffield J C 1973 Rev.Sci.Instrum. 44 183
Dietz L A and Sheffield J C 1975 J.Appl.Phys. 46 4361
Saltek Services Van Nuys Ca 91409 USA
Dahl D A and Delmore J E 1988 EGG-CS-7233 Informal Report

Paper presented at RIS 90, Varese, Italy, 16–21 September 1990

The use of autoionizing states for efficient detection of Sr in resonance ionization mass spectrometry

Lars Bengtsson, Ove Axner, Halina Rubinsztein-Dunlop*

Department of Physics, Chalmers University of Technology, S-412 96 Göteborg, Sweden and
*Department of Physics, University of Queensland, Queensland 4072, Australia

ABSTRACT

An efficient excitation scheme for RIMS of Sr has been investigated. Two - colour excitation has been used for excitation of the atoms to an autoionization level with both transitions in the visible region. The first step excitation was at 460.733 nm ($5s^2\ ^1S_0 \rightarrow 5s5p\ ^1P_1$) and the second step leading to ionization was at 405.32 nm which connects the first excited state with a $^1D_2$ autoionizing state at 46380 cm^{-1}. This type of RIMS excitation scheme ensures selectivity in both steps as well as an efficient degree of ionization. The choice of the visible wavelengths helps also to minimize the influence of spectral interferences as well as wing excitations of matrix species.

1.INTRODUCTION

Resonance Ionization Mass Spectrometry (RIMS) is a spectroscopic technique for elemental analysis which is characterized by an extremely high selectivity. In RIMS the sample has to be first atomized. The second step is to excite the created free atoms by resonant laser light and subsequently ionize them. The final step of the technique is the detection of the created ions in the mass spectrometer.

The high selectivity of RIMS is obtained by optimal conditions for each step of the technique. The furnace in which the sample to be analysed is placed is optimized to ensure the highest possible atomization efficiency by a proper choice of vaporization temperature. The proper choice of the excitation scheme for each individual element is of great importance for the best selectivity as well as the choice of the ionization step (unless the atoms are ionized by photoionization to the continuum).

Although the method in itself possesses a high selectivity, it is of importance to strive for as high efficiency as well as as high selectivity as possible in each separate step. In the present report a very useful excitation scheme for most efficient detection of Sr in RIMS is presented in which the ionization step is performed to an autoionizing state giving the selectivity not only in the excitation step but also in the ionization step.

The experimental set-up is described below. In this set-up the atoms under study can be ionized either through photoionization to the continuum or by being excited to an autoionizing state.The ionization from an autoionization state has several advantages as compared to photoionization to the continuum. The most important is that the cross sections for excitation to autoionizing states are significantly larger (orders of magnitude) than those to the continuum. This implies that the ionization step does not need such high intensity as when excited atoms are being photoionized. The need for moderate intensities in the ionization step is vital in order maintain the selectivity of the excitation step - which otherwise might decrease if immense intensities are being used (excitation of matrix elements on the wings of ordinary transitions, by two-step excitation, or by direct photoionization).

2. EXPERIMENTAL

The RIMS apparatus has been described in details in ref.(Bengtsson 1990). It consists of a closed tantalum furnace with a small outlet channel. This arrangement produces a collimated atomic beam. The furnace is heated by electron bombardment from surrounding tungsten filaments. Typically, the electrons impact on the furnace with a kinetic energy of ≈ 1 keV, and the current is continuously adjustable up to ≈ 100 mA.

The laser light interacts orthogonally with the atomic beam 20 cm above the furnace, see figure 1. The produced ions are accelerated by electric fields into the field free region of a reflectron type Time -of - Flight mass spectrometer. The ion detector is a channel electron multiplier capable of an amplification of 10^8. The ion signal from the channel electron multiplier may be connected via to different devices for further signal analysis. The signal can be analysed either by a gated boxcar integrator or in a digital sampling oscilloscope equipped with optional signal averaging modes.

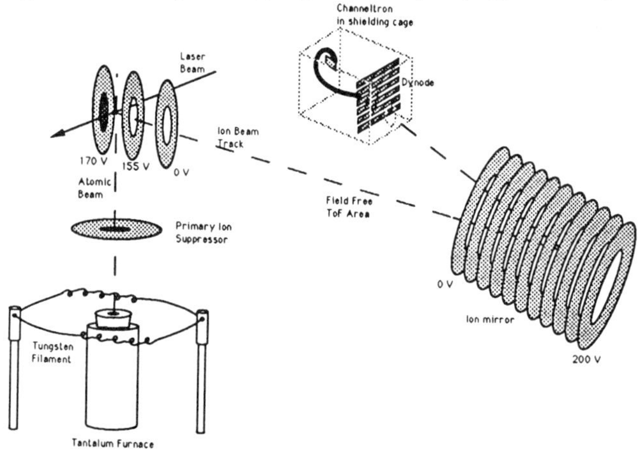

Fig. 1. A schematic diagram of the RIMS apparatus.

3. RESULTS AND DISCUSSION

Strontium is an element which has a rather low ionization limit (45925 cm^{-1}). In addition, Sr can be very conveniently excited from its ground state (5s^2 $^1S_0$) to its first excited level (5s5p $^1P_1$ at E = 21698 cm^{-1}) using visible light (460.733 nm) which is easily accessible by most modern laser systems.

Esherick (1977) has previously found a strong J = 2 autoionizing resonance approximately 450 cm^{-1} above the ionization limit (most probably the 4d^2 $^1D_2$ state at E ~ 46380 cm^{-1}). See Fig. 2. This resonance can be reached from the first excited state by using light of 405.32 nm wavelength. Also this wavelength can be produced rather conveniently by most dye-laser systems. Hence, an efficient excitation scheme which combines a high ionization efficiency with remained high selectivity has been found.

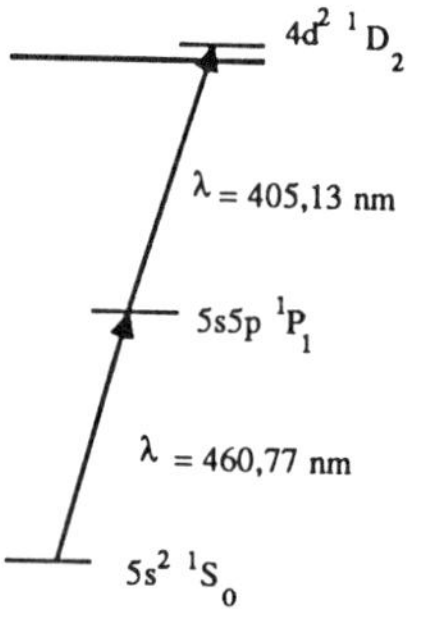

Fig. 2. Simplified energy level diagram of Sr.

This autoionizing state has previously been observed by Hart et al. (Hart 1985) in a study of laser induced stepwise and two-photon ionization schemes for strontium in an air acetylene flame. In that study, however, the resonance was detected by a direct two-photon excitation from the ground state at 431.10 nm [as was done by Esherick (1977)]. There are, however, several disadvantages when performing two-photon excitations as compared to two-step (step-wise) excitations.

The most efficient way of ionizing atoms by a direct two-photon excitation to an autoionizing state is to focus the laser in the atomic beam. (Since the excitation rate for each individual atom has a quadratic dependence on the laser intensity and the number of atoms illuminated decreases linearly with the area of the laser beam, the total number of ions produced increases linearly with focussing). There is, however, an optimum degree of focussing (corresponding to a certain diameter of the beam) which occurs when a significant part of the atoms in the interaction region is being ionized or when the transition is becoming saturated. This optimum intensity is often found to be rather high which implies that additional matrix interferences, i.e. ionization of concomitant species through either resonant or non-resonant processes, might take place.

Similarly, since the degree of ionization in a two-photon process is quadratically dependent upon the light intensity, the actual beam profile must be known, or at least must be known to be unchanged during the experiment, since otherwise fluctuations in the signal will occur as a consequence of spatial inhomogeneities or spatial fluctuations of the laser beam.

When the atoms are excited by a two-step process, such as in the present case, none of the above-mentioned drawbacks will influence the signal.

Since the first step can be easily saturated using modern pulsed dye-laser systems (g_2A_{21} = 0.85 10^8 s^{-1}), the signal will be linearly dependent upon the laser light intensity of the second step. Hence, normalization of the measured signal with respect to long term drifts in the laser power output can be performed by simply monitoring the intensity of the second step laser light (e.g. by a simple photo diode). In addition, since the signal is linearly dependent upon the second step laser, spatial inhomogeneities and spatial fluctuations in the second step laser light beam will not affect the signal (assuming an homogeneous distribution of atoms in the interaction region). Furthermore, the signal will be independent of focussing (as long as the first step laser is saturating the first transition) due to the same reasons.

All these aspects are of high importance for the RIMS technique since quantifications of the signals are most often done by integrating (summing) the signal measured during one whole vaporization cycle of the sample [which might take a considerable amount of time (several minutes or tens of minutes)].

There is even more advantages connected with using this set of transition for detection of Sr in the RIMS (or RIS) technique. Since the excitation scheme only consists of visible (or near visible) wavelengths (as opposed to UV light which most elements require for excitation either of the first step or for reaching the ionization limit), a minimum of spectral interferences with other elements will occur. Moreover, the use of transitions in the visible also reduces the possibilities of wing excitation of many matrix species since most elements need a UV photon in order to be excited in the first step.

As has been studied previously in Laser-Enhanced Ionization (LEI) technique,which is closely related to RIMS, considerable spectral interferences can occur whenever UV light with shorter wavelengths than needed for direct photoionization of elements from their ground state or from low lying thermally populated states is being used for excitation or ionization of the element under study (Axner 1986, Axner 1990) - Li might give considerable background signals when light below 230 nm is being used; Ba below 238 nm; Na below 241 nm; K below 285 nm; Rb below 297 nm, Cs below 318 nm, etc due to direct ionization from their ground states. Also these types of potential sources of interferences are minimized by employing a set of visible transitions in Sr.

Figure 3 shows two scans of the second step laser across the autoionizing state ($4d^2$ $^1D_2$ state) in a RIMS experiment. In this experiment the output signal from the mass-spectrometer was measured by a box-car integrator with its window positioned at the appropriate position of time for probing of the Sr peak. The second step laser was scanned here from 404 to 407 nm. It is possible to deduce from this scan that the width of the autoionizing state is in the order of 1 nm which is a rather narrow bandwidth for an autoionizing state. (The decrease of the signal with time is due to a change of the degree of vaporization due to slowly changes in the resistance, and hence heating, of the furnace with time. The narrow bandwidth of this state, together with its property of being a $^1D_2$ state which can be very efficiently reached from the intermediate state ($^1P_1$), makes this choice of excitation/ionization wavelengths very useful in RIMS, RIS or other ionization analyses.

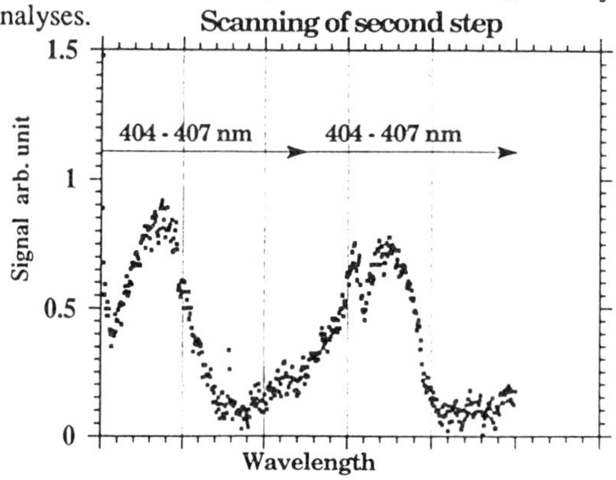

Fig. 3. The signal from a RIMS experiment in Sr scanning the second step across the autoionizng state from 404 to 407 nm.

Figure 4, shows a similar scan of the first step laser over its 460.73 nm transition. This time, the laser was scanned only from 460.6 to 460.9 nm, indicating that the width of the first step transition is considerably smaller than that of the second step, as was expected.

No quantitative measurement of Sr has been performed so far but will take place in various samples in the authors' laboratory in the closest future.

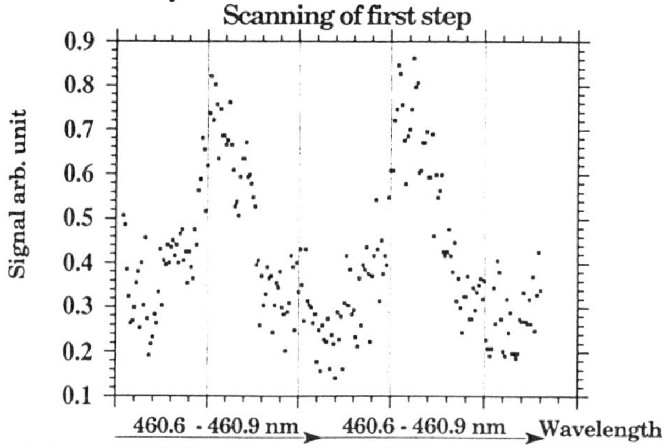

Fig. 4. The signal from a RIMS experiment in Sr scanning the first step across from 460.6 to 460.9 nm.

REFERENCES

Axner O, Berglind T, and Sjöström S (1986) *Physica Scripta* **34** 18

Axner O, Norberg M, Rubinsztein-Dunlop H, and Persson M (1990) *Applied Spectroscopy* **44,** August. ...

Bengtsson L (1990) *Thesis, ISBM-91-7032-512-X, Göteborg*

Esherick P (1977) *Physical Review A* **15** 1920

Hart L P, Smith B W and Omenetto N (1985) *Spectrochimica Acta* **40B** 1637

AMS an ultra sensitive method for the detection of long-lived radioisotopes

Martin Suter
Paul Scherrer Institut, c/o HPK, ETH-Hönggerberg, CH-8093 Zürich, Switzerland

Abstract. The present status of accelerator mass spectrometry (AMS) is summarized in term of the sensitivity, efficiency and accuracy which are attained. The discussion will concentrate on the detection of long-lived cosmogenically produced radioisotopes such as ^{10}Be, ^{14}C, ^{26}Al, ^{36}Cl ,^{41}Ca and ^{129}I.

1. Introduction

Accelerator mass spectrometry (AMS) of long-lived radioisotopes has become an important tool in environmental sciences. World wide, several thousand measurements are performed every year. Radiocarbon studies make up the majority of applications, for which reduction of sample size is the main advantage compared to conventional counter techniques. In addition, the efficient detection of radioisotopes with longer half-lives, such as ^{10}Be, ^{26}Al, ^{36}Cl, ^{41}Ca and ^{129}I has opened up a large variety of new applications which could not have been undertaken with decay counting techniques. These radioisotopes, which are primarily produced by cosmic rays, are used both to date geophysical and geochemical events and as tracers of geophysical processes. In addition, ^{14}C has many application in archeometry and for dating of art objects.

In this paper the present status of the AMS-technique is summarized in terms of the sensitivity, efficiency, accuracy and precision. This information should enable the reader to estimate the potential of this technique and allow a comparison with the high sensitive LASER techniques discussed at this conference. More detailed reviews on AMS are found in Proceedings of AMS Conferences (Litherland 1984, Wölfli 1987, Suter 1990).

2. Techniques

Conventional mass spectrometry is based on the fact that ions with the same energy are deflected in a magnetic field according to their masses. It is limited in sensitivity due to the background caused by molecules and isobars, which have

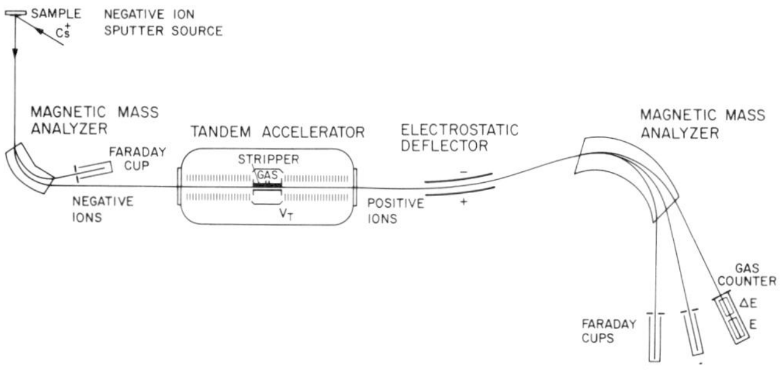

Figure 1: Schematic diagram of an AMS facility

masses extremely close to that of the nuclide to be determined. Other limitations arise from background counts caused by scattering on walls, appertures and residual gas of the mass spectrometer. Accelerator mass spectrometry is a special kind of mass spectrometry which overcomes most of these problems and therefore extends the sensitivity of conventional mass spectrometry by many orders of magnitude. Particle accelerators, originally developed for nuclear physics research, are used to accelerate the the ions of interest to energies high enough to destroy molecules and to identify isobars and neighboring masses by nuclear detection techniques. In addition, scattering cross sections are much smaller at high energies leading to a further background reduction. Because interferences are removed by techniques other than deflection in magnetic and electric fields, relatively low mass resolution spectrometer can be used as components of AMS systems. Therefore the transmission and thus the efficiency can by kept very high, as required to detect the low content of rare radiosotopes in small samples.

Figure 1 shows a typical set-up of a tandem accelerator mass spectrometer arrangements, as used at most places. The purified and pretreated material under investigation is loaded into a sputter ion source and bombared with positive Cs ions to produce a negative ion beam. Because in several cases interfering elements do not form stable negative ions, the source can act as a first filter. This effect is certainly responsible for the large success of AMS for ^{14}C dating since the interfering $^{14}N^-$ ion is not stable. Other examples for this type of suppression technique are the following isobar pairs: ^{26}Al-^{26}Mg, ^{36}Cl-^{36}Ar and ^{129}I- ^{129}Xe. For ^{41}Ca a high discrimination of ^{41}K is possible when the CaH_3^- moleculues are extracted from the ions source .

After the extraction of the negative ions from the source, the mass of interest is selected with a first magnetic spectrometer (second filter). All ions which have passed this filter including molecules and isobars with nearly the same mass are accelerated to the high voltage terminal of the tandem accelerator. There they pass through a thin carbon foil or a narrow canal filled with gas. This causes several electrons to be stripped away and ions to end up in a positive charge state. This stripping process, which causes molecules to be destroyed because multiply charged molecules ($q \geq 3+$) are not stable, constitutes the third filter. The positive ions are then accelerated back to ground potential and analysed in a high energy mass spectrometer, which normally consists of an electrostatic and a magnetic deflector (fourth filter). In some arrangements also Wien filters (crossed electric and magnetic fields) are used. Finally the ions are identified in a detector systems (fifth filter). The detector consists in most cases of a gas counter giving a signal proportional to the total energy of the ion and one or more Δ E signals, which provide information on the specific energy loss of the ion. These energy loss signals are related to atomic number of the ion, allowing an isobar separation. In some cases time-of-flight detectors are added to provide a velocity determination, which allows a further mass identification and thus an additional background reduction. With all these filtering stages a discrimination of neighboring masses in the 10^{-15} range is possible while still maintaining a high transmission.

Essential for reaching precise results is an appropriate normalizing procedure. In most cases radioisotope concentrations are calibrated with standards. Because sputter rates and thus the ion beam current may vary from sample to sample and do not necessarily remain constant during the analysis, an adequate beam currents measurement of the abundant isotope(s) is needed. Various procedures have been developed to accomplish this. In the simplest arrangement, the currents of the stable isotope(s) are measured at the low energy side of the accelerator after the first magnet (see figure 1). The disadvantage of this technique is that there is no direct way to estimate possible variations in the transmission through the accelerator to the final detector. Therefore most systems monitor the abundant isotope(s) on the high energy side of the accelerator either sequentially or simultaneously with the rare isotope measurement.

3. Performance

The concept of tandem-AMS has not much has changed, since it was introduced in 1977. But the performance and reliability have been improved enormously, either by modifying and upgrading existing tandem accelerators and the associated equipment or by the construction of new dedicated AMS facilities.

3.1. Background and sensitivity

The background measured in the final detector limits the lowest isotopic ratios which can be determined. It also limits the minimum number of atoms detectable. The background can be divided into two categories: contamination and instrumental background.

Contamination can be introduced into the sample during sample collection, sample pretreatment of final target preparation. Preparation contamination can come from labware or from the general environment. This type of background presently limits the ^{14}C detection, in the best cases to about 2 10^{-15}, where as instrumental background measured with unprocessed graphite can be a factor of 5 to 10 lower.

The rare isotope can also be introduced when carrier has to be added to samples to provide sufficient material for processing. This has been reported for ^{10}Be: normal carrier has a ^{10}Be/^{9}Be ratio of 10^{-14} and by selecting special material a reduction to the 10^{-15} level is possible. The background of ^{129}I may also be restricted by this type of background to a few times 10^{-14}.

Instrumental background can be divided in two groups: Scattered ions with a wrong mass may reach the final detector or particles with the right mass but from an other element (isobars) are not properly identified. Background from neighboring isotopes can largely be reduced with approriate designs of mass spectrometers at low and high energy side of the accelerator. In addition, the vacuum in the electric and magnetic deflectors as well as in acceleration tubes has to be good to avoid scattering and charge-changing processes. Isobaric interferences are more severe, if they can not be suppressed in the ions source. The separation has then to performed exclusively in the final detector system. The separating power in this case increases with beam energy and decreases with the atomic number (Suter 1990). For lighter isotopes up to ^{26}Al small accelerator operating with terminal voltages around 2 MV give sufficient isobar separation to reach background levels of 10^{-15}. For ^{36}Cl and ^{41}Ca larger accelerators (6-10 MV) are needed to reach similar sensitivities. Heavier radioisotopes such as ^{53}Mn, ^{60}Fe, ^{59}Ni, ^{126}Sn and ^{205}Pb, which all have strong isobaric interferences could so far only be detected with the largest available AMS facilities which are using LINAC-configuration to post accelerate the ions to energies above 100 MeV. In addition, some special techniques such as complete stripping (Korschinek et al 1897) or gas-filled magnets (Paul et al 1989) had to be applied to reach background levels in order of 10^{-10} to 10^{-13}. Especially for the detection of these heavier isotopes it would be desirable to have element selective ion sources, as might be provided by LASER-techniques. Since LASER sources would produce positive ions, an additional charge exchange canal would be needed to form the negative ions.

Recently initial experiments have been reported, the goal of which is to reduce the isobaric background of ^{36}Cl by the electron detachement of the interfering S^- in a LASER-beam. The technique works in principle, but due to the low duty cycle of the LASER, the efficiency is presently too low by orders of magnitudes for practical applications (Berkovits 1989).

3.2. Efficiency and throughput

A high detection efficiency is important when working with small samples and very low concentrations. It also enhances the throughput. High counting rates are needed to reach low statistical errors and thus a high accuracy. Neglecting losses during sample preparation, the overall detection efficiency can be divided in two parts: the yield of the ion source and the transmission to the final detector (including stripping efficiency). The efficiency improvements made during the last 13 years and the higher beam current available today reduced the measuring time per sample from hours to minutes and the annual throughput could be enhanced to more than 2000 samples for some facilities.

The Cs-sputter sources have been improved in geometry and concept. By using large spherical Cs-surface ionizers, the currents could be enhanced to the 100 μA range for some elements (C, Cl) and for others from 100 nA range to several μA (Be,Al,CaH_3). At at the same time these new sources are characterized by higher yields for the negative ion formation (Middleton 1984). In addition, appropriate multy sample loading systems through a vacuum lock and automatic sample changing capabilities improved throughput considerably.

Also the beam transmission to the final detector has been enhanced. Stripping yields of radioisotopes were studied very carefully as function of ion energy, stripper media and thickness (Suter 90). Based on these data it is now possible to optimize stripping conditions. Important are terminal pumping systems which allow to reach the stripper densities required, while still maintaining good vacuum conditions in the acceleration tubes. This avoids charge changing processes which would lead to counting losses and background. For light elements such as (^{10}Be and ^{14}C) stripping yields for the charge state 3+ or 4+ of more 60% are possible by the appropriate choice of energy and stripper gas. For heavier ions yields of 30 % to 40% can be reached. In general, the optical transmission in the older AMS system was very low. Beam losses of 50 % or higher were not uncommon. With newer acceleration tube designs, computer-optimized injection optics and well matched stripper arrangements, transport losses smaller than 10 % seem possible.

Based on the current experience, one can characterize the overall detection efficiency for existing system and, by including theoretical limits on optics and stripping yield one can estimate maximum efficiencies which might be reached under optimal conditions. For ^{14}C typical efficiencies are around one percent for well designed systems and might reach a few percent under favorable conditions. For Be efficiencies range from several times 10^{-4} up to a few times 10^{-3} with larger accelerators. The possible maximum would be about 1 %, which could be reached with optimized systems operating at terminal voltages of 8-10 MV. Due to the low sputter yield of Al^- and CaH_3^- ions the overall efficiency is low for these elements, namely in the 10^{-4} range. High efficiency, at the percent level, is in principle possible for Cl, but in practice, much lower levels are achieved, since the count rate of the interfering S increases enormously when the Cs beam hits the target holder. High efficiency is advantageous, but in many applications efficiency is not the limiting factor. Sample preparation and handling of milligram quantities is in general much easier, more reliable and more efficient to perform than dealing with smaller quantities. In many cases the background limits the smallest number of atoms detectable to about 10^5 to 10^6.

3.3. Uncertainties, precision, accuracy

The source efficiency and transmission improvements made have enhanced the count rates in the final detector and thus reduced the statistical uncertainties. With improved stabilities of power supplies for the ion source and the beam guiding elements as well as with the development of appropriate normalization procedure the systematic errors could also be significantly reduced. The developement toward high accuracy measurement was stimulated by the conventional ^{14}C decay counting techniques which has reached a accuracy around 2 permil on large samples and allowed to derive precise calibration curves for dating based on tree ring studies. Today AMS is approaching the same accuracy level for ^{14}C on mg samples as indicated by several studies. For other isotopes, an accuracy of a few percent in the isotopic ratios is normally sufficient.

4. Summary and conclusions

Today more than 30 facilities are involved in AMS-related research. By far the largest fraction of facilities (ca 20) is using older tandems originally designed for nuclear physics research. The maximum terminal voltages for these facilities range from (6-16 MV). They are therefore very versatile. For light ions (Be, C) very high stripping yields can be obtained and for isotopes up to ^{41}Ca good isobar suppression is possible. They have the disadvantage that maintenance and operation cost are high, and to get good performance significant upgrading is required. A few large facilities with post accelerators make available the range above 100 MeV, thus opening the possibility to separate some heavy isobar pairs, which otherwise would not be accessible to AMS.

New systems have been designed and built for AMS. All these new accelerator operate around 2 MV. They have mostly been optimized for precise ^{14}C measurements. They have also successfully been applied for ^{7}Be, ^{10}Be, ^{26}Al and ^{129}I measurements. These facilities are also well suited for stable isotope applications in material and geosciences. They require relatively low maintenance and are easy to operate.

By looking at the present performances of AMS facilities and comparing them with the limits given by the physics or by the best devices in operation, one can say that the limits have not yet been reached. Significant improvements can be expected in the next few years with respect to efficiency, accuracy and throughput. For the isobar separation new techniques would be needed to enhance the sensitivity much further.

References

Berkovits D, Boretta E, Hollos G, Kutschera W, Naaman R, Paul M, and Vager V 1989, Nucl. Instr. and Meth. **A281** 663

Korschinek G, Morinaga H, Nolte E, Preisenberger E, Ratzinger U, Urban A, Dragovitsch P and Vogt S 1987, Nucl. Instr. and Meth. **B29** 67

Litherland A E 1984, Nucl. Instr. and Meth. **B5** 100

Middleton R 1984, Nucl. Instr. and Meth. **B5** 193

Paul M ,Glagola, Henning W, Keller J G, Kutschera W, Lia Z, Rehm K.E., Schneck B and Siemsson R H 1989, Nucl. Instr. and Meth. **A277** 418

Suter M 1990, Nucl. Instr. and Meth., Proceedings of the 5th International AMS Conference, to be published

Wölfli W 1987, Nucl. Instr. and Meth. **B29** 1

Paper presented at RIS 90, Varese, Italy, 16–21 September 1990

Isotopically selective RIMS of rare radionuclides by double-resonance excitation with single-frequency cw lasers

B. A. Bushaw and J. T. Munley

Pacific Northwest Laboratory, Richland, WA, 99352, USA

ABSTRACT: Double-resonance RIMS using two single-frequency dye lasers and a CO_2 laser for photoionization has been shown to be both extremely sensitive and highly selective. Measurements on the radioisotope ^{210}Pb have demonstrated optical selectivity in excess of 10^9 and detection limits of less than 1 femtogram.

1. INTRODUCTION

There are a number of radionuclides that occur at relative abundances in the range of 10^{-8} to 10^{-13} with respect to other isotopes of the same element. Measurement of isotopes at these low abundances is usually beyond the capabilities of resonance ionization mass spectrometry (RIMS) using pulsed lasers because the relatively broad laser linewidths limit selectivity and duty cycle constraints limit sensitivity. Thus, we have been investigating the use of high-resolution cw dye lasers to obtain improved isotopic selectivity, beyond that which can be provided by a mass spectrometer, for the detection of these rare isotopes (Cannon 1985, Bushaw 1987, 1988, 1989). In this paper we will discuss, as an example, measurements on the isotope ^{210}Pb, which is of interest both as an indicator of radon exposure and in the microelectronics industry.

In our experiments, the analyte atoms of interest are generated with a vacuum-compatible graphite furnace and detected using the scheme depicted in Figure 1: a single-frequency cw dye laser is tuned to a resonance of the atom originating in the ground state, raising it to a low-lying excited state, a second single-frequency cw dye laser is tuned to promote the initially excited atom into a high-lying Rydberg state, the Rydberg state atom is then ionized by a cw CO_2 laser, and the ions produced are analyzed with a quadrupole mass spectrometer. There are several features of this scheme that make it attractive for the extremely low-level measurements being addressed. The graphite tube furnace is a self collimating beam source providing good transport of the atoms to the excitation region. All three steps of the excitation process can be saturated using cw lasers, leading to detection limits in the attogram range. The two dye lasers are counterpropagated to reduce residual Doppler broadening, and optical isotope selectivity can be attained in both resonance excitations. The Rydberg states are long-lived, providing a long access time for the CO_2 ionization laser as well as very narrow linewidths with small Lorentzian tails, enhancing selectivity. There is essentially no background ionization because the low-energy photons from the CO_2 laser can ionize only those atoms that have been specifically prepared in high-lying Rydberg states.

In previous studies on barium (Bushaw 1988), we have shown that this approach is capable of yielding detection limits near 10^{-17} grams. However, Pb is a more difficult case and thus a more stringent test of the capabilities of the technique. First, one of the dye lasers must be frequency-doubled to reach the first step excitation wavelength of 283.3 nm. Second, while the first excitation in barium involves a two-level system and any atoms not promoted to the Rydberg state will return to the ground state where reexcitation can occur, in Pb the first excitation is very susceptible to optical pumping: approximately 85% of the atoms excited to the 6p7s $^3P_1$ state will decay into the metastable $^3P_1$ and $^3P_2$ fine structure components of the ground state, where they are not accessible for further excitation.

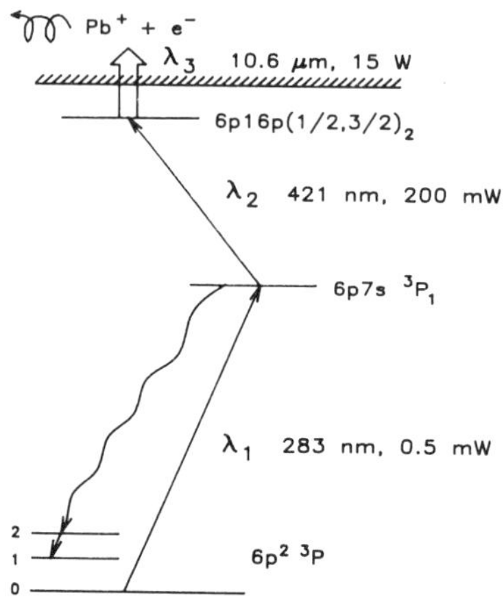

Fig. 1. Energy level diagram for double-resonance excitation of Pb isotopes, followed by photoionization with a CO_2 laser.

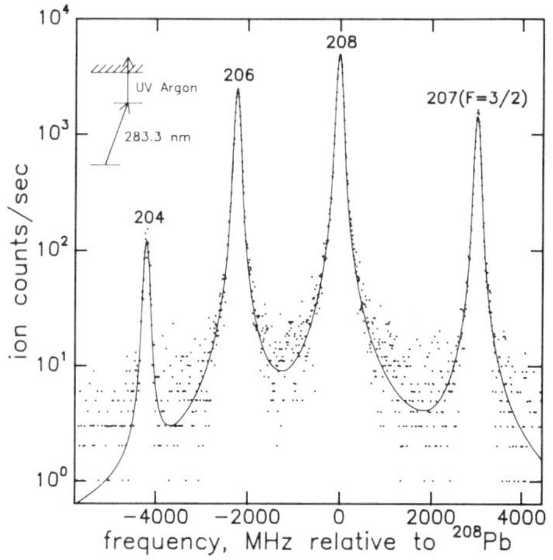

Fig. 2. Isotopic structure for the stable isotopes of Pb in the 6p7s $^3P_1$ ← $6p^2$ $^3P_0$ resonance transition.

2. OPTICAL ISOTOPE SHIFTS AND SELECTIVITY

Isotope shifts in the 6p7s $^3P_1$ ← $6p^2$ $^3P_0$ resonance for the stable, naturally occurring isotopes of Pb were determined using a 1+1 photoionization scheme in which the frequency-doubled cw dye laser excited the resonance line (35287.23 cm^{-1}) and the atoms were then directly photoionized out of the 6p7s $^3P_1$ state with the ultraviolet output of an argon ion laser. For this measurement, the mass spectrometer was not used so that all the Pb isotopes could be observed. The resulting spectrum is shown in Figure 2, where the points represent the actual data and the solid line is nonlinear least squares fit of a numerically generated Voigt line profile to the data. The fitted Gaussian component width of 99 MHz FWHM is consistent with a residual Doppler width for an atomic beam collimation factor of 14.7, provided by the tube furnace geometry. The Lorentzian component of 56 MHz FWHM corresponds to a 2.8 nanosecond lifetime for the 6p7s $^3P_1$ state. The second hyperfine component for ^{207}Pb, F = 1/2, is off scale at -10166 MHz. Our measured isotope shifts agree to within 0.3% with prior fluorescence measurements (Thompson 1983). Extrapolating the tails of one isotope to the position of its nearest neighbors, it can be seen that there is optical selectivity greater than 10^3 in this transition. We did not measure the isotope shift for ^{210}Pb by this method since it has relatively low sensitivity and would require working with high-level radioactive samples. Because of the excellent agreement for the stable isotopes, we have accepted and used Thompson's value of +3974 MHz for ^{210}Pb.

Measurement of the isotope shifts in Rydberg states used the full excitation scheme shown in Figure 1, however, the mass spectrometer was still not used. Figure 3 shows the results of these measurements for the $6p16p(1/2,3/2)_2$ Rydberg state. In each of the cases shown the dye laser pumping the first transition was tuned to resonance for a specific isotope, as labelled in the figure, while the second laser was scanned, but with an offset opposite to that of the first laser such that the total two-photon energy was the same in all four scans. While the positions of all the isotopic peaks remains constant, the intensity patterns vary greatly depending on the tuning of the first laser. For instance, the ^{208}Pb peak in (B), where the first laser is tuned for ^{206}Pb, is reduced by a factor of 3.3×10^3 from that in (D). These intensity patterns simply reflect the selectivity of the first excitation step.

Selectivity in the second step, excitation to the Rydberg state, can be estimated from the lineshapes shown in Figure 3. For example, in (D) the intensity of the ^{208}Pb tail at the ^{206}Pb position is reduced by a factor of $3x10^5$ from the peak value. This implies an overall two-step laser selectivity near 10^9, which may also be seen by overlaying the lineshape for ^{208}Pb from (D) with its intensity in (B) and extrapolating the tail intensity to under the ^{206}Pb position.

3. MASS SPECTROMETRY AND SENSITIVITY

Figure 4 compares mass spectra of the stable Pb isotopes obtained by conventional electron impact ionization, double-resonance excitation using two cw dye lasers where absorption of an additional photon from the second laser causes photoionization, and double-resonance excitation using a cw CO_2 laser (15 watt) for photoionization. In the case shown the lasers were tuned for the minor (1.5% abundance) ^{204}Pb. The dye lasers alone improve ionization efficiency by a factor of 60 over electron impact, while the CO_2 laser increases the signal by another factor of 670 to produce an overall increase in efficiency of about 40,000 over electron impact ionization.

The isotopic selectivity of the laser approach is also apparent in Figure 4. Integrating the counts over each of the mass peaks and normalizing for abundances yields a nearly constant selectivity value of $2x10^7$ for ^{204}Pb versus ^{206}Pb, ^{207}Pb, and ^{208}Pb. This is considerably less than the value of 10^9 predicted above from lineshape analyses. The fact that the selectivity remains constant for all isotopes suggests that the ^{206}Pb, ^{207}Pb, and ^{208}Pb ions observed result from charge transfer, not a limitation of laser selectivity. Further, it should be noted that this charge transfer background will be inconsequential, less than the laser selectivity limit, when ionizing rarer isotopes at abundances of less than 0.01%.

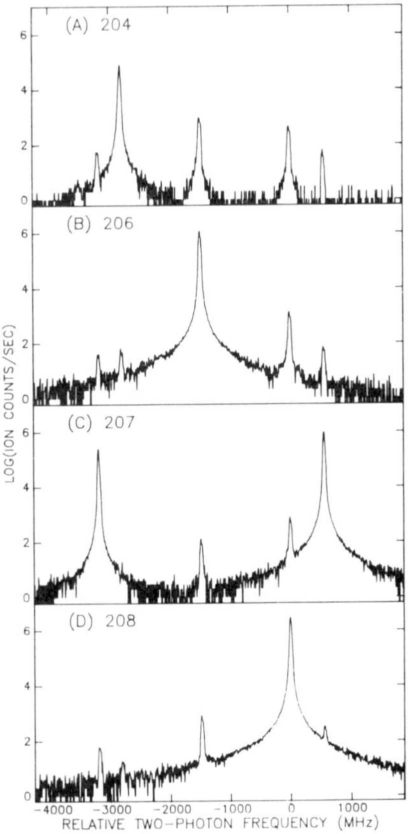

Fig. 3. Isotope shifts of the stable Pb isotopes in the 6p16p(1/2,3/2)$_2$ Rydberg state. Details of the measurements are given in the text.

It can be also seen from the shape of the high side of the mass 204 peak that, when measuring ^{210}Pb, the mass spectrometer will provide selectivity against the nearest stable isotope, ^{208}Pb, of greater than 10^7. In conjunction with the laser selectivity, the overall isotopic selectivity is greater than 10^{16}! It is apparent that actual measurements of ^{210}Pb will not be limited by the abundance of other Pb isotopes, but rather by other sources of background such as dark counts and nonresonant ionization of other materials such as residual pump oils in the vacuum system.

Absolute sensitivity was determined by holding the laser frequencies and mass spectrometer tuning constant, set for the isotope of interest, and evolving a small discrete sample with a ramp of the furnace temperature. Integrating observed ion counts over the sample evolution period and subtracting the integrated counts for a blank sample load yields a measure of the absolute efficiency, conveniently expressed as observed ion counts per femtogram of the isotope placed in the furnace. This is shown in Figure 5, where a 2 microliter aqueous sample of 14.7 ppm total Pb, containing 432 picograms of ^{204}Pb, was placed in the furnace. The majority of the sample is evolved in a broad peak between 500 and 1300°C. This is attributed to the Pb on the graphite surface, while the continuing evolution and structure at higher temperatures is probably from Pb which has penetrated into the graphite matrix. Even though the furnace is

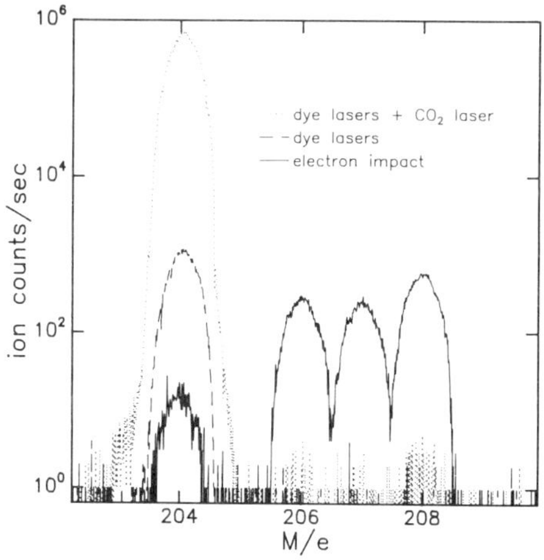

Fig. 4. Mass spectra of Pb with electron impact ionization and laser ionization tuned for ^{204}Pb, with and without a CO_2 laser.

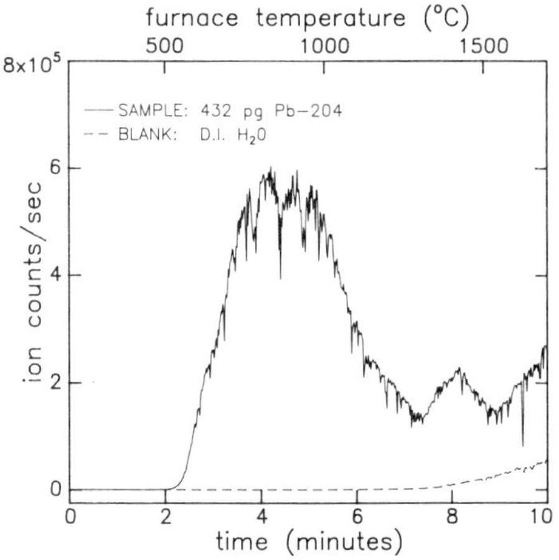

Fig. 5. Measurement of absolute sensitivity by evolution of a small discrete sample during temperature ramp of the graphite furnace.

"cleaned" by heating to 2100°C for one minute between runs, some of this deep matrix contamination is always present, as can be seen in the blank signal. Fortunately, the temperature profile for the matrix contamination is quite different from that of the sample and does not significantly interfere with the measurements. Integrating the main peak over the 500-1300°C range yields 1.06×10^8 ion counts, while a deionized water blank produced 1.5×10^5 counts over the same region. The net integrated counts due to the sample yields a sensitivity figure of 244 counts per femtogram, or an absolute efficiency of 1 of every 12,000 atoms placed in the furnace is detected as an ion.

4. MEASUREMENTS ON ^{210}Pb

The first task in applying the method to the actual measurement of the radioisotope ^{210}Pb is to determine its isotope shift in the selected Rydberg state. Fortunately, for the even isotopes (those without nuclear spin) the shift can be predicted with reasonable accuracy from the simple proportionality: $S_R(210)/S_R(i) \simeq S_1(210)/S_1(i)$, where $S_R(210)$ is the unknown shift for ^{210}Pb with respect to ^{208}Pb in the Rydberg state, $S_R(i)$ is the known shift for the other even isotopes, either ^{204}Pb or ^{206}Pb, and $S_1(210)$ and $S_1(i)$ are the known shifts in the 6p7s $^3P_1$ state. The average of the values for the ^{210}Pb shift in the $6p16p(1/2,3/2)_2$ Rydberg state, predicted from the known ^{204}Pb and ^{206}Pb shifts, is +2653 MHz. Thus, we were able to constrain our search for the actual resonance to a limited range in this vicinity. The measurement was accomplished by tuning the first dye laser to the known ^{210}Pb frequency, setting the mass spectrometer for mass 210, and then repeatedly scanning the second dye laser over the predicted region while a sample containing 203

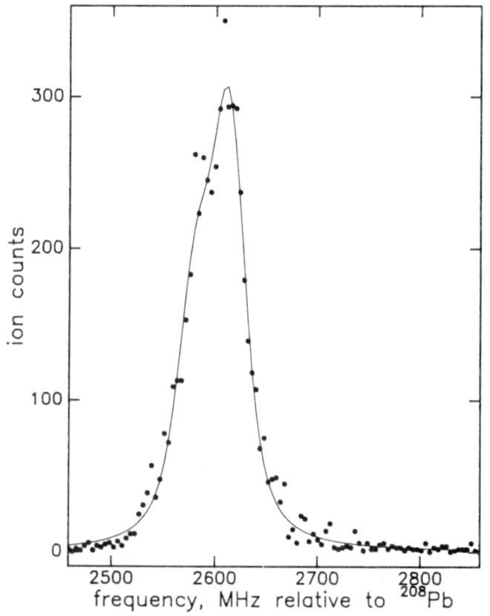

Fig. 6. Isotope shift for ^{210}Pb measured with a sample containing 203 femtograms. Sum of 20 laser scans repeated during sample evolution.

femtograms of ^{210}Pb slowly evolved from the furnace. These repeated scans were then summed to yield the spectrum shown in Figure 6. The observed isotope shift is +2622 ± 4 MHz, in good agreement with the predicted value. The slight asymmetry in the peak is caused by the AC stark effect and a slight detuning (60 MHz) of the first dye laser, which is used to reduce metastable state trapping.

With the isotope shift known, all the information necessary for test analytical measurements is available. These were performed in a manner analogous to that described previously for the absolute sensitivity measurements with ^{204}Pb: both dye lasers were tuned for excitation of ^{210}Pb, the mass spectrometer was set at mass 210, and ion counts were recorded as the furnace temperature was ramped. Figure 7A shows the result for a prepared sample which contained 18.4 femtograms of ^{210}Pb in the presence of 2.95 micrograms of the stable Pb isotopes. This was compared to a "blank" containing only the 2.95 micrograms of the stable isotopes. The ^{210}Pb signal is clearly observed, producing an average of about 15 counts per second at the peak of the temperature profile, while the blank produced no observable increase in the count rate above the constant background of approximately 0.3 counts per second. The rapid increase in signal above 1400°C is present without laser excitation and thus is not the matrix contamination discussed earlier. It is attributed to the onset of blackbody ultraviolet photons generated by the furnace, which can be detected with the channeltron detector. Figure 7B shows the result for a sample having an even lower ^{210}Pb content of 1.6 femtograms, but still containing 2.95 micrograms of the stable isotopes: a relative ^{210}Pb concentration of 5.6×10^{-10} with respect to the stable isotopes. Because the expected count rate at the peak of the temperature profile is less than 2 per second, the signal has been summed over one minute intervals. The signal from the 1.6 femtograms of ^{210}Pb is still clearly observed, and, with the blank sample, there is still no apparent increase in signal, due to the stable isotopes, above the constant background. The results appear to be statistically limited: over the period including the fourth through seventh minutes, 200 total ion counts were recorded for the sample containing ^{210}Pb, while 64 counts were recorded for the sample without ^{210}Pb. This corresponds to a net signal of 136 ion counts from the 1.6 fg of ^{210}Pb and a signal-to-background noise ratio $[(S-B)/\sqrt{B}]$ of 17. Thus, the 3σ detection limit is slightly less than 3×10^{-16} grams.

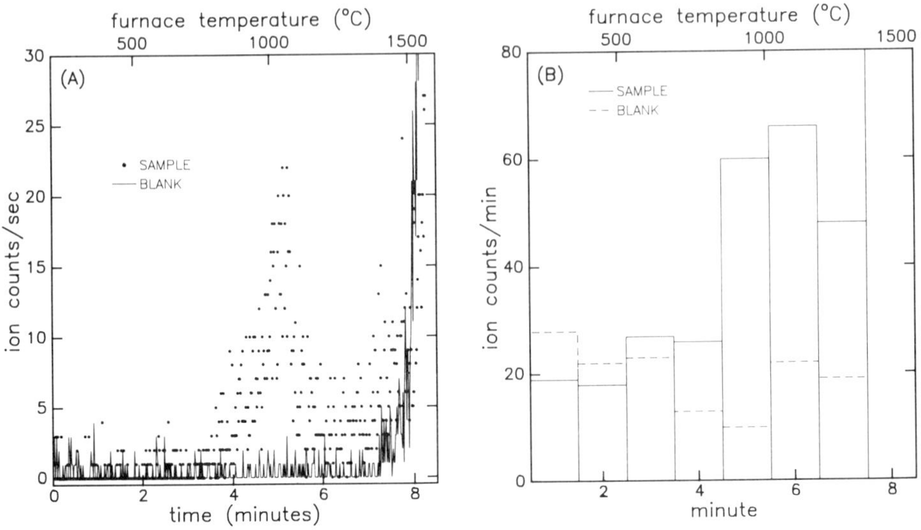

Fig. 7. Analytical measurements on prepared samples containing traces of ^{210}Pb in the presence of a large excess of the stable isotopes. The sample in (A) contained 18.4 femtograms of ^{210}Pb while (B) contained 1.6 femtograms. Both samples and the blanks contained 2.95 micrograms of the stable isotopes. The relative isotopic concentration of ^{210}Pb in (B) is 5.6×10^{-10}.

5. CONCLUSIONS

We have shown that double-resonance RIMS, using single-frequency dye lasers and an auxiliary cw CO_2 laser for photoionization, can effectively measure extremely rare isotopes at relative abundances of 10^{-10} or less. Detection limits for ^{210}Pb of 0.3 femtograms have been demonstrated with actual sample measurements made at levels within a factor of 5 of this detection limit. It is expected that these detection limits will be improved by another one or two orders of magnitude with continued work and a cleaner vacuum system to reduce nonresonant background ionization. Further, it has been shown that these techniques are applicable for complicated spectroscopic systems where optical pumping and metastable trapping occur, and where ultraviolet wavelengths are required for the excitation scheme. Thus, it should be possible to make measurements for most of the elements in the periodic table (Bushaw 1989) with similar isotopic selectivity and subfemtogram detection limits.

6. ACKNOWLEDGEMENT

This work is supported by the Office of Health and Environmental Research of the US Department of Energy (DOE) under contract DE-AC06-76RLO 1830. Pacific Northwest Laboratory is operated for DOE by Battelle Memorial Institute.

7. REFERENCES

Bushaw B A, Cannon B D, Gerke G K and Whitaker T J 1987 *Inst. Phys. Conf. Ser.* **84** 103

Bushaw B A and Gerke G K 1988 *Inst. Phys. Conf. Ser.* **94** 277

Bushaw B A 1989 *Prog. Analyt. Spectrosc.* **12** 247

Cannon B D, Bushaw B A and Whitaker T J 1985 *J. Opt. Soc. Am. B* **2**, 1542

Thompson R C, Anselment M, Bekk K, Goring S, Hanser A, Meisel G, Rebel G, Schatz G and Brown B A 1983 *J. Phys. G* **9** 443

Inst. Phys. Conf. Ser. No 114: Section 10
Paper presented at RIS 90, Varese, Italy, 16–21 September 1990

High ionization efficiency techniques for CW RIMS

B. L. Fearey, S. G. Johnson, and C. M. Miller
Los Alamos National Laboratory, Los Alamos, New Mexico 87545

ABSTRACT: The demand to measure high dynamic range isotope ratios on small samples with RIMS continues to increase. This paper discusses high ionization efficiency methods which can be applied to CW RIMS to potentially achieve several tens of percent ionization efficiencies for certain elements. The primary technique under development to achieve this is an external laser cavity which can generate very high circulating laser powers.

1. INTRODUCTION

The ability to measure large isotope ratios for rare isotopes with small samples continues to increase (Hurst 1986, Miller *et al.* 1986, and Blum *et al.* 1990). The major problem which has limited the accomplishment of this task has been to achieve high overall ionization efficiency with good sample utilization.

Some aspects of this problem are apparent in the choice of excitation and ionization source for the RIMS process. While pulsed lasers have high peak powers and broad tuning ranges, several characteristics limit their general utility: low duty cycle (low efficiency) [largely solved by kHz repetition rates], pulse pile-up difficulties (limited dynamic range), poor stability (poor precision), and laser pulse spectral and temporal concerns (ratio biases). In contrast, CW lasers offer 100% effective duty cycles, easily controlled laser profiles (spectral, spatial and temporal), as well as excellent power stability. The major obstacle which has precluded the utility of CW lasers has been power. While sufficient intensity is available to saturate the resonant transition, efficient promotion of the excited atoms above the ionization potential is difficult.

High dynamic range measurements ($\leq$1:10^5) have been demonstrated previously with CW RIMS , primarily through its high duty cycle and power stability versus pulsed sources (Miller and Nogar 1983, Miller *et al.* 1985). We have also demonstrated that a secondary, non-resonant laser can substantially increase ionization to yield efficiencies of ~0.1% in lutetium (Fearey *et al.*, 1989). However, further increases through more powerful lasers are not available.

Therefore, our primary effort is to significantly increase signal levels through an external cavity technique which has the potential to achieve overall ionization efficiencies of tens of percent for several elements (*e.g.*, Th). Such a technique is anticipated to yield a 100-fold increase in coherent circulating laser power. This paper will focus on the theoretical and experimental aspects of such a system, as well as some of our most recent results.

2. EXPERIMENTAL

Our general CW RIMS experimental setup has been described previously (Fearey *et al.* 1988). Specifically, the external cavity RIMS set-up consist of an Ar^+ laser pumped standing wave dye laser coupled into the external laser cavity which will be constructed around a 90° single sector magnetic mass spectrometer (see Figure 1). The external cavity consists of a 98% reflective input coupler mirror, a ~100% high reflector, a polarization selective element and active-feedback stabilization lock electronics. The stabilization scheme (Hansch and Couillaud 1980) consists of a piezo-electrically driven high reflector mirror whose drive voltage is generated from an error signal derived from the polarization dependent element. The purpose of the stabilization circuits is to maintain a phase matching condition within the cavity such that the external cavity has maximum gain (*i.e.*, the length is $n*\lambda/2$).

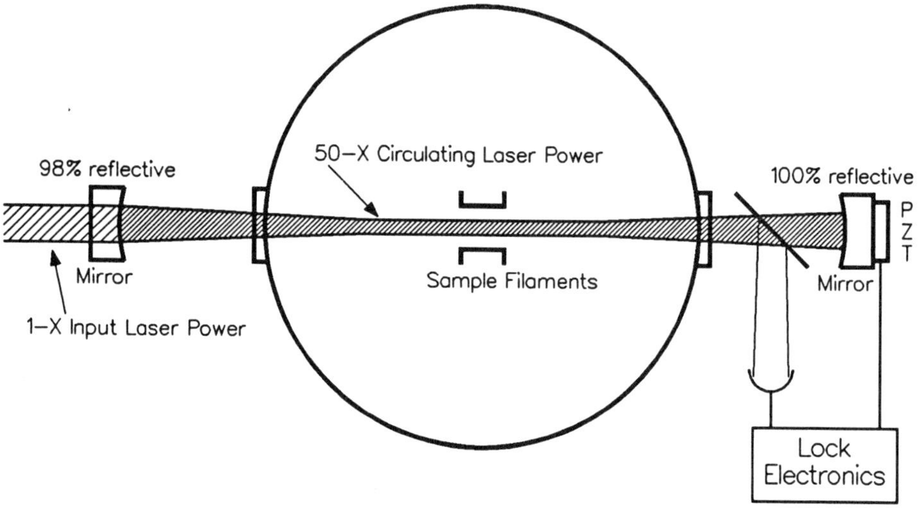

Fig. 1. Mass spectrometer source and external laser cavity configuration to be used in RIMS experiments. With mirror reflectivities shown, a 50-fold circulating laser power increase over the input laser power is calculated.

3. BASIC PRINCIPALS

Although the basic theory in the properties of external laser cavities has been thoroughly discussed before (Hansch and Couillaud 1980 and Helmcke *et al.* 1977), it is perhaps useful to briefly discuss some of the important basic principals and properties involved. The essential gain function can be described as follows: if the reflectivity of the input coupler is R, and the round-trip attenuation is V, the enhancement factor, E, of the input power by the cavity is given by: $E = (1 - R)/(1 - sqrt(RV))^2$. Upon examination, the maximum enhancement occurs when the input coupler reflectivity exactly matches the round-trip attenuation (*i.e.*, $R = V$, such that $E = (1-\mathrm{R})^{-1}$).

In addition, several other factors must be controlled and optimized. These include the external cavity length, mode matching to the cavity, and variable beam waists (Corney 1977). Importantly, the capability to vary beam waists allows the optimization of the cavity gain with the atom reservoir for maximum efficiency.

4. RESULTS

The experimental apparatus is in the construction and testing stage. The present external laser cavity (in its simplest form) consists of two spherical mirrors (100% high reflector and a 98% reflective input coupler). With optimal conditions and different reflectivity mirrors, a ≥100-fold increase in circulating power can be achieved (Berquist 1988).

We have recently completed prototype testing of the cavity in an unstabilized configuration (*i.e.*, without feedback electronics). Under these conditions, we have demonstrated extremely promising values of an average gain factor of ~20 (corresponding to 4 watts of circulating laser power) and a peak (instantaneous) enhancement of ~45-fold (9 watts). In an active feedback stabilization set-up, we expect to be able to match or exceed the peak enhancement factor, approaching the theoretical value of ~50-fold increase in input laser power. In addition, the importance of mode matching and variable external cavity beam waists has been verified.

The actively-stabilized cavity configuration is presently in testing. It has been shown in our bench-top experiments that the PZT motion is intimately involved in the gain (as expected). Our first-stage electronics have been shown to be performing essentially as expected. The more comprehensive second-stage electronics are presently in fabrication. An actively phase-locked cavity should be operational in the immediate future.

5. DISCUSSION

The potential for a two order of magnitude (or more) gain in the circulating laser power and the corresponding increase in ionization efficiency make the external laser cavity approach to RIMS extremely attractive. Figure 2 shows theoretical modeling calculations (similar to Miller and Nogar 1983) of the relative RIMS signal enhancement (ionization efficiency) under various conditions. The plots show the dependence of relative ionization efficiency as a function of laser beam diameter. The lower curve shows this dependence for a single dye laser with 200 mW of power. The other three curves indicate conditions where a 50-fold gain external cavity is utilized for both/either the dye laser and/or argon ion laser.

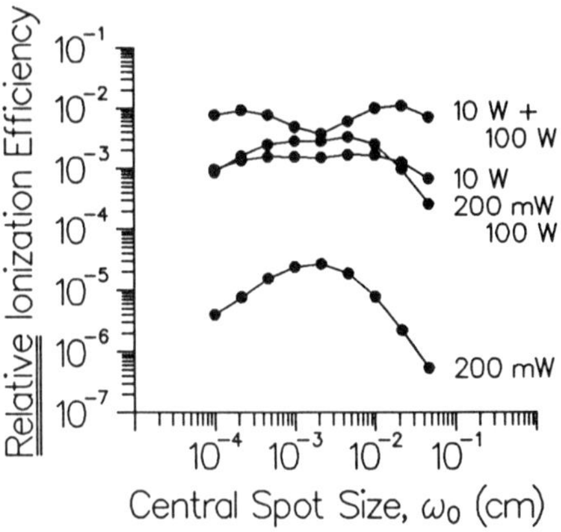

Fig. 2. Detailed calculation of relative ionization efficiencies as a function of beam diameter with differing dye and argon ion laser powers.

Amplification of a single laser leads to a ≥60-fold improvement in the ionization efficiency. That the dependence on laser power is greater than linear, points to increased sample utilization or, equivalently, ionization volume. Figure 3 shows a calculational comparison of the increase in sample utilization between without and with the benefit of an external cavity.

From these calculations, the promising results on our unstabilized external cavity and with our current ionization efficiency of ~1% for Th using simple CW RIMS (described elsewhere in this volume (Fearey *et al.* 1990) — the utilization of the external cavity could potentially lead to overall ionization efficiencies for Th of several tens of percent. This increased efficiency should allow for sub-nanogram isotopic ratio analyses of thorium at the 5:10^6 level ($^{230/232}$Th) which are critical for internal isochron measurements important for geochronology (Fearey *et al.* 1990).

Finally, a second laser, such as an Ar^+ ion laser, can also be used for the ionization step in a 1+1 RIMS scheme (Fearey *et al.* 1989). This second laser can also be amplified via another external cavity which then allows for even further increases in the the volume of atoms addressed and hence, ionization efficiency. Such increases could lead to ionization efficiencies ≥2.5 orders of magnitude (see Fig. 2) greater than single-pass dye configurations. As laser powers continue to increase, saturation conditions will become important; however, this is not necessarily a limitation because the laser beam diameter can be increased. Obviously, there will be some optimum combination of laser power and beam diameter.

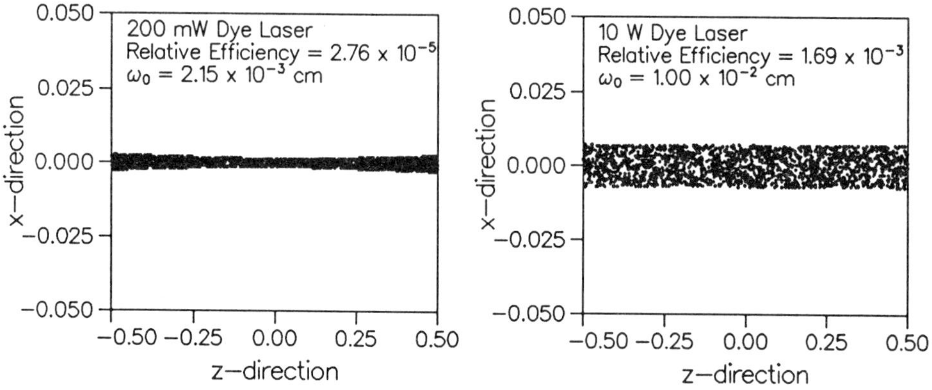

Fig. 3. Ionization calculation showing the comparison in ionization volume both without and with the operation of the external cavity. The x-direction is across the sample filament (cross-section of laser beam) and the z-direction is along the active length of the filament (along the laser beam).

6. SUMMARY

Theoretical calculations and preliminary data have been presented that demonstrate that the utilization of an external laser cavity to boost the circulating laser power leads to substantially increased ionization efficiencies for CW RIMS. An average gain of ~20-fold and an instantaneous gain of ~45-fold have been observed. A gain factor of 100 is expected for our final configuration utilizing a 99% reflective input coupler. Incorporation of the external cavity technique into CW RIMS should lead to ionization efficiencies as high as tens of percent for certain elements. This will largely remove present limitations of CW lasers for RIMS research and analysis.

References

Berquist J C private communication

Blum J D, Pellin M J , Calaway W F , Young C E , Gruen D M, Hutcheon I D and Wasserburg G J 1990 *Anal. Chem.* **62** 209

Corney A 1977 *Atomic and Laser Spectroscopy* (Oxford: Oxford Univ. Press) 355

Fearey B L, Parent D C, Keller R A and Miller C M 1989 *Resonance Ionization Spectroscopy 1988* eds T B Lucatorto and J E Parks (Bristol: Inst. of Physics) 263

Fearey B L, Miller C M, Rowe M W, Anderson J E and Nogar N S 1988 *Anal. Chem.* **60** 1787

Hansch T W and Couillaud B 1980 *Opt. Comm.* **35** 441

Hurst G S 1986 *Resonance Ionization Spectroscopy* eds G S Hurst and C G Morgan (Bristol: Inst. of Physics) 1

Helmcke J, Lee S A and Hall J L 1982 *Appl. Opt.* **21** 1686

Miller C M, Engleman, Jr. R and Keller R A 1985 *J. Opt. Soc. Am.* **2** 1503

Miller C M and Nogar N S 1983 *Anal. Chem.* **55** 481

Miller C M, Nogar N S, Apel E C and Downey S W 1986 *Resonance Ionization Spectroscopy* eds G S Hurst and C G Morgan (Bristol: Inst. of Physics) 109

Inst. Phys. Conf. Ser. No 114: Section 10
Paper presented at RIS 90, Varese, Italy, 16–21 September 1990

RIMS measurement of odd-even isotope ratios with wide-band lasers on gadolinium: Influence of the polarization, energy and delay of the laser pulses in the presence of an earth-like magnetic field

- E. LE GUYADEC
C.E.A. - C.E.N. VALRHO - DCC/DTE/SLC, BP 171, 30205 BAGNOLS SUR CEZE Cedex, FRANCE.

ABSTRACT : A three-step photoionization scheme is studied on gadolinium. It allows the obtention of a large variety of odd-even isotope signal ratios.
The influence of parameters as laser polorization, pulse energy, inter-pulse delay, magnetic field intensity on this photoion signal ratio is discussed.

I. INTRODUCTION

Fairbank et al./Fairbank et al. 88/ have shown that the result of an isotope ratio measurement made by a Resonance Ionization Mass Spectroscopy (R.I.M.S) method can depend on the Laser polarizations or on the photoionization scheme.

A study on gadolinium with wide-band polarized pulsed lasers (6 GHz F.W.H.M., larger than isotope shifts and hyperfine splittings / Comaskey et al. 86/) was planned in order to confirm these results. The natural gadolinium is contituted of five predominant isotopes (^{155}Gd (14,7 %) ; ^{156}Gd (20,5 %) ; ^{157}Gd (15,7 %) ; ^{158}Gd (24,9 %) ; ^{160}Gd (21,9 %)). The nuclear spin value is I = 3/2 for the odd isotopes and I = 0 for the even ones. This fact induces an odd-even differentiated behaviour when excited by polarized light / Balling-Wright 76, Zeldovich-Sobelman 75/. A photoionization scheme and an experimental set-up previously described/ Le Guyadec et al. 90/ is used. It allows the observation of the odd-even isotope ratio over a large range (from 48 to 0.4), the observed result depending on the following experimental conditions : laser polarizations and energies, time delays between the optical pulses and magnetic field intensity and orientation.

After a review of the magnetic field effect, we will interest in the evolution of the isotope ratio with the rotation of one of the polarization axis for different laser pulse energies.

2. MAGNETIC FIELD AND TIME-DELAY EFFECTS

This Hanle-like effect has been previously experimentally and theoretically described for pulsed excitation conditions / Le Guyadec et al. 90/ and for CW ones / Slijkhuis et al. 86/.

All Gd isotopes are photoionized through a 3-step scheme (Figure.1) by 3 pi-polarized photons. The ^{156}Gd signal (as all the even isotope signals) appears periodically when the two synchronous probe optical pulses λ2 and λ3 are delayed from λ1 (Figure.2). The ^{157}Gd signal (as the other odd isotope signals) exhibits only the radiative lifetime decay curve of the $^9F_2$ probed level. These oscillations are the result of the periodic transfer of population between the magnetic sublevels induced by the residual $B_\perp$ component (Figure.3) provided principally by the earth magnetic field, by the oven heating device and by compensation Helmholtz coils. The shape of the periodical ^{156}Gd ion signal depends on the angle (Figure. 3) and it oscillates with the Larmor period of the probed level.

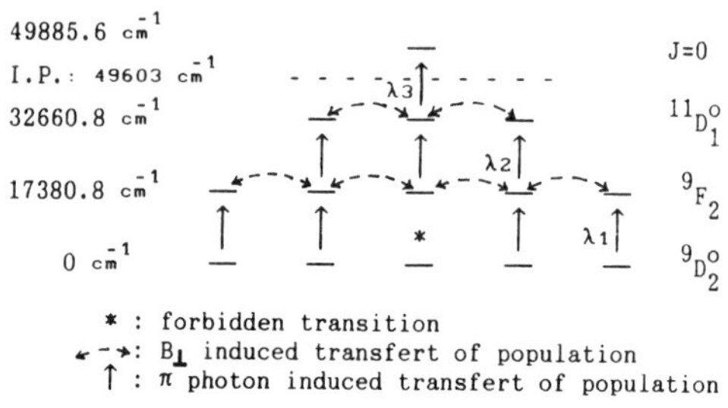

Fig.1 : Gd photoionization scheme and level coupling structure.

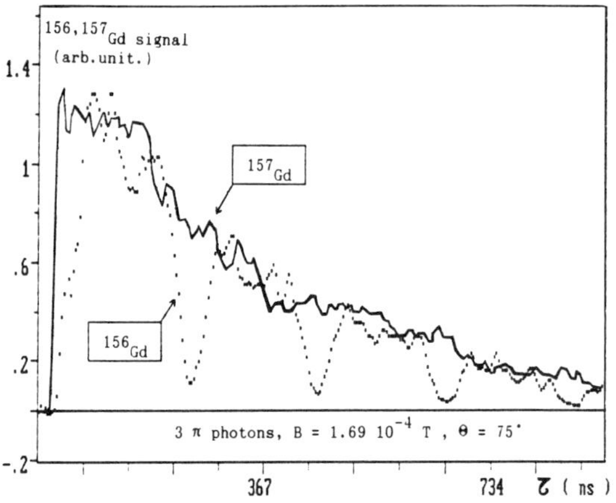

Fig.2 : ^{156,157}Gd photoion signal versus the λ_1-(λ_2,λ_3) interpulse delay time τ.

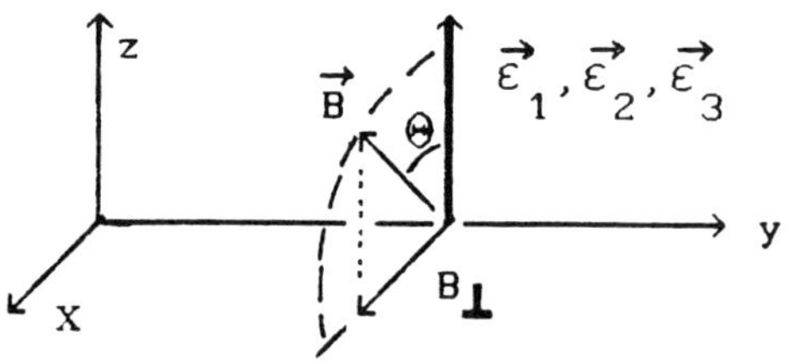

Fig.3 :
Laser and magnetic fields geometry

For such a photoionization scheme, the observed odd/even signal ratio varies between 48 (high selectivity (Figure. 4a)) and 0.4 (natural ratio (Figure.4b)) when the λ1-λ2 interpulse delay is changed of few 10 ns.

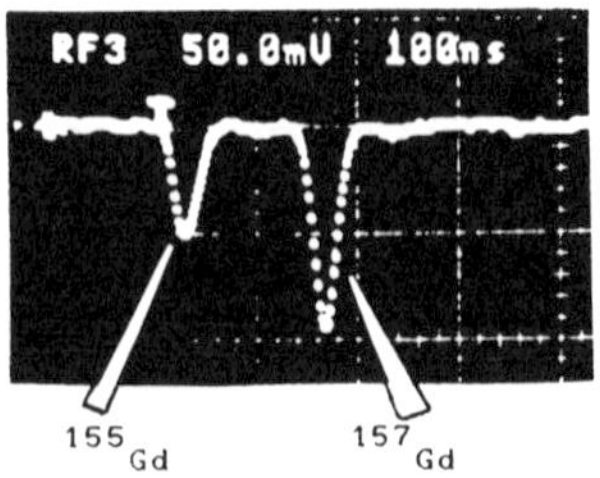

Fig.4a : Mass spectrum with synchronous pulses.

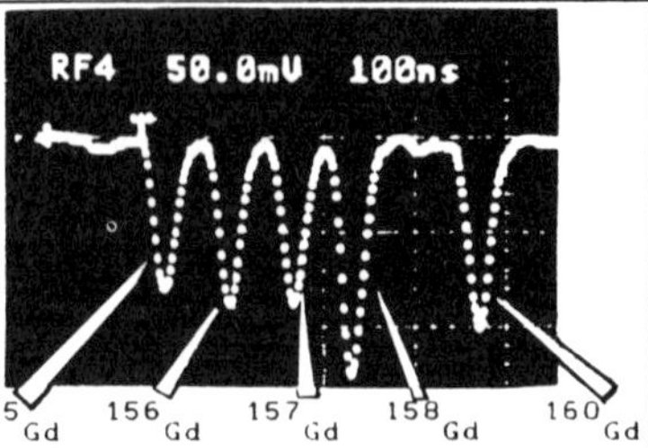

Fig.4b : Mass spectrum with a λ_1-λ_2 delay time of 80 ns.

3. POLARIZATION AND ENERGY EFFECTS

An other type of experiments was performed. It consisted in observing the evolution of the odd-even isotope photoion ratio when rotating one polarization of the 3 ionizing laser pulses (Figure.5).

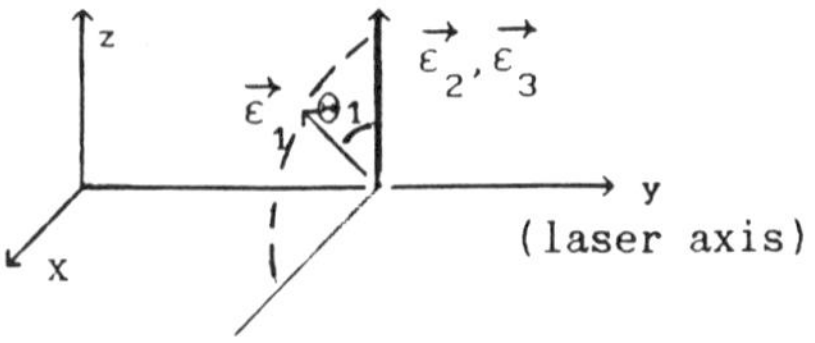

Fig.5 : Laser fields geometry

This was repeated for different values of the rotated pulse energy. Here the evolution of the $^{156}Gd/^{157}Gd$ photoion signal ratio is displayed when the $\lambda 1$ polarization axis is turned from $\Theta 1 = 0°$ to 90° and for 3 different energies of the $\lambda 1$ laser pulse (Figure. 6).

a/ 0,5μJ : 1 st transition not saturated

b/ 3 μJ : Just saturated.

c/ 25μJ : well saturated.

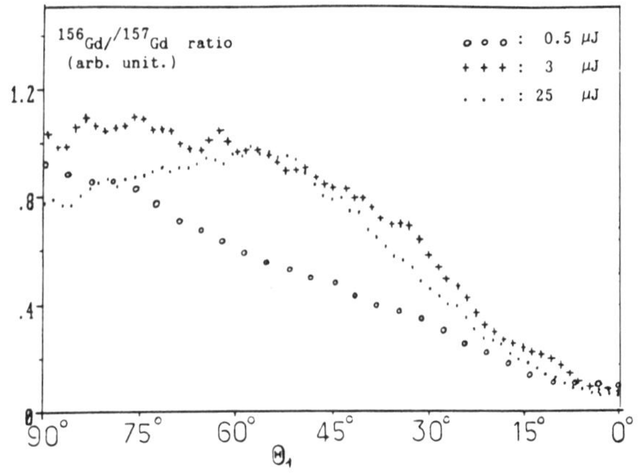

Fig.6 : $^{156}Gd/^{157}Gd$ photoion signal ratio versus θ_1 for three λ_1 laser pulse energies.

In any case the odd-even isotope ratio depends highly on the angle $\Theta 1$. For a given $\Theta 1$ angle the ratio can vary with the saturation of the transitions. A modelisation of this result is given /Le Guyadec, Thesis/. Then an estimation of the cross-section of the studied transition is possible.

4. CONCLUSION

It is shown that the measurement of an isotope ratio made by the R.I.S. technique depends greatly on the spectroscopic, laser and electromagnetic experimental conditions : photoionization scheme (autoionizing level included), laser polarization - laser energy, laser pulse delay time - external magnetic field.

In the presented experiments, odd Gd isotopes are generally not sensible to those coupled parameters and this is interpreted as the averaging of these effects on their complex hyperfine structure. But for laser pulses shorter than the evolution time of the hyperfine coherences, this would not yet occur as demonstrated by T.W. Ducas et al./ Ducas et al. 75/. Therefore the use of ultrashort synchronous pulses could increase the accuracy of isotope ratio measurements.

REFERENCES

- Balling L.C., Wright J.J., 1976 Appl. Phys. Lett., 29, 411.
- Ducas T.W., Littman M.G., Zimmerman M.L., 1975, Phys. Rev. Lett, 35, 1752.
- Fairbank W.M., Spaar M.T., Parks J.E., Hutchinson J.M.R., 1986, Resonance Ionization Spectroscopy 1986, eds T.B. Lucatorto and J.E. Parks (Bristol : Inst. of physics) 293.
- Le Guyadec E., Ravoire J., Botter R., Lambert F., Petit A., 1990. Optics comm., 76, 34.
- Le Guyadec E., 1990, Thesis, Univ. d'Orsay (Paris XI), France.
- Slijkhuis S., Nienhuis G., Morgenstern R., 1986, Phys. Rev. A., 33, 3977.
- Zeldovich Y.B., Sobelman I.I., 1975, J.E.T.P. Lett., 21, 168.

Inst. Phys. Conf. Ser. No 114: Section 11
Paper presented at RIS 90, Varese, Italy, 16 21 September 1990

Depth profiling resonance ionization mass spectrometry of compound semiconductors

S.W. Downey, A.B. Emerson and R.F. Kopf

AT&T Bell Laboratories, 600 Mountain Ave., Murray Hill NJ USA 07974

ABSTRACT: Depth profiling resonance ionization mass spectrometry (DPRIMS) is used to help elucidate dopant incorporation mechanisms, dopant migration and SIMS matrix effects in III-V materials. Quantitative measurements of dopants in Si are presented.

1. INTRODUCTION

Depth-resolved information about the elemental composition of semiconductors like Si and GaAs is of extreme importance to electronics technology. The measurement of elemental distributions of optical and electronic devices made of III-V compound semiconductors such as GaAs and AlGaAs is difficult because such devices use layers of these materials. The compositional changes between these layers are often large and usually have abrupt interfaces. Techniques such as secondary ionization mass spectrometry (SIMS) are very mature and are routinely used to characterize starting materials and processed products for major and minor constituents. However SIMS requires extensive use of relative sensitivity factors (RSFs) for each matrix studied.(1) Unfortunately, even with careful calibration, these matrix effects can be problematic, especially at interfaces where sputtering mixes the matrices. Changing matrices exhibit sputter rate changes and may also have additional problematic mass interferences.

In an effort to minimize matrix effects and mass interferences, we have been using resonance ionization mass spectrometry (RIMS) as a sputtered neutrals probe for compound semiconductor depth profiling. (2) In RIMS, the atomization and ionization steps occur in two separate steps. Sputtering is still used to sample the solid, while laser light ionizes the gas-phase neutrals. This two-step approach to the creation of ions for mass spectrometry allows optimization of both processes for sensitivity and/or selectivity and relative immunity to secondary ionization matrix effects.(3) Other groups have performed depth profiling with resonant (4,5) and non-resonant (6) ionization demonstrating excellent sensitivity. If the photoionization process is demonstrated to be saturated, RIMS signals can be related to the absolute concentrations of many elements in matrices such as Si, GaAs or other electronic materials. Here, RIMS signals are demonstrated to be element-independent, that is, equal concentrations of different impurities give equivalent signals. In this ideal case, only one RSF is needed. Comparisons between matrices is possible if the relative sputtering yields (atoms sputtered per incident ion) are known.

2. EXPERIMENTAL

The RIMS/SIMS instrument has been described previously, a VG - Microtrace IX-70S, UHV magnetic sector SIMS is modified to admit laser light.(7) RIMS and SIMS can both be performed under nearly identical conditions and thus the two techniques are easily compared. The incident primary ion beam energy is typically 2 keV and currents of 500 to 1000 nA are contained in a spot of 100 um (FWHM) diameter. The flat-bottomed portion of the craters formed by the beam is 250 to 400 um. Xe^+ sputtering enhances the neutral secondary atom yield.

An excimer-pumped, frequency-doubled, pulsed dye laser creates ions at wavelengths between 234 and 245 nm. The laser light has an intensity of up to 2 x 10^7 W/cm^2, and is directed into the sputtering region just in front of the sample to optimize sample-laser beam overlap. The laser is synchronized to the voltages controling the rastering of the primary beam and fires when the primary beam is in the center of the crater. In this way, the laser samples atoms sputtered from the crater bottom for optimal depth resolution, which is about 5 nm for 2 keV ions. The laser repetition rate is 41 Hz.

Sample high-voltage bias and time-gated detection are used to select the RIMS ions and reject SIMS ions. In DPRIMS mode, the output of a Daly detector is monitored with a time-gated integrator and computer. Data acquired as a function of time and depth scales are later added from profilometry of the sputtered craters. The time-gated useful RIMS yield of the instrument has been determined to be 2%.(2)

A variety of III-V samples grown by molecular beam epitaxy (MBE) are examined with SIMS and DPRIMS. The MBE is performed in a Varian Gen II system. Alternating layers of GaAs and AlAs 100 nm thick, were grown on a GaAs substrate at 400°C, and doped with Be to nominal uniformity at approximately 2 x 10^{19} cm^{-3}. Some ion implanted Si, GaAs, and polycrystalline Al are used in quantitative experiments. The implanted doses are all 1 x 10^{14} cm^{-2}. The accuracy of the dose measurement is better than 10%. The peak concentration of these elements is about 50 to 200 ppm, depending on the matrix and the energy of the implant.

3. RESULTS AND DISCUSSION

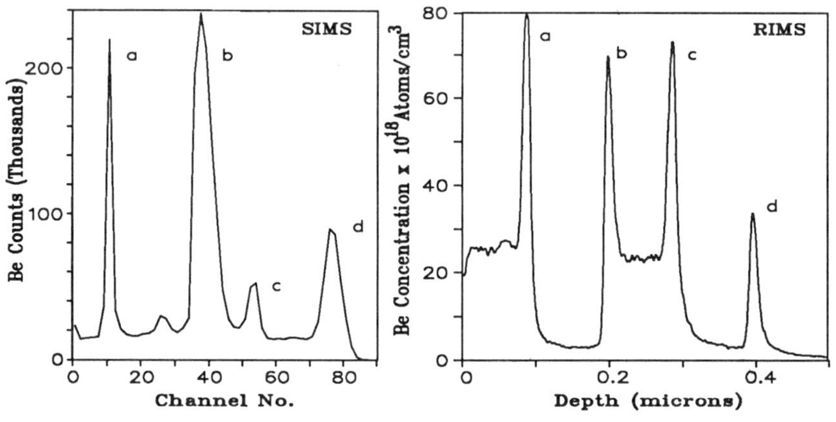

Figure 1. Depth profiles for Be-doped $(GaAs/AlAs)_{x2}$ sample. O_2^+ sputtering was used to obtain the SIMS profile. The letters in the figure mark the same interface in each trace. The top layer and the layer between b and c are GaAs.

Figure 1 shows the raw SIMS and DPRIMS Be data for the $(GaAs/AlAs)_{x2}$ structure. Two main features in the DPRIMS data are easily noticed. Be is about ten times more soluble in GaAs than in AlAs under these growth conditions. (A factor of two is used to correct for the sputter rate change between GaAs and AlAs.) Also, Be is noted to pile-up at the interfaces in these data. These spikes may be the result of Fermi-level pinning or diffusion or both occurring during growth. Because the Be flux and growth rates are constant and the integrated areas contained in the spikes very nearly matches the signal missing from the AlAs layers, the dominant mechanism of the Be pile-up is probably diffusion in the AlAs layers. This apparent diffusion is unusual at this growth temperature where diffusion coefficients are relatively low. However, the Be can exist in either substitutional or interstitial sites in the crystal, the latter having extremely large diffusion as compared to the former. The data suggest that at 400°C, the Be substitutionally incorporates into the GaAs but interstitially incorporates into the AlAs. Notice that in the DPRIMS data, the widths of the spikes at the interfaces are all about the same. The full width at half maximum of the peaks is about 5 nm, which is the same as for delta-doped samples measured previously on this instrument.(2) This result suggests that the interstitial Be in the AlAs diffuses to the interfaces and is trapped in defect sites.

In contrast, the SIMS data show several matrix effects. First, the Be signals from the GaAs and AlAs are about the same, although the concentrations are quite different as shown in the DPRIMS data. This effect could be offset through the use of RSFs and standards. The spike widths are different depending upon the ordering of the layers. Dynamic corrections for this effect are possible, but not desirable for analysis of dopants located close to interfaces.

We have examined some As atomization matrix effects in $Al_xGa_{1-x}As$.(2,3) A loss in As signal was determined to be due to a sizeable fraction of sputtered neutral AlAs. The atomization of As from the Ga-only layers is more complete than in the Al-containing layers, consistent with bond strength data for these compounds. When compared to the As RIMS signal from GaAs, the As RIMS signal from the AlAs is 40% too low after correction for changes in sputtering rates. The effect is roughly linear with Al concentration; in $Al_{0.5}Ga_{0.5}As$ 20% of the As signal is missing as compared to GaAs. Atomization matrix effects have also been noted in previous work involving uranium oxides (8) and $TiSi_2$.(9)

Figure 2 shows the ionization saturation curves for a variety of elements in Si. The data are obtained at the peak concentrations of the ion implanted standards. The curves show that the one-photon resonant, two-photon ionization (1+1) process is very efficient (100%) for all elements. The slopes of the curves suggest the resonance step is saturated easily and the ionization step depends linearly upon laser pulse energy. The remarkable saturation of the signals above 1 mJ (about 4 x 10^6 W/cm^2 inside the chamber) is probably due to a number of instrumental factors reducing edge effects. The laser beam (diameter 400 um) is larger than the Xe^+ sputtering beam (diameter 100 um) and intercepts a large fraction of the sputtered atoms. We have previously estimated this to be 25 to 30% for a 1 us sputtering pulse. In addition, the secondary ion collection optics may not be accepting the outer portions of the RIMS excitation volume.

Note at high pulse energies, the relationship between atomic concentrations and RIMS signals is element-independent. That is Be, Co and Al give signals proportional to their concentration. Only one response factor is needed to quantify those elements that behave ideally, once saturation of an analyte is ensured. From this data, the sensitivity factor is about 16 mV/ppm with 1 uA beam sputtering current. RIMS's ability to easily achieve saturation for many elements is unique from other post-ionization mechanisms where the ionization efficiency of the elements is high, but not constant.

For adequate comparison of signals, the instrument must be tuned properly. The transmission of a matrix isotope produced by off-resonance excitation at an analyte wavelength is maximized through the spectrometer before a profile is obtained. Since the wavelength used in the tuning procedure changes for each analyte and the off-resonance ionization efficiency of the matrix is not known, the absolute signal is not constant for all wavelengths. However, the goal of maximizing the matrix signal before depth profiling produces the constant interelement response factor for resonance ionization of analytes.

TABLE I. Comparison of normalized integrated signals for 1 x 10^{14} cm^{-2} Be implants into several matrices. The Be RIMS wavelength is 234.8 nm.

Energy (keV)	Matrix	Signal
100	GaAs	.89
70	Al	1.0
70	Si	.96

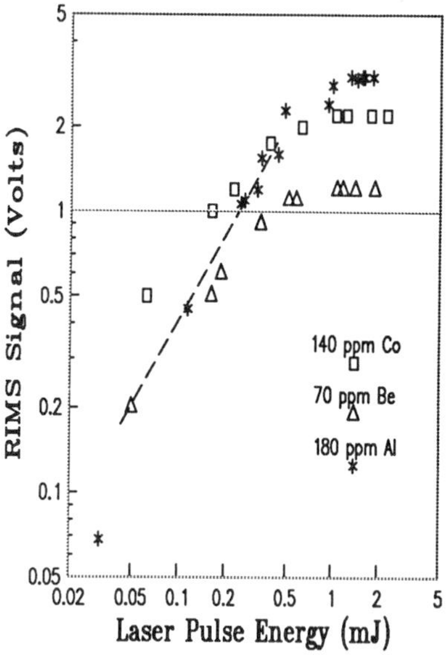

Figure 2. (Right) Demonstration of RIMS saturation for Al, Be, and Co implanted into Si. The data are taken at depths corresponding to the implants' peak concentration. Above 1 mJ, each element responds similarly and the signals are proportional to concentration. The dotted line marks a slope of one on the log-log plot. Co: 240.7nm; Al: 236.7 nm.

Table I shows the integrals for Be implants into Si, GaAs and Al. Because the implanted doses are the same (1 x 10^{14} cm^{-2}), the integrated signals should be similar if the entire implant is sampled under similar conditions. The integrated signals have a 11% relative standard deviation (RSD). Similar RSDs are obtained for replicate profiles of one implant, taken over several days. Consecutively acquired profiles have RSDs of 5%.

These results demonstrate the matrix independence of the RIMS response under ideal circumstances. After calibration with any one standard and insurance of saturation of the ionization, only the bulk sputtering yield, and the primary ion current is needed to convert the measured voltage to atomic concentration. If not known, the sputter yield is determined from crater size, beam current and sputtering time.

4. ACKNOWLEDGEMENTS

The authors are indebted to F.A. Stevie, A.E. White and K. Short for providing implanted Si samples and RBS information about a variety of samples.

5. REFERENCES

(1) R.G. Wilson, F.A. Stevie and C.W. Magee, Secondary Ion Mass Spectrometry: *A Practical Handbook for Depth Profiling and Bulk Impurity Analysis*, Wiley, New York, 1989.

(2) S.W. Downey, R.F. Kopf, E.F. Schubert and J.M. Kuo, Appl. Optics, in press.

(3) S.W. Downey, A.B. Emerson, R.F. Kopf and J.M. Kuo, Surface and Int. Anal., accepted.

(4) H.F. Arlinghaus, M.T. Sparr, and N. Thonnard, J. Vac. Sci. and Technol. A., **8** (1990) 2318.

(5) M.J. Pellin, C.E. Young, W.F. Calaway and D. M. Gruen, Nucl. Instr. Meth. Phys. Res. **B13** (1986) 653.

(6) J.B. Pallix, C.H. Becker and K.T. Gillen, Appl. Surf. Sci., **32**, (1988) 1.

(7) S.W. Downey and R.S. Hozack, J. Sci. and Technol. A., **8** (1990) 791-796.

(8) W.H. Christie and D.E. Goeringer, Inst. Phys. Conf. Ser. No. 84 (1986) 169.

(9) J.E. Parks, M.T. Spaar, D.W. Beckman, L.J. Moore and P.J. Cressman, Inst. Phys. Conf. Ser. No. 94 (1988) 197.

Paper presented at RIS 90, Varese, Italy, 16–21 September 1990

Comparison of resonant and nonresonant ionization in sputtered initiated laser ionization spectrometry

G. J. Havrilla, M. Nicholas, S.R. Bryan, J.G. Pruett
BP Research Warrensville Research Center, Cleveland, Ohio 44128-2837 USA

ABSTRACT: Resonant and nonresonant laser ionization of sputtered neutrals are compared for silver/gold alloys and silica/silicon wafer implants. The information from these two different ionization schemes results in complementary information on the sputtered species. Results indicate that there are chemical and photofragmentation effects that must be considered if quantitative determinations are to be made.

1. INTRODUCTION

Surface analyses using either resonant or nonresonant laser ionization of sputtered neutrals have been demonstrated by a number of groups (review by Young 1987). The objective has been to overcome the inherent limitations of SIMS, particularly the matrix effects of the secondary ion yield which limits general quantitative applications of SIMS. In each case the advantages of either resonant or nonresonant ionization have shown that matrix effects are indeed reduced because the ionization process is decoupled from the sputtering process. However, direct comparisons of both resonant and nonresonant ionization on the same samples have not appeared. This work presents comparisons of the two processes in two widely different matrices and the different kinds of information that can be obtained from each. In each case there is unique information that complements the information of the other. This work indicates that for complete surface analytical characterization, both laser ionization processes should be used.

2. EXPERIMENTAL

The instrument is a prototype SIRIS TOF system custom designed and built by Atom Sciences, modified by us for both resonant and nonresonant laser ionization. The resonant ionization is provided by a Molectron Nd:YAG (model MY 34-10) operating at 533 nm with 250 mJ and at 355 nm with 125 mJ. Nonresonant ionization is provided by a Lambda Physik excimer laser (EMG 103-MSC) operating at 193 nm with 50-120 mJ. The Nd:YAG pumps a Molectron dye laser (model DL-18) with either DCM or Rhodamine 610 and a Quanta-Ray dye laser (PDL-2) with coumarin 485. The resonant beams are directed into the chamber via alignment holes which cross the beams at the center of the chamber approximately 1 mm above the sample surface. The excimer is focussed with a 30 cm focal length lens. The lasers are operated at 10 Hz.

The ion gun is a VG model DP-51 with a beam diameter of around 400 μm at an ion current of approximately 5 μA. The ion gun is operated in the pulsed mode with a pulse width of 400-800 ns. The signal is detected with a Daly detector/microchannel plate combination. The resulting time-of-flight spectrum is collected by a LeCroy model 7200 DSO and then processed using LABCALC software. The depth profiling is under computer control and has flexibility to allow up to five different sputtering rates, different accumulation times and number of sampled depths.

The SIMS spectra were obtained with a Cameca IMS 3f operated with a 12.5 keV oxygen positive ion primary beam at around 100 nA. The two types of samples studied are silver/gold alloys and silica/silicon wafers with implanted elements. The silver/gold alloys are NIST SRM 481 and the SiO_2/Si samples are thickness standards manufactured by Surface Interface Inc. (surface etch rate monitor model Si400) which has a 45 nm thick silicon dioxide layer on a silicon wafer. Both silver and iron were implanted into this material.

3. RESULTS

Figures 1 and 2 show the depth profile of the 20% Ag/80% Au alloy for resonant and nonresonant ionization respectively. The silver resonance ionization is achieved by stepwise excitation using one photon at 328.068 nm and two photons of 546.549 nm. Both silver and gold require two 193 nm photons for nonresonant ionization.

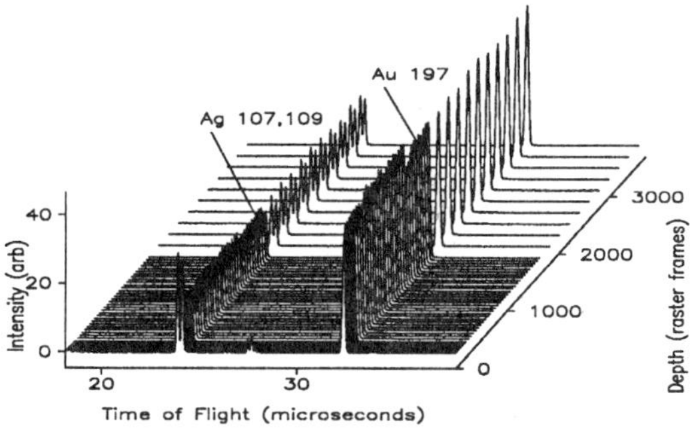

Figure 1. Nonresonant depth profile of 20%Ag/80%Au alloy.

The striking difference between the two spectra is the enhanced elemental silver signal for nonresonance ionization, while in resonance ionization the signal builds up to a steady level. The reason for this is the presence of a silver oxide layer on the surface of the sample. Apparently the 193 nm photon from the nonresonant ionization process both fragments and photoionizes the silver oxide layer.

This is evident from the small silver oxide peaks seen in the first layer of the depth profile between the silver and gold peaks. Once the silver oxide is sputtered away the silver "bulk" signal is lower than that observed at the surface with the oxide layer.

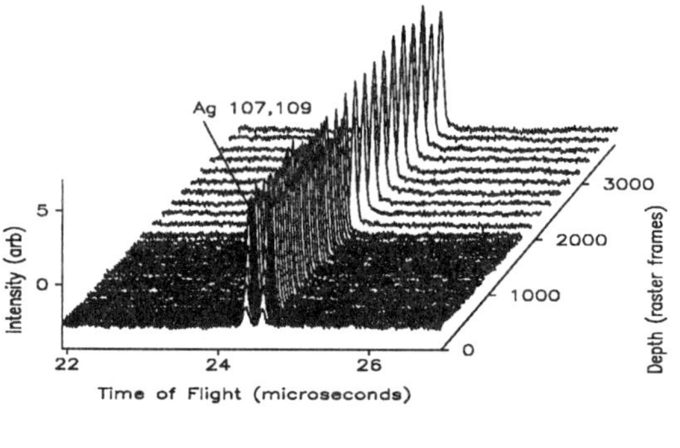

Figure 2. Resonant depth profile of 20%Ag/80%Au alloy.

This is confirmed by the gradual increase of the silver resonance signal with increasing depth. These results indicate that a significant portion of the silver sputtered from the oxide layer is tied up a molecular neutrals, which are not fragmented and ionized with the resonance scheme. This was seen with the entire series of the alloys. The composition varied from 100% silver to 80%, 60%, 40%, 20% and 0% silver (100% Au).

An interesting phenomenon was observed when silver foil was coated with gold and then depth profiled. The nonresonance depth profile is shown in Figure 3. Silver is present on the surface of the gold coating. This indicates room temperature diffusion of the silver through the gold layer. The results in Figure 3 were produced several months after the gold coating. A depth profile taken just after gold coating exhibits a high silver signal at the surface, then the silver completely disappears in the gold overcoat. Evidently there is some room temperature diffusion of silver through the gold overcoat occurring. This is similar to what has been observed by Bukaluk (1990) in gold/silver multilayers using AES. Although their work was done using above ambient temperatures, our work supports their observation that silver diffusion is quite rapid at elevated temperatures.

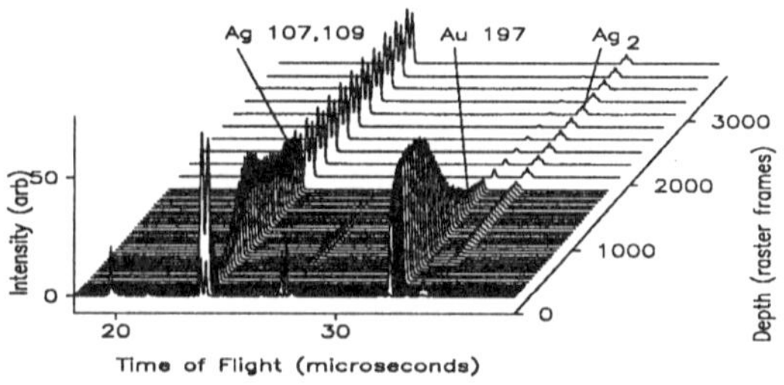

The sample in Figure 3 exhibits silver throughout the gold coating. Another interesting point is the silver oxide coating below the gold coat. We can see that it is several layers thick, with a corresponding enhancement of the elemental silver signal. Once the bulk silver is reached we also see silver dimers appearing to the right of the gold peak.

Figure 3. Nonresonant depth profile of gold coated silver foil.

The resonance depth profile is similar to the nonresonance profile except there is no enhanced elemental silver signal from the oxide at either the gold surface or at the gold/silver interface.The resonance signal does indicate the same silver diffusion through the gold layer.

A nonresonant depth profile of the SiO_2/Si sample is shown in Figure 4. We would expect that the elemental Si signal would be low in the oxide layer and high in the silicon layer. What we observe is that the intensity of the detected species is dependent on the laser power. Figure 4 shows that under high excimer laser power (>60 mJ) the elemental signal is high in the oxide layer, enhanced by the oxide due to photofragmentation.

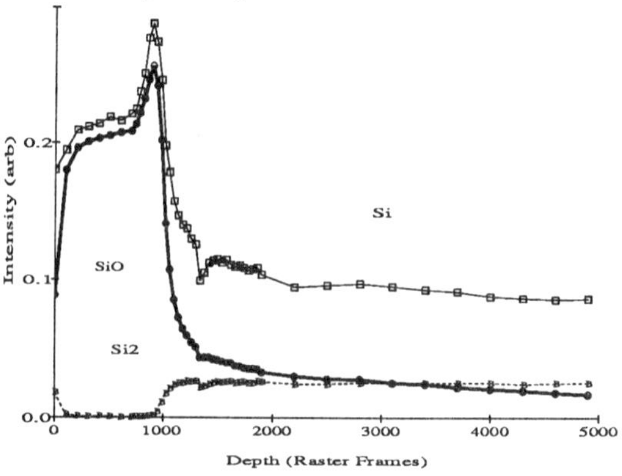

This is seen by the apparent decrease of the elemental Si signal after sputtering through the oxide layer. The onset of the interface region is exhibited by the appearance of the silicon dimer signal. In some aspects this depth profile is similar to what is obtained from SIMS. The elemental Si signal decreases after the interface region. This is due to the change in the matrix from oxide to element with a corresponding change in secondary ion yield. Both the elemental silicon and silicon oxide signals exhibit increases at the interface region.

Figure 4. Depth profile of the SiO_2/Si wafer sample without any implants.

The reason for this is unclear, however it was apparent in all samples under a variety of conditions. Although at lower laser power (<50 mJ) we were able to achieve a steady elemental Si signal through the oxide layer. This was evidently below the fragmentation threshold for the oxide. These results indicate that the laser power has a large effect on the shape of the profile and the resulting species observed in the mass spectrum.

The iron 54 and 56 isotope implants offer some interesting comparisons among the different ionization methods used. The SIMS profile is similar to that observed for the nonresonant data shown in Figure 5. In the SIMS profile the ion yields from the oxide layer are much higher than from the bulk Si layer as indicated by the fact that the Si signal is 3 times higher in the oxide. Presumably the iron signal undergoes a similar change in ion yield. Although similar, the nonresonant data exhibits some subtle differences. First of all in the nonresonance profile, the leading edge is lower than the trailing edge while in the SIMS profile the leading edge is higher than the trailing edge. We also observe a factor of 4 decrease in signal in the nonresonant profile while the SIMS profile shows only a factor of 2 dip.

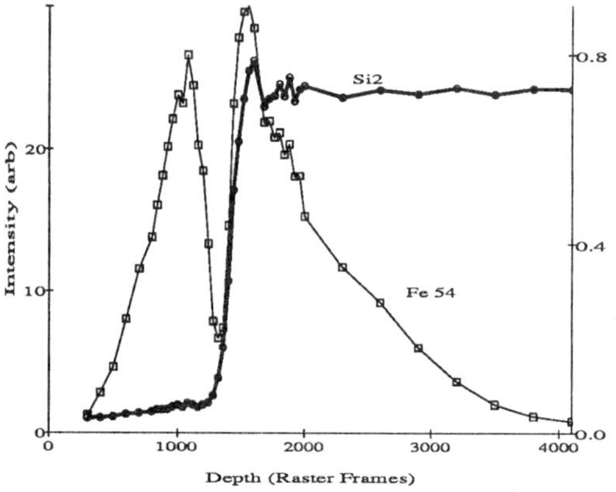

Figure 5. Nonresonant Depth Profile of Iron Implant in SiO_2/Si wafer.

In the resonant depth profile (Figure 6) there is no apparent iron signal prior to the interface region. This seems to suggest that the iron does not sputter as ground state atoms from the oxide layer and is not available for resonant photoionization, while the excimer is able to dissociate the iron and produce an elemental signal. It may be that the iron, although implanted in the atomic state, during the sputtering process undergoes excitation by the ion beam and reacts with the oxygen in the oxide layer to sputter off as iron oxide.

The iron oxide cannot be resonantly photoionized but is easily fragmented and ionized by the excimer laser (Christie 1987). This would account for the apparent differences in the three profiles. The curious aspect is that there does appear to be a depletion zone of iron atoms at the interface. At this point it is not clear which side the depletion zone is on.

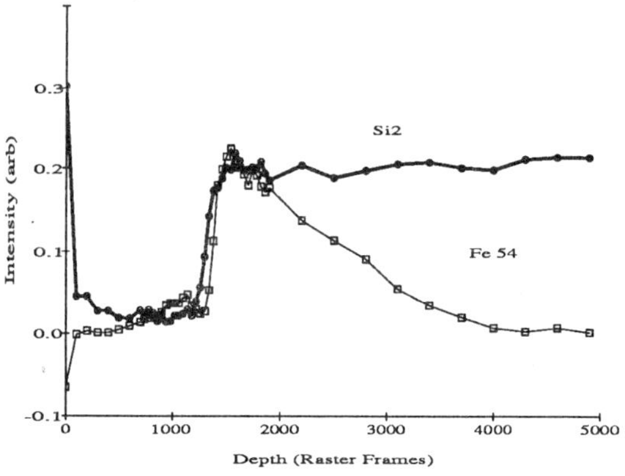

Figure 6. Resonant Depth Profile of Iron Implant in SiO_2/Si Wafer.

Additional work such as RBS is being done to confirm the presence of the depletion zone and more detailed ionization studies are needed to explain and understand the mechanisms involved in sputtering process.

4.REFERENCES

Bukaluk, A. 1990 Appl. Surf. Sci. **45** 57-64
Christie, W.H. 1987 SIMS VI ed. A. Benninghoven (Chichester:Wiley) pp 805-808
Young, C.E. 1987 Nucl. Instr. Meth. Phys. Rev. **B27** 119-129

Inst. Phys. Conf. Ser. No 114: Section 11
Paper presented at RIS 90, Varese, Italy, 16–21 September 1990

Development of an instrument for laser-assisted resonance ionization of sputtered atoms

P De Bisschop[1], D Huyskens[1], W Vandervorst[1], B Rasser[2], F Costa de Beauregard[2]

[1] Imec, Kapeldreef 75, B-3030 Leuven, Belgium; [2] Cameca, F-92403 Courbevoie, France

ABSRACT: We are developing an instrument for resonant post-ionization of sputtered neutrals in which a photoionization mode and a field-ionization mode have been implemented. We have demonstrated the viability of the latter and showed that it offers advantages with respect to suppression of background ions: the detection limit for Ge in Si was found to be an order of magnitude better than with photoionization.

1. INTRODUCTION

Resonant laser-induced post-ionization of sputtered neutrals is generally viewed as a means to overcome some of the limitations of Secondary Ion Mass Spectrometry (SIMS). It aims at an improvement of detection limits, not only through an increase of the sensitivity, but in the first place because of the low background level that follows from its inherent selectivity of the technique. A second issue on which an improvement is expected is the quantitativity of the technique, as matrix effects should in general be absent or largely reduced with respect to SIMS. These advantages have been proven in a number of cases. We are developing a laser-assisted post-ionization instrument to further explore the potential of the method in the field of impurity analysis in semiconductors. We decided that our instrument should permit good SIMS analysis as well, next to post-ionization. The idea here is that a preanalysis of the sample with SIMS will permit a more "efficient" use of the post-ionization capability.

2. THE INSTRUMENT

The present version of our instrument has been described before (De Bisschop 1989). The mass spectrometer is of the classical double-focusing magnetic type; we typically use it at a mass resolution of 200, but higher values are possible. The spectrometer has the property of conserving ion images. Although this is of no direct use if the instrument operates in its post-ionization mode, it permits a very good alignment of the primary column and the spectrometer by tuning to the SIMS mode. At the same time it improves the SIMS performance considerably. Although a pulsed primary column has been developed and tested in a "stand-alone" configuration, it has not yet been mounted on the apparatus: we still use a continuously operating column, equipped with a duoplasmatron source (angle of incidence 45°).

In all other existing instruments using resonant laser-induced post-ionization of sputtered atoms, the atoms are ionized directly by the lasers through successive photoabsorption, i.e. RIS. In a number of cases the occurrence of isobaric background photoions has been reported (Kimock 1984, Pellin 1987, Downey 1990) in spite of the elements selectivity of RIS. As there is no possibility to suppress this type of background it could severely curtail the detection limit. This problem could be circumvented in some cases by using another excitation sequence but such an alternative may not always be available. Most elements that are important to semiconductor technology require the use of UV radiation, suggesting to us that this background problem could prove to be a relatively important one. We have therefore been looking for an alternative approach to RIS that would offer the possibility to suppress such background ions, and we realized that the use of electric field ionization of Rydberg atoms instead of photoionization might well be a viable alternative.

Field ionization of laser-excited Rydberg atoms (with dc or pulsed fields) was already used in the early seventies for selective detection of atoms in vapors or atomic beams. (A good account can be found in Letokhov 1987.) To our knowledge it has however never been applied to sputtered particles. The Rydberg atoms can easily be ionized simply by exposing them to a sufficiently high electric field (DC or pulsed). The Rydberg atoms have a relatively long lifetime if the field is only small and sputtered particles typically have eV energies, which implies that sputtered particles that have been excited to a Rydberg state can travel over an appreciable distance before decaying radiatively. On this property we based the idea of a "double-zone secondary ion source" as an implementation of the field-ionization technique in our instrument. The gist of this approach is to keep the electric field low in the region where the laser-atom interaction takes place and only apply the high ionizing field in a second region, situated somewhat further on the flight path of the Rydberg atoms. (The transition from low to high field is achieved by applying appropriate dc potentials on one or more source electrodes and does not require the use of voltage pulses.) In this way, the actual field-ionization process takes place in a region that is spatially separated from the region through which the lasers are sent. Due to the variation of the electrostatic potential applied in the source, the ion energy will then depend on the mechanism with which it was formed (SIMS, photoionization or field ionization). This offers the possibility to distinguish them from one another and hence suppress SIMS and photoions when using field ionization. For convenient reference, we use the acronym LEFI to refer to this method combining Laser Excitation and Field Ionization.

When considering LEFI as an alternative to RIS, one should be aware of the fact that ion yields with RIS should be higher than with LEFI. We argued before (De Bisschop 1989) that the Rydberg-excitation probability will always be smaller than 1 due to the occurrence of stimulated emission from the Rydberg state. At the same time, efficient ion collection from a large interaction volume will be easier with RIS due to the larger focusing action of the accelerating field. RIS and LEFI can therefore be viewed as complementary techniques, and in designing our instrument we decided that it should be possible to apply either of them, and that switching from one to the other should be easy.

The secondary ion source, i.e. the region near the sample in which post-ionization occurs, was designed to permit LEFI, RIS and SIMS. It contains an electrode on which a variable potential can be applied, situated between the sample and the extraction electrode. Which source mode is selected depends on the value of this potential and on the sample potential. A variable potential can be applied on the extraction electrode too, but so far, we have usually kept it at ground potential. In operating the instrument in any of the source modes, we always keep the ion energy for which the spectrometer is optimized fixed (at 3 keV) and vary the source until the potential at the position where the ions are generated is 3 kV. For RIS, the 3 kV equipotential plane hence has to lie at the position of the laser beam; for LEFI it must coincide with the position where the field reaches the critical value for field ionization of the particular Rydberg state that was excited.

3. SPECTROSCOPIC RESULTS ON GE

The experimental work done so far has been mainly devoted to Ge. The ground state of Ge is a $4p^2\ {}^3P_J$ triplet, and we used one of the $4p5s\ {}^3P_J$ levels as intermediate state of a 1+1 two-color excitation scheme. The wavelength of the $J_g = 0 \rightarrow J_e = 1$ transition is $\lambda_1 = 265.16$ nm. The threshold wavelength for photoionization of the 4p5s $^3P_1$ state, λ_{thr}, is 384.34 nm.

With the source in RIS mode, we made a number of λ_2 scans above the ionization threshold (1+1 two-color RIS). We observed a considerable number of resonances, which could be attributed to states belonging to the second Rydberg series of Ge (converging to the 4p $^2P_{3/2}$ excited state of Ge^+). Apparently, these states behave as autoionizing states, an assumption that was confirmed by measuring the energy distribution of these ions. A comparison with the energy distribution found when doing 1+1 single-color RIS (λ_1 only), showed that in the two-color experiment ions are created not only in the laser-beam volume but also outside this volume, further away from the sample. We observed states of this second series with effective principal quantum number n_2* from 7.56 to 13.67.

The location of the free-atom Rydberg states (of the first series), with effective principal quantum number between 15.6 and 25, was determined by scanning λ_2 below threshold, with the source now tuned to LEFI (i.e. with the 3 kV plane in the high-field region of the source). As these states are easily perturbed by any residual electric field in the excitation region, the source potentials had to be adapted so as to reduce this field to a value not larger than a few tens of V/cm. At higher fields, we found a completely perturbed spectrum consisting of a large amount of peaks at an almost constant spacing.

4. ANALYTICAL RESULTS ON GE

After demonstrating LEFI and localizing the Rydberg states, we tried to assess the major analytical characteristics of our instrument for an analysis of ^{74}Ge in Si, choosing three post-ionization modes. First of all we looked at one LEFI mode without an electric field in the interaction region and one with adapted source potentials to increase the potential barrier in the source ('non-zero field' case). Further we investigated an autoionizing scheme. The results are summarized in Table 1.

With our continuous primary beam, the classical useful yield notion has no meaning, and we therefore use the sensitivity S to quantify the number of ions that our instrument can produce (S ≡ counts/pulse devided by the primary current and the atomic concentration). The use of the autoionizing state clearly offers the best sensitivity. Comparing our sensitivities with data in the literature, however, indicates that values up to ~ 1 cpp/μA·ppm are sometimes attained in other instruments, being much better than our present values. It should be mentioned that the present version of our instrument was mainly intended for demonstrating LEFI, assessing

Table 1: Comparison of analytical characteristics for analysis of ^{74}Ge in Si with LEFI and RIS, in or present instrument

Quantity	LEFI non-zero field	LEFI zero field	RIS autoionization
S (cpp/μA·ppm)	5.2 10^{-4}	2.5 10^{-4}	5.2 10^{-3}
BRF: $^{74}Si_2O$ SIMS	2 10^6	3 10^5	1.8 10^4
BRF: photoions	1.2 10^3	2.7 10^2	—
S/N_{SIMS}	1.0 10^7	7.3 10^5	1.2 10^6
c_{min} (ppm)	0.21	2.7	1.7

its performance characteristics and comparing it to RIS; no real optimization of the yield was attempted. In particular the secondary source - having an extraction hole of only 1 mm - puts a curb on the useful interaction volume, and hence the sensitivity. A new version of the source offering an expected increase in sensitivity of one order of magnitude has been designed.

The main item to be verified was the background-suppression capability of the source in both post-ionization modes. SIMS analysis of Ge in Si under O_2^+ bombardment suffers from an interference of Si_2O. In our experiments we use Ar sputtering, but the vacuum level in the machine is only $\sim 10^{-7}$ Torr, and there is apparently enough oxygen adsorption on the sample surface to produce an appreciable Si_2O SIMS signal: the ratio of the $^{74}Si_2O/^{30}Si$ SIMS signal was found to be ~ 9. (The ratio observed in an IMS 4f Cameca SIMS instrument, with O_2^+ bombardment, is ~ 10.) We quantify the background suppression potential of a given source mode through the "background reduction factor" (BRF), defined for example in case of the SIMS background as the ratio of the optimum secondary ion signal (instrument in SIMS mode) with the residual secondary ion signal when the source and transfer optics are tuned to the particular post-ionization mode. The best LEFI BRF for secondary ions is two orders of magnitude better than the BRF for the autoionization scheme. The detection limit, c_{min}, for LEFI is therefore one order of magnitude lower, in spite of the higher sensitivity obtained when using autoionization. (At present, our detection limit is imposed by the secondary-ion background.)
Suppression of photoion background is only possible in the LEFI mode. The BRF was measured by optimizing the 1+1 single color RIS signal and then tuning the source back to LEFI: a value of $\sim 10^3$ was found in the non-zero field LEFI case. This means that an isobaric background generated through e.g. non-resonant photoionization or photodissociation can be suppressed with three orders of magnitude.

5. CONCLUSION

We have implemented field ionization (LEFI) in an instrument for post-ionizing sputtered neutrals, and showed that it can be a viable technique. It was found to be complementary to RIS. Our first experimental results, on Ge, have confirmed the expectation that RIS is more sensitive, but that the use of LEFI can lead to better detection limits in cases where the detection limit is determined by a secondary-ion or photoion-background source. We also found that states of higher-order Rydberg series, lying above the first ionization limit, show an autoionization behaviour for Ge. The use of such states might be a general way to enhance the photoionization cross section in RIS, as such higher-order Rydberg series exit in many atoms.

We acknowledge the EEC for financial support (contract Esprit 1056). DH is endebted to the Belgian IWONL for a scholarship. We aknowledge J.B. Clegg (Philips, Redhill, UK) and R. von Criegern (Siemens, Munich, FRG) for fruitful discussions while developing our instrument.

De Bisschop P, Huyskens D, Meuris M, Vandervorst W, Rasser B and Costa de Beauregard F 1989 *Proc. of the VIIth Int. Conf. on SIMS* eds A Benninghoven, C A Evans, K D McKeegan, H A Storms and H W Werner (Wiley) pp 907-910

Downey S W and Hozack R S 1990 *J Vac Sci Tech* A8 791-6

Kimock F M, Baxter J P, Pappas D L, Kobrin P H and Winograd N 1984 *Anal Chem* **56** 2782-91

Letokhov V S 1987 *Laser Photoionization Spectroscopy* (Academic Press)

Pellin M J, Young C E, Calaway W F, Burnett J W, Jϕrgensen B, Schweitzer E L and Gruen D M 1987 *Nucl Inst Meth* **B18** 446-51

Inst. Phys. Conf. Ser. No 114: Section 11
Paper presented at RIS 90, Varese, Italy, 16–21 September 1990

Charge emission from silicon and germanium surfaces irradiated with KrF excimer laser pulses: Applications to the atom bank

G. S. Hurst, M. M. Bialkowski, and J. E. Parks
Institute of Resonance Ionization Spectroscopy, University of Tennessee, Knoxville, Tennessee 37996 USA

D. H. Lowndes, and G. E. Jellison, Jr.
Oak Ridge National Laboratory, Oak Ridge, Tennessee 37831 USA

ABSTRACT: Atoms can be stored in materials such as semiconducting crystals and recalled for counting or other analytical purposes by the application of laser pulses. Such a concept has already been used to isotopically enrich and to count noble gas atoms. In some cases it is desirable to know the neutral and the ionic environment associated with the process of laser annealing of the storage medium. Some investigations of this type are discussed for electronics grade silicon and germanium.

1. INTRODUCTION

It is interesting to speculate on the uses that an atom bank would have if one existed with certain characteristics. In connection with the development of the Maxwell Demon for sorting and counting noble gas atoms with isotopic selectivity, we have already shown that Krypton atoms can be implanted into silicon and then recovered for further isotopic enrichment or repeated counting by the application of laser pulses to anneal the crystal (Hurst *et al* 1985). Other conceivable uses of the atom bank could include the preparation of low-level standards for atom counting. In addition it would be desirable to count and save atoms for long-term archival purposes. Surely the ability to deposit, save, transport, and withdraw atoms from an atom bank would be of interest in basic sciences such as particle physics, and we would not be surprised to see entirely new applications in the future.

If such a concept is to be truely useful, the atom bank must meet certain rigorous requirements: (1) the implantation or deposit of atoms must be done with essentially unit efficiency; (2) the atom must be counted with near unit efficiency as it enters the bank; (3) the bank must be a chemically safe place to store the atom for an indefinite period; (4) the bank must be portable; and (5) it is essential that the atoms be withdrawn on request without either losses or gains. From these it follows that atoms can be sent from one bank to another and that some of the transactions involved in monetary practices are possible.

Presently it is not possible to meet all of the requirements listed above, but the gap is being closed in the case of noble gas atoms. In the next section, we examine the technical basis for realizing an atom bank. In section 3 we will examine a special case of the recovery of atoms, to prove that it is now possible to transfer atoms from bank-to-bank on a microsecond time scale.

2. REVIEW OF THE TECHNICAL REQUIREMENTS FOR AN ATOM BANK

The technical basis for achieving an atom bank has already been discussed (Hurst *et al* 1988a) and will be reviewed here only briefly.

When an ion, for example, of 10 keV strikes a material, a neutral atom is stored in the material at a depth of about 10 nm. For certain materials, the probability that the ion will be implanted, called the sticking coefficient, approaches unity at 10 keV. Special experiments were done by Hurst *et al* (1985) to determine that the sticking coefficient for 10 keV ions of ^{85}Kr was 0.9 ± 0.2. Therefore, there is reasonable certainty that Kr atoms can be stored in a semiconducting material with near unit probability at a depth of several nm in the material, thus satisfying requirement (1).

The absolute counting of atoms as they enter the bank is somewhat more difficult. Normally, this is accomplished by counting the secondary electrons emitted when an ion is implanted. The number of electrons emitted per ion, known as the gamma coefficient, needs to be 3 or greater to ensure that more than 95% of the ions are counted, assuming the usual Poisson statistics. This requirement can be met with certain materials, and in particular in Cu-Be targets for noble gas ions above 3 keV. Whether this requirement can be met with the same material needed for the storage medium is still open to question.

The use of small targets made of electronics grade silicon or germanium would appear to satisfy requirements (3) and (4), and we need to discuss the recovery of atoms from these materials. We have already shown that essentially all of the Kr atoms implanted into silicon can be recovered by laser annealing. But since this is such a key idea for our discussions below, we wish to review the concept in a bit more detail. Figure 1 illustrates the redistribution of atoms following laser annealing of silicon. Two examples are shown to illustrate two clearly delineated cases. In the first case, the implanted atom is segregated to the surface as the liquid-solid interface propagates back to the annealed surface, see the Cu illustration in Figure 1. In the second case, the annealing process merely diffuses the implanted atoms. Thus the atom bank will only work whenever we have met certain conditions on bonding energies. In the case of noble gases, the annealing process results in atoms in the gas phase whereas in other cases the atoms are attached to the surface of the bank. In both cases the atoms are most likely expelled from the bulk of the material in a very short time corresponding to the time of propagation (ns) of the liquid-gas interface back to the surface.

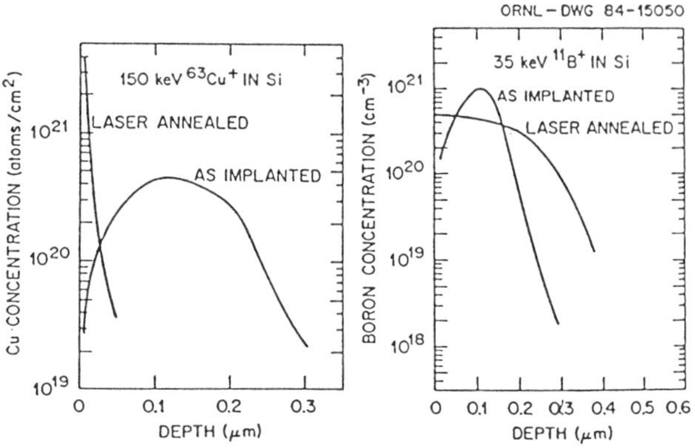

Figure 1

3. TIME-RESOLVED STUDIES OF LASER ANNEALING OF Si AND Ge CRYSTALS

We have already shown (Hurst *et al* 1985) that Kr atoms can be removed from Si targets by laser annealing. Applications of this have been made to process the atoms through another step of isotopic enrichment. But in these cases the time between laser annealing to return the atoms to the gas phase and the RIS process required for subsequent acceleration and reimplantation can be quite long (minutes or hours). On the other hand, if it is required to reduce this time span to a microsecond time scale, as in the RISTRON concept (Hurst 1988b), then some new considerations must be addressed. Namely, on a microsecond time scale, is there space charge associated withthe laser annealing step which could interfere with the ion optics required for further analysis of the ions? That was the motivation for a recent study reported by Bialkowski *et al* (1990) which will be briefly discussed.

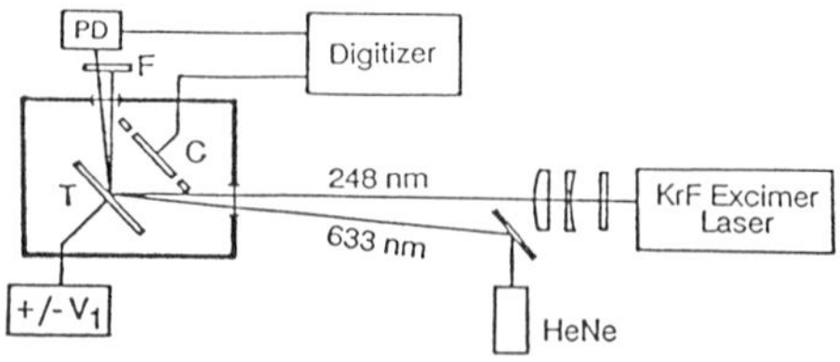

Schematic of Annealing Study

Figure 2

A schematic of the experiment where both Si and Ge were studied to assess the charge generated as a by-product of the laser annealing process is shown in Figure 2. A KrF excimer laser was used to anneal the targets (T), and a time-resolved reflectivity method using a He-Ne laser was used to determine the melting threshold. Time-resolved studies of the electrical charge, as measured with the collector (C), were recorded with a two-channel transient digitizer. Thus one channel recorded the reflectivity while the other channel recorded the charge released from the targets, which were studied both below and above the melting threshold.

An example of the time-resolved emission from Ge is shown in Figure 3 for positive as well as negative potentials on the Ge target. We note that just below the melting threshold the positive ion emission from Ge is collected in about 60-ns which was about the width of the KrF laser pulse. The negative carriers also show a strong component corresponding to the same time behavior of the positive emission, suggesting that the dominant mechanism showing up in Figure 3 is a recombination of positive and negative carriers.

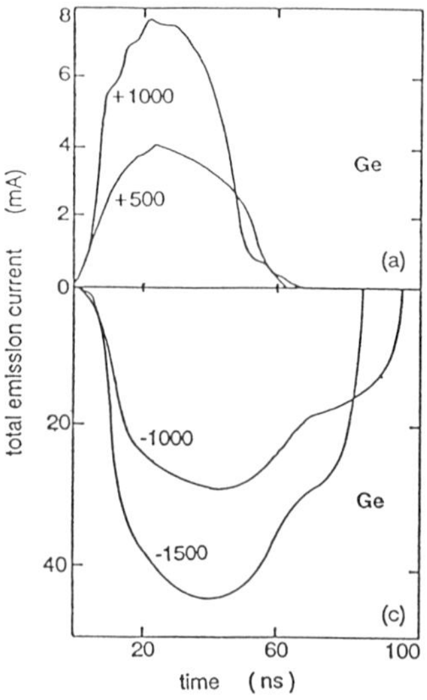

Ge: Below Melting Point

Figure 3

Figure 4 illustrates the results obtained with Si targets just above the melting threshold; however there is a very interesting process occurring in the positive emission from Si targets. Note that this emission persists for more than 3 microseconds while the negative carriers are collected more quickly, still showing the 60-ns laser pulse width.

The long collection time of positive ions in Si is not a deterrent to the use of this material in an atom bank since a few microseconds are required for the neutral atoms to travel to the RIS beam. Further, there is no such time delay in the case of the Ge target. Thus, in either material there is no charge emission process that would interfere with ion optics in applications such as the RISTRON. Our requirement that withdrawals from the atom bank be made on a microsecond time scale have been met.

We wish to comment on a possible physical mechanism to account for the time delay in the positive ion emission in Si and it's absence in Ge. In reviewing the physical properties of these materials it is seen that at the melting points the vapor pressure is much higher in Si compared to Ge. This suggests that the neutral emission from Si when it is annealed could play some role in delaying the collection of the positive ions. The importance of Knudsen layer formation during laser annealing has been stressed by Kelly and Dreyfus (1988). In this highly non-equilibrium process, particles of widely differing velocities are colliding to produce a more uniform velocity distribution in the forward direction.

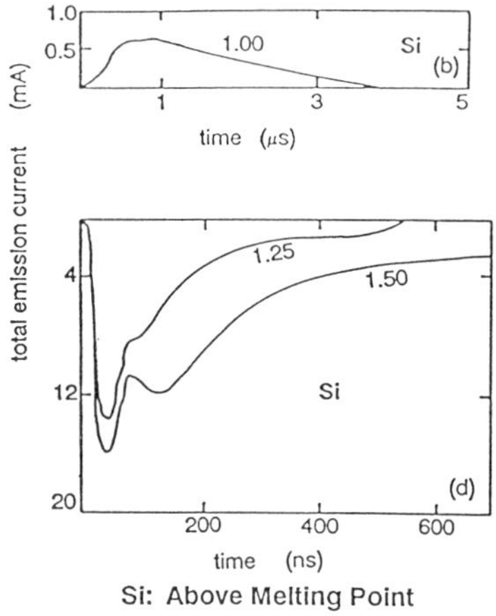

Figure 4

The effect of this Knudsen layer could be to provide a dense layer of atoms with which the positive ions must collide and thus accounts for a time delay. The thickness of this layer is just the product of the mean velocity and the time duration of the laser pulse (since the material cools rapidly after the laser pulse). The density of neutral Si atoms can be estimated to be about 10^{17} cm^{-3}, assuming that one monolayer of Si is ablated in one laser annealing pulse. This is such a high pressure (3 torr) that one can approximate the motion with an ion mobility. Thus the time delay is just the thickness of the Knudsen layer divided by the mean velocity of the positive ions determined from mobility estimates. A reasonable thickness is 10^{-2} cm and an estimated ion drift velocity at an E/P value of 200 volts cm^{-1} torr^{-1} is 10^4 cm s^{-1}, thus delays of the order of microseconds can be accounted for. Since the vapor pressure at the melting point is much less for Ge, it is reasonable to assume that less than a monolayer is removed during laser annealing, and this accounts for the absence of the Knudsen layer for this material.

Further studies of this interesting phenomena are planned in which RIS would be used to study the neutral species themselves, and thus provide much more definitive information on densities, angular distributions, and velocities of neutral species released from semiconducting materials during laser annealing.

4. REFERENCES

Bialkowski, M. M., Hurst, G. S., Parks, J. E., Lowndes, D. H., and Jellison, Jr., G. E., J. Appl. Phys. (accepted for publication November 1, 1990).

Hurst, G. S., Payne, M. G., Kramer, S. D., Chen, C. H., Phillips, R. C., Allman, S. L., Alton, G. D., Dabbs, J.W.T., Willis, R. D., and Lehmann, B. E., 1985 Rep. Prog. Phys. **48**, 1333.

Hurst, G. S. and Payne, M. G., 1988a *Principles and Applications of Resonance Ionization Spectroscopy* (Bristol: Adam Hilger Ltd.).

Hurst, G. S., Bialkowski, M. M., and Lowndes, D. H., 1988b *Resonance Ionization Spectroscopy 1988*, eds. T. B. Lucatorto and J. E. Parks [Bristol: Institute of Physics (Conference Series No. 94)] pp. 229-232.

Kelly, R. and Dreyfus, R. W., 1988 Surface Science **198**, 263.

Inst. Phys. Conf. Ser. No 114: Section 11
Paper presented at RIS 90, Varese, Italy, 16–21 September 1990

Surface sensitive detection of organic molecules using ion beam induced desorption

M. H. Ervin, D. M. Hrubowchak, M. C. Wood and N. Winograd

Department of Chemistry, The Pennsylvania State University
152 Davey Laboratory, University Park, PA 16802, U.S.A.

Abstract: The ability to use energetic ion bombardment coupled with multiphoton resonance ionization to quantitatively detect molecules on surfaces has been investigated. Benzo(a)pyrene has been used to establish the resonant nature of the ionization process and to demonstrate the tunable fragmentation available with this type of ionization. Serotonin was then used to demonstrate the sensitivity of the method.

1. Introduction

Surfaces play a vital role in many important processes such as heterogeneous catalysis, friction, corrosion, adhesion, etc. One of the techniques commonly used to characterize surfaces is secondary ion mass spectrometry (SIMS). This technique uses the interaction of an energetic ion beam with a solid to cause ejection of secondary ions from the topmost atomic layer. These ions are then detected using a mass spectrometer.

While the SIMS technique has made many contributions to the understanding of various surface processes it has significant limitations when it comes to the quantitation of trace analytes. First, the secondary ions are only a small fraction of the material desorbed by the primary ion impact. The majority of this material is ejected as neutral species. For SIMS, this leads to inefficient sampling of the ablated material so that sensitivity is lost. Second, the probability that a particle will leave the surface as an ion, may vary by orders of magnitude depending on the surface composition, so that it is very difficult to obtain quantitative results (Kimock et al. 1984). In order to overcome these limitations, multiphoton resonance ionization (MPRI) has been combined with ion induced desorption. Laser ionization allows the sampling of the more abundant and less variable neutral component of the ejected flux which increases the analytical signal and makes it more quantitative. The selectivity of resonant ionization reduces noise, thereby enhancing sensitivity. It also is a very efficient means of ionization, allowing relatively large ionization volumes for a given laser power. Using this method, a detection limit of 9 parts per trillion for In atoms uniformly doped into Si has been obtained (Pappas et al.1989). It is not obvious, however, that similar results can be achieved when this technique is applied to the measurement of molecules on surfaces. For one thing, molecules have many rotational and vibrational levels so that there are larger densities of states with more populated levels than are observed for atoms. This reduces the selectivity of the ionization. Although the desorbed molecules could be cooled by entaining them in a supersonic jet expansion in order to reduce the number of populated levels (Grotemeyer et al. 1986), this entrapment would sacrifice sensitivity. Sensitivity can also be lost due to the fragmentation of the analyte. For this reason the resonant nature of the ionization is even more important when studying molecules, since the lower power densities that can be used will reduce the absorption of photons in excess of those needed for ionization, which frequently result in fragmentation. Single photon ionization can also avoid this fragmentation but it lacks selectivity, and typically only small ionization volumes can be used which reduces sensitivity (Schule et al. 1988).

In this work, we examine polycyclic aromatic hydrocarbons (PAH's) as initial specimens for molecular detection because they absorb strongly in the near UV at easily attained energies. These energies are sufficient such that two photons will exceed the ionization potential of the molecule. PAH's are also important pollutants with real world trace measurement problems. Initial experiments involve the investigation of the power dependence of the signal. Moreover, the increased sensitivity relative to SIMS, is demonstrated for benzo(a)pyrene. Finally, the linearity and limits of detection are investigated for serotonin, a biologically important molecule.

2. Experimental

The apparatus has been described in detail elsewhere (Kimock et al. 1984, Pappas et al. 1989). All experiments are performed in an ion pumped Perkin-Elmer Utek TNB-X ultra high vacuum chamber with a base pressure of 5 x 10^{-9} torr after bakeout. The system is equipped with a load-lock pumped by a Balzers Model TSU 170 liter/sec turbomolecular pump for fast sample insertion. A 5.6 µs pulse of primary Ar^+ ions (34 µA, 10 keV) generated by a Physicon Model DP10-01 duoplasmatron source is focused to a 1 mm diameter spot on the sample at a 45° angle of incidence. The laser system consists of a Quanta-Ray Model PDL-2 dye laser pumped by the second harmonic of a Quanta-Ray Model DCR-2A Nd:YAG laser with a 30 Hz repetition rate. The dye laser produces visible 590 nm light with Exciton Rhodamine 590/methanol dye solutions and this light is frequency doubled with a Quanta-Ray Model WEX-1 wavelength extension unit to produce 280 nm UV light. While the output characteristics vary from alignment to alignment, some typical values are: 5 mJ/6 ns pulses of light with a beam area of 0.5 cm^2 (1.6 x 10^6 W/cm^2). The laser is triggered ~150 ns after the end of the primary ion pulse, with the resultant ions being extracted into a reflecting time-of-flight (TOF) mass spectrometer with a pulsed extraction field. The reflecting potentials can be optimized to pass either photoions or secondary ions to a Galileo Electro-Optics Corp. Model FTD 2002 dual microchannel plate (MCP) detector. The output signal from the MCP is processed with a Stanford Research Systems Model SR 250 gated integrator for power dependence studies and ion fraction measurements. Mass spectra are recorded by sending the MCP signal to a 100 MHz Digital Signal Processing Technology (DSP) Model 2001AS transient recorder with summing memory. The instrument is interfaced to a Digital Equipment Corp. MicroVax II computer using CAMAC technology.

For the experiments involving pyrene and benzo(a)pyrene, bulk samples were used since sensitivity was not an issue. These chemicals were obtained from Aldrich Chemical Company and used without further purification. Milligram quantities were dissolved in benzene and deposited onto an ultrasonically cleaned substrate attached to a Cu backplate. The pyrene samples were evaporated onto polycrystalline Au foils whereas the benzo(a)pyrene was deposited onto pieces of {100} Si wafers. These samples were allowed to air dry before being inserted into the chamber. Serotonin was obtained from Sigma Chemical Company and used without further purification. The serotonin samples were prepared in HPLC grade water and deposited onto Si substrates at which point the solvent was vacuum evaporated in the load-lock. For the quantitative studies a series of standards ranging in concentration from 10^{-4} M to 10^{-7} M were prepared by quantitative dilution of a 10^{-3} M stock solution. 10 µl aliquots of these standards were analyzed where the amount of serotonin in the analysis zone varied between 5 x 10^{-11} to 5 x 10^{-14} moles. Each concentration was repeated three to four times to provide independent analysis.

3. Results

Previously we have demonstrated that carbazole gives a large molecular ion signal when desorbed as neutral molecules that are ionized with two 287 nm photons (Winograd 1988).

Similar results were obtained with pyrene and benzo(a)pyrene at 280 nm (Hrubowchak et al. 1990). When the power dependence at 280nm of the benzo(a)pyrene signal was investigated, results very similar to those for pyrene were obtained. Below 5 x 10^5 W/cm^2 the dependence is linear indicating a one photon dependence, presumably because the resonant excitation step is saturated. At power densities above 1 x 10^6 W/cm^2 the signal drops with increasing power due to fragmentation caused by the absorption of additional photon(s) by the parent ion. The ability to control the amount of fragmentation by varying the laser power allows one to choose between optimum sensitivity or the extraction of structural information simply by changing the laser power (Boesl et al. 1987). This tunable fragmentation is illustrated in Figures 1 and 2. In Fig. 1 the laser power density is 3 x 10^5 W/cm^2. This yields a clean spectra that is dominated by the parent ion. Conversely, in Figure 2 where the power density is 2 x 10^6 W/cm^2, the molecular ion peak is diminished in favor of different fragment peaks characteristic of the structure of the parent molecule. When we compare the benzo(a)pyrene MPRI signal in Figure 1, which is in excess of 250 mV, to the less than 0.1 mV SIMS signal obtained on the same instrument, we see that it is advantageous to sample the neutrals when sensitivity is important. This is not surprising since we have previously shown that at least 98% of desorbed pyrene leaves the surface as neutral molecules (Hrubowchak et al. submitted).

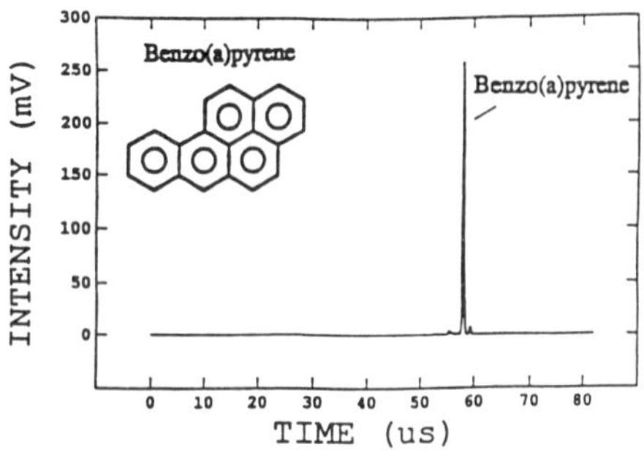

Fig. 1. Time-of-flight MPRI spectrum of benzo(a) pyrene taken at 280 nm with a laser power density of 3 x 10^5 W/cm^2.

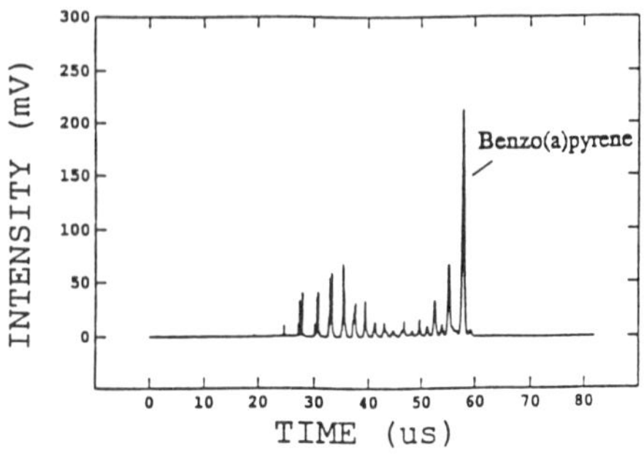

Fig. 2. Time-of-flight MPRI spectrum of benzo(a)pyrene taken at 280 nm with a laser power density of 2 x 10^6 W/cm^2.

When MPRI with 280 nm light is applied to serotonin, a very important biomolecule, the mass spectrum in Figure 3 is observed. As can be seen from the spectrum, the base peak is the parent ion less CH_2NH_2 from the side chain. It is interesting to note that this is also the base peak in the electron impact mass spectrum of serotonin. The quantitative nature of the laser signal is demonstrated in Figure 4. Here the log of signal intensities are plotted against the log of the number of molecules sampled by the ion beam. Before the individual signal intensities at a given concentration are averaged, they are normalized to the laser power and primary ion current used during their measurement. The result is a line of slope 0.946 ± 0.031 and a correlation coefficient of 0.996. A detection limit is calculated by measuring the signal to noise ratios (S/N) from spectra and extrapolating to a concentration that would give a S/N of 2. This extrapolated concentration is then normalized to 100 μA of obtainable ion current with the result being averaged with the results from other runs and other concentrations. This yields a detection limit of 5.2 x 10^{-15} moles ± 20% for serotonin. This compares to a detection limit of 6.4 x 10^{-16} moles for pyrene (Hrubowchak et al. submitted). The difference between the two probably results from differing ionization efficiencies and the fragmentation that takes place in the serotonin's case.

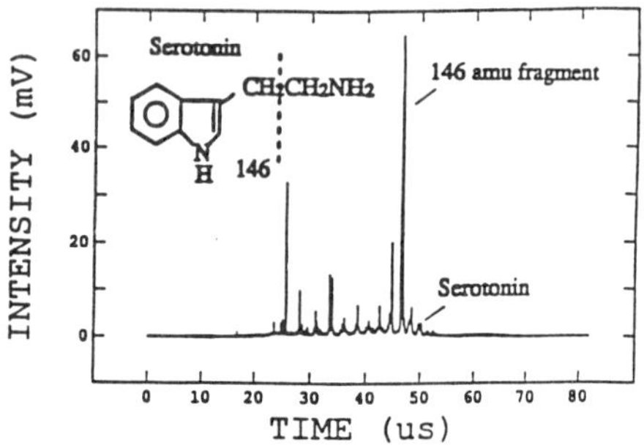

Fig. 3. Time-of-flight MPRI spectrum of serotonin taken with 280 nm light.

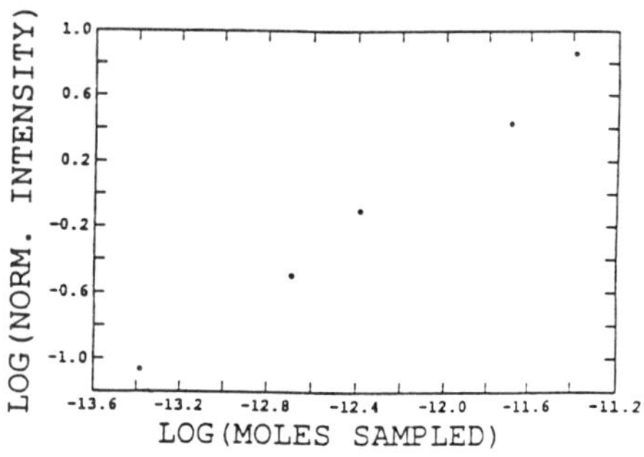

Fig. 4. Log-log plot of 146 amu serotonin fragment intensity versus the number of moles of serotonin sampled by the ion beam.

4. Conclusion

We have demonstrated that energetic ion bombardment coupled with MPRI can be used to sensitively and quantitatively detect submonolayer quantities of molecules on surfaces. The ionization process has been shown to be resonant by the power dependences observed. The resonant nature of the ionization process is important because it increases the sensitivity of the method, as compared with nonresonant ionization, through the increased spacial overlap and decreased fragmentation that comes from efficient ionization at low power densities. Resonant ionization also increases sensitivity by lowering noise that would degrade the S/N ratio. In addition, we have shown that this method can be applied to biological molecules which are typically fragile molecules. Serotonin, which served as a test case for biomolecules, yielded a linear calibration curve with a detection limit of 5.2 x 10^{-15} moles. We are currently applying this method to the detection of molecules in complex matrixes.

5. Acknowledgement

The authors are grateful to the National Science Foundation, the Department of Energy and the Office of Naval Research for the financial support of this work.

6. References

Boesl U, Grotemeyer J, Walter K and Schlag E W 1987 *Anal. Instrum.* **16** 151

Grotemeyer J, Boesl U, Walter K and Schlag E W 1986 *Org. Mass Spectrom.* ***21*** 645

Hrubowchak D M, Ervin M H and Winograd N 1990 *Proc. 7th Int. Conf. on Secondary Ion Mass Spectrometry (SIMS VII)* eds A Benninghoven, C A Evans, K D Mckeegan, H A Storms and H W Werner (New York: Wiley) pp 805-8

Hrubochak D M, Ervin M H and Winograd N, submitted

Kimock F M, Baxter J P, Pappas D L, Kobrin P H and Winograd N 1984 *Anal. Chem.* ***56*** 2782

Pappas D L, Hrubowchak D M, Ervin M H and Winograd N 1989 *Science* ***243*** 64

Schüle U, Pallix J B and Becker C H 1988 *J. Am. Chem. Soc.* ***110*** 2323

Winograd N 1988 *Resonance Ionization Spectroscopy 1988*- Institute of Physics Conference Series number 94 T B Lucatorto and J E Parks (Bristol:Institute of Physics) pp 183-7

Inst. Phys. Conf. Ser. No 114: Section 11
Paper presented at RIS 90, Varese, Italy, 16–21 September 1990

Two-colour ionization dip spectroscopy of size assigned hydrogen bonded clusters

M. Schmitt, M. Pohl, K. Kleinermanns

Physikalisch Chemisches Institut, Im Neuenheimerfeld 500
D-6900 Heidelberg

ABSTRACT: Vibrational spectra of supersonically cooled complexes of p-cresol with H_2O were analyzed by mass-resolved two-photon ionisation, dispersed fluorescence and stimulated emission, detected by two colour ionisation dip. In p-cresol·$(H_2O)_1$ progressions of the intermolecular cluster stretch vibration were observed with similar frequencies in the S_0 - and S_1 - state. The intermolecular stretch of p-cresol · $(H_2O)_3$ again has similar frequencies in the S_0-and S_1- state but has a considerably higher frequency than in p-cresol·$(H_2O)_1$. In p-cresol·$(H_2O)_3$, further intense intermolecular bands arise beside the totally symmetric stretch vibration, namely the hydrogen-bridge bending and torsion vibration.

1. Introduction

The development of sensitive, high resolution laser spectroscopic methods over recent years in combination with adiabatic cooling in supersonic expansions has yielded much information on electronically excited states of clusters and their intra- and intermolecular vibrations. Relatively little is known about the vibrational states of the electronic ground state of clusters. Only the ground state information, however, can be compared with ab initio calculations of cluster potentials and vibrations on the Hartree-Fock level (Kleinermanns et al. ; 1990). For large organic molecules and their clusters, only single determinant ground state calculations can be performed with reasonably sized basis sets and tolerable computer time.
For (aromatic) organic molecules and clusters with chromophoric groups, dispersed laser fluorescence can be used with no restriction in the attainable vibrational states so long as the Franck-Condon overlap between the specific (ro-) vibronic level in S_1 and S_0 exists. The spectral resolution of the method is, however, limited by the monochromator resolution which is normally a few cm^{-1} - considerably larger than the laser resolution.
An interesting method for obtaining cluster ground state vibrational frequencies is stimulated emission. Here, two laser colours ω_1 and ω_2 are used to first pump a molecule to a level in S_1 with ω_1 and then dump it with ω_2 via stimulated emission to a selected vibrational level in S_0. High resolution (that of the laser used) can be achieved here, which is important for cluster spectroscopy due to the low frequency intermolecular vibrations of the clusters, which may lead to a rather high density of vibrational states already at comparable low energies. Furthermore, high number densities of ground state molecules in a selected vibrational state (up to 40-50% of the excited molecules in our experiment) can be produced because the pump / dump process consists of two sequential one-photon transitions. It should be possible to achieve a

high population of clusters vibrationally excited above their predissociation limit for time - and state - resolved observation of the dissociation by laser spectroscopy of the dissociation products.

Stimulated emission spectroscopy (SEP) of stable molecules has been pioneered by Hamilton et al. (1986) who used the dip in the fluorescence from the pumped S_1 level to monitor a SEP resonance. The dip of the ion signal in one - color multiphoton ionization was used by Cooper et al. (1981) as monitor. Ito et al. (1986) used the two - colour ionization dip for monitoring stimulated emission (Fig. 1). Here, ω_2 induces stimulated emission and ionization, while ω_1 is weakened to avoid one - colour ionization. Temporal and spatial mismatching between the pump and dump laser can be minimized easily here by aligning the dump laser for maximum ion signal. This is automatically the best alignment for the pump / dump ion dip signal also. The alignment is easier to achieve because the two - colour ion signal is larger and has a less noisy baseline than the dip signal and furthermore is independent of the dump laser frequency (as far as $hc(\omega_1 + \omega_2) \geq IP$). The one - and two - colour ionization dip methods have no problems with scattered light from the dump laser and this is another reason for prefering it to fluorescence dip spectroscopy.

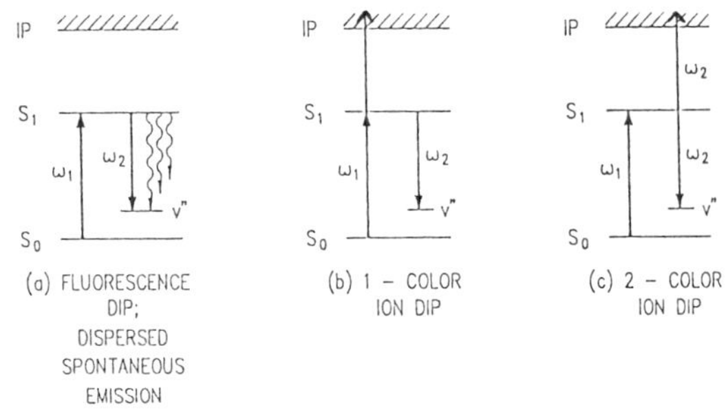

2. REMPI AND TWO COLOUR ION DIP SPECTRA OF P-CRESOL·$(H_2O)_1$

Fig. 2 shows the two - photon ionization spectra of p-cresol·$(H_2O)_n$ (n = 0-3) after S_0-S_1 (p-p*) excitation as measured by tuning the boxcar gate to the time-of-flight of the individual clusters. Fig. 2a shows the REMPI spectrum of free p-cresol (m/e = 108) seeded in 4 bar of Helium between 35000 and 36200 cm^{-1}. The band at 35334 cm^{-1} is the lowest frequency band of bare p-cresol and can be assigned to the electronic origin 0,0 (free). The bands at 35334 + 419 and 35334+803 cm^{-1} can be assigned to the totally symmetric a_1 (S_1) state fundamental vibrations $6a_0^1$ and 12_0^1 which are localized mainly in the aromatic ring. The very weak low frequency bands in fig. 2a arise from the hindered internal rotation of the CH_3 group of p-cresol. The bands generally show efficient rotational cooling with band widths < 3.0 cm^{-1} FWHM at 4 bar Helium. Fig. 2b shows the REMPI spectrum of p-cresol·$(H_2O)_1$ at mass number 126 (108 + 18). The electronic origin 0,0 of the monohydrated p-cresol is located at 34976 cm^{-1} with a red shift of 358 cm^{-1} relative to bare p-cresol. From the complex origin, a vibrational progression starts with a frequency interval of 152 cm^{-1}. From the isotope shift of the corresponding band in $C_7H_7OD \cdot D_2O$ and from comparison with ab initio SCF calculations of the cluster potential along the OH...O- coordinate we can assign this transition to the intermolecular stretch vibration. The dispersed fluorescence spectrum from excitation of (0,0) p - cresol·$(H_2O)_1$ shown in fig. 3 indicates a slightly lower frequency for the stretching vibration in the electronic ground state. A progression with 146 cm^{-1} spacing can be observed starting from the origin as well as from some intramolecular vibrations of the cluster. Fig. 4 shows typical two - colour ionization dip signals after excitation of (0,0) p- cresol·$(H_2O)_1$ which can be associated with the corresponding bands in the dispersed emission spectrum . From the dip spectra quite precise ground state vibrational frequencies can be obtained with an uncertainty of ± 1 cm^{-1} compared to ± 10 cm^{-1} in the dispersed emission spectra.

Both methods yield a value of 146 cm^{-1} for the intermolecular stretch vibration. The red shift of p-cresol·$(H_2O)_1$relative to p-cresol shows the hydrogen bond to be stronger in the excited S_1 state than in the electronic ground state. Hence a higher stretch frequency in the S_1 state (152 cm^{-1})compared to the S_0 state (146 cm^{-1}) is reasonable.The resonances observed in our first ion dip experiments agree well with the frequencies obtained from the corresponding dispersed fluorescence spectrum.This comparision shows directly that the observed ion dip resonances indeed arise from dumping to the S_0 state of the cluster.In our experiments we found the very low frequency range $(\omega_1 - \omega_2) < 300$ -400 cm^{-1} of the p-cresol·$(H_2O)_1$ cluster hard to obtain by two - colour ionization dip spectroscopy. The high intensity of the dump laser can lead to non-resonant multiphoton ionization of the cluster as background signal if ω_2 is near to ω_1 and 2 $hc\omega_2$ > IP. Furthermore, the ionization dip signal generally increases with increasing intramolecular vibrational relaxation (IVR) rate of the dumped vibrational state allowing dips of more than 30% which would be the saturation limit in a three level system. We did indeed observe dips up to 50% . The IVR rate however,increases linearly with the state density of the bath in accordance with Fermi´s golden rule and is small at low vibrational energy. Many of the intermolecular cluster vibrations are situated in the range 0 - 300 cm^{-1} and may be hard to obtain with ion dip spectroscopy despite the generally high state density of clusters with their low frequency intermolecular vibrations. In case of cresol, the state density is further enlarged by the low fre quency of the hindered internal rotation of the methyl group . Another possibility is to obtain the intermolecular vibrations from combination bands at higher frequencies.Two-colour ionization dips cannot be used for detection of SEP of very high vibrational states with $E_{vib} > 2\,hc\omega_1 - I_p$, because then $hc(\omega_1 + \omega_2) < I_p$. In principle one-colour ionization with an optimized laser alignment from two-colour ionization is the alternative.This is especially important for vibrational predissociation spectroscopy of hydrogen bonded clusters. OH stretch vibrations in the range 3500-3800 cm^{-1} for example lie sufficiently above the hydrogen bond energy to in-

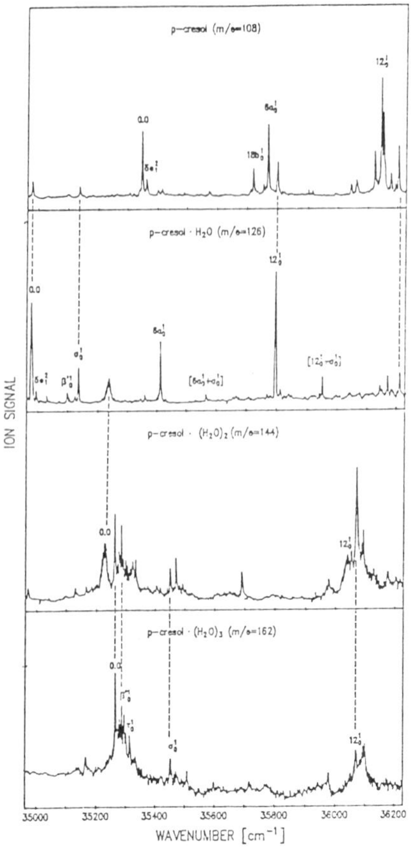

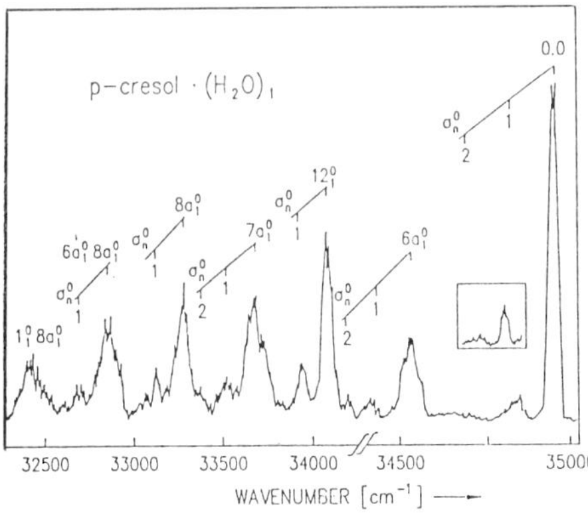

duce efficient cluster dissociation but their energy is often >2 $hc\omega_1$ -IP. Hence their SEP excitation can only be detected by a one-colour ion or a fluorescence dip
not by the two-colour ion dip. A rise of the ω_1 power to the one-color ionization limit, however, changed our dip signals for the worse so that high frequency vibrations were not easier to detect. While the absolute ion signal increased drastically, the ion noise at the top of the resonance also increased, whereas the dip did not change much because of its limitations by the power and spatial and temporal characteristics of the dump laser.

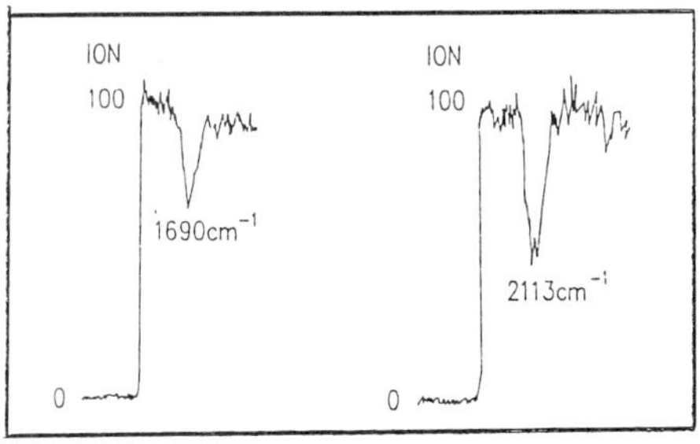

3. REMPI AND DISPERSED EMISSION SPECTRA OF P-CRESOL·$(H_2O)_{2,3}$

Fig. 2c shows the broad bands of the n = 2 cluster, namely its electronic origin and the intense 12_0^1 band. As pointed out by Kleinermanns et al.; (1990), the blueshift of the n = 2 relative to the n = 1 spectrum can be quantitativly explained with the second H_2O acting as proton donor to the hydroxyl group of p-cresol. Fig. 5 shows the dispersed emission spectrum after excitation of (0,0) p-cresol·$(H_2O)_3$. Starting from the electronic origin of this cluster and from several intramolecular vibrations, a progression with a frequency of 185 cm^{-1} is observed which can be assigned to the intermolecular stretch vibration.

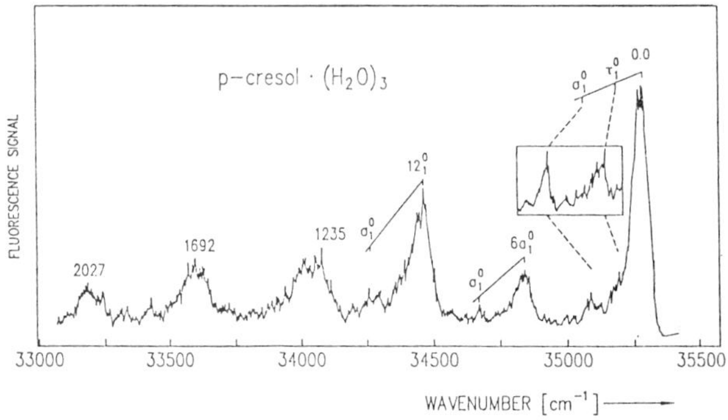

In the REMPI-spectrum in fig.2d we see a prominent band at 187 cm^{-1}, which is the corresponding intermolecular stretch vibration in the S_1 state, . Neither in the dispersed emission nor in the absorption spectrum of p-cresol·$(H_2O)_3$ did we find indications of an intermolecular stretch attributable to the second water subunit. This stretch vibration may be optically inactive or too weak to be observable here.

REFERENCES

Cooper D. E., C.M. Klimcak, J.E. Wessel, Phys. Rev. Lett. 46, 324 (1981)
Hamilton C. E., Kinsey, R.W. Field, Ann. Rev. Phys. Chem. 37, 493 (1986)
Suzuki T., N. Mikami, M. Ito, J. Phys. Chem. 90, 6431 (1986)
Pohl M., M. Schmitt, K. Kleinermanns, "Microscopic shifts of p-cresol/H_2O cluster spectra", submitted to J. Chem. Phys. (1990)

Inst. Phys. Conf. Ser. No 114: Section 11
Paper presented at RIS 90, Varese, Italy, 16–21 September 1990

Laser applications for high-T_c superconductors

C. H. Chen, R. C. Phillips, and M. P. McCann*

Chemical Physics Section, Oak Ridge National Laboratory
Oak Ridge, Tennessee 37831-6378

ABSTRACT: Laser ablation and resonance ionization have been used to monitor superconductive film preparation. From the desorption spectra, trapped O_2 in high-T_c superconductors was found to play an important role in superconductivity.

1. INTRODUCTION

During the past four years, the research and development of high-T_c superconductors has aroused unprecedented attention, not only from the scientific community, but also from various industries and governments. However, because these new superconducting materials are in general very "brittle" and difficult to machine, it is extremely difficult to fabricate them into thin wires. Thus thin film preparation is particularly important for many applications of superconductors, such as electronics for the computer industry. Films with high superconducting transition temperatures have been prepared by electron beam evaporation, organometallic chemical vapor deposition, DC sputtering, molecular beam epitaxy, and laser evaporation and deposition processes. However, the right ratio of various metal elements is one of the key factors in achieving superconductivity. Superconductive films are usually obtained based on repeated trials and adjustments. The quality of the superconductive film depends on controlling the film stoichiometry. Thus, the need for a monitor for high-T_c thin film preparation is quite critical.

In general, the preparation of a superconductive thin film involves the vaporization of various chemicals. The species deposited on substrates include neutral atoms or molecules in ground states, atoms and molecules in excited states, ions in ground states and ions in excited states. An instrument which has the capability of the quantitative determination of atoms, molecules, and ions in various electronic states should provide very detailed information on the mechanism of formation of superconductive films. An ideal method of monitoring the deposition process is to continuously track the deposition *in situ* and automatically maintain optimum film growth. In this work, fluorescence spectra are used for monitoring excited atoms, molecules, and ions; mass analysis of laser desorbed ions is obtained; and the application of resonance ionization spectroscopy (RIS) to monitor desorbed neutral atoms and molecules is discussed.

There are still no agreeable theories to explain high-T_c materials, at present. One of the "peculiar" features of these high-T_c materials is the noninteger number of oxygen atoms. The oxygen content is extremely critical to the superconductive properties. Take $YBa_2Cu_3O_{7-x}$ as an example. Its superconductive properties disappear whenever x is larger than 0.5.

The crystal structures of $YBa_2Cu_3O_{7-x}$ have been studied very extensively by X-ray diffraction and neutron diffraction. It was confirmed as orthorhombic for x = 0 in $YBa_2Cu_3O_{7-x}$. However, the oxygen content is more or less determined by the thermogravimetric method. It cannot distinguish trapped O_2 from oxide. It is well known that high-T_c superconductors in an oxygen-free atmosphere will lose oxygen. It has been assumed that the oxide anion (O^{-2}) in the CuO plane is in equilibrium with oxygen molecules in the gas phase at the same temperature and a fixed partial pressure. It is known that these metal oxide superconductors are, in general, very porous and that there are many deficit sites in the crystal. However, the possible existence of trapped O_2 molecules in the bulk or thin film materials has been ignored. During the development of the present monitor for superconductive films preparation, we observed the existence of trapped O_2 in superconductors. The trapped O_2 gas can possibly have significant implications for the nature of the superconductivity of these high-T_c materials.

2. EXPERIMENTAL RESULTS AND DISCUSSION

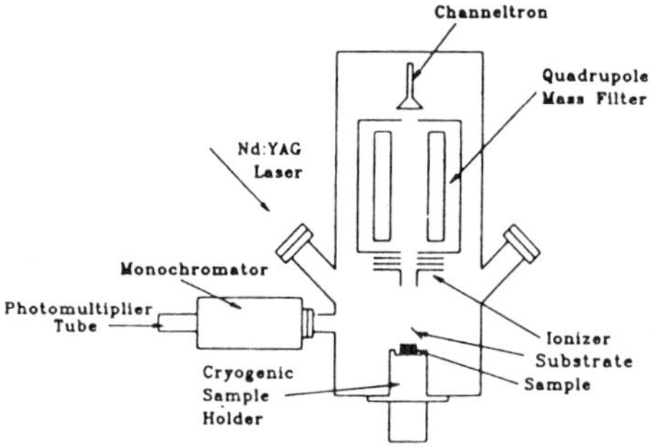

Fig. 1: Experimental schematic for fluorescence spectra, desorbed ion spectra, and resonance ionization spectra for laser ablation of the superconductors.

A Nd:YAG laser beam was used to do laser ablation of samples. A quadrupole mass spectrometer was used to monitor the desorbed ions directly. The desorbed neutral atoms or molecules were ionized by electron impact or by a laser resonance ionization method. Fluorescence spectra were obtained from excited atoms and molecules.Most samples were made by a standard solid-state chemistry approach. Transition temperatures of samples for Y-Ba-Cu-O and Bi-Sr-Ca-Cu-O were measured as $\sim 92°K$ and $\sim 125°K$, respectively. Most fluorescence data were taken using laser ablation of the superconductor sample by the fundamental of a Nd:YAG laser beam. Most secondary ion spectra were obtained using the second harmonic of the Nd:YAG laser beam. The laser beam power density on the superconductor target was varied from 5×10^5 W/cm^2 to 1×10^9 W/cm^2. The laser beam pulse duration was 5–7 ns which was measured by a fast response photodiode. Fluorescence spectra were obtained by using a GCA McPherson spectrometer. A photomultiplier (RCA 31034) which has an almost flat response for visible and near-infrared (IR) photons was used to detect the wavelength-resolved photons.

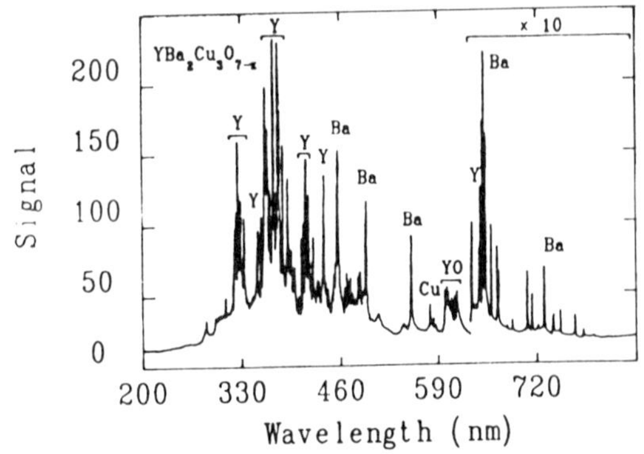

Fig. 2: Emission spectrum of laser ablation of $YBa_2Cu_3O_{7-x}$

The emission due to different metal elements was identified from atomic spectra. Many different transitions for Y, Ba, and Ca were observed. However, no definite indication of transitions due to copper atoms, which had been suggested as the key element for superconductive properties, was observed. No clear spectra corresponding to CuO transitions were identified. The complete fluorescence spectra of high-T_c superconductors are quite complicated. However, the sharp transition lines, especially in the red and near-IR spectral regions, should be convenient for use in monitoring the concentration of most metal elements, since the broad continuum emission due to the plasma can be neglected in this region. The fluorescence band for YO was also observed in the Y-Ba-Cu-O system.

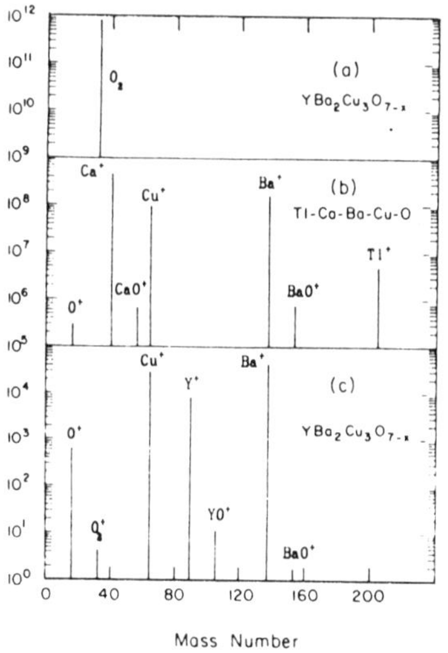

Fig. 3: Desorbed mass spectra from a laser ablation process of superconductor targets. (a) represents mass spectrum from $YBa_2Cu_3O_{7-x}$ target with laser density at 2 x 10^6 W/cm^2, the desorbed neutral O_2 molecules were ionized by electron impact ionization. No other desorbed ions or molecules were observed. (b) and (c) represent desorbed ion spectra from $YBa_2Cu_3O_{7-x}$ and Tl-Ca-Ba-Cu-O targets, respectively. The ablation laser power density was 2 x 10^7 W/cm^2. Filament was off.

Elements which provide low emission intensity such as Cu and Tl can be monitored by desorbed ion spectra. Ion spectra of laser-ablated Y-Ba-Cu-O and Tl-Ca-Ba-Cu-O are shown in Fig. 3(b) and Fig. 3(c). No CuO^+ signals were observed for the Y-Ba-Cu-O or Tl-Ca-Ba-Cu-O systems.

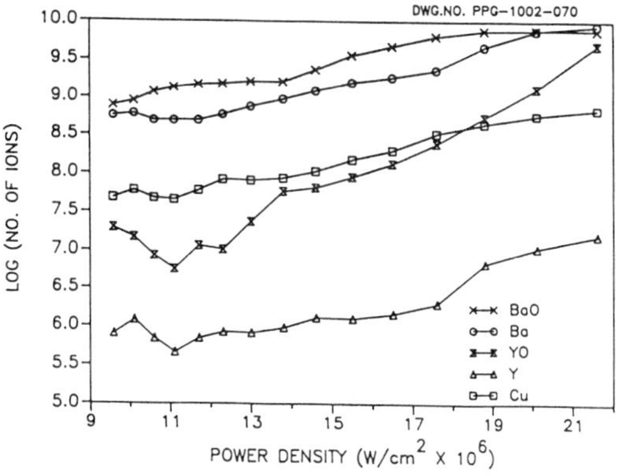

Fig. 4: Number of desorbed ions at various ablation powers

Quantitative determination of desorbed ions versus laser power density is shown in Fig. 4. The crystal structure of Y-Ba-Cu-O has been extensively studied by X-ray diffraction and neutron scattering methods. There were no oxide atoms on the Y plane. Thus the formation of YO^+ was thought to be either from peroxide or from trapped oxygen molecules. For the Tl-Ca-Ba-Cu-O system, CaO^+ was observed, although X-ray diffraction data indicate that there are no oxides on the Ca plane. The formation of YO^+ and CaO^+ can possibly come from the chemical reaction between Ca or Y atoms with oxygen in the plume just above the surface.

The major species detected following laser ablation of superconductors are neutral atoms and molecules in the ground state. RIS has been demonstrated to have the capability of detecting a single neutral metal atom. Different laser excitation schemes for various metal elements are given in Table 1. It shows that commercially available lasers can be used to detect all desorbed neutral atoms and molecules.

Table 1
Resonance ionization schemes for various metal elements in metal oxide superconductors

					λ
Ba ($6s^2$)	$\xrightarrow{h\nu}$	Ba (6s 7p)	$\xrightarrow{h\nu}$	Ba^+	324.5 nm
Tl ($6s^2$ 6p)	$\xrightarrow{h\nu}$	Tl ($6s^2$ 7s)	$\xrightarrow{h\nu}$	Tl^+	377.6 nm
Pb ($6s^2$ $6p^2$)	$\xrightarrow{h\nu}$	Pb ($6s^2$ 6p 7s)	$\xrightarrow{h\nu}$	Pb^+	283.3 nm
Bi ($6p^3$)	$\xrightarrow{h\nu}$	Bi ($6p^2$ 7s)	$\xrightarrow{h\nu}$	Bi^+	306.8 nm
Ca ($4s^2$)	$\xrightarrow{h\nu}$	Ca (4s 5p)	$\xrightarrow{h\nu}$	Ca^+	273.6 nm
Y (4d $5s^2$)	$\xrightarrow{h\nu}$	Y (4d 5s 5p)	$\xrightarrow{h\nu}$	Y^+	359.4 nm
Cu ($3d^{10}$ 4s)	$\xrightarrow{h\nu}$	Cu ($3d^{10}$ 4p)	$\xrightarrow{h\nu}$	Cu^+	324.8 nm
Sr ($5s^2$)	$\xrightarrow{h\nu}$	Sr (5s 6p)	$\xrightarrow{h\nu}$	Sr^+	295.2 nm

Experimental results on quantitative determination of desorbed neutral atoms and molecules for various laser power have also been obtained.

It was observed that a large amount of O_2 was desorbed during the laser ablation process. the amount of desorbed O_2 was estimated as $\sim 10^{10}$ molecules per pulse from laser ablation of a $YBa_2Cu_3O_{7-x}$ target at 22°K with a very modest laser power density of 1×10^6 W/cm^2. With this laser power level, there should be no significant destruction of any chemical bonds. No significant numbers of atoms, molecules, or ions other than O_2 were observed, as shown in Fig. 3(a). This process of desorption should be due to a heating effect. Since the desorption of O_2 was observed for laser ablation by the fundamental Nd:YAG laser beam at 1064 nm, it is very unlikely the desorption was due to bond breaking of a metal oxide by the one-photon absorption process. Thus, the present results indicated that O_2 molecules were loosely trapped in these superconducting materials. Even a mild heating process is enough to drive out significant quantities of oxygen molecules trapped in these materials. There was a concern that the observed neutral O_2 molecules could be due to the adsorption of O_2 on the surface of the superconductor sample. To minimize this concern, a power density of 1×10^9 W/cm^2 was used to vaporize the top layers of material for several minutes before making measurements of desorbed O_2. We also carefully checked to see that there was no H_2O, N_2, or CO desorbed from the target to assure that the laser was probing below the original surface layer. These procedures rule out the possiblity of O_2 being adsorbed just on the surface. Therefore, the fact that oxygen molecules are trapped in $YBa_2Cu_3O_{7-x}$ and Tl-Ca-Ba-Cu-O superconductors is confirmed.

From our results, the existence of trapped O_2 in the bulk of these superconductive materials is confirmed. The implication of trapped O_2 inside of the bulk of superconductors can be quite significant. It has long been considered that the existence of Cu^{+3} accounts for the excess of oxygen in $YBa_2Cu_3O_{7-x}$. The nominal composition suggests that 33% of the copper is in the +3 state. However, no Cu^{+3} ions were observed by X-ray photoelectron spectroscopy. With the trapped molecular oxygen, there is no need to have Cu^{+3} account for the number of oxygen atoms to be higher than 6.5. This should have an important effect on the calculation of the electronic structure of these crystals. Since the trapped O_2 is critical for the superconductivity, the conservation of O_2 in superconductive film becomes very critical for any practical application for these newly invented superconductors.

3. ACKNOWLEDGEMENTS

This research was sponsored by the Office of Health and Environmental Research, U.S. Department of Energy, under contract number DE-AC05-84OR21400 with Martin Marietta Energy Systems, Inc. *M. P. McCann is currently at University of Central Florida, Chemistry Department

Paper presented at RIS 90, Varese, Italy, 16–21 September 1990

Development of an internal calibration method in RIMS and application to the analysis of iron in fluoride glasses

T. Gibert[1], M.F. Barthe[2], B. Dubreuil[1], J.L. Debrun[2]

1 GREMI Université d'Orléans / CNRS, BP 6759, 45067 Orléans - France
2 CERI - CNRS, 3A rue de la Férollerie, 45071 Orléans Cédex 2 - France

ABSTRACT : Interesting properties of the fluoride glasses are their ultratransparency around 2.5 μm. Preparation of high-quality glasses requires ultrapure chemicals because metal impurities strongly increase the absorption losses. A method based on S.I.R.I.S. has been developed to perform quantitative analysis of Fe in these glasses. Fluoride glasses are insulators . To overcome charging problems the ion gun is pulsed and the absolute concentration of Iron is obtained by comparison of the Fe RIMS signal with the signal of a major element such as Zr during the same run.

1. INTRODUCTION

Infrared transmitting glasses are of interest as materials for applications in optical communications. New glassy materials based on some specific fluorides (Fluoride glasses) exhibit very interesting properties for applications in mid-infrared (near 2.5 μm) fibre optics (Lucas 1989). Their potential ultratransparency is about hundred times better than that of silicate-based glasses. The glass called ZBLAN (ZrF_4,BaF_2, LaF_3, AlF_3, NaF) appears as a good candidate for such applications. However, preparation of high-quality fluoride glasses with low absorption is a difficult challenge. In addition to intrinsic absorption losses (Rayleigh scattering, multiphonon absorption), absorption by cations such as transition metals (Fe, Co, Ni) drastically increases the losses. The impurity level should be limited to the ppb range to keep the expected performances. This requires synthesis of ultrapure fluoride chemicals and precise control of the impurity level in the different steps of the glass elaboration and fibre fabrication.

The work presented in this paper deals with the development of a method based on S.I.R.I.S. (Parks et al. 1983) for the quantitative analysis of iron in ZBLAN glasses. For this purpose, we have extended and adapted our analytical studies on metals and semiconductors (Gelin et al. 1989, Gobert et al. 1990) to the more complicated case of insulating materials.

2. EXPERIMENTAL

The S.I.R.I.S set-up developed in Orléans has been previously described (Gobert et al. 1990). The ZBLAN sample is mounted on a (x,y,z,ϕ) manipulator in a U.H.V. chamber. Atoms are sputtered from the sample surface by an Ar^+ beam produced by a Riber Dig 1 duoplasmatron ion gun (beam diameter 300 μm, current 2 μA). Sputtered atoms are resonantly photoionized by the beam of a tunable Nd-Yag pumped dye laser (Quantel Datachrome 5000) operating at 10 Hz. Photoions are filtered and detected by a quadrupole mass spectrometer (Riber SQ 156). Measurements can be performed either in analog or counting mode.

In our previous works on elemental analysis in metals and semiconductors, it was shown that S.I.R.I.S. is a method practically free from matrix effects (Gelin et al. 1989 b) allowing easy calibration, exhibiting good linearity, detection limits at the ppb level and limited mass interference effects (Gelin et al. 1989 b, Gobert et al. 1990). However, in the case of ZBLAN glasses some of these points have to be reconsidered due to the insulating nature of these materials. As a matter of fact, primary Ar ion bombardment induces uncontrolled charging of the glass sample. These charging effects change the electric field map in a very complicated way so that it becomes impossible to fit the parameters of the ion optics to the charged surface potential. The result is a severe or total loss of Fe photoion signal. To reduce this effect by minimizing the net charge deposited in the material, the Ar ion gun is pulsed (15 μs time duration) at the laser repetition rate. Stable and reproducible Fe photoion signals are then obtained and they are comparable to those measured in conductor materials.

However, the unknown and heterogeneous charged surface potential depends strongly on the specific structure of the sample under study and particularly on the defects present or induced in the bombarded volume. The experimental conditions have to be adjusted for each type of sample. This may lead to a large variation of the mass spectrometer transmission factor which a priori prevents any reliable comparison between the measurements performed on different samples.

The presence of a large amount of fluorine in the glass is also supposed to give rise to specific sputtering effects like those encountered with oxide materials (particularly a large increase of the secondary ions and molecular species could be responsible of significant matrix effects).

For these reasons, the absolute concentration of Fe in different ZBLAN samples cannot be simply deduced from the comparison with a standard. To overcome this problem of quantitativity, an internal calibration method has been developed, based on the comparison in the same experimental conditions of the Fe RIMS signal with the signal of a major element of the matrix such as Zr (10.3 % at.). Indeed, Fe and Zr atoms can be resonantly photoionized with the same dye laser at slightly different wavelengths and switching from one to another is straightforward.

3. RESULTS

1+1 resonant multiphoton ionization of Fe has been previously studied (Gelin et al 1989 a). For the present study, the a $^5D_4$ ---> y $^5F_4$ resonance (λ = 293.69 nm) was selected. With the laser fluence available in the interaction region (10 mJ/cm^2), the photoion yield is not fully saturated.

The ionization spectrum of Zr has been recorded in the 290 - 300 nm range. It is shown on the figure below.

Resonances corresponding to highly excited states of Zr are clearly identified on a non resonant background. These resonances are of small amplitude compared to the background. The ionization yield is indeed not saturated and varies almost linearly around 10 mJ/cm^2.

The a $^3F_2$ ---> u $^3D_3$ transition (λ.= 292.06 nm) is used as reference signal.

To deduce the photoionization yield ratio Zr/Fe in the conditions of glass analysis, a ZBLAN standard doped with 11.5 ppm at. of Fe has been analyzed. The Zr/Fe signal ratio for this standard is 85 $\pm$ 5 for 3 mJ of laser energy. It is much lower than the concentration ratio due to the low ionization efficiency of Zr at this wavelength.

Using this value, the absolute Fe concentration of any ZBLAN sample can be easily deduced from the ratio of the Fe/Zr photoion signals measured during the same run.

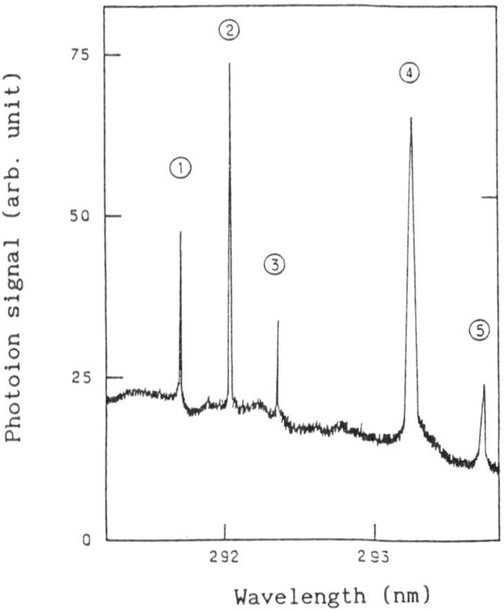

Transitions corresponding to the observed resonances :

1 - $4d^2\ 5s^2\ ^3F_2$ ---> $4d^3(b\ ^4P)5p\ ^5D^o_4$

2 - " " " ---> $4d^2\ 5s(a\ ^2F)5p\ ^3D^o_3$

3 - " " " ---> $4d^3(b\ ^4P)5p\ ^5F^o_2$

4 - " " " ---> $4d^3(a\ ^2H)5p\ ^3H^o_5$

5 - " " " ---> $4d^3(b\ ^4P)5p\ ^5P^o_1$

Ionization spectrum of Zr

By this internal calibration method, real analyses of Fe in several undoped ZBLAN samples have been performed using the counting mode. The results (in the ppm range) present good statistics (15 - 25 % dispersion) whereas calibration based on comparison of the different Fe signals lead to slightly different but more dispersed results. Indeed, the internal calibration method is practically sample-independent and is free from variations of the sputtering yields, of the detection efficiency and of the long term variations of the laser and ion beam characteristics. Finally, from these results, a conservative detection limit of about 3ppb can be espected for an incident ion beam current of 0.5 μA and a counting time of 30 minutes.

References

Gelin P, Gobert O, Dubreuil B, Debrun JL, Inglebert RL 1989 a "Resonance Ionization Spectroscopy 1988", edited by Lucatorto TB and Parks JE, Institute of Physics Conference Series, N° 94 (Institute of Physics)

Gelin P, Debrun JL, Gobert O, Inglebert RL, Dubreuil B 1989 b Nucl. Instrum. Methods B **40/41** 290

Gobert O, Dubreuil B, Inglebert RL, Gelin P, Debrun JL 1990 Phys. Rev A **41** 6225

Lucas J 1989 J. Mat. Science **24** 1

Parks JE, Schmitt HW, Hurst GS, Fairbank Jr WM 1983 Thin Solid Films **108** 69

Inst. Phys. Conf. Ser. No 114: Section 11
Paper presented at RIS 90, Varese, Italy, 16–21 September 1990

Three-colour resonance ionization of sputtered Ti for isotopic analysis of meteoritic samples

C. E. Young,[1] D. R. Spiegel,[2] M. J. Pellin,[1] W. F. Calaway,[3] S. R. Coon,[4] J. W. Burnett,[5] D. M. Gruen,[1] A. M. Davis,[2] and R. N. Clayton[2]

ABSTRACT. Progress in the development of RIS techniques for the determination of isotopic abundances in meteorites is reported. Measurement of isotopes of calcium and titanium in meteoritic hibonite [$CaMg_xTi_xAl_{12-2x}O_{19}$; $0 \leq x \leq 0.3$] crystals was chosen as a test case, where it is necessary to distinguish between overlapping Ca and Ti mass peaks at positions 46 and 48. The SARISA (surface analysis by resonance ionization of sputtered atoms) instrument used in this work is an energy- and angle-compensated time-of-flight mass spectrometer, providing high useful yield (several percent) at reasonable mass resolution (≥200). Two distinct 3-step RIS schemes for Ti, utilizing only visible wavelengths and autoionization lines in the final transition, were investigated. Data for several autoionization lines in each scheme are tabulated, and saturation behavior on all transitions is given. Selectivity factors were determined for Ti in the presence of Ca in test samples of perovskite, sphene, and hibonite, a value of 500 being achieved for the latter sample.

1. INTRODUCTION

Resonance ionization mass spectrometry (RIMS) is a technique that is finding increasing application in the analysis of the isotopic composition of geological and meteoritic samples [Belal et al. (1990), Blum et al. (1990b)]. The intrinsic elemental selectivity of resonance ionization spectroscopy (RIS) is a powerful technique to apply to the problem of eliminating isobaric interferences. There is also great interest in utilizing the RIMS method with high spatial resolution in a microprobe instrument, where domains of unique sample composition could be interrogated directly, conserving valuable samples, and eliminating many preliminary chemical processing steps [Blum et al. (1990a)].

The present paper reports progress in our laboratory in the application of RIMS techniques to the analysis of the isotopic composition in meteorites. In such samples, variation of isotopic abundances in calcium-aluminum-rich inclusions (CAIs) is the subject of considerable current interest. At mass positions 46 and 48, the ability to distinguish between the isotopes of Ca and Ti is needed [terrestrial abundances(%): ^{46}Ca 0.004, ^{48}Ca 0.187, ^{46}Ti 8.0, ^{48}Ti 73.8]. Isotopic anomalies for these elements at the percent level are known in meteoritic samples. Because the concentration of Ca can greatly exceed that of Ti in mineral samples, RIS schemes must be able to discriminate strongly against either element. For example, a hibonite inclusion within CAIs is stoichiometrically $CaO.6(Al_2O_3)$ with the substitutions of $Ti^{4+} + Mg^{2+} = 2Al^{3+}$, resulting in atomic concentrations of Ti at the level of a few percent. Hibonite inclusions are of

*Work supported in part by the U.S. Department of Energy, BES-Materials Sciences, under Contract W-31-109-ENG-38, by NASA Grants NAG 9-51 and NAG 9-111, and a NSF Graduate Fellowship.

[1]MSD/CHM, Argonne National Laboratory, Argonne, IL 60439, USA
[2]Enrico Fermi Institute, University of Chicago, Chicago, IL 60637, USA
[3]MSD/CMT, Argonne National Laboratory, Argonne, IL 60439, USA
[4]Lab Grad Participant, University of Texas at Austin
[5]Lab Grad Participant, University of Pittsburgh

considerable interest to cosmochemists [Hinton et al. (1987)]. Part of our program is aimed at establishing the accuracy, reproducibility, and sensitivity to machine parameters of the isotopic composition of pure metallic and mineral test samples. That data will be reported elsewhere.

2. EXPERIMENTAL

The studies reported here were performed with the SARISA (surface analysis by resonance ionization of sputtered atoms) instrument, in which samples are sputtered by a pulsed (≤ 1 μs at 2 μA) 4 keV argon ion beam and photoionized by excimer-pumped dye lasers. The energy- and angle-refocusing properties of this type of time-of-flight (TOF) instrument allow a good combination of useful yield and mass resolution (≥1% and ≥200, respectively), with strong rejection of secondary ions from the sample surface [Pellin et al. (1988), Young et al. (1987)]. For isotope studies, instrumental and procedural details, which are discussed in the references cited above, are largely applicable. Characterization of multistep RIS excitation schemes was performed with pure metallic Ti samples. Selectivity data were then obtained through the use of a specially prepared sample consisting of Ti metal, for reference measurements, and the minerals perovskite, rutile, hibonite and sphene, with a thin carbon coating to minimize charging effects. Typical areas of the samples were 2x2 mm^2, much larger than the ion beam spot (diameter ~ 150 μm). TOF spectra were acquired with a transient recorder data system having 5 ns time resolution and the capability of averaging over laser shots in either analog or counting modes. The laser system comprised a set of three dye lasers with excimer pumping, typically operated at low repetition rate (23 Hz), for the present study.

The atomic transitions employed are indicated in Fig. 1. Both excitation schemes 1 and 2 avoid the use of ultraviolet light, in order to minimize the photoionization of excited species falling within the range of nonresonant ionization through 1-photon absorptions. Scheme 2, utilizing three green colors, comes very close to dividing the ionization energy into three equal packets. In the choice of dye wavelength regions, consideration was also given to facilitating a convenient switch to Ca transitions.

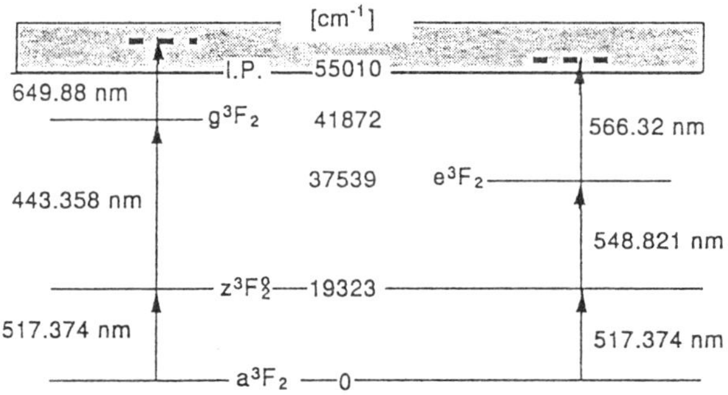

Fig.1. RIS schemes for selective ionization of Ti. Left side: scheme 1; right side: scheme 2.

3. RESULTS AND DISCUSSION

Autoionizing lines were sought for the third step in both schemes as a means of minimizing the required saturation intensity and consequent interference due to nonresonant ionization of elements other than Ti. Tables 1 and 2 contain the results of a search in the wavelength regions 632 to 652 nm and 530 to 578 nm, respectively. The latter interval crosses the energy threshold for Ti ionization. Assignment of spectral position utilized a 0.75 m grating spectrometer as a calibrated reference for combining several laser scans, with an estimated uncertainty of ~1 cm^{-1}. The strongest autoionization feature in each table, which was used in subsequent studies, is underlined in each table. Dye gain-curves were taken into account in estimating relative intensities within each scheme; no direct comparison among signal levels between tables is implied beyond the inferences available from saturation data, below. Large autoionization linewidths were observed but significantly less so for the second scheme, which lies closer to the energetic threshold. Nominal laser bandwidths were 0.2 cm^{-1} on step 3 and about 0.06 cm^{-1} on steps 1 and 2. No clear pattern of Rydberg series was apparent in the autoionization data, not an unexpected result in Ti which has many low-lying excited states available for the ion core.

Table 1. Ti autoionization lines. Scheme 1.

Energy Over Ionization limit (cm^{-1})	Ionization-Step Wavelength (nm)	Relative ^{48}Ti Signal	Autoionization Linewidth (cm^{-1})
2221.0	650.89	0.6	1.2
2230.8	650.45	0.8	2.5
2245.2	649.88	1.0	2.4
2401.0	643.36	0.4	3.4
2407.2	643.11	0.4	4.2
2479.5	640.13	0.6	6.3
2516.3	638.62	0.4	8.6
2612.2	634.74	0.2	8.6
2630.4	634.00	0.8	12.2
2677.6	632.11	0.5	9.8

Table 2. Ti autoionization lines. Scheme 2.

Energy Over Ionization limit (cm^{-1})	Ionization-Step Wavelength (nm)	Relative ^{48}Ti Signal	Autoionization Linewidth (cm^{-1})
42.2	570.83	0.1	1.6
53.4	570.47	0.1	1.3
128.2	568.04	0.1	0.8
165.4	566.85	0.2	3.4
170.5	566.68	0.6	0.5
181.8	566.32	1.0	0.6
253.3	564.04	0.04	1.0
379.1	560.06	0.02	2.4
748.1	548.72	0.06	1.8
861.1	545.34	0.01	1.3

The effectiveness of employing autoionization states in reducing intensity requirements for laser saturation of the RIMS signal may be estimated from Fig. 2, which gives data for all three steps of scheme 2 as well as the final step in scheme 1. (Saturation on the first two steps of scheme 1 was comparable to the corresponding scheme 2 behavior shown.) Laser beam profiles, controlled by cylinder lens telescopes, were Gaussian (FWHM ~ 0.6 mm) in the dimension perpendicular to the sample and extended (~ 4 mm) in the parallel direction. Metallic Ti was the sample in these experiments. On the third step, a small detuning shift caused a much stronger signal decrease in the saturation regime in scheme 2 than in scheme 1. It was also subsequently discovered, on going to mineral samples, discussed below, that most of the nonresonant ionization of Ca was attributable to radiation at λ_3, giving incentive for reducing the intensity at that wavelength.

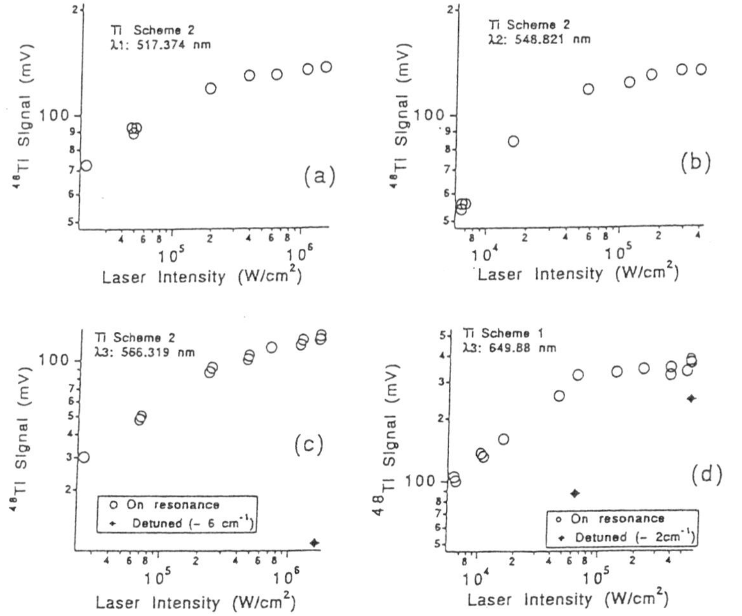

Fig. 2. Laser-intensity saturation curves. Panels a-c give the results for steps 1-3 of RIS scheme 2; panel d is for the 3rd (ionization) step in scheme 1.

Selectivity tests were performed on samples of hibonite, perovskite, and sphene via the RIS scheme 2. Step 1 was attenuated below the maximum intensities shown in Fig. 2 by a factor varying from 3 to 10 with little effect on Ti signal. Results are summarized in Table 3. The selectivity parameter is defined as the ($^{48}Ti/^{40}Ca$) signal divided by the corresponding ratio in the sample, known very accurately by electron microprobe analysis. The resulting value would be directly applicable to isobaric interference, with the assumption of identical RIS behavior for all isotopes. Surface deviations from bulk composition would be expected to be similar in meteoritic and test samples. The selectivity of 500 observed with hibonite is already very satisfactory for determination of significant isotope anomalies in meteoritic samples. The observed selectivity with perovskite (100) can be compared with the best value obtained via scheme 1 (~ 6). A Ti/Ca selectivity factor of 58 has been obtained by a group using a RIS scheme, employing 300 nm region UV photons and a sample comprising various elemental impurities in a glass matrix [Willis and Thonnard (1990)].

Extensive investigations of the nature of the Ca-containing species, which give rise to the Ca^+ photoion observed, have not yet been carried out. On perovskite, in RIS scheme 2, it was observed that over 30% of the nonresonant Ca signal remained when only step 3 laser light was present, and 40 % remained after detuning step 3 by 3 cm^{-1}. By comparison, the Ca^+ photoion signal from the combination of steps 1 and 2 was a factor of 30 less than induced by the step 3 beam. In general, if molecular species are dissociating to result in atomic photoions, the optimal approach may require a search for a "quiet" region of molecular absorptions.

Table 3. Selectivity test ($^{48}Ti/^{40}Ca$), RIS scheme 2.

Material	Formula	Selectivity
Hibonite[a]	$CaO.6(Al_2O_3)$	500
Perovskite	$CaTiO_3$	100
Sphene	$CaTiSiO_5$	200

a. With substitutional Ti ~2.7 wt.% (see text).

4. REFERENCES

Belal I K, Xiong X, Fassett J D and Moore L J 1990 *Bull. Am. Phys. Soc.* **35** 1498

Blum J D, Pellin M J, Calaway W F, Young C E, Gruen D M, Hutcheon I D and Wasserburg G J 1990a *Geochim. Cosmochim. Acta.* **54** 875

Blum J D, Pellin M J, Calaway W F, Young C E, Hutcheon I D and Wasserburg G J 1990b *Anal. Chemistry.* **62** 209

Hinton R W, Davis A M and Scatena-Wachel D E 1987 *Astrophys. J.* **313** 420

Pellin M J, Young C E and Gruen D M 1988 *Scanning Microsc.* **2** 1353

Willis R D and Thonnard N 1990 *SBIR Phase 1 Report #ISI-8860854* Atom Sciences, Inc.

Young C E, Pellin M J, Calaway W F, Jørgensen B, Schweitzer E L and Gruen D M 1987 *Inst. Phys. Conf. Ser.* **84** 163

Inst. Phys. Conf. Ser. No 114: Section 11
Paper presented at RIS 90, Varese, Italy, 16–21 September 1990

On-line studies of implanted isotopes with RIMS: The PILIS systems

F. Le Blanc, J. Arianer, P. Kilcher, J. Obert, J. Oms, J.C. Putaux, B. Roussière, J. Sauvage
Institut de Physique Nucléaire, Orsay

J.E. Crawford, J.K.P. Lee, G. Savard, H. Dautet, G. Thekkadath
McGill University, Montréal

H.T. Duong, J. Pinard, S. Liberman‡
Laboratoire Aimé Cotton, Orsay

and the ISOCELE Collaboration

ABSTRACT: PILIS II, a new experimental system for the spectroscopic investigation of isotope shift and hyperfine structure of laser-desorbed samples of collected isotopes has been constructed at the ISOCELE mass separator (IPN, Orsay). Since operation of this system is completely on-line, measurements on isotopes with $T_{1/2} < 1mn$ are possible. A system for collinear laser spectroscopy called COMPLIS, which uses a version of the the PILIS system to produce a pulsed atomic beam, is under construction at the CERN ISOLDE separator.

1. INTRODUCTION

Several versions of a system called PILIS (Post-ISOCELE Laser Isobar Separator) have been used successfully at the Institut de Physique Nucléaire, (Orsay, France) for the study of isotope shift (IS) and hyperfine structure (HFS) of series of radioactive isotopes. The spectroscopic system of all of the PILIS systems is essentially the same. Isotopes produced by the ISOCELE on-line mass separator are implanted in a graphite substrate, and are subsequently desorbed by pulses from a Nd:YAG heating laser. In front of the graphite, the desorbed atoms are resonantly ionized by overlapping beams from pulsed tunable dye lasers. The ions produced are detected by microchannel plate (MCP) detectors mounted at the end of a time-of-flight (TOF) line. The transition from the atomic ground state of the isotope studied yields information for the IS and HFS analysis. For this transition, a swept-frequency CW dye laser is used to injection-lock the frequency of a pulsed laser beam established in the same cavity, yielding a single-mode output with relatively narrow linewidth (<100 MHz). The YAG heating provides a bunched atomic beam from the graphite reservoir; synchronous pulsing of the YAG with the RIS beams is an effective way to overcome the problem of the low duty cycle of the pulsed lasers. In a second version of this system, PILIS II, the collection and desorption techniques have been redesigned to adapt the apparatus for the study of short-lived isotopes.

‡ deceased

2. Design features of PILIS I and PILIS II

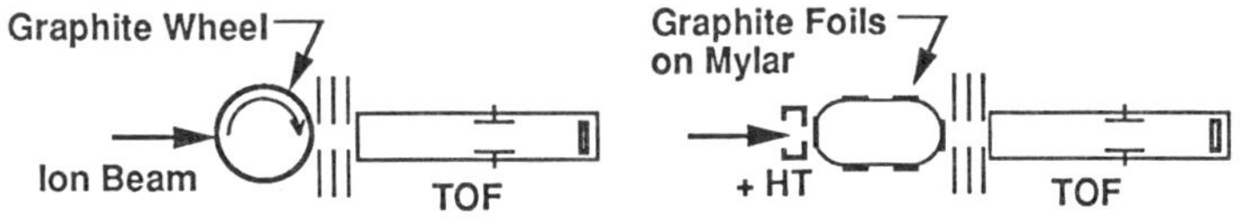

Fig. 1 (a) Original PILIS I system (b) With incident beam deceleration

The original PILIS I systems have been described previously (Lee, 1988a,b). The essential features are shown schematically in fig. 1. For gold, which is easily desorbed, implantation at the full incident ion beam energy of 45 keV is done over an area of ~16 mm^2 on a graphite wheel, which is rotated 180° to align the implant with the electrodes of the TOF line. The YAG pulses (λ=532 nm., τ=10 ns, E~1 mJ) are focused to a line of width ~50μ and area ~0.1 mm^2 on the graphite, which is slowly displaced to provide a new source of the implanted isotope. In principle, the technique may also be used for platinum, the radioactive daughter of the implanted Au. However platinum proved much more difficult to desorb, and the system was modified to retard the incident ion beam, thus reducing the implant depth. The graphite wheel is replaced by thin (~20μ) graphite disks mounted on an insulating mylar tape (fig. 1b). During implantation, the disk is held 500 V below the ion beam potential; following implantation, the disk is moved to the TOF electrode assembly, where its voltage is reduced to 1.5 kV for the TOF acceleration. Both versions of the PILIS I system use deflecting plates which are pulsed to prevent low mass parasitic ions produced directly by the YAG laser from saturating the MCP detector. The main limitation of the PILIS I scheme is the relatively long time (~10 mn) required to load and scan the graphite reservoir—this effectively limits the range of studies to isotopes with $T_{1/2} > 10$ m. This was the main motivation for the PILIS II design, which can be operated in a true on-line mode. This system uses a stationary graphite collector: implantation and desorption occur at the same position.

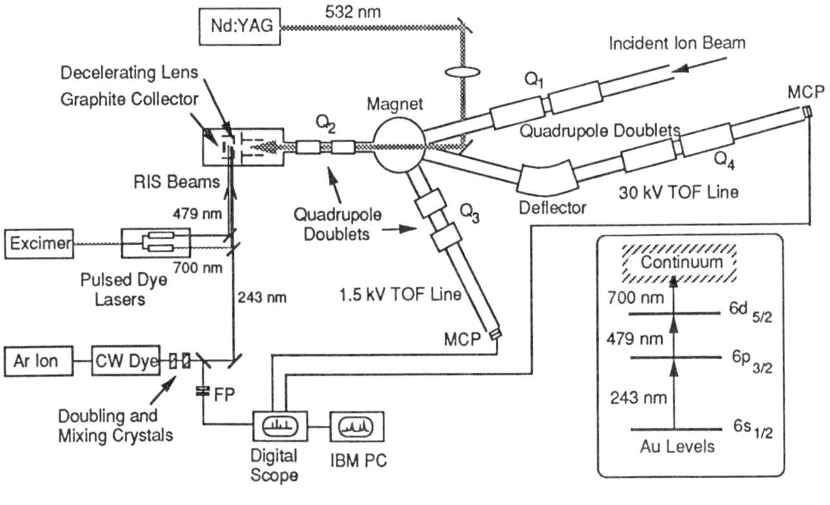

Fig. 2 The PILIS II system. The inset shows the RIS steps for Q_s studies of Au

Fig. 2 shows the main features of the PILIS II design. The incident ion beam is deflected by the magnet, and focused by two sets of electrostatic quadrupole doublets. During implantation, a retarding electrostatic lens held 500 V below the 30 kV acceleration potential of the ISOCELE beam focuses the slowed ions to a spot of diameter ~2 mm on the collector. The YAG beam passes through the magnet, and strikes the graphite at normal incidence. The RIS beams are incident parallel to the graphite surface, and about 10 mm in front of it. These are arranged so that beams 2 and 3 are collinear and perpendicular to beam 1, defining a narrow interaction volume in the desorbed atomic cloud (see fig. 3). Since the aim is to collect the sample in a short time, and to extract it in one or a brief burst of desorption pulses, the YAG energy is much higher than in the PILIS I system (~40 mJ) and its line image covers about 17% of the implanted region. Under these conditions, large numbers of parasitic low-mass ions are directly produced by the heating; however, the magnet prevents most of these from reaching the MCP detectors. Further suppression is achieved by keeping the target at a potential ~50V lower than the extraction electrode.

Two different TOF lines, using acceleration voltages of 1.5 and 30 kV, were used for various experiments with the PILIS II system. With the 1.5 kV line, immediately after implantation the deceleration voltage of 29.5 kV is reduced to 1.5 kV, and the Q_2 quadrupole set is switched off. The RIS ions are then deflected into the shorter TOF line. Following each desorption and collection cycle, the retarding voltage is reapplied to prepare a new collection for the next point on the HFS scan. With the 30 kV line, no switching is required. The graphite remains at the full deceleration voltage during collection and desorption; the TOF line is mirror-symmetric with the input beam line.
This system also has improved mass resolution (M/ΔM~660, compared to 360 for the 1.5 kV line). The electrostatic deflector shown was required only because of space limitations in the ISOCELE beam hall.

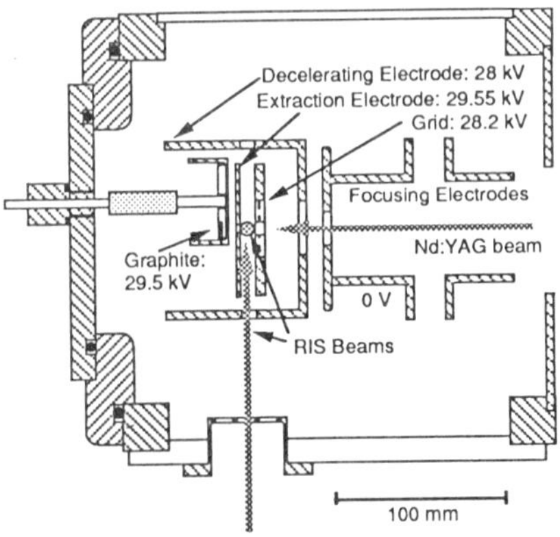

Fig. 3 The PILIS II Interaction Region

3. System Performance

To compare the various versions of the systems tested, we may define the detection efficiency as N_d/N_0, where N_d is the number of ions detected at the MCP for one YAG pulse, and N_0 is the estimated number of ions desorbed from the graphite. This measurement is made at the peak of a resonance (in ^{186}Au in the present comparison). Table 1 shows efficiency estimates for PILIS I (without incident beam retardation) and for the two TOF lines of PILIS II.

	PILIS I	PILIS II (1.5 kV)	PILIS II (30 kV)
Ions collected	10^{10}	6×10^8	3.8×10^7
Atoms in YAG beam	5×10^7	1.1×10^7	6×10^6
Atoms desorbed (N_0)	2×10^6	2.2×10^6	1.2×10^6
Ions detected (N_d)	100	50	150
Detection Efficiency	5×10^{-5}	2×10^{-5}	1.3×10^{-4}

Table 1. Detection efficiencies for different versions of PILIS

In this comparison, it should be noted that in the PILIS I system, the number of ions collected represents the total required for a complete spectral collection in PILIS I, and the number per pulse for PILIS II. For the 30 kV TOF line of PILIS II, we make the following rough estimates of contributions to the overall system efficiency: about 20% of the implanted atoms can be desorbed in each pulse, 1% of the atomic cloud is probed by the RIS beams, and the transport efficiency of the TOF line (reduced by grids in the line) ~0.5. The lower two RIS transitions are saturated, and the final ionizing step has an efficiency of about 10%. The figures in the table show that the 30 kV version of PILIS II was clearly better than the 1.5 kV system; we believe that improved ion optics in the laser-atom interaction region at the higher voltage are responsible for the observed difference.

Both PILIS II TOF lines have been used for spectroscopic measurements and tests. The 1.5 kV TOF line was used for the determination of the spectroscopic quadrupole moments of ^{191}Au and ^{186}Au (Le Blanc,1989). Here the spectroscopic resolution of ~800 MHz was somewhat poorer than for the PILIS I measurements: for these measurements, it was necessary to produce the first laser excitation step of 243 nm by frequency tripling. The 30 kV line has been used to obtain partial hyperfine spectra for $^{186,185,184}Au$ and ^{183}Au ($T_{1/2}$ = 42 s). From these measurements, the spectroscopic quadrupole moment of ^{185}Au was determined .

4. The COMPLIS System

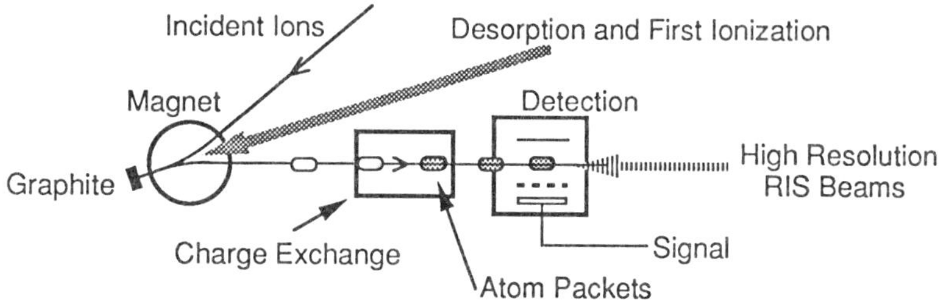

Fig. 4 Concept of the COMPLIS system for pulsed fast-beam collinear spectroscopy

A version of the PILIS II system is presently under construction at the ISOLDE III mass separator at CERN. Initially, spectroscopic studies will be done using the technique just described. Eventually, it is proposed to use the system as a pulsed atomic source for collinear fast-beam spectroscopy. This idea for the COMPLIS (COllinear Measurements with a Pulsed Laser Ion Source) is shown in fig. 4. To achieve this, the RIS interaction region will be designed not for high spectroscopic resolution, but for maximum yield of ions of the isotope studied. A charge-exchange cell will be used to produce the atomic beam, and detection may be done using collinear resonant excitation and field-induced ionization. It has been estimated that overall detection efficiencies may be improved by about two orders of magnitude by increasing the area bombarded by the desorbing laser, and the overlap of the RIS beams with the desorbed cloud.

REFERENCES

Le Blanc, F. et al 1989 *Proc. XII Workshop on Nuclear Physics, Iguatzú Falls, Argentina,* ed M.C. Cambiaggio, A.J. Kreiner, E. Ventura (World Scientific) p 174

Lee, J.K.P. et al 1988a *Nucl. Inst. and Meth.* **B34** 252

Lee, J.K.P. et al 1988b *Phys. Rev.* **C38** 2985

Inst. Phys. Conf. Ser. No 114: Section 11
Paper presented at RIS 90, Varese, Italy, 16–21 September 1990

Investigation of the surface impurity content of silicon samples by resonance ionization spectroscopy

J S Bakos, I B Földes, P N Ignácz, M Á Kedves

Central Research Institute for Physics, H-1525 Budapest P.O.B.49. Hungary

ABSTRACT: Resonance ionization spectroscopy with laser atomization is applied for the investigation of silicon surfaces. The atomization process was studied and the ablation intensity was optimized. The measuring set-up was used for plasma transport analysis.

1. INTRODUCTION

The surface sodium contamination density of silicon samples exposed to sodium impurity fluxes is investigated by laser ablation followed by resonance ionization spectroscopy (RIS). The atomization process was studied by determining the velocity distribution of the ablated neutrals. The measuring setup has been used for the analysis of collector probes exposed to high temperature tokamak plasmas. Transport processes of the tokamak plasma can be investigated by the analysis of the silicon probes. The application of RIS for this analysis, due to its high sensitivity, enables the study of single discharges.

2. EXPERIMENTAL ARRANGEMENT

The RIS experimental arrangement can be seen in Figure 1. The surface of the silicon sample was atomized by the focussed beam of a Q–switched ruby laser with a pulse duration of 40 ns and intensity between $5 \times 10^7 - 10^8$ W/cm^2 forming a crater with a diameter of 0.7 mm and depth of 10 nm. The sodium content of the resulted atomic cloud was analysed by a two step resonant photoionization at a distance of about 4 cm from the surface. The Na atoms in the ground state were excited through the $3^2S_{1/2} \rightarrow 3^2P_{3/2}$ transition by the pulse of a dye laser (λ_e = 589 nm) and ionized by the pulse of the former's pumping XeCl excimer laser (λ_i = 308 nm). The energy of the exciting and ionizing laser pulses was 0.1 and 1 mJ, respectively, the pulse durations were 20 ns. The exciting and ionizing laser beams crossed the atomizing laser beam at about right angles. The ions formed in the resonance ionization process were accelerated by a static electric field ($U = -2.5$ kV) to an electron multiplier. The output signal of the electron multiplier was digitized by a transient recorder and the data were processed by a personal computer. The synchronization was provided by an adjustable electronic delay between the atomizing and analysing pulses and by an optical delay between the exciting and ionizing pulses.

3. VELOCITY DISTRIBUTION MEASUREMENT

The velocity distribution of the evaporated cloud of particles propagating perpendicularly to the surface of the sample has been determined by applying different delay times between the atomizing and analysing laser pulses. Assuming that the expansion of the

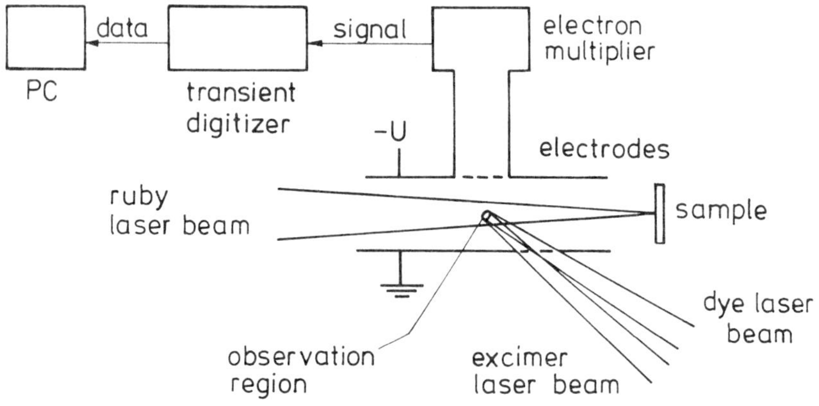

Figure 1. Scheme of the RIS apparatus.

vaporized substance can be described by a Maxwellian velocity distribution function (Beekman et al, 1984), on which a center-of-mass velocity (v_{cm}) is superimposed in the direction perpendicular to the surface (x), the density of atoms versus time and spatial coordinates can be given as:

$$\rho(x,y,z,t) = N_0 \left(\frac{m}{2\pi kT}\right)^{3/2} t^{-3} exp\left[\left(\frac{-m}{2kT}\right)\left(\frac{(x-v_{cm}t)^2+y^2+z^2}{t^2}\right)\right],$$

where N_0 is the number of sodium atoms removed from the surface by the laser shot, m is the mass of a sodium atom, k is the Boltzmann constant and T is the temperature. After integration over the observation volume (x_0, x_1, $-y_0, y_0$ and $-z_0, z_0$ are the limits in the three directions) the number of illuminated atoms versus delay time can be obtained:

$$N_i(t) = N_0 \frac{1}{2}\left[erf\left(\frac{x_1 - v_{cm}t}{t\sqrt{2kT/m}}\right) - erf\left(\frac{x_0 - v_{cm}t}{t\sqrt{2kT/m}}\right)\right] \times$$

$$erf\left(\frac{y_0}{t\sqrt{2kT/m}}\right) erf\left(\frac{z_0}{t\sqrt{2kT/m}}\right).$$

By fitting this function to the measured data (Figure 2.) the temperature and translational velocity of the atomic packet can be calculated. In the intensity range of $5 \times 10^7 - 10^8$ W/cm^2 the temperature was found to be between 1100 - 1800 K with a practically zero center-of-mass velocity. If we calculate the intensity limits of the evaporation and ablation regimes for silicon applying the model of Krokhin (1971), it is found that below 10^8W/cm^2 only heating of the surface occurs, between $10^8 - 8\times10^8$W/cm^2 an evaporation wave is produced and over 8×10^8W/cm^2 starts the ablation. So the result $v_{cm} << v_T$ is in accordance with this model ($v_T = \sqrt{3kT/m}$), and the slight degree of ionization is advantageous from the point of view of resonance ionization detection.

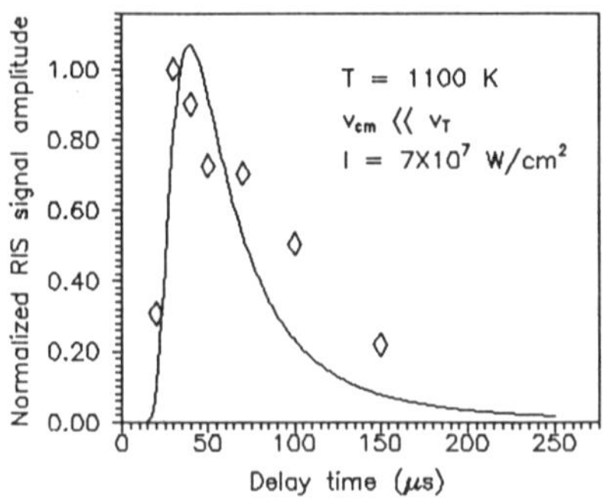

Figure 2. A set of measured data with fitted distribution function.

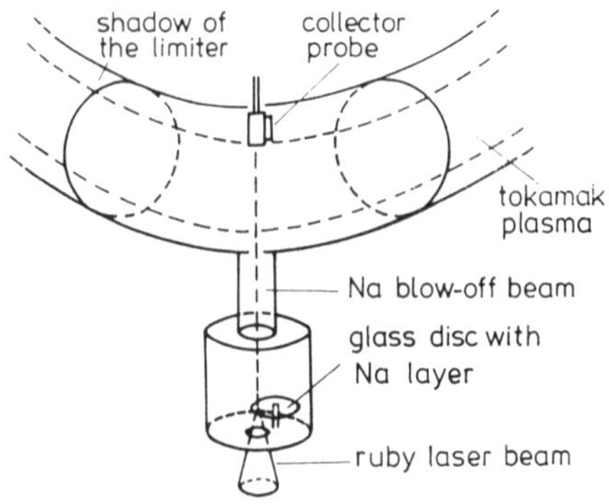

Figure 3. The tokamak blow-off arrangement.

4. APPLICATION FOR PLASMA DIAGNOSTICS

The experimental arrangement was used for the analysis of tokamak collector probes. The deposition probes located at the top of the torus (see Figure 3.) in the edge plasma collected the toroidally propagating intrinsic and injected impurities between radial positions of 75 and 90 mm during the discharge. Sodium atomic beam was injected into the plasma by irradiating a fresh evaporated 0.4 μm thick sodium film from the side of the glass substrate by the focussed beam of a Q–switched ruby laser (laser blow-off beam; Marmar et al, 1975), with an energy density of 50 J/cm^2, pulse duration of 30 ns and spot diameter of 1 mm. By the analysis of the probes the radial distribution of the toroidal flux of the injected sodium atoms can be determined.

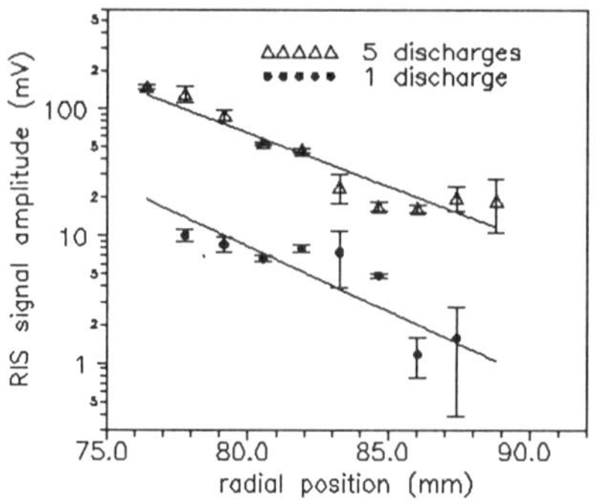

Figure 4. RIS signal amplitudes proportional to the sodium fluxes versus radial position.

In our measurements the probes contained the sodium impurities after 1 and 5 tokamak discharges corresponding to the same number of impurity injections. The radial dependence of the impurity concentration can be seen in Figure 4. It shows exponential decay with an e-folding length of about 5 mm (4.3 mm and 5.2 mm for 1 and 5 discharges, respectively). This result is in reasonable agreement with that obtained in earlier measurements on the same tokamak but with different plasma parameters and surface analytical methods after 20 impurity injections (approximately 10 mm; Bakos et al, 1989), considering that this decay length may depend on the actual plasma parameters and no detailed study has been carried out on this dependence.

Using simple approximations the surface concentration of the sodium can be roughly estimated. The sodium blow-off beam consisting of approximately 10^{16} atoms entered the tokamak after a flight of 70 cm through a hole of diameter of 3 mm. So the number of the particles actually entering the plasma can be estimated as 3×10^{12}, as it can be calculated from the velocity distribution of the blow-off beam (Utterback et al, 1976) with the actual center-of-mass velocity and temperature data (obtained in earlier experiments; Bakos et al, 1990). If we assume uniform distribution of sodium on the 15400 cm^2 internal surface of the tokamak, approximately 2×10^8 cm^{-2} average surface concentration is obtained. The analysed probe layer with 10 nm thickness contains approximately 5×10^{16} atoms per cm^2, so the average Na concentration after one discharge can be estimated as 4×10^{-9}, the number of the detected Na atoms was approximately 8×10^5.

5. TOWARDS DEPTH ANALYSIS

The optimum intensity for the evaporating laser was found to be less then $10^8 W/cm^2$ as mentioned earlier. In this case the RIS signal and the sensitivity was high enough and the ionic background could be kept still low corresponding to a good selectivity.

The laser crater (as checked by electronmicroscope and talystep) was shallow with a depth of approximately 10 nm. Several shots on the same spot will allow us to carry out depth analysis. It is, however, necessary to improve the beam quality which must be homogenous on the whole spot.

Summarizing the results, silicon surface analysis by laser atomization and resonance ionization has been realized. Our method has been used for plasma diagnostics, i.e. sodium surface impurity distribution on tokamak collector probes was analysed after a single tokamak discharge. Depth analysis and absolute calibration of the method is under development.

REFERENCES

Bakos J S, Bürger G, Ignácz P N, Szigeti J, Petravich G, Zoletnik S and Hildebrandt D 1989 J. Nucl. Mater. **162-164** 381

Bakos J S, Földes I B, Ignácz P N, Kocsis G, Szigeti J and Kovács J 1990 Opt. Commun. **74** 374

Beekman D W and Callcott T A 1984 Proc. 2nd Int. Symp. on Resonance Ionization Spectroscopy and its Applications eds. G S Hurst and M G Payne (Bristol: The Institute of Physics) pp 143-150

Krokhin O N 1971 Proc. Int. School of Physics "Enrico Fermi" Course XLVIII ed P Caldirola (London: Academic Press) pp 278-305

Marmar E S, Cecchi J L and Cohen S A 1975 Rev.Sci.Instrum. **46** 1149

Utterback N G, Tang S P and Friichtenicht J F 1976 Phys. Fluids **19** 900

Inst. Phys. Conf. Ser. No 114: Section 11
Paper presented at RIS 90, Varese, Italy, 16–21 September 1990

UV-laser desorption of neutral organic molecules: Speed distributions and fragmentations

K Dreisewerd, U Bahr, F Hillenkamp

Institut für Medizinische Physik, Universität Münster,
Hüfferstr. 68, D-4400 Münster, FRG

ABSTRACT: The distributions of the kinetic energy of neutral organic molecules, desorbed by the pulse of a KrF-excimer laser at 248 nm and (resonantly) postionized by a second excimer or a dye laser are presented. The influence of the desorption irradiance is reported. By comparision of the distributions of parent and fragment molecules fragmentation patterns are determinded indicating whether dissociation is mainly due to the desorption or the postionization step.

1. INTRODUCTION

Distributions of the kinetic energy of desorbed molecules give insight into the fundamental mechanisms of UV-laser desorption. In contrast to desorbed ions, the neutrals are not influenced by external electric fields or field inhomogeneities in the probe which may possibly distort or cover the aspired information. Speed distributions of organic neutrals therefore give a more direct picture of the desorption process. Energy distributions of desorbed neutrals differ distinctly from those of desorbed ions. While Spengler et al. (1988) determined energy distributions for the latter with a width of up to several tens of eV, Antonov et al. (1982) and Spengler et al. (1988) showed that the maximum energies of neutral desorbed organic molecules are in the sub-eV-region. Further inside into the desorption process can be obtained by investigations on the influence of the desorption irradiance on the energy distributions, but up to now no such experiments have been published for UV-laser desorbed neutral organic molecules, with only a few experiments related to other systems, like e.g. metals, semiconductors and polycondensates reported. A general trend in these systems is that the mean kinetic energy of the desorbed particles increases with growing irradiance as was e.g. shown by Schueler et al. (1987) for Ga- and As-atoms, desorbed from a GaAs-semiconductor surface, and Danielzik et al. for MMA-monomers, stemming from irradiated PMMA. In this paper it is shown that the influence of the desorption irradiance in UV-laser desorption on the kinetic energy distribution of neutral organic molecules differs distinctly from the former results. In postionization mass spectra besides the molecular ions in many cases fragment molecules are also observed. The precursors of these fragments can in principle be intact desorbed neutral molecules which are photoionized and dissociated by interaction with the postionization laser pulse or neutral fragment molecules formed in the condensed phase during the desorption step and ionized later. In this paper this question will be answered for two selected substances.

2. EXPERIMENTAL

For the investigations a laser desorption time-of-flight mass spectrometer was used that facilitates the photoionization of desorbed neutrals.
Desorption was achieved by the use of an excimer laser with a wavelength of 248 nm and a temporal pulse width of 10 ns After a certain time of flight the neutrals reach the postionization region where they are photoionized either by a second excimer laser operating at 193 nm or by the output pulse of a frequency doubled dye laser (FL 2001, Lambda Physik) with a pulse width of 15 ns and output wavelengths between 270 and 320 nm. The dye laser is tuned such that resonant absorption by the analyte molecules takes place. The time the neutral molecules need to get from the sample surface to the postionization region reflects their initial kinetic energy. At different times with respect to the desorption pulse molecules with different velocities are located inside the postionization region. Thus by variating of the time delay between the desorption and the postionization laser pulse molecules out of different velocity intervals are detected. By scanning the delay between the two laser pulses and determining the signal intensities at every delay time, intensity-time distributions are achieved. A further transformation of the experimental results into velocity distributions is not referred to in this paper because the influence of effects such as cluster desorption and dissociation are more obvious with this kind of representation. A comparision between the experimental and the simulated intensity-time distributions - calculated under the assumption of a desorption of the molecules with Maxwell-velocity- and Knudsen-cos-distributions - is given instead.

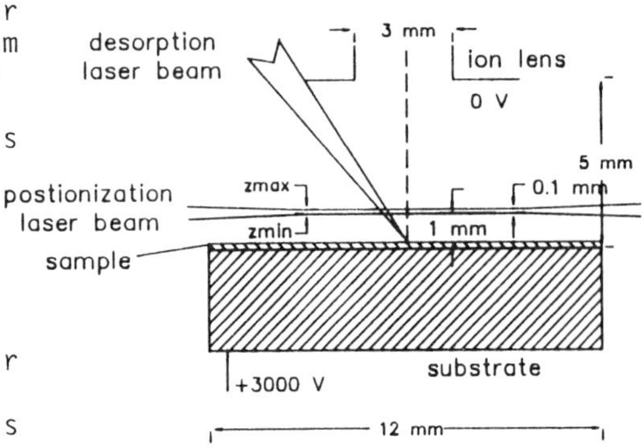

Fig. 1: Postionization area

3. RESULTS AND DISCUSSION

Figures 2 and 3 present the intensity-time distributions of neutral tryptophane and nicotinic acid molecules near threshold and 20-fold threshold irradiance for the desorption of ions which are approximately the same for both substances.

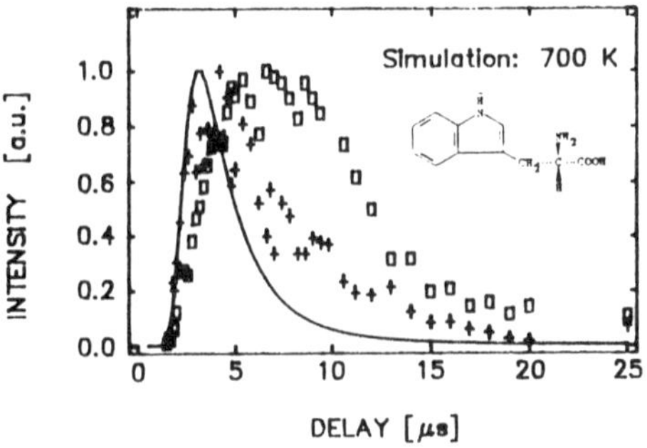

Fig. 2: Intensity-time distributions of tryptophane molecules near threshold (+) and 20fold desorption irradiance

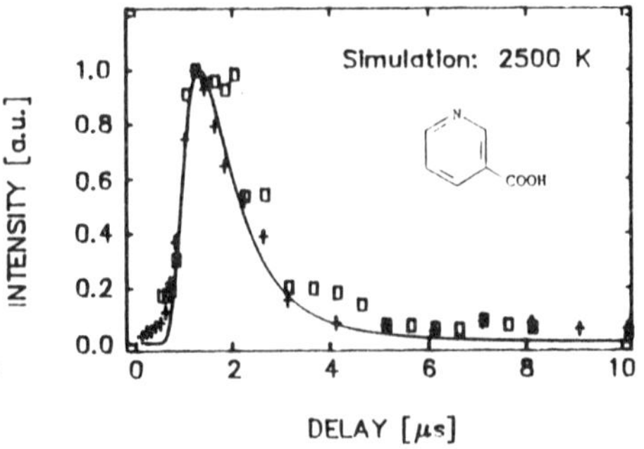

Fig. 3: Intensity-time distritions of nicotinic acid molecules near threshold (+) and 20fold desorption irradiance

Although in the case of nicotinic acid the experimental results can be well described by a Maxwell velocity distribution with a mean kinetic energy of $E_{Trans} = 2k_BT_{Trans}$ of 430 meV, corresponding to a translational temperature, T_{Trans}, of 2500 K, it is not possible to do so in the case of tryptophane. The reason for the broadening of the intensity-time distribution of tryptophane is most likely the desorption of neutral tryptophane dimers and oligomers and their later dissociation into monomeric untis on their way to or inside the postionization region. If oligomers and monomers start with similar kinetic energies, due to their higher mass the former will be slower, leading to the signal intensities at great delay times. As can be seen in figure 4 the desorption of oligomers is enhanced by an increase in desorption irradiance. In the case of nicotinic acid only a few oligomers are desorbed and no desorption irradiance effect observed. The different behaviors are likely due to different intermolecular bond strengths which allows the desorption of greater units of tryptophane molecules and prohibits it in the case of nicotinic acid. This explanation is supported by the difference in vapor pressures of the two substances. The front branchs of the experimental curves represent the fastest molecules reaching the postionization volume first. Possible decay reactions of greater units lead only to minor modifications of the distributions in this time region. Therefore the simulation for tryptophane as a fit to the front branch gives the best approximation to an intensity-time distribution, stemming from molecules which are desorbed exclusively as monomers. The translational temperature of 700 K, corresponding to a mean kinetic energy of 120 meV, is distinctly below that of nicotinic acid molecules although the desorption irradiances are approximately the same. The reason for the difference in mean kinetic energy may again be the difference in intermolecular bonding strengths. In the case of tryptophane more energy is needed to dissolve the bindings of the molecules to the surrounding whereas there is more energy left over to be converted into kinetic energy in the case of

nicotinic acid. In contrast to the results at other systems mentioned earlier no desorption of faster molecules with higher irradiance takes place as it would e.g. be expected from a thermal desorption mechanism.
Figures 4 and 5 show the intensity-time distributions of the two indole derivatives tryptophane and indole acetic acid and those of the typical indole fragment with a mass of 130 Dalton which is observed in the mass spectra of both substances. The postionization wavelength is 290 nm, the desorption irradiance near threshold.

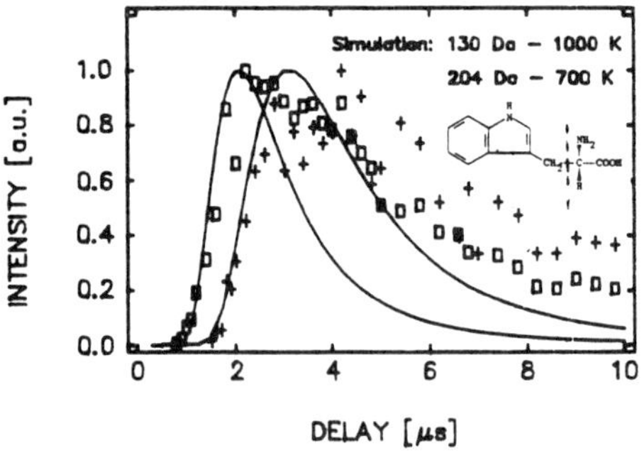

Fig. 4: Intensity-time distributions of tryptophane molecules (+) and fragments of mass 130 Da (□), $E_{Des}=E_0$

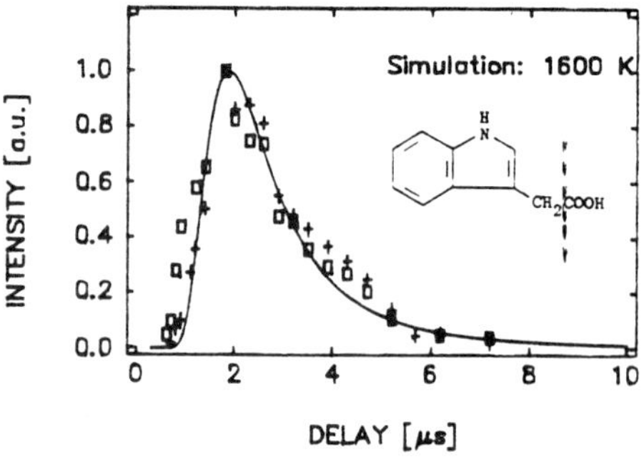

Fig. 5: Intensity-time distributions of indole acetic molecules (+) and fragments of mass 130 Da (□), $E_{Des}=E_0$

Whereas in the case of indole acetic acid the distributions of parent and daughter molecules are the same, faster tryptophane fragments are desorbed. Due to their lower mass fragments from molecules that dissociate already during the desorption step are generally faster than the parent molecules and have therefor intensity-time distributions shifted to shorter delay times compared to the parent molecules. This leads to the conclusion that in the case of indole acetic acid no neutral indole fragments are desorbed and the observed fragment ions stem exclusively from a dissociation reaction due to the interaction with the postionization laser pulse. In contrast trpyptophane fragmentation occurs in the condensed phase prior to photoionization. An increase of the desorption irradiances neither leads to the desorption of neutral indole acetic fragments nor to an increase in the maximum velocity of tryptophane fragment molecules.

REFERENCES

Antonov V S, Lethokov. V S, Matveyets Y A, Shibanov A N 1982 Laser Chemistry **1** 1 pp 37-
Danielzik N, Fabricius N, Röwekamp M, von der Linde, D 1986 Appl. Phys. Lett. **48** pp 3-
Schueler B, Odom, R W 1987 J. Appl. Phys. **61** 9 pp 4652 -
Spengler B, Bahr U, Hillenkamp F 1988 Inst. Phys. Conf. Ser. **94** pp 231-

The support of the Deutsche Forschungsgemeinschaft and the Bundesminsterium für Forschung und Technologie is gratefully acknowledged.

Paper presented at RIS 90, Varese, Italy, 16–21 September 1990

Surface characterization and adsorbate-site determination using multiphoton resonance ionization detection of desorbed particles

M. El-Maazawi, Z. Postawa, R. Maboudian, B. J. Garrison, and N. Winograd

Department of Chemistry, The Pennsylvania State University
University Park, PA 16802, U.S.A.

Abstract: The high efficiency, sensitivity and selectivity of multiphoton resonance ionization (MPRI) coupled with energetic ion-bombardment and with a specially designed position-sensitive detector enabled us to simultaneously measure energy and angular distributions of neutral Rh atoms ejected from clean and oxygen covered Rh{001}. The results taken along φ=0° (open) and φ=45° (close) azimuthal directions are well reproduced by molecular dynamics computer simulations and are explained in terms of the geometrical structure of the very near-surface region. It is also found that oxygen adsorbs in the fourfold hollow sites of Rh{001}.

1. Introduction

Angular distributions of particles desorbed from ion-bombarded single crystals show characteristic anisotropies corresponding to preferred ejections along certain low-indexed directions. These anisotropies are perturbed by the presence of adsorbate atoms or molecules on the surface. By carefully studying these perturbations one can learn a great deal about the atomistic nature of the binding sites, coverages and possible ordering of the adsorbates. Computer simulations of the ion-impact process have offered a detailed picture of the mechanisms leading to the angular anisotropies. However, the progress has been hindered due to the lack of detailed experimental measurements. Although there have been a variety of techniques, such as LEED and SEXAFS, that have been used to study surface structures, our group has shown over the last few years that ion beam technique can be employed to get a detailed structural information through the simple concepts of atom channeling and blocking (Chang *et al* 1987, Blumenthal *et al* 1988, Reimann *et al* 1988 & 1989 and Maboudian *et al* 1990). We have also employed multiphoton resonance ionization (MPRI) process to simultaneously measure energy and angular distributions of desorbed species. This allows us to make direct and reliable comparisons to computer simulations using the molecular dynamics (MD) approach (Garrison *et al* 1988).

Despite the importance of oxygen in promoting the catalytic activity of Rh surfaces, only few investigations of oxygen adsorption on Rh exist (Fisher *et al* 1983 and Oed *et al* 1988). There has not been any detailed study of the interaction of oxygen with Rh surfaces at room temperature. In this paper we show that the angular distribution of the ejected atoms from clean Rh{001} is determined mainly by the relative positions of surface atoms which influence the trajectory of an ejecting particle via channeling and blocking. Furthermore, upon adsorption of oxygen, the angular distribution of the desorbing Rh atoms is affected in such a way as to suggest that oxygen adsorbs in the fourfold hollow site, in agreement with the LEED experiments done by Oed *et al* (1988).

2. Experimental Setup and Computer Simulations

The experimental arrangement is described in detail elsewhere (Kobrin *et al* 1986). Briefly, it consists of a UHV chamber (2×10^{-10} Torr base pressure) equipped with LEED and Auger spectroscopy, a Nd:YAG pumped-dye laser with a frequency doubler to resonantly ionize the desorbed neutrals, a position-sensitive microchannel plate (MCP) detector with a phosphor screen in the back of it and an image processing system which includes a CCD camera interfaced to a micro-VAX station II computer via a real-time frame grabber. Molecular dynamics simulations of the ion-impact event are performed using 3 keV Ar^+ impact at normal incidence on a Rh{100} micro-crystallite. The interaction among the target atoms is described using the many-body potential based on the embedded-atom method (EAM) (Daw and Baskes 1984). It is already shown that the EAM potential developed originally for the Rh{111} surface by Garrison *et al* (1988), successfully accounts for the distributions obtained from the Rh{331} stepped surface (Reimann *et al* 1988). Here, the same many-body potential is further used to simulate the desorption process for Rh{001}.

3. Results and Discussion

The experimental polar-angle (θ) distributions of Rh atoms desorbed from Rh{001} surface along $\varphi=0°$ (open) and $\varphi=45°$ (close) azimuthal directions are shown in Fig. 1(b). The off-normal intensity for the 0° azimuth is higher than for the 45° direction indicating that there is a strong preference for ejection along the open channels of the crystal. In addition, ejection is focussed into a polar angle of about 50° along the open direction and 35° for the close-packed direction. The results of the classical molecular dynamics simulation of the desorption process are presented in Fig. 1(b). As can be seen, there exists a very good overall agreement between the measurement and the calculation in the high kinetic energy range. However, in the low energy regime, the agreement with the experimental data is less satisfactory. At this time, the reason for this discrepancy is not clear. The characteristics of the angular distributions along the two azimuthal directions emphasize the importance of the relative positions of surface atoms (Maboudian *et al* 1990). For an open surface such as {001}, the peak position in the angular distribution directly correlates with the nearest-neighbor distance. The larger this separation is, the larger is the peak angle, as is evident when comparing the distribution along $\varphi=0°$ to that for the 45° ejection. Based on the above observations, one can conclude that the relative positions of surface atoms are the dominant factor in determining the angular distributions of desorbed atoms from an ion-bombarded single crystal.

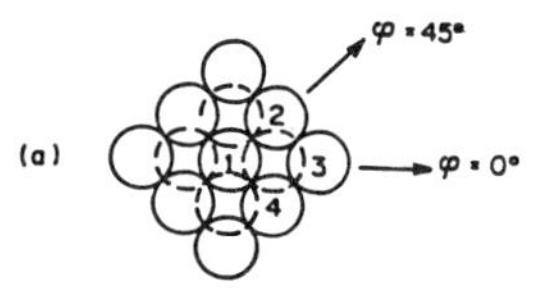

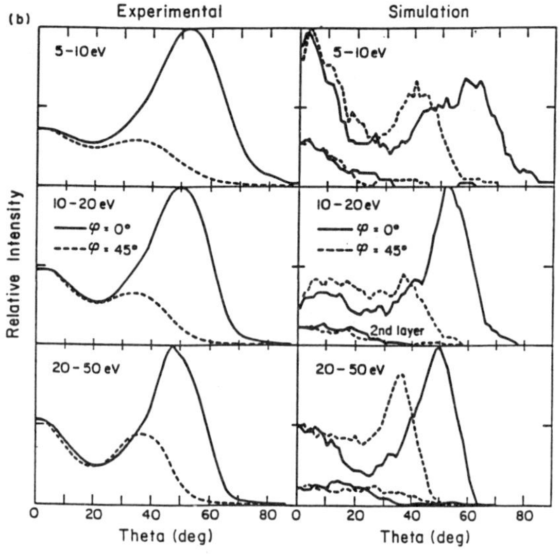

Fig. 1. (a) Rh{001} crystal face with $\varphi=0°$ (open) and $\varphi=45°$ (close) azimuthal directions shown. The dashed circles denote the positions of the second layer atoms.
(b) Experimental (left) and calculated (right) angular distributions of Rh atoms desorbed atoms desorbed from Rh{001}. The curves marked "2nd layer" are the distributions obtained from simulations corresponding to the second layer atoms only.

Kinetic energy distributions of Rh atoms emitted from clean and oxygen covered Rh{001} surface are shown in Fig. 2 for φ=0° and φ=45° azimuthal directions. Generally, no significant differences are visible between distributions taken before and after oxygen exposure. The small broadening of the spectrum for oxygen-covered surface is within the uncertainty of our measurements. Potentially, two processes can influence the energy spectra of emitted substrate atoms in the presence of oxygen overlayer. Oxygen can increase the surface binding energy which will shift the peak in the energy distribution to higher values (Thompson 1968). It can also cause an exiting particle to lose part of its kinetic energy due to collision with the adsorbed oxygen, moving the energy distribution towards lower values. These two processes compete with each other. The energy distribution then obtained is determined by the relative strength of these two processes. The lack of any change in the shape of the kinetic energy distributions differs from our previous measurements for Rh{111} crystal where a shift toward lower kinetic energies was observed after oxygen adsorption and explained by the dominance of the second process (Reimann *et al* 1989). One should, however, remember that the {111} surface has higher surface atomic density and higher oxygen coverage than the {001} and collisions between exiting Rh atoms and adsorbed oxygen atoms are much more probable and effective. More pronounced changes are visible in the energy-resolved polar angle distributions. In Fig. 3, such distributions are shown for atoms emitted along the open (φ=0°) and close (φ=45°) azimuthal directions. The position of the off-normal peak for the oxygen-covered surface is shifted closer to the surface normal for the open azimuth but is not altered in the spectrum measured along the φ=45° azimuth. The magnitude of the shift depends on the energy of the emitted particle, being almost 10° in the 5-10 eV range and decreasing for higher kinetic energies. This trend is easy to be understood as Rh atoms with higher kinetic energies are less sensitive to the collisions with much lighter O atoms.

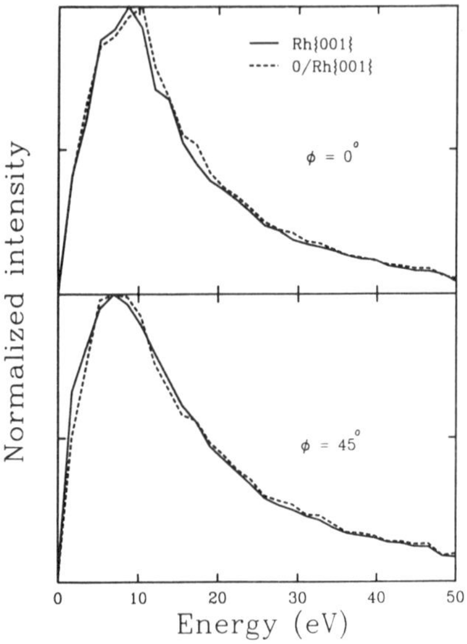

Fig. 2. Energy distributions of Rh atoms emitted along both azimuthal directions from clean and oxygen covered Rh{001}.

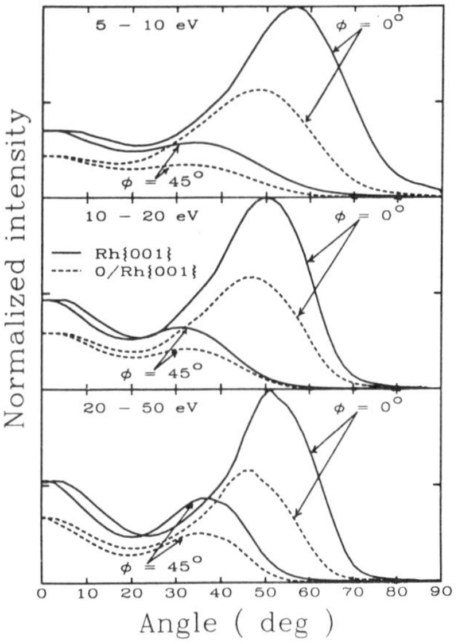

Fig. 3. Energy resolved polar angle distributions of Rh atoms emitted from clean and oxygen covered Rh{001}.

A careful inspection of Fig. 3 reveals an additional feature. The off-normal emission along the φ=0° direction is suppressed more than the normal ejection whereas for the φ=45°, both components are reduced roughly by the same amount. The three possible high symmetry p(2x2) oxygen overlayers on the Rh {001} surface are the A-top site, the twofold bridge site, and the fourfold bridge site. If oxygen atoms are located directly above the first layer atoms (A-top bonding site) they should equally influence the emission of Rh atoms. However, as we can see from Fig. 3, the angular distribution along $\varphi=0°$ is suppressed more than that along $\varphi=45°$. The total Rh signal is decreased roughly by 30%. Computer simulations for the O/Cu{001} system, which is analogous to the case discussed in this paper, suggest that in the case of the A-top oxygen adsorption the total sputtering yield of the emitted copper atoms is nearly unchanged whereas both twofold- and fourfold-bridge sites reduce the number of ejected copper atoms (Garrison *et al* 1978). Both arguments disqualify the A-top site as a possible bonding site for the adsorbed oxygen. Oxygen in both two- and fourfold bridge sites will partially block emission of second layer atoms. However, if the twofold-bridge sites are occupied, the emission along the $\varphi=45°$ azimuth should be shifted toward the surface normal since in this direction an additional atom is placed in the way of the exiting Rh atom, whereas the emission along the open direction should not be changed. The situation reverses if oxygen atoms are bonded in fourfold sites. As is shown in Fig. 3, the second alternative dominates and therefore we conclude that oxygen adsorbs in the fourfold bridge sites on Rh {001} at room temperature.

4. Summary and Conclusions

Energy-resolved angular distributions of Rh atoms ejected from ion-bombarded Rh{001} single crystal have been measured. Using the same many-body interaction potential developed originally for Rh{111}, a good agreement between the experimental results and the classical dynamics calculation of the desorption process has been obtained. By extracting the information available from the molecular dynamics simulations, we have shown that despite the complexity of the ion-bombardment process, the anisotropies observed in the angular distributions are mainly determined by the extreme local environment experienced by a surface atom. This causes the particle to preferentially desorb along certain crystallographic directions. Furthermore, the energy and the angular distributions of Rh atoms ejected from ion-bombarded oxygen-covered Rh{001} surface have been measured and used to identify bonding sites of adsorbed oxygen atoms. These observations are significant in terms of providing a simple and comprehensive picture of the mechanisms leading to the observed distributions. Moreover, they provide an important frame-work for future developments of any analytical models of the ion-bombardment process.

Acknowledgement: The financial support of the U. S. Office of Naval Research and the U. S. National Science Foundation is greatly acknowledged.

References

Blumenthal R, Donner S K, Herman J L, Trehan R, Caffey K P, Furman E, Winograd N and Weaver B D 1988 J. Vac. Sci. Technol. **B6** 1444

Chang C C, Malafsky G P and Winograd N 1987 J. Vac. Sci. Technol. **A5** 981

Daw M S and Baskes M I 1984 Phys. Rev. **B29** 6443

Fisher G B and Schmig S J 1983 J. Vac. Sci. Technol. **A1**(2) 1064

Garrison B J, Winograd N, Deaven D M, Reimann C T, Lo D Y, Tombrello T A, Harrison D E,Jr. and Shapiro M H 1988 Phys. Rev. **B37** 7197

Garrison B J, Winograd N and Harrison D E 1978 Phys. Rev. **B18**(11) 6000

Kobrin P H, Schick G A, Baxter J P and Winograd N 1986 Rev. Sci. Instrum. **57** 1354

Maboudian R, Postawa Z, EL-Maazawi M, Garrison B J and Winograd N 1990 Phys. Rev. **B42**(12)

Oed W, Dötch B, Hammer L, Heintz K, Müller K 1988 Surf. Sci. **207** 55

Reimann C T, EL-Maazawi M, Walzl K, Garrison B J and Winograd N 1989 J. Chem. Phys. **90**(3) 2027

Reimann C T, Walzl K, EL-Maazawi M, Garrison B J and Winograd N 1988 J. Chem. Phys. **89**(4) 2539

Thompson M W 1968 Philos. Mag. **18** 377

The detailed characteristics of resonant laser ablation

L.Wang, I.S.Borthwick, R.Jennings, P.T.McCombes, K.W.D.Ledingham, R.P.Singhal
Department of Physics & Astronomy, University of Glasgow, Glasgow G12 8QQ

C.J.McLean
Department of Electronics & Electrical Engineering,University of Glasgow,Glasgow G12 8QQ

ABSTRACT: This paper deals with the characteristics of RLA in detail for a GaAs sample, e.g.,the angular dependence,laser power dependence and the widths of the resonant signals.

1. INTRODUCTION

Laser ablation has been used extensively to vaporize target materials and has given rise to the analytical instruments LIMA (Southon et al. 1984) and LAMMA (Krier et al. 1985), which use a quadrupled Nd:YAG laser in conjunction with a reflectron time of flight mass spectrometer to analyse the ions produced in the ablation process.It is believed that the ion yield is dominated by plasma processes, and in most cases the yield of ions is two or three orders of magnitude lower than that of neutrals, estimated by the Saha equation (Fürstenau 1981). Thus greater sensitivity can be achieved by efficiently ionising the ablated neutral species with a delayed laser pulse tuned to a transition wavelength of the specific atoms. This is the basis of the one of the variants of the RIMS process (Towrie et al. 1990), namely laser ablation together with resonant post ablation ionisation.

Recent studies(Verdun et al 1987; Ho-ming Pang et al. 1989) have suggested that both ablation and resonant ionisation may be performed with a single tunable laser pulse to provide signal enhancement and higher selectivity. However, considerable broadening of resonant signals was observed (~nm) in both studies, as well as strong matrix dependence, although enhancements up to 70 were observed. It should be pointed out that in both cases the ablating laser was directed normally to the sample, in transmission mode in one case (Verdun et al. 1987) and in conjunction with 50 Torr of He buffer gas in the other (Ho-ming Pang et al. 1989).

In a recent paper McLean et al.(1990) have shown that if a moderately focused resonant laser beam is directed at grazing incidence to the target, signal enhancements of greater than two orders of magnitude can be obtained. In addition the resonant widths obtained are less than 0.05 nm at wavelengths corresponding to known atomic transitions. It was felt that the analytical potential of these observations was of considerable value and thus a detailed study of grazing incidence resonant laser ablation was undertaken.

2. EXPERIMENTS AND RESULTS

The mass spectrometer used here has been described in detail elsewhere (McLean et al,1990). A GaAs sample with a smooth mirror finish was used in these experiments. The laser was directed at grazing incidence to the surface and moderately focused by a lens (f=50cm) to give a beam diameter of 0.5mm. The UV output of frequency-doubled dye laser, with pulse length being 5ns and bandwidth 0.003nm, was tuned to around the $^2P_{1/2}$-$^2D_{3/2}$ transition (287.5 nm) of Ga and together with the absorption of a fundamental photon resulted in the ionisation of the excited atoms.

Experiment 1) The grazing incident angle on the sample was kept constant at 5° and the total laser pulse energy was varied up to 1mJ. At each energy a wavelength scan was taken . The spectra are shown in Fig.1. A resonant enhancement of several hundred is clearly seen although as the pulse energy increases, the resonant width increases asymmetrically, being greater at longer wavelengths. At low pulse energies the width was about 0.1nm. On resonance it is assumed that the laser first desorbs neutrals which are subsequently resonantly excited and ionised in gas phase by the same pulse. This model was supported by the close correspondence between the known atomic transition wavelength and the laser wavelength at resonance calibrated with a uranium hollow cathode lamp.

The resonant widths observed at high laser energies cannot be attributed to the interaction of isolated atoms with the laser field. Taking into account the short duration of the laser pulse (~5ns) and the average velocity of the particles leaving the surface of the sample (Schueler et al. 1987), the interaction occurs within a few μm of the surface. Since the density of the neutrals increases rapidly with the laser fluence, it is expected that collisional processes give rise to the observed broadening. It is worthwhile noting that no significant resonant shifts were observed at the different laser power levels. Such shifts usually appear with collisional broadening processes and increase with increasing pressure. Power broadening is small because the laser intensity is relatively low, and thus it is expected that the main contribution to the width comes from inelastic collisions, which cause broadening but no shift of the resonant lines (Demtröder 1982). It is also possible that inelastic collisions would directly ionise some of the excited atoms.Fig.2 shows the dependence of the Ga^+ yield on the ablation laser energy on resonance with a power dependence of approximately 4~5.

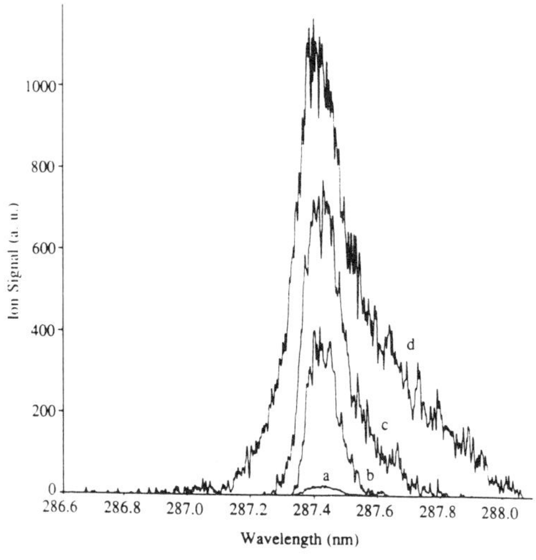

Fig.1 The ion signal of Ga as functions of the ablation laser wavelength,λ, and energy E. The grazing incident angle of the laser beam on the sample surface is θ=5°. The corresponding total laser energy is (a) E=0.39mJ, (b) E=0.66mJ, (c)E=0.79mJ and (d) E=0.98mJ.

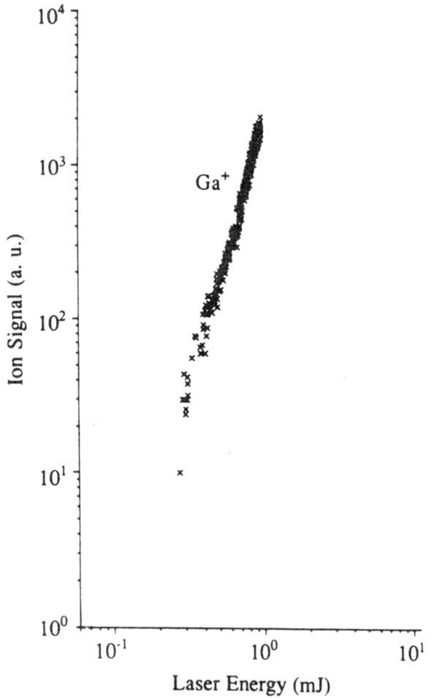

Fig.2 Dependence of Ga^+ yield on ablation laser energy at the resonant wavelength. The power dependence is approximately 4~5.

The ablation yield, Y, of neutrals is proportional to, among other things, the laser fluence to some high power, n, the ablated area A ($\propto 1/\sin\theta$), and the energy absorbed (1-R), i.e.,

$$Y \propto F^n A(1-R) \propto E^n \sin^{n-1}\theta(1-R) \quad (1)$$

where F and E are the laser pulse fluence and energy, θ is the angle of grazing incidence and R is the reflectivity of the surface. Although R is a function of fluence (Ready 1971) it is assumed to be constant over the small range of fluences in the present experiment. Thus at constant θ and hence reflectivity, the neutral yield is proportional to a high power of the pulse energy which was shown in Fig.1.

Experiment 2) To investigate the effect of target geometry on RLA, the laser pulse energy was kept constant at fixed values and a wavelength dependence of the ion signal was taken for several angles between 5° and 8° by rotation of the precision manipulator. Fig.3 shows the data at total laser energy E=0.26mJ (UV energy was 56μJ) while Fig.4 was taken at E=0.61mJ (UV energy was 110μJ). The right inset in Fig.3 is the Ga^+ time of flight signal showing the two isotopes ^{69}Ga and ^{71}Ga well resolved.

From Eq.(1) at constant laser energy the neutral yield is proportional to $\sin^{n-1}\theta(1-R)$, and since R decreases as θ increases the ion yield will increase with some high power of $\sin\theta$, which is in qualitative agreement with what was observed. At low pulse energy (Fig.3) the peaks remain narrow as θ increases but at high energy (Fig.4) the peaks become broad asymmetrically at longer wavelength, similar to Fig.1. This observation supports the view that the broadening of the resonant peaks resulted from the increasing density of the neutrals near the surface.

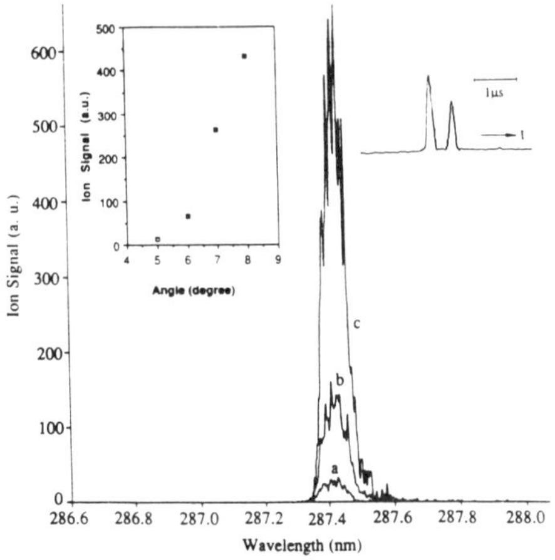

Fig.3 RLA spectra of Ga at relatively low laser energy, E=0.26mJ, with different incident angles of laser beam: (a) θ=5°, (b) θ=6° and (c) θ=7°. Upper right insert is the Ga^+ signal with ^{69}Ga and ^{71}Ga well resolved.

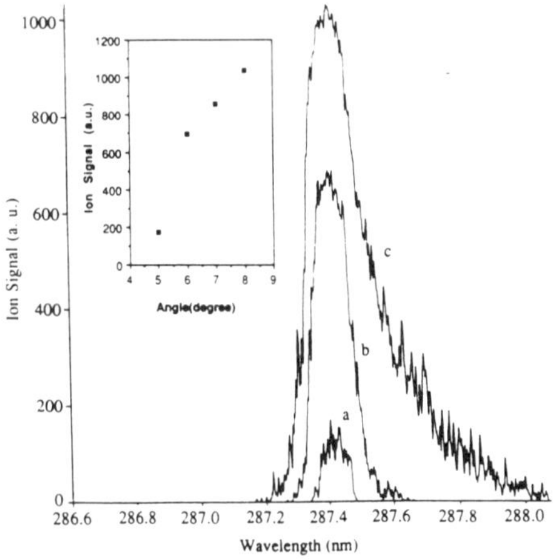

Fig.4 RLA spectra of Ga at slightly higher laser energy, E=0.61mJ, with different incident angles:(a)θ=5°, (b) θ =6° and (c) θ=8°. In comparison with Fig.3, broadening occurs at larger incident angles.

Experiment 3) Fig.5 shows the results where both the laser energy E and angle of incidence θ were varied such that the fluence was kept constant and the ion signal was again plotted as a function of wavelength. The ion yield still rises with increasing angle when laser energy was reduced, although rather slowly, and the peak shape remains relatively narrow. At constant fluence the neutral yield according to Eq.(1) varies as A(1-R) or (1-R)/sinθ. The ion yield depends on the number of neutrals Y and on E, the energy available for ionisation. Y increases,but E decreases, as θ increases in this experiment. Thus the angular dependence of ion yield is uncertain although it is certainly smaller than in experiment 2).

3. CONCLUSIONS

Resonant laser ablation has been extensively demonstrated at different powers and several different angles at grazing incidence for Ga from a GaAs sample. At grazing incidence the resonant widths are fairly narrow at low fluence but increase with both fluence and angle of incidence, and at the high fluence tend to approach the values reported in references 5 and 6, where the normal incident configuration must produce more neutrals near the surface of sample resulting in broad resonant peaks. This unusual broadening was attributed to inelastic collision due to the high density of neutrals near the sample surface. The asymmetric profile of the resonant peak in the present experiment remains to be explained.

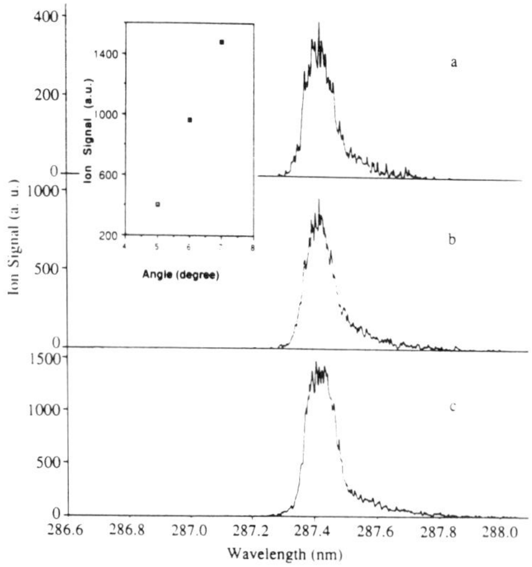

Fig.5 RLA spectra of Ga with laser fluence kept constant by holding $E\sin\theta$=constant for different laser energy E and incident angles θ: (a) E=0.51mJ,θ=5°,(b) E=0.43mJ,θ=6° and (c) E=0.39mJ,θ=7°.

Since the reflectivity of the sample plays an important role in RLA, further work is at present being carried out to determine quantitatively the angular dependence observed in Figs.4-6. In addition RLA at large angles will also be investigated.

The work described in this paper was carried out on Ga, a major constituent of GaAs. Further work will be carried out on RLA of trace atomic levels of gold in copper matrices to determine the sensitivity of RLA in comparison with RIMS.

Although RLA has been investigated in this paper with respect to atom detection its importance may well be greater in connection with trace molecular detection.

ACKNOWLEDGEMENTS

L.W. wishes to thank the Royal Society for a fellowship while I.S.B., P.T.McC. and C.J.McL acknowledge receipt of SERC research studentships.

REFERENCES

Demtröder W 1982 *Laser Spectroscopy* (Berlin:Springer-Verlag) pp92.
Fürstenau N 1981 Fresenius Z.Anal.Chem. **308**, 201
Krier G, Verdun F R and Muller J F 1985 Fresenius Z. Anal. Chem. **322**, 379
McLean C J, Marsh J H, Land A P, Clark A, Jennings R, Ledingham K W D,McCombes P T, Marshall A, Singhal R P and Towrie M 1990 Int. J. Mass Spectrrom. Ion Proc. **96** R1
Pang H and Yeung E S 1989 Anal. Chem. **61**, 2546
Ready J F 1971 *Effect of High-Power Laser Radiation* (NewYork:Academic Press)
Schueler B and Odom R W 1987 J. Appl. Phys. **61**, 4652
Southon M J, Witt M C, Wallach E R and Myatt J 1984 Vacuum **34**,903
Towrie M,Drysdale S L T,Jennings R,Land A P, Ledingham K W D, McCombes P T, Singhal R P, Smyth M H C and McLean C J 1990 Int. J. Mass Spectrom.Ion Proc. **96**, 309
Verdun F R, Krier G and Muller J F 1987 Anal. Chem. **59**, 1383

Paper presented at RIS 90, Varese, Italy, 16–21 September 1990

A time of flight mass spectrometer for the analysis of solids, gases and liquids

M. Andersson, H. Fallgren and A. Rosén

Department of Physics, Chalmers University of Technology and University of Göteborg, S-412 96 Göteborg, Sweden

ABSTRACT: We present a time-of-flight spectrometer for the analysis of solid surfaces, gases and volatile liquids. We have analysed a sample of $Ga_{0.99}In_{0.01}As$ with one-photon ionization as well as resonant and non-resonant two-photon ionization to study the prospects for sensitive and quantitative analysis using these ionization methods.

1. INTRODUCTION

The technique to combine sputtering, laser ionization and mass spectrometry to study surfaces has been developed during the last decade (Winograd et al 1982, Parks et al 1983, Becker and Gillen 1984 and Pellin et al 1987). Both resonant and non-resonant multiphoton ionization (MPI), and lately also one-photon ionization with VUV-light (Schühle 1988), has been used. The main reasons for using resonant MPI are a higher degree of selectivity (isobaric and state selectivity), probably higher sensitivity for specific elements and normally less fragmentation of molecules. Non-resonant MPI on the other hand has its advantages in multi-element analysis, as all elements and compounds can be probed simultaneously with the same laser light.

We have constructed a time-of-flight spectrometer for surface analysis with sputtering and MPI. With our available lasers we can perform both resonant and non-resonant MPI. In this paper we shortly describe the apparatus and present the first results, which is a comparison between these two ionization methods for the analysis of a GaAs sample containing a small amount of In.

2. EXPERIMENTAL

Our apparatus is primarily designed for analysis of various solid samples, but it has been modified to enable analysis of gases and liquids as well. A schematic of the experimental set-up is shown in figure 1, and a more extensive description has been made by Fallgren (1989).

The experiments are performed in a UHV-system pumped by turbomolecular pumps. Sputtering is performed with 4-keV Ar^+-ions from an ion gun (Physical Electronics 6033) directed to the sample with an incidence angle of 65° to the

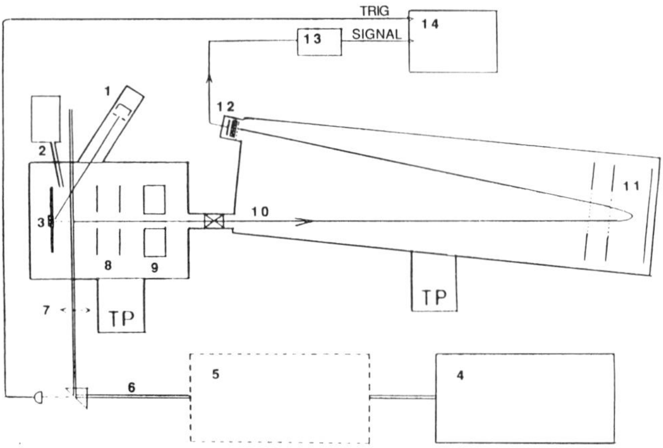

Figure 1. Schematic of the experimental set-up. Legend: 1 - Primary ion gun, 2 - Gas inlet, 3 - Solid sample, 4 - Excimer laser, 5 - Dye laser (only by resonant MPI experiments), 6 - Excimer or dye laser beam, approaching a prism and a photodiode, 7 - Focusing lens, 8 - Extraction and acceleration ion optics, 9 - Electrostatic lens, 10 - Beam of ionized species, 11 - Reflector, 12 - Detector, 13- Preamplifier, 14 - Digital storage oscilloscope or boxcar integrator, TP-Turbomolecular pumps

surface normal. Between the ion gun and the sample we have placed an aperture, on which we can apply a pulsed high voltage in order to chop the primary ion beam. We have used primary ion pulses of 3-4 µs duration. The sputtered atoms and molecules are ionized with laser light passing a few mm above the sample surface. The laser light pulse is delayed a few µs from the primary ion pulse. The lasers we use are two excimer lasers (Lambda Physik LPX210i and EMG102E), operating on ArF (193 nm, 6.4 eV) and XeCl (308 nm, 4.0 eV) respectively, and a dye laser (Lambda Physik FL2002E) with a frequency-doubling crystal. The photoions are detected in a time-of-flight spectrometer with a two-field reflector. The acceleration voltage is approx. 800 V and the drift length is kept at about 2 m. Secondary ions are discriminated by the reflector. The signal from the dual microchannel plate detector is recorded on an averaging storage oscilloscope (Le Croy 9400) or a boxcar integrator (Stanford Research SR 250). To analyse gases and volatile liquids there are two gas inlet systems into the ionization region. One is a regulated flow from an intermediate pressure reservoir containing the gas or vaporized liquid to be analysed. The other gas inlet is through a fused silica capillary from ambient pressure.

3. RESULTS AND DISCUSSION

We have analysed a sample of GaAs containing approx. 0.5 % In, and especially we have determined the detected amounts of Ga and In using three different ionization methods: one-photon ionization, resonant and non-resonant two-photon ionization.

Figure 2a shows a mass spectrum where we have used the light from the ArF-excimerlaser for ionization. The ionization potential (IP) of Ga is 6.0 eV and of In 5.8 eV, so absorbing one photon with λ=193 nm is sufficient for ionization. However, only a limited number of elements are possible to ionize with one photon from the ArF-laser and most elements like As (IP=9.8 eV) demand two or in a few cases even more photons for ionization from the ground state. The number of ionized and detected Ga-atoms is thus much larger (30 times) than the number of As-atoms. The number of detected In-atoms is 1.8 % of the number of Ga-atoms, which reasonably well corresponds to the estimated composition of the sample. The estimated intensity of the laser light for the spectrum in figure 2a is 60 MW/cm^2, and the pulse energy is 10 mJ.

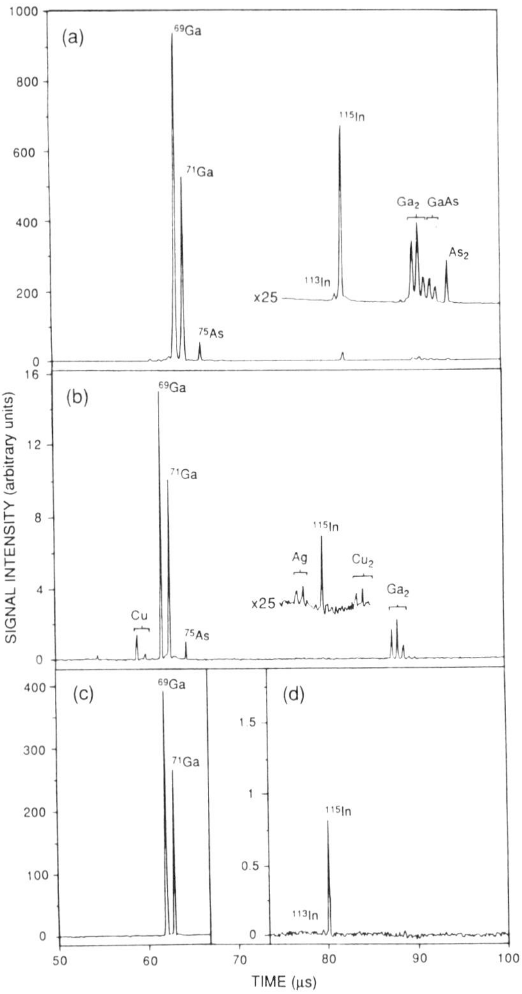

Figure 2a-d. Mass spectra recorded using laser light with wavelengths: (a) 193 nm, (b) 308 nm, (c) 287.4/574.8 nm and (d) 303.9/606.9 nm for ionization. All spectra have the same, arbitrary, signal intensity units. The Cu and Ag originate from the sample holder.

When the light from the XeCl-excimer laser is used for non-resonant ionization, two photons are required for the ionization of Ga and In, and three photons for As. From the spectrum in figure 2b, where the same signal intensity units as in figure 2a has been used, it can be seen that the number of detected atoms is much lower when we use 308 nm-light instead of 193 nm-light. The laser light power density used for the spectrum in figure 2b is approx. 60 MW/cm^2, but the pulse energy is 50 mJ and the ionization volume is larger than in figure 2a. The number of detected Ga-atoms is 1.2 % of the number detected using the ArF-laser. In figure 2b the number of detected In atoms is 0.9 % of the number of detected Ga-atoms. This is lower than in figure 2a, but still close to the estimated 1 %. The lower In/Ga-ratio probably depends on a different laser beam geometry causing different ionization and transmission probabilities. Local variations of the sample composition or different relative one-photon and two-photon ionization cross-sections might also influence the ratio. Although As is ionized in a non-resonant three-photon process it gives a significant peak that represents 3.5 % of the Ga-peaks.

Figures 2c and d display mass spectra of resonantly ionized Ga and In. To ionize Ga we use light with λ=574.8 nm, which is frequency doubled to λ=287.4 nm. The Ga-atoms are excited from the $4p^1\ ^2P_{1/2}$ ground state to the $4d^1\ ^2D_{3/2}$ state with a UV-photon and then ionized with a photon of yellow light. The pulse energies used are 4 mJ at 574.8 nm and 0.2 mJ at 287.4 nm. The number of Ga-atoms detected by resonant two-photon ionization is 20 % of the number detected using the 193-nm light for one-photon ionization, but more than an order of magnitude larger than by non-resonant two-photon ionization with the light from the XeCl-excimer laser. We have also ionized Ga-atoms from the $4p^1\ ^2P_{3/2}$ state via the $4d^1\ ^2D_{3/2}$ or $4d^1\ ^2D_{5/2}$ states with wavelengths 294.4/588.8 nm or 294.3/588.7 nm. The number of atoms in the $^2P_{3/2}$ state is estimated to be 40 % of the number in the $^2P_{1/2}$-state. In-atoms have been ionized according to the corresponding ionization scheme: excitation from the $5p^1\ ^2P_{1/2}$ ground state to the $5d^1\ ^2D_{3/2}$ state with a 303.9 nm-photon and then ionization with the fundamental 606.9 nm-light. The number of detected In-atoms is only 0.2 % of the number of Ga-atoms detected by resonant two-photon ionization with the same pulse energy. Although these spectra have not been recorded simultaneously, possible changes in the experimental parameters cannot explain the low In/Ga-ratio. A possible explanation is different ionization cross-sections, though the ionization schemes are similar.

4. CONCLUSIONS

In our apparatus and with available lasers we have found that one-photon ionization with the light from an ArF excimer laser is the most efficient ionization method for measurements of Ga- and In-atoms. It is about five times more efficient than resonant two-photon ionization, which on the other hand is an order of magnitude better than non-resonant two-photon ionization. The relative efficiency of the different ionization methods vary however from element to element. The efficiency of the non-resonant MPI principally depends on how many photons that are demanded, while the resonance ionization schemes of the various elements may be more or less difficult to carry out. Our conclusion is that both methods can be used for sensitive surface analysis and which method to choose should be governed by the purpose of the measurement. Non-resonant MPI provides mass spectra that are suitable for quantitative multi-element analysis. Resonant MPI gives a higher degree of selectivity. The state selectivity gives more information, but also complicates a quantitative analysis.

ACKNOWLEDGEMENTS

The authors want to thank Dr. Thorwald Andersson for providing us with samples. We are grateful for financial support from Knut and Alice Wallenberg's Foundation for the ArF excimer laser, and from Carl Trygger's Foundation and the Swedish Board for Technical Development (STU project no. 86-03067P) for the construction of the time-of-flight spectrometer.

REFERENCES

Becker C H and Gillen K T 1984 Anal. Chem. **56** 1671

Fallgren H 1989 Licentiat thesis Göteborg University

Parks J E, Schmitt H W, Hurst G S and Fairbank jr W M 1983 Thin Solid Films **108** 69

Pellin M J, Young C E, Calaway W F, Burnett J W, Jørgensen B, Schweitzer E L and Gruen D M 1987 Nucl. Instr. and Meth. **B18** 446

Schühle U, Pallix J B and Becker C H 1988 J. Vac. Sci. Technol. **A6** 936

Winograd N, Baxter J P and Kimock F M 1982 Chem. Phys. Lett. **88** 581

Paper presented at RIS 90, Varese, Italy, 16–21 September 1990

Laser induced fluorescence in plasma plumes produced by laser ablation of solid samples

W. Sdorra and K. Niemax
Institut für Spektrochemie und angewandte Spektroskopie(ISAS)
Dortmund, Federal Republic of Germany

ABSTRACT: Laser ablation of solid samples (Fe, Al, Cu and glass) into a low-pressure argon gas has been studied applying a Nd:YAG laser. Elemental analysis was performed by laser induced fluorescence in the plasma plume. The application of internal standardization gave matrix independent, precise data with very low detection limits.

1. INTRODUCTION

The process of laser ablation and atomization is a powerful technique for direct spectrochemical analysis of solid samples without chemical preparation. With laser induced fluorescence (LIF) for selective excitation of analytes the method is well suited for very sensitive quantitative micro and trace analyses. In contrast to the ion microprobe technique laser ablation is independent of the electric porperties of the sample. Conductors, semiconductors and insulators (for example, ceramics and glass) can be investigated. Papers on analytical applications of laser ablation using optical emission spectroscopy (OES) and LIF have been published in the last two years by our group (Leis et al. 1989, Sdorra et al. 1989, Ko et al. 1989, Quentmeier et al. 1990, Sdorra and Niemax 1990, Niemax and Sdorra 1990).

2. EXPERIMENT

A pulsed Nd:YAG laser beam (Spectron SL 801, 1064 nm, 10 ns pulse length, power range: 5 - 10 mJ) was focussed on the sample placed in a chamber filled with buffer gas (argon, 140 hPa). The focus diameter on the sample surface was typically 100 μm. The atomic fluorescence of analyte atoms in the laser-generated plasma plume was excited by a tunable pulsed dye laser pumped by an excimer laser. The fluorescence radiation was detected by a gated optical multichannel analyzer system

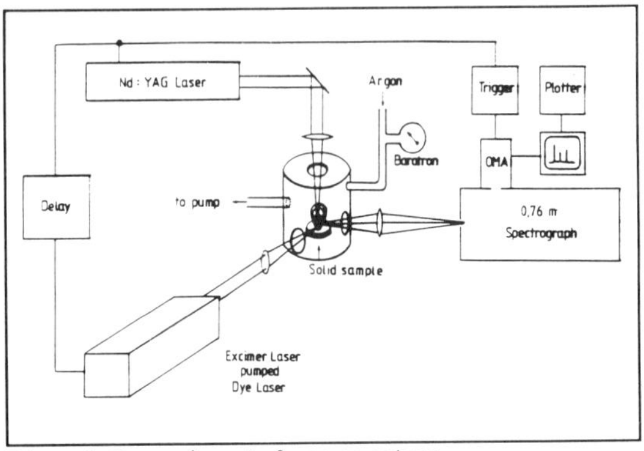

Fig. 1 Experimental apparatus

(OMA III, Eg & G/Par) adapted to a medium dispersion 0.76 m spectrograph (Fig. 1).

3. ABLATION AND ATOMIZATION PROCESS

If the Nd:YAG laser pulse hit the solid the surface is heated rapidly and electrons are released from the surface. These electrons are heated by absorption and a plasma breakdown takes place in the buffer gas. Sample material is ablated in form of atoms, ions, molecules, droplets and particles. The matrix determines the temperature of the plasma as well as the total amount of ablated material (Tab. 1). The Ar plasma serves as an energy reservoir for the complete evaporation of penetrating sample material (Fig. 2). In the first microseconds, the plasma temperature is very high and free atoms are completely ionized. But within 10 - 15 μs the temperature decreases rapidly due to the atomization process (Fig. 3). The electronic recombination of ions is relatively slow. Fig. 4 shows the LIF signals of the atomic and ionic resonance lines of Mg in dependence on time after the plasma production. From the ratio of the signals the degree of ionization can be deduced (Sdorra and Niemax 1990). Optimum analytical conditions were found at reduced argon pressure of about 140 hPa. At vacuum conditions the atomization is not efficient. Because of the missing hot buffer gas the atomization is not complete. At atmospheric pressure only a small fraction of the ablated particles and droplets can penetrate into the dense and viscose plasma.

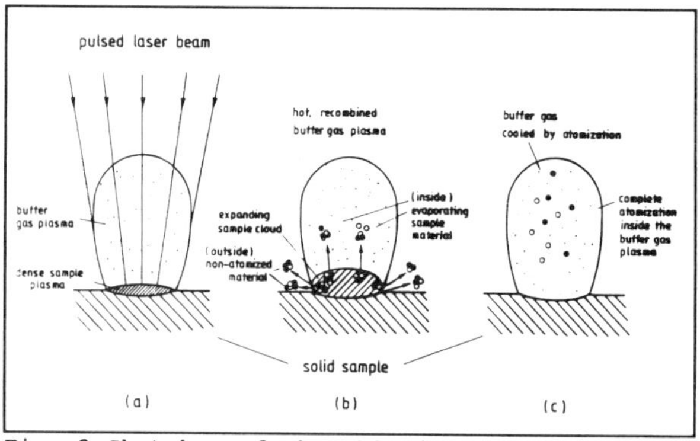

Fig. 2 Sketches of the ablation and atomization process by laser radiation

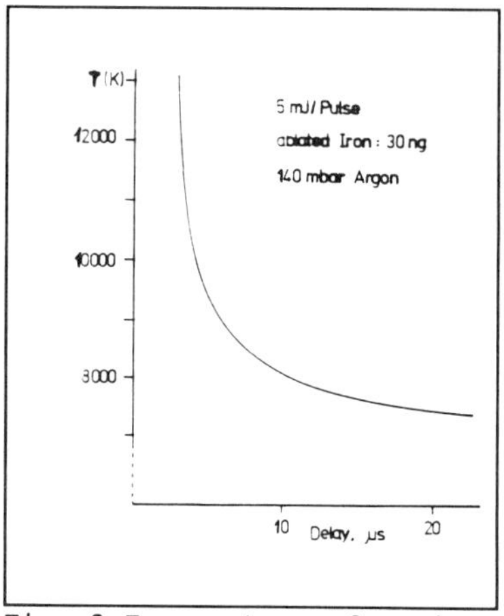

Fig. 3 Temperature of an iron sample plasma in dependence on time

Table 1: Amount of ablated material by one laser shot for different samples

sample	mass [ng]	
Al	170	
Cu	84	
steel	30	
borax-glass	3	($Na_2B_4O_7$)
ceramics	4	(Al_2O_3)

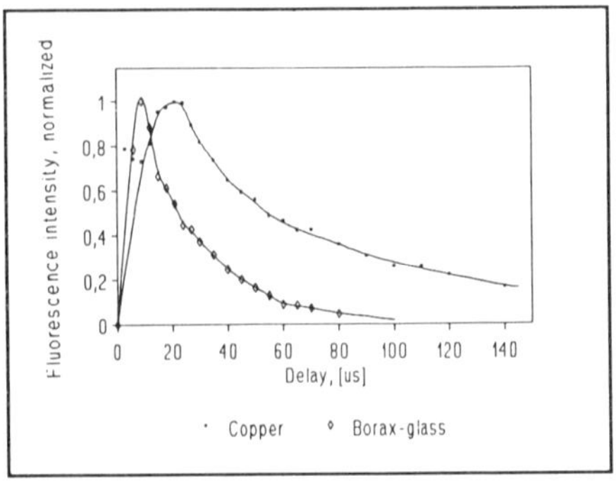

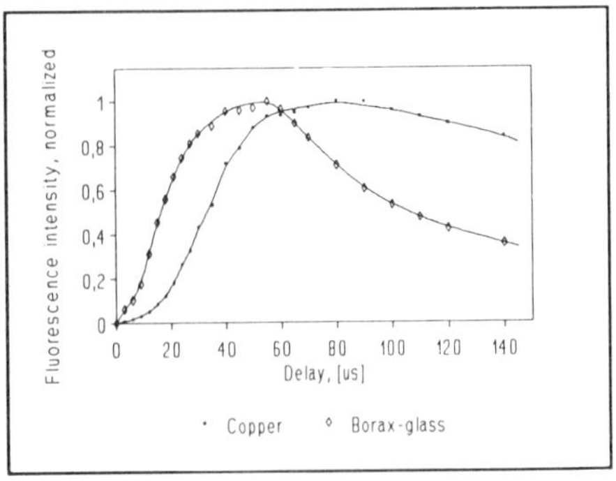

Fig. 4 Fluorescence signals of the ionic and atomic resonance lines of Mg in dependence on time after the plasma production (left: Mg II 280.27nm, right: Mg 285.21nm). From the ratio of the signals the degree of ionization can be deduced (Sdorra and Niemax 1990).

4. LASER INDUCED FLUORESCENCE (LIF) AND DETECTION LIMITS

The atoms and ions in the plasma plume were excited by the dye laser beam 55 μs after the ignition of the plasma about 1 mm above the sample surface. Then the atomization process is completed and the concentration ratios of elements in the vapor plume should be representative for the element ratios in the bulk. If the concentration of analytes in the sample is homogeneous, the reproducibilities of LIF are good and the standard deviations were 5 to 10 % by one laser shot for steel and borax-glass ($Na_2B_4O_7$) samples. The detection limits (DL) were determined on the basis of the 3 σ rule (σ: standard deviation of the background noise).

Table 2: Relative and absolute detection limits of different elements by one laser shot for low-alloyed steel and borax-glass ($Na_2B_4O_7$) samples

sample	element	rel.DL [μg/g]	abs.DL [fg]	transition line [nm]
steel	Mg	0.15	4.5	285.213
	Si	2.1	63	251.61
	Mn	2.2	66	280.106
	B	1.1	33	249.77
	Pb	9	270	283.3/405.8
	Sn	21	630	284.0
	W	41	1230	255.135
borax-glass	Mg	1.0	3	285.213
	Si	14.7	44	251.61
	Mn	9.1	27	280.106

5.INTERNAL STANDARDIZATION

The technique of internal standardization could be successfully applied. The elements Mg, Si and Mn were probed by the dye laser beam after a time delay of 80 μs. Steel, copper, aluminium and borax-glass ($Li_2B_4O_7$) samples were used. Although the matrices are very different, the ratios of the LIF signals were on common straight calibration lines (Fig. 5). This can be expected since the investigated elements have nearly the same ionization energies.

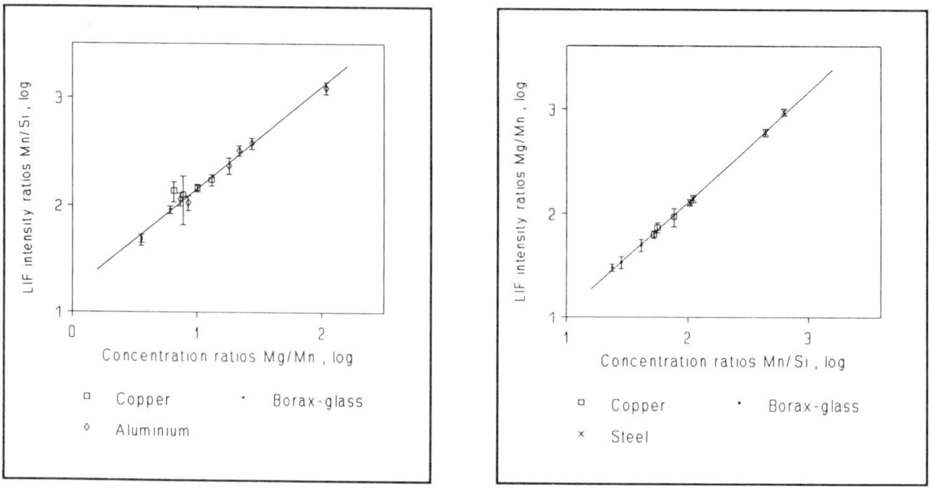

Fig. 5 Fluorescence intensity ratios plotted against the ratio of the concentrations (left: Mg/Mn, right: Mn/Si)

CONCLUSION

Laser induced fluorescence in plasma plumes produced by laser ablation is a promising analytical technique for solid samples. If sample material is ablated into a reduced pressure atmosphere, complete atomization can be achieved. The detection limits are low and the precision of the results is good. Matrices with unknown composition can be calibrated by certified samples using the method of internal standardization.

REFERENCES

Ko JB, Sdorra W, Niemax K 1989 Fresenius Z Anal Chem 335 648
Leis F, Sdorra W, Ko JB, Niemax K 1989 Mikrochim Acta II 185
Niemax K, Sdorra W, 1990 Appl Opt in press
Quentmeier A, Sdorra W, Niemax K 1990 Spektrochim Acta 45B 537
Sdorra W, Quentmeier A, Niemax K 1989 Mikrochim Acta II 201
Sdorra W, Niemax K 1990 Spektrochim Acta 45B 917

Paper presented at RIS 90, Varese, Italy, 16–21 September 1990

Determination of radionuclides in a variety of matrices by sputter initiated resonance ionization spectroscopy (SIRIS)

ADAM W. McMAHON and SIMON A. MOUNTFORT

AEA Technology, Harwell Laboratory, Didcot
Oxfordshire, OX11 0RA

and

HEINRICH F. ARLINGHAUS and MICHAEL T. SPAAR

Atom Sciences Inc.,
114 Ridgeway Center, Oak Ridge, TN 37830

ABSTRACT: Sputter Initiated Resonance Ionisation Spectroscopy has been used to measure known, low levels of uranium, technetium and plutonium in a variety of matrices. The freedom of the technique from many of the interferences associated with other analytical methods is demonstrated, especially in complex biological matrices.

1. INTRODUCTION

Radiometric methods for the determination of radionuclides become less sensitive (in mass terms) as the half-life of the isotope of interest is increased. Non-radiometric analytical methods including SIRIS and secondary ion mass spectrometry (SIMS) are therefore being considered as potential alternatives to radiometric techniques in order to improve the sensitivity for radionuclide determinations and to provide further complementary information such as the concentration of stable isotopes.

An intercomparison of these techniques has involved a preliminary review of the literature (McMahon 1989) and the methods of interest are currently being assessed using a set of well-characterised materials, containing known levels of radionuclides. Three radionuclides will be highlighted in this paper, uranium, plutonium and technetium. The sample types include electrodeposits, doped gelatins and ion implants which have been chosen to represent a much larger range of materials. The electrodeposits for example represent the end step of a chemical separation and preconcentration procedure which might be used in the analysis of waters, soils, sediments, foodstuffs etc.

2. EXPERIMENTAL.

SIMS measurements were performed using the Cameca 3F instrument at the Harwell laboratory. The SIRIS instrument at Atom Sciences Inc., Oak Ridge, Tennessee was used for the resonance ionisation work. The SIRIS instrument sputters material with a focused argon ion beam. Analyte atoms are ionised using two tunable dye lasers and mass analysis is performed using a sector mass spectrometer. The instrument has been described in more detail elsewhere (Arlinghaus et al. 1990). The ionisation scheme for technetium utilised two visible photons at 313.3 nm and 554.2 nm with a 1064 nm infrared photon for the ionisation step. Uranium was ionised by a visible photon at 591.5 nm an ultraviolet photon at 319.9 nm and a second 591.5 nm visible photon for the ionisation step. Plutonium was ionised by a single colour three photon scheme using 586.4 nm light.

Electrodeposits were prepared by a method described by Burton and Yarnold (1988). Gelatin samples were prepared from standard solutions set in BDH pigskin gelatin. The gelatin blocks were snap frozen, cryosectioned and dried onto pieces of silicon wafer. Ion implanted materials were prepared on the Mark IV ion implanter at the Harwell laboratory, the implantion energy used for uranium-238 implants was 10 keV. Uranium was implanted into pieces of silicon wafer, tantalum metal and capton (a polyimide).

3. RESULTS AND DISCUSSION

3.1 Electrodeposits.

The uranium and plutonium content of the electrodeposits were measured by alpha spectroscopy and are given below:

TABLE1:	Activity	Mass	Number of Atoms Sampled by Beam
U-238	918 mBq	7.4×10^{-5} g	3.0×10^{12}
U-235	39 mBq	4.9×10^{-7} g	2.0×10^{12}
U-234	750 mBq	3.3×10^{-9} g	1.3×10^{8}
Pu-242	119 mBq	8.2×10^{-10} g	2.5×10^{8}

To be accessible to radiometric determination the mass of uranium 238 and 235 on the deposits is relatively high and gives rise to a strong SIRIS as well as measurable SIMS signals.

Sequential measurements using the Atom Sciences instrument in SIMS and SIRIS mode, showed that a signal enhancement by a factor of 380 could be obtained by the use of SIRIS rather than SIMS.

Plutonium-242 and uranium-234 were present in much smaller concentrations due to their shorter half-lives but also gave rise to a detectable SIRIS signal whereas these isotopes could not be detected by SIMS. A depth profile through the plutonium-242 deposit is shown in Figure 1.

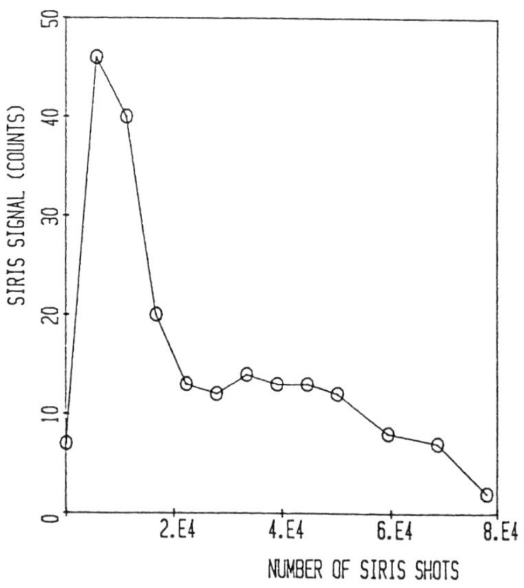

Figure 1: Depth profile of Pu-242 in an electrodeposit.

A 100 um diameter area of the electrodeposit was sampled by the SIRIS instrument. The number of atoms sampled in the generation of the depth profile in Figure 1 is thus only 2.5×10^{8}. This clearly demonstrates the potentially very high sensitivity of SIRIS although it would be necessary to concentrate the sample into a 100 um diameter spot to attain such sensitivity in a real analysis.

The single colour scheme used to ionise plutonium also ionises uranium. This lack of selectivity is a problem for actinides where the density of atomic states is high. A vast excess of uranium has been seen to give rise to an interference at m/z = 239, due to the tail of the 238 peak. In such a situation, it would be preferable to use a two or three colour scheme to attain the degree of selectivity required.

3.2 Gelatins.

The gelatin standards used are uniformly doped solids which are close in physical and chemical form to a range of biological soft-tissues. SIMS analysis of such materials is limited by the presence of molecular interferences which contribute to a background signal level worsening the detection limit of the method. This is demon-strated in the analysis of technetium 99 in gelatin (see Figure 2). The calibration curve for SIRIS analysis is linear down to the 10 ppb level whereas the SIMS signal levels off at the 1 ppm level, indicating the presence of a matrix interference at m/z=99.

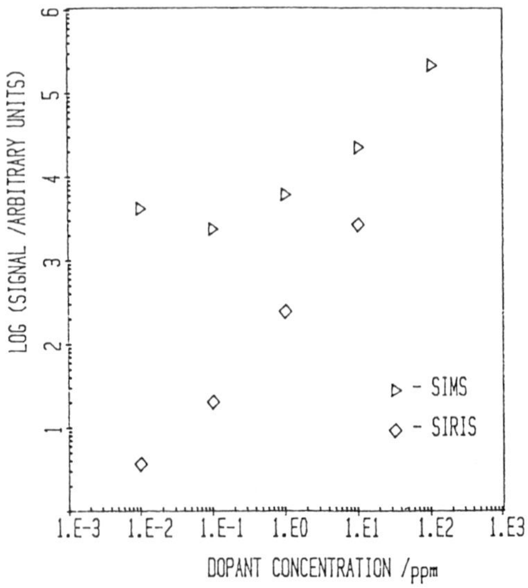

Figure 2 Calibration plots for the determination of technetium-99 in gelatin by SIMS and SIRIS

3.3 Ion Implants.

SIRIS depth profiles of uranium 238 implants (10^{13} cm^{-2}, 10 keV) in silicon, tantalum and capton, were measured. The profiles all tail to zero signal level, implying that there are no interferences at m/z = 238 arising from matrix components. The implants are at different depths and the sputter rates for the three matrices are different, so it is unlikely that the profiles would be identical. However, the uranium concentration integrated over the depth of the profile is the same in all samples. This is not true of the integrated uranium SIRIS signal:

Matrix	Integrated SIRIS (U) Signal
Silicon	101.6
Tantalum	81.5
Capton	17.2

The integrated signals are within an order of magnitude of one another but are obviously different. It is suggested that this arises from the powerful gettering abilities of uranium atoms. Formation of uranium oxides, carbides and other molecular species in the sputtering process will diminish the atomic signal observed in the SIRIS experiment. The low signal observed in the polyimide sample may be due to the ready availability of carbon, nitrogen and hydrogen in this matrix, relative to the tantalum and silicon matrices.

This would imply that matrix matched standards are required to perform any quantitative determination.

4. CONCLUSIONS

This study has demonstrated that SIRIS is more sensitive than alpha spectroscopy for the determination of uranium 238 and 235 and for plutonium 242. The high selectivity of the technique allows interference free determinations to be performed even in complex biological matrices. However, this selectivity may be significantly reduced for actinides which have high densities of atomic states. It should also be noted that despite the high selectivity, matrix-matched standards will be required to perform quantitative determinations by SIRIS.

5. ACKNOWLEDGEMENTS

This work has been funded jointly by the following U.K. organisations: the Department of Energy, the Department of the Environment, the Department of Health, the Ministry of Agriculture, Fisheries and Food, and the United Kingdom Atomic Energy Authority.

6. REFERENCES

A.W. McMahon
"A Critical Intercomparison of Techniques for the Determination of Radionuclides."
Harwell Report, AERE R 13617, July 1989, HMSO, London.

P.J. Burton and L.P. Yarnold,
Sci. Total Environ. 69, 239-260, (1988)

H.F. Arlinghaus, M.T. Spaar and N. Thonnard,
J. Vac. Sci and Technol. A8, 2318, (1990)

Author Index